C–H and C–X Bond Functionalization

Transition Metal Mediation

RSC Catalysis Series

Series Editor:
Professor James J Spivey, *Louisiana State University, Baton Rouge, USA*

Titles in the Series:
1: Carbons and Carbon Supported Catalysts in Hydroprocessing
2: Chiral Sulfur Ligands: Asymmetric Catalysis
3: Recent Developments in Asymmetric Organocatalysis
4: Catalysis in the Refining of Fischer–Tropsch Syncrude
5: Organocatalytic Enantioselective Conjugate Addition Reactions: A Powerful Tool for the Stereocontrolled Synthesis of Complex Molecules
6: *N*-Heterocyclic Carbenes: From Laboratory Curiosities to Efficient Synthetic Tools
7: *P*-Stereogenic Ligands in Enantioselective Catalysis
8: Chemistry of the Morita–Baylis–Hillman Reaction
9: Proton-Coupled Electron Transfer: A Carrefour of Chemical Reactivity Traditions
10: Asymmetric Domino Reactions
11: C–H and C–X Bond Functionalization: Transition Metal Mediation

How to obtain future titles on publication:
A standing order plan is available for this series. A standing order will bring delivery of each new volume immediately on publication.

For further information please contact:
Book Sales Department, Royal Society of Chemistry, Thomas Graham House, Science Park, Milton Road, Cambridge, CB4 0WF, UK
Telephone: +44 (0)1223 420066, Fax: +44 (0)1223 420247
Email: booksales@rsc.org
Visit our website at www.rsc.org/books

C–H and C–X Bond Functionalization
Transition Metal Mediation

Edited by

Xavi Ribas
Institut de Química Computacional i Catàlisi (IQCC) &
Departament de Química, Universitat de Girona, Girona, Catalonia-Spain
Email: xavi.ribas@udg.edu

RSC Publishing

RSC Catalysis Series No. 11

ISBN: 978-1-84973-570-4
ISSN: 1757-6725

A catalogue record for this book is available from the British Library

Published by The Royal Society of Chemistry,
Thomas Graham House, Science Park, Milton Road,
Cambridge CB4 0WF, UK

Registered Charity Number 207890

For further information see our web site at www.rsc.org

Printed in the United Kingdom by CPI Group (UK) Ltd, Croydon, CR0 4YY

Dedicated to my daughter Laia, my son Arnau and my wife Brenda.

Preface

Metal-mediated organic synthesis has revolutionized the way organic chemists design their retro-synthetic strategies to prepare new compounds unachievable only a few years ago. Among those strategies, carbon–carbon and carbon–heteroatom cross-coupling reactions mediated by palladium and copper emerge as the most reliable ones to effect the desired C–X bond functionalizations. More recently, an unstoppable rush in metal-mediated direct C–H bond functionalizations is occurring, which will probably be the major advance in organic synthesis in the first decades of the 21st century. In both of these synthetic strategies, a profound mechanistic understanding is crucial to design new metal catalysts with improved performance.

This book gives an extensive overview of the different metal-catalyzed methodologies to achieve the desired C–X and C–H bond functionalizations to produce a myriad of new organic compounds, giving a prominent relevance to the molecular mechanistic aspects of the reactions with special attention to the organometallic species and different metal oxidation states involved. The first two chapters deal with copper-catalyzed C–heteroatom cross coupling reactions, the so-called Ullmann-type reactions. Chapter 1 (Ma) covers the key role of the commonly used dinucleating ancillary ligands, whereas Chapter 2 (Casitas) discusses extensively the different mechanistic proposals for these Cu-catalyzed transformations, nowadays under intense debate.

Chapter 3 (Fairlamb) begins a block about palladium-catalyzed reactions, starting with the fundamental Pd^0/Pd^{II} redox steps in C–C cross-coupling reactions, with special emphasis to the everyday most recognized role of the heterogeneous pathways due to nanoparticle formation, which competes with the classical homogeneous proposals. Chapter 4 (Vedernikov) then continues with an overview of different C–heteroatom bond formation reactions by reductive elimination involving organopalladium(IV) species, which are involved in Pd^{II}/Pd^{IV} catalytic cycles. Moreover, Chapter 5 (Mirica) closes this

RSC Catalysis Series No. 11
C–H and C–X Bond Functionalization: Transition Metal Mediation
Edited by Xavi Ribas

Published by the Royal Society of Chemistry, www.rsc.org

block on Pd-based mechanistic understanding by showing the connections between the recently discovered mononuclear and dinuclear organopalladium(III) species and the Pd^{II}/Pd^{IV} redox chemistry, in C–C, C–heteroatom bond formation reactions, as well as in C–H functionalization.

The two following chapters of the book deal with metal-catalyzed defluorination and fluorination reactions. Chapter 6 (Love) describes in detail the strategies used to effect aromatic C–F functionalization mediated by metals ranging mainly between group 7 and group 11 of the periodic table (including Fe). Then follows Chapter 7 (Gouverneur), which reports on the importance of fluorination in pharmaceutical industries and medical diagnosis, covering aromatic, allylic and alkylic fluorinations catalyzed mainly by Pd, but also by Au, Ni and Pt.

The next four chapters are devoted to the recent inrush of direct C–H activation and functionalization, mainly mediated by palladium catalysts. Chapter 8 (Urriolabeitia) covers many examples of the coordination-directed metallation strategy to effect C–H functionalization at vicinal positions, with Pd as the most prominent metal, but also including examples with Cu, Rh and Ru. It is followed by Chapter 9 (Lei), which includes direct arylation of arenes with aryl halides under the catalysis of heavy-metals such as Rh, Ir and Pd, but also first-row transition metals such as Fe, Co, and Ni. In Chapter 10 (Su) the palladium-catalyzed double C–H activation processes under oxidative conditions are highlighted, being somehow the culmination of the cross-couplings described in this book in terms of using non-activated arenes as substrates. The last chapter of this block, Chapter 11 (Liu), is devoted to aerobic oxidative C_{sp2}–H functionalizations catalyzed by either palladium or copper catalysts, in which the use of molecular oxygen to reactivate the catalyst is highlighted.

Finally, the last chapter of the book, Chapter 12 (Fairlamb), is devoted to the synthesis of modified biomolecules by means of palladium-catalyzed cross-coupling chemistry. Small nucleoside derivatization by C–X activation is reported, as well as C-functionalization of larger biomolecules such as nucleic acids, peptides and proteins. The use of biomolecules as substrates opens the door to the synthesis of highly complex biomolecule targets.

Finally, I want to sincerely thank all the authors for their excellent contributions, and for giving me the opportunity to tackle in this book some of the hottest topics in metal-catalyzed C–X and C–H functionalization strategies, which in my view will completely alter the strategies in organic synthesis in the next years.

Xavi Ribas

Contents

RSC Catalysis Series No. 11
C–H and C–X Bond Functionalization: Transition Metal Mediation
Edited by Xavi Ribas

Published by the Royal Society of Chemistry, www.rsc.org

CHAPTER 1

Cu-Catalyzed Ullmann-Type C–Heteroatom Bond Formation: The Key Role of Dinucleating Ancillary Ligands

YONGWEN JIANG,[a] LANTING XU,[b] CHENGGANG ZHOU[c] AND DAWEI MA*[a]

[a] State Key Laboratory of Bioorganic & Natural Products Chemistry, Shanghai Institute of Organic Chemistry, Chinese Academy of Sciences, 354 Fenglin Lu, Shanghai 200032, China; [b] Department of Chemistry, Fudan University, Shanghai 200433, China; [c] Roche R&D Center (China) Ltd., 720 Cai Lun Road, Building 5, Pudong, Shanghai 201203, China
*Email: madw@sioc.ac.cn

1.1 Introduction

Copper-catalyzed coupling reactions of aryl halides with nucleophiles, so called Ullmann-type reactions, are well-established methods for preparing pharmaceutically and materially important compounds. The traditional version of these coupling reactions normally requires harsh reaction conditions, and the reactions have a reputation for erratic yields. Since the late 1990s, a great number of ligands, particularly *N,N*-, *N,O*- and *O,O*-bidentate ligands, have been revealed to have the ability to promote Ullmann-type reactions.[1] The combination of copper salts and these ligands allow the coupling reactions to be conducted in milder conditions and with dramatically enhanced yields.

RSC Catalysis Series No. 11
C–H and C–X Bond Functionalization: Transition Metal Mediation
Edited by Xavi Ribas

Published by the Royal Society of Chemistry, www.rsc.org

Under these mild reaction conditions, an increasing number of nucleophiles have been found to be applicable for copper-catalyzed arylation reactions, therefore leading to structural moieties that are prevalent in building blocks of functional molecules in the life sciences and material sciences.

As a general trend, ligand promoted Ullmann-type reactions are not too sensitive to the choice of the copper source, but the choice of ligands is often crucial for different coupling reactions.[1] In most cases ligands play an essential role for obtaining improved results, speeding up cross-coupling under much milder conditions, widening the substrate scope, and enhancing chemoselectivity and enantioselectivity. These ligands were believed to coordinate with Cu^{I}, thereby changing the electronic nature of this active species. Some proposed intermediates formed from Cu^{I} and bidentate ligands are outlined in Figure 1.1. The complexes **1–4** are formed by coordinating with ionic ligands,[2–4] while complexes **5–8** result from the coordination of Cu^{I} with nonionic ligands.[5]

The structures of some useful *N,O*-, *N,N*-, *O,O*-bidentate ligands as well as copper-complexes are outlined in Figure 1.2. In this chapter we wish to summarize their applications in promoting Ullmann-type coupling reactions, which can be classified into C–N, C–O, C–S and C–P bond formation reactions.

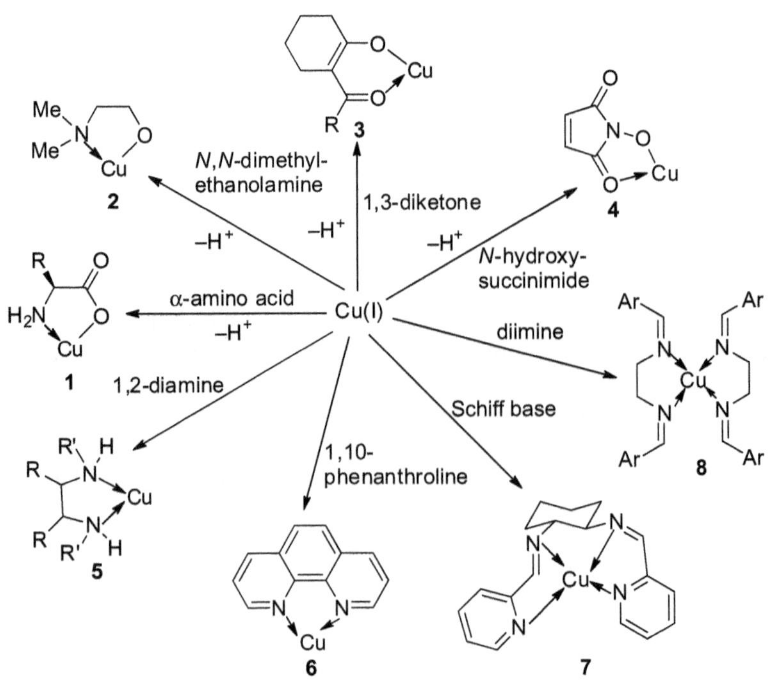

Figure 1.1 Possible intermediates resulting from coordination of Cu^{I} with bidentate ligands.

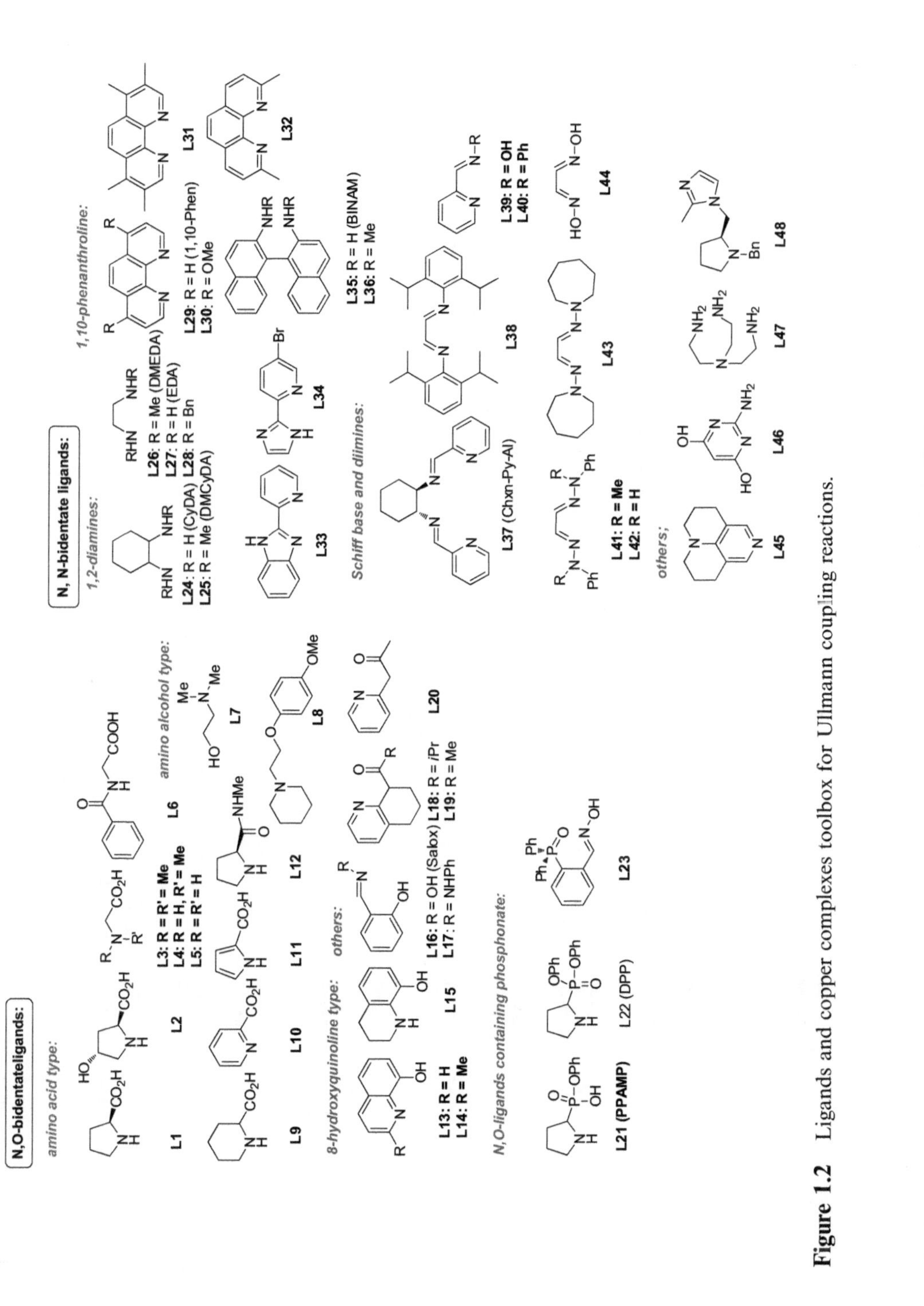

Figure 1.2 Ligands and copper complexes toolbox for Ullmann coupling reactions.

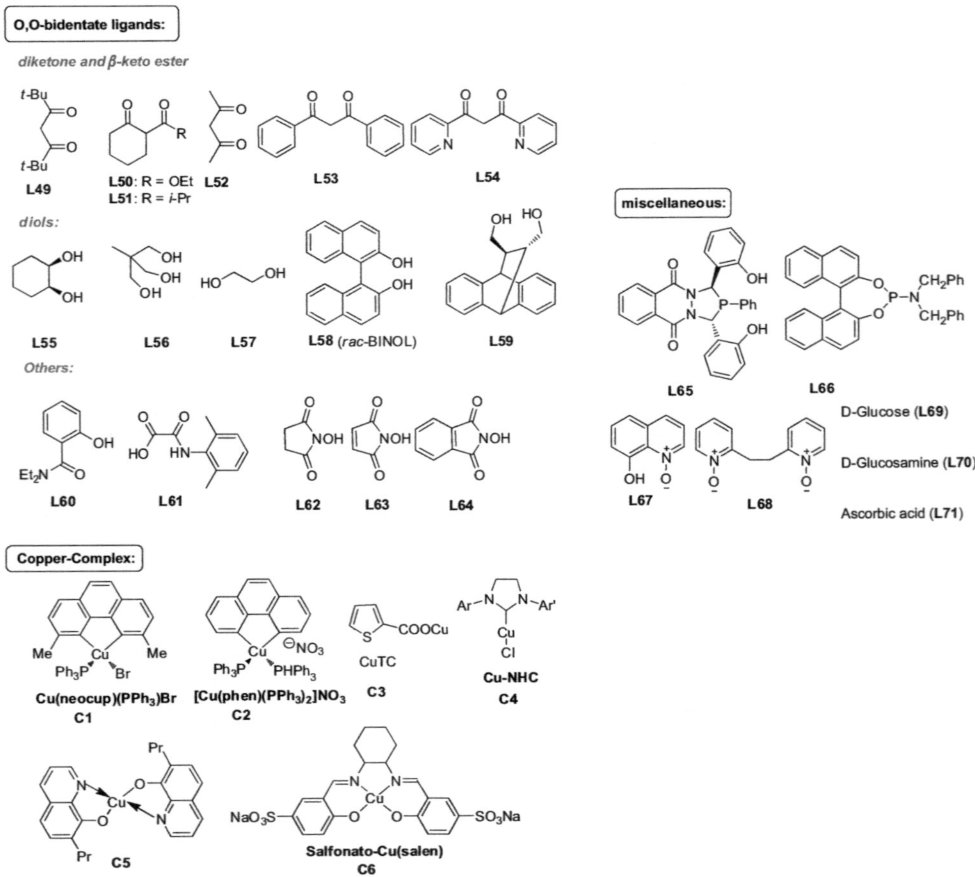

Figure 1.2 (Continued)

1.2 C–N Bond Formation

1.2.1 Arylation of Amines

1.2.1.1 Arylation of Aliphatic Primary and Secondary Amines

Copper catalyzed coupling between aryl halides and aliphatic primary amines is a typical transformation for checking the efficiency of newly developed ligands. Many ligands have proven effective for this reaction. The reaction conditions employed with these ligands are summarized in Table 1.1. As a class of cheap and conveniently available *N*,*O*-bidentate ligands, amino acids have been revealed to have a strong accelerating effect for Ullmann arylation amination.[2] L-Proline (**L1**) was shown to be applicable for both aryl iodides and bromides with a working temperature range from 60 to 90 °C. In some cases *N*-methylglycine (**L4**) gave similar results (entry 1).[6,7] 2-*N*,*N*-dimethylaminoethanol (deanol, **L7**) was another useful *N*,*O*-bidentate ligand for the amination of aryl halides although excessive addition was required (entry 2).[8] Notably, the coupling reaction works well in aqueous media, and various amines (including simple primary amines, amino acids, amino alcohols and even peptides) could couple with aryl iodides at 80–90 °C. Two other *N*,*O*-bidentate ligands, 2-hydroxylbenzaldehyde *N*-phenylhydrazone **L17** and 1-(5,6,7,8-tetrahydroquinolin-8-yl)ethanone **L19** showed the powerful acceleration effect, which made the coupling reaction of aryl iodides with primary amines proceed at room temperature (entries 3 and 4).[9,10] However, in this case the higher reaction temperature was still required if aryl bromides were used (entry 4).[10]

The first efficient *O*,*O*-bidentate ligand for this transformation was ethylene glycol (**L57**), which was reported by Buchwald and coworkers, but only aryl iodides were suitable for the coupling (entry 5).[11] Subsequently, they found that diethylsalicylamide (**L60**) was a powerful ligand for accelerating the aryl amination of aryl bromides (entry 6).[12] Some years later, 2-isobutyrylcyclohexanone (**L51**) was found to be another powerful ligand, leading to the coupling reaction of aryl iodides with amines occur at room temperature, and aryl amination with aryl bromides proceed at 90 °C (entry 7).[13] Additionally, a combination of CuCl as the catalyst, ethyl 2-oxocyclohexanecarboxylate (**L50**) as the ligand, and tetramethylammonium hydroxide as the base has been showed effective for aryl amination (entry 8).[14]

rac-BINOL (1,1′-binaphthyl-2,2′-diol, **L58**) is another important and interesting *O*,*O*-bidentate ligand for the amination of aryl halides. The copper source was found to have a marked influence to this coupling reaction. The combination of CuBr and *rac*-BINOL (entry 10)[15] could make amination of aryl iodides work at room temperature, while the combination of other copper sources (Cu, CuI, CuO/$FeCl_3$) and *rac*-BINOL was not so effective (entry 9, 11).[16,17] Additionally, with the assistance of *ortho*-carbonyl acid, coupling of aryl bromides and amines could be carried out at room temperature (entry 12).[18]

A few *N*,*N*-bidentate ligands have been used to promote the coupling of aryl halides and primary amines (entries 13 and 14).[19,20] Using diimine-type ligand

Table 1.1 Copper/ligand catalyzed coupling between aryl halides and aliphatic primary amines.

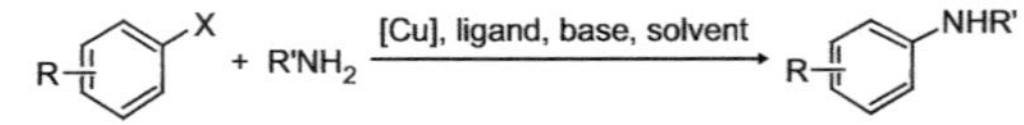

Entry	*X*	*[Cu] (mol%)*	*Ligand (mol%)*	*Reaction conditions*	*Yield (%)*	*Number of examples*	*Ref.*
1	I	CuI (10)	**L1 or L4** (20)	K_2CO_3, DMSO, 40–60 °C	64–91	28	6,7
	Br			K_3PO_4, DMSO, 80–90 °C	81–98	15	
2	I	CuI (10)	**L7** (150–300)	K_3PO_4, H_2O, 80–90 °C	32–92	23	8
	Br				62–90	4	
3	I	CuI (10)	**L17** (20)	K_3PO_4, DMF, 25–35 °C	56–90	31	9
4	I	CuI (10)	**L19** (20)	Cs_2CO_3, DMF, rt	80–97	20	10
	Br			Cs_2CO_3, DMF, 65 °C	83–95	7	
5	I	CuI (1–10)	**L57** (200)	K_3PO_4, ^{i}PrOH or *n*BuOH, 80–100 °C	58–91	35	11
6	Br	CuI (5)	**L60** (5–20)	K_3PO_4, DMF or solvent-free, 90–100 °C	71–95	23	12
7	I	CuI (5)	**L51** (20)	Cs_2CO_3, DMF, RT–50 °C	79–99	17	13
	Br			Cs_2CO_3, DMF, 90–100 °C	87–94	5	
8	I	CuCl (10)	**L50** (10)	$(CH_3)_4{}^+NOH^-$, DMSO, 80 °C	56–87	14	14
9	I	Cu (5)	**L58** (10)	Cs_2CO_3, DMSO or DMF, 90 °C	77–99	5	16
	Br	Cu/CuI (5/5)	**L58** (20)	Cs_2CO_3, DMF, 110 °C	62–96	13	
10	I	CuBr (20)	**L58** (20)	K_3PO_4, DMF, RT–50 °C	38–94	17	15
11	I	CuO/$FeCl_3$ (10/10)	**L58** (20)	Cs_2CO_3, DMF, 80–90 °C	72–89	7	17
	Br			Cs_2CO_3, DMF, 90–100 °C	76–88	4	
12	Br	CuI (10)	**L58** (20)	K_3PO_4, DMF, rt	40–96	20	18
	Cl				16–20	2	
13	I	CuBr (5)	**L44** (10)	*n*Bu_4NBr, NaOH, H_2O, 100 °C	74–91	8	19
	Br				72–91	4	
	Cl				87–89	2	
14	Br	CuI (10)	**L29** (20)	Cs_2CO_3, DMSO, 150 °C (Mw)	69–96	12	20
15	I	Cu_2O (1–5)	**L23** (20)	Cs_2CO_3, MeCN or Toluene or solvent free, 60–90 °C	64–95	17	21
16	I	CuBr (2.5)	**L66** (5)	Cs_2CO_3, DMF, 90 °C	96–99	9	22
	Br	CuBr (5)	**L66** (10)		89–97	14	23
17	I	CuBr or CuCl (10)	**L65** (12)	Cs_2CO_3, DMSO, 55–80 °C	51–95	20	24

L44 ((1E,2E)-oxalaldehyde dioxime (OADO)) and phase-transfer catalyst made the reaction work well in water (entry 13).[19] Two activated aryl chlorides, 2-chlorobenzoic acid and 2-chloronicotinic acid, could give good results in this case.

Some phosphine ligands were also efficient for the amination of aryl halides although the relatively high reaction temperatures were required (entries 14–17).[20–24]

Unlike Pd-catalyzed aryl amination, ligand-promoted Ullmann amination is very sensitive to steric hindrance of the amines. For example, when secondary amines were employed, the reaction became sluggish and only less hindered cyclic amines could give satisfactory conversions. Some ligand-promoted arylation of cyclic secondary amines are indicated in Table 1.2.

When acyclic secondary amines were used as the coupling partners, low conversions were observed in most cases. Twieg and coworkers reported the first useful example for arylation of acyclic secondary amines (Scheme 1.1). When *N*,*N*-dimethylaminoethanol (**L7**) was used as both the ligand and the solvent, 2- and 3-bromothiophenes could be coupled with some acyclic secondary amines under the catalysis of Cu/CuI. In this case complete conversion was only observed when amino alcohols were employed, presumably because amino alcohol could also accelerate this coupling reaction.[3]

Table 1.2 Copper salt/ligand catalyzed coupling between aryl halides and aliphatic cyclic secondary amines.

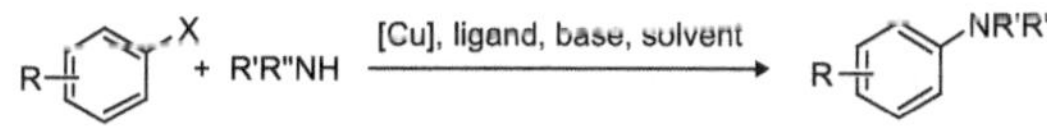

Entry	*X*	*[Cu] (mol%)*	*Ligand (mol%)*	*Reaction conditions*	*Yield (%)*	*Number of examples*	*Ref.*
1	I	CuI (10)	**L1** (20)	K_2CO_3, DMSO, 65–90 °C	21–100	5	7
	Br			K_2CO_3, DMSO, 90 °C	81–99	8	
2	I	CuI or Cu (10)	**L7** (>100)	K_3PO_4, RT–75 °C	34–91	8	25
	Br	Cu (10)		K_3PO_4, 80 °C	28–81	3	
3	I	CuI (1–10)	**L57** (200)	K_3PO_4, ^{i}PrOH, 80 °C	71–90	6	11
4	I	Cu (5)	**L58** (10)	Cs_2CO_3, DMSO or DMF, 90 °C	75–96	4	16
5	I	CuBr (20)	**L58** (20)	K_3PO_4, DMF, RT–40 °C	54–91	7	15
6	I	CuO/$FeCl_3$	**L58** (20)	Cs_2CO_3, DMF, 80–90 °C	75–90	6	17
	Br	(10/10)		Cs_2CO_3, DMF, 100 °C	75–89	2	
7	I	CuBr (2.5)	**L66** (5)	Cs_2CO_3, DMF, 90 °C	78–99	10	22
	Br	CuBr (5)	**L66** (10)		78–86	7	23
8	I	Cu_2O (5)	**L23** (20)	Cs_2CO_3, MeCN or Toluene or solvent-free, 80–90 °C	60–90	10	21

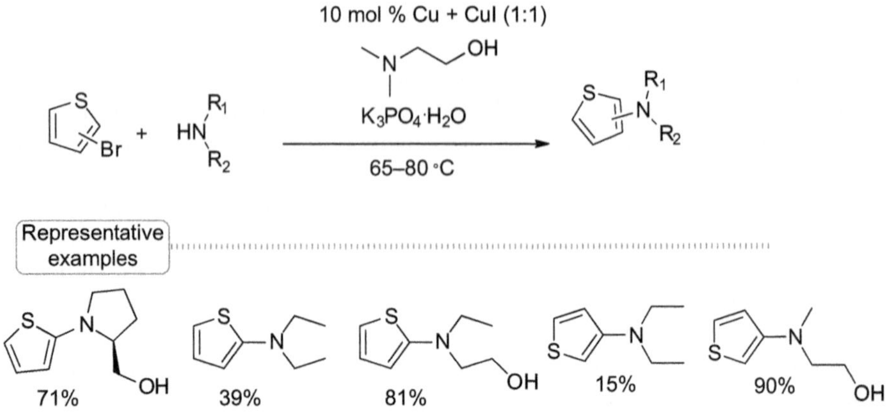

Scheme 1.1 Copper-catalyzed coupling of halothiophenols and acyclic secondary amines in *N*,*N*-dimethylaminoethanol.

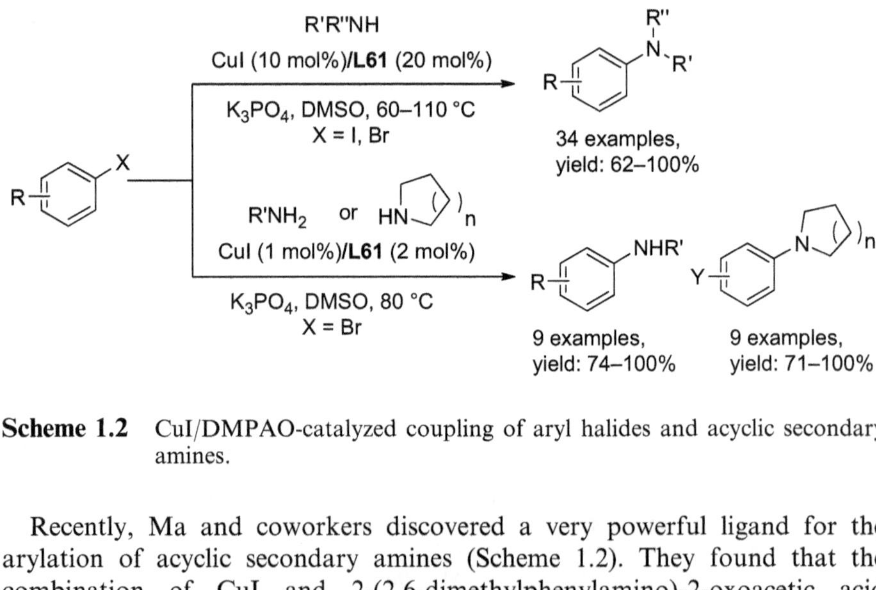

Scheme 1.2 CuI/DMPAO-catalyzed coupling of aryl halides and acyclic secondary amines.

Recently, Ma and coworkers discovered a very powerful ligand for the arylation of acyclic secondary amines (Scheme 1.2). They found that the combination of CuI and 2-(2,6-dimethylphenylamino)-2-oxoacetic acid (DMPAO, **L61**) could make the coupling reaction of aryl halides with various aliphatic acyclic secondary amines work well at 60–110 °C. Notably, for the coupling of aryl bromides with primary amines and cyclic secondary amines, the catalytic system was still very efficient even when decreasing the catalyst loading to 1 mol%.[26]

1.2.1.2 Arylation of Aryl Amines

Due to poor reactivity, aryl amines normally required higher reaction temperatures to ensure good conversion compared with aliphatic amines.

In early studies, phenanthroline as well as its Cu^{I}-complex (**C1**) were used in the arylation of aryl amines,[27,28] but were only applicable for the synthesis of triarylamines from secondary aryl amines (Table 1.3, entry 1).[28] Two other *N*,*N*-bidentate ligands, *N*,*N′*-bis(2,6-diisopropylphenyl)-1,4-diaza-1,3-butadiene (DAB) **L38** and 9-azajulolidine **L45**, were also efficient for the arylation of diarylamines when a strong base like ^{t}BuOK and ^{t}BuONa was used (entries 2 and 3).[29,30] DMEDA (**L26**) was found to be a better ligand for the *N*-arylation of heteroarylamines (entry 4).[31]

For *N*,*O*-bidentate ligands, L-proline (**L1**) could promote CuI-catalyzed arylation of primary aryl amines and the working temperature was 90 °C (entry 5).[6,7] But only electron-rich anilines gave complete conversions, while electron-deficient anilines provided low yields. Fu found that this drawback could be overcome by heating the reaction mixture at 110 °C and using pipecolinic acid (**L9**) as a ligand (entry 6).[32] Widening the substrate scope was possible when pyrrole-2-carboxylic acid (**L11**) was used, as evident from the fact that various aryl amines and some hindered aryl halides could give satisfactory coupling yields (entry 7).[33]

rac-BINOL was the only *O*,*O*-bidentate ligand that has been used for the monoarylation of anilines (entry 8).[17] Additionally, pyrrolidine-2-phosphonic acid phenyl monoester (**L21**) and pyrrolidine-2-phosphonate (**L22**) were also effective for promoting Cu-catalyzed coupling of aryl halides with anilines (entries 9 and 10).[34,35]

1.2.1.3 Arylation of ammonia

The first copper/ligand catalyzed coupling reaction of aryl halides with ammonia was reported by a Merck group (Scheme 1.3, eqn (1.1)).[36] In this case ethylene glycol was believed to serve as both ligand and solvent, and the reaction worked at 80 °C under an elevated pressure. These reaction conditions are suitable for substituted pyridine bromides and aryl bromides (eqn (1.2)).

Subsequently, several more powerful ligands for promoting this special coupling reaction were discovered, and the results are summarized in Table 1.4. Mild conditions were discovered by Kim and Chang, in which L-proline (**L1**) was employed as the ligand (entry 1).[37] Both NH_4Cl and $NH_3{\cdot}H_2O$ could be used as the nitrogen source, and the reaction proceeded smoothly at room temperature in the case of aryl iodides. But for aryl bromides, only substrates with an electron-withdrawing group gave good yields even increasing reaction temperature to 80 °C. To solve this problem, another ligand, *trans*-4-hydroxy-L-proline **L2**, was tested. This promoter was found to be applicable for both electron-deficient and electron-rich aryl bromides, could make their coupling with ammonia complete at 50 °C (entry 2).[38]

Taillefer and coworkers reported that some 1,3-diketones are powerful ligands for copper-catalyzed coupling of aryl halides with ammonia. The optimized combination was $Cu(acac)_2$ and acetylacetone **L52**, which makes this transformation complete at 60–90 °C in DMF. Both aryl iodides and aryl bromides are suitable coupling partners (entry 3).[39]

Table 1.3 Copper/ligand catalyzed coupling between aryl halides and aryl amines.

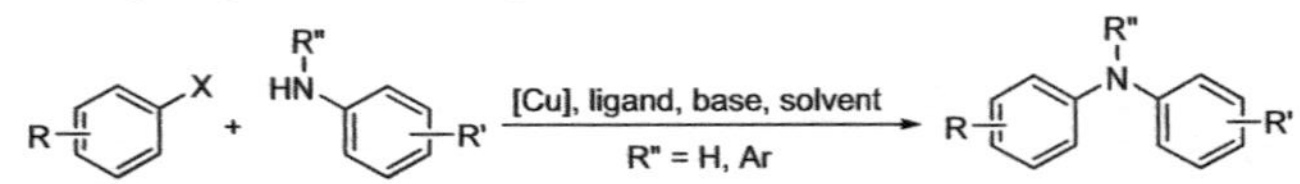

Entry	*X*	*Copper Source (mol%)*	*Ligand (mol%)*	*Reaction conditions*	*Yield (%)*	*Number of examples*	*Ref.*
1	I, Br, Cl	Cu(neocup)(PPh_3)Br (10)		*t*-BuOK, Toluene, 110 °C	49–88	6	28
2	I	CuI (5)	**L38** (6)	*t*-BuOK, Toluene, 120 °C	38–97	19	29
3	I	CuI (5)	**L45** (10)	*t*-BuONa, Toluene, 110 °C	60–96	7	30
4	I	CuI (10100)	**L26** (10–100)	K_2CO_3, dioxane, 100 °C	62–84	3	31
	Br					8	
5	I, Br	CuI (10)	**L1** (20)	K_2CO_3, DMSO, 90 °C	51–97	11	6,7
6	I	CuI (10)	**L9** (20)	K_2CO_3, DMF, 110 °C	79	1	32
	Br				29–93	13	
	Cl				15–31	4	
7	I	CuI (10)	**L11** (20)	K_3PO_4, DMSO, 70–90 °C	50–82	10	33
	Br			K_3PO_4, DMSO, 100 °C	51–76	9	
8	I	CuO/$FeCl_3$ (10/10)	**L58** (20)	Cs_2CO_3, DMF, 80–100 °C	68–89	9	17
	Br			Cs_2CO_3, DMF, 110 °C	76–86	5	
9	I	CuI (10)	**L21** (20)	K_3PO_4, DMF, 100–110 °C	75–90	10	34
	Br			K_3PO_4, DMF, 90–110 °C	71–98	4	
10	I	CuI (10)	**L22** (20)	K_3PO_4 , DMF, 100–110 °C	57–80	3	35
	Br					10	

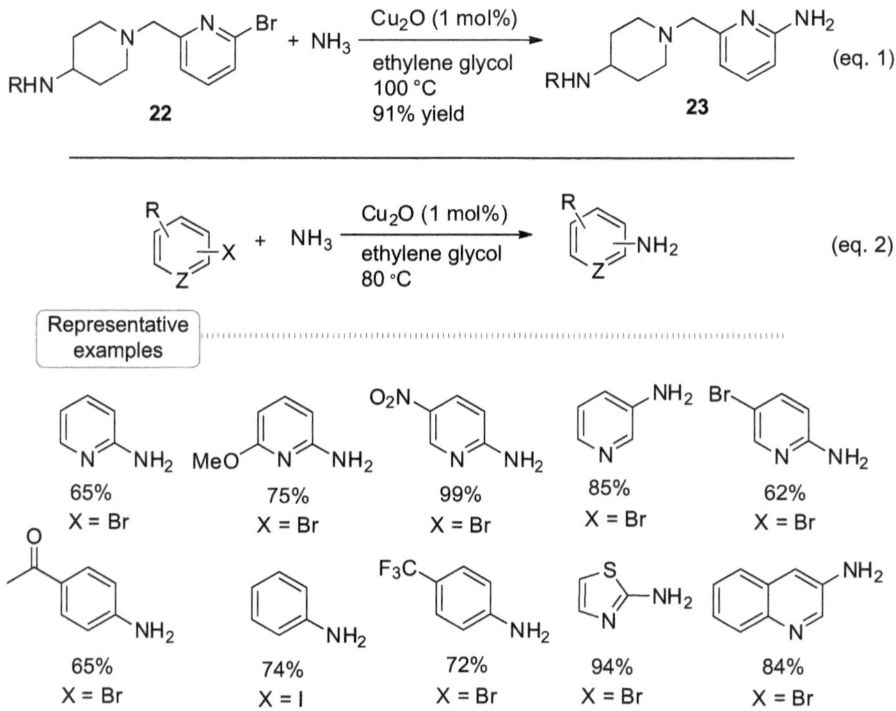

Scheme 1.3 Cu_2O-catalyzed synthesis of primary aryl amines from aryl halides and ammonia.

Copper-carbene complexes **C4,** 1-(5,6,7,8-tetrahydroquinolin-8-yl)-2-methylpropane-1-one **L18**, sulfonato-Cu(salen) complex **C6** and D-glucosamine **L70** could also affect this reaction (entries 4–7).[40–43] But in one case aqueous ammonia was not a suitable coupling partner and a saturated ammonia solution in MeOH/NMP had to be used.[40]

Recently, Page and coworkers reported that amination of aryl iodides could work well in liquid ammonia at room temperature under the catalysis of 1 mol% CuI and 1 mol% ascorbic acid **L71**. Increasing the reaction temperature to 100 °C could lead to complete conversion in case of aryl bromides as substrates (entry 8).[44]

By employing some synthetic equivalents of ammonia, other alternative methods for preparing primary aryl amines have been reported. During the studies on the CuI/L-proline catalyzed coupling of aryl halides with amidine hydrochlorides, Fu and coworkers discovered that primary anilines were directly isolated. This transformation was believed to undergo a coupling/hydrolysis process as indicated in Scheme 1.4.[45]

The CuI/L-proline or CuI/*N*,*N*′-dimethylethylenediamine (**L26**) catalyzed coupling of aryl halides with sodium azide generally provided the

Table 1.4 Copper/ligand-catalyzed preparation of aryl amines from aryl halides and ammonia.

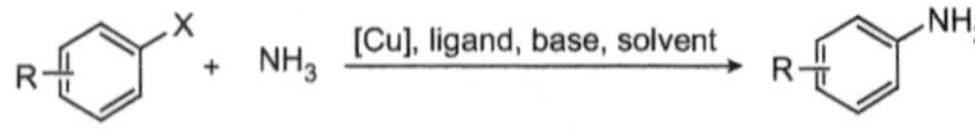

Entry	X	[Cu] (mol%)	Ligand (mol%)	Nucleophile	Reaction conditions	Yield (%)	Number of examples	Ref.
1	I	CuI (20)	**L1** (40)	NH_4Cl or	K_2CO_3, DMSO or DMSO/H_2O, rt	7–97	22	37
	Br			$NH_3{\cdot}H_2O$	K_2CO_3, DMSO or DMSO/H_2O, rt–80 °C	10–91	3	
2	Br	CuI (20)	**L2** (40)	$NH_3{\cdot}H_2O$	K_2CO_3, DMSO, 50–70 °C	55–92	15	38
3	I			$Cu(acac)_2$ (10)	**L52** (40)			
				$NH_3{\cdot}H_2O$	Cs_2CO_3, DMF, 60–90 °C	23–99	5	39
	Br					41–92	10	
4	Br	**C4 (5)**		NH_3 in MeOH	K_2CO_3, MeCN or NMP, 90–120 °C	7–95	17	40
5	I	CuBr(5)	**L18** (10)	$NH_3{\cdot}H_2O$	K_3PO_4, DMSO, rt-70 °C	60–95	19	41
	Br	(10)	(20)		K_3PO_4, DMSO, 110 °C	52–95	13	
6	I	**C6** (5)		$NH_3{\cdot}H_2O$	NaOH, H_2O, 120 °C	71–95	9	42
	Br	**C6** (10)				64–95	14	
7	I, Br	CuI (10)	**L70** (10)	$NH_3{\cdot}H_2O$	K_2CO_3, acetone/H_2O, 90 °C	50–96	14	43
						57–98	7	
8	I	CuI (1)	**L71** (1)	NH_3	rt	57–99	16	44
	Br				100 °C	63–99	16	

corresponding aryl azides.[46,47] Surprisingly, the Helquist group observed that primary anilines could be produced exclusively if the reaction was carried out by using 100 mol% CuI, 130 mol% L-proline and 2 equiv. of sodium azide (Scheme 1.5).[48] Although the detailed mechanism is not clear yet, this reaction provides an alternative approach for assembling primary anilines from aryl halides.

The CuI/*N*,*N*′-dimethylethylenediamine (**L26**)-catalyzed coupling of trifluoroacetamide with aryl halides could proceed smoothly in DMF (for aryl iodides) or dioxane (for aryl bromides) to deliver the corresponding amidation products, which could be easily converted into primary aryl amines *via* hydrolysis in a one-pot manner (Scheme 1.6).[49]

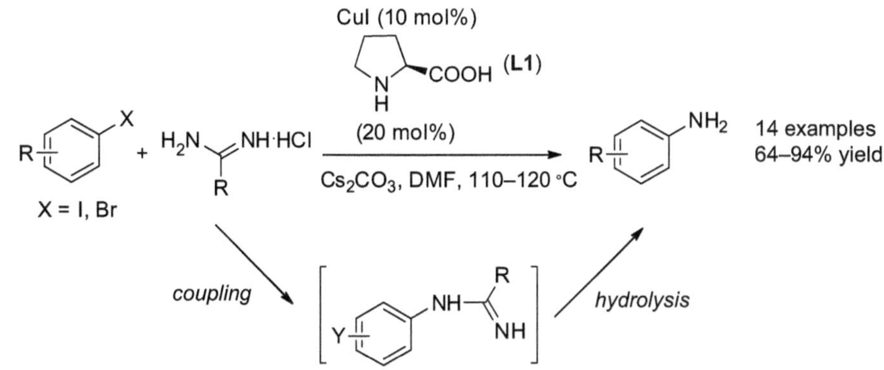

Scheme 1.4 CuI/L-proline-catalyzed synthesis of anilines from aryl halides and amidines.

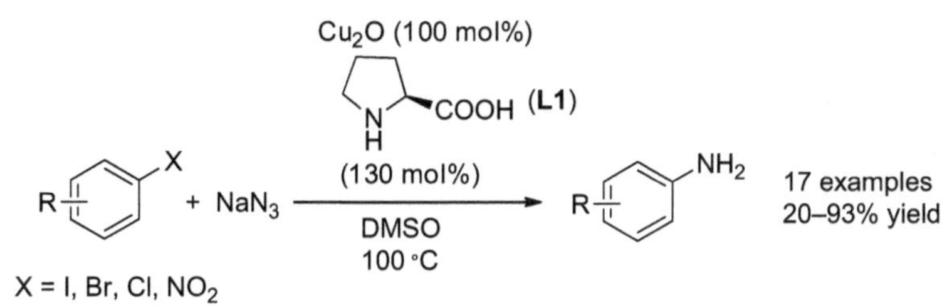

Scheme 1.5 CuI/L-proline-catalyzed synthesis of anilines from aryl halides and sodium azide.

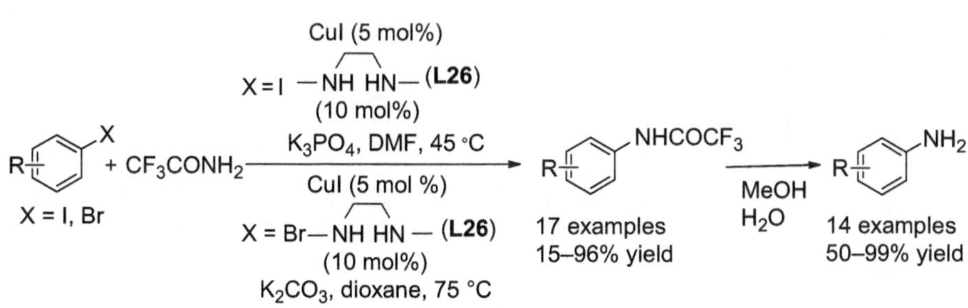

Scheme 1.6 CuI/*N*,*N*′-dimethylethylenediamine-catalyzed one-pot synthesis of anilines from aryl halides and 2,2,2-trifluoroacetamide.

1.2.2 Arylation and Vinylation of *N*-Heterocycles

1.2.2.1 Coupling of Aryl Halides and N*-Heterocycles*

N-Aryl heterocycles are frequently found in bioactive molecules and are useful precursors for material science. Therefore, method development for synthesis of *N*-aryl heterocycles has become a hot area in organic synthesis.

Table 1.5 Copper/ligand catalyzed coupling between aryl halides and *N*-heterocycles amines.

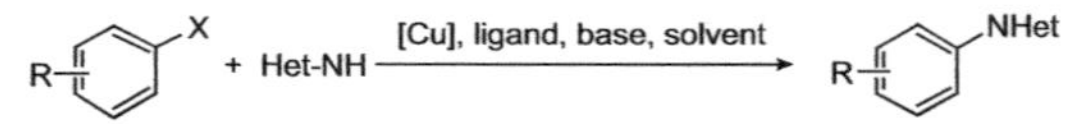

Entry	*X*	*[Cu] (mol%)*	*Ligand (mol%)*	N-*Heterocycle*	*Reaction conditions*	*Yield (%)*	*Number of examples*	*Ref.*
1	I	$(CuOTf)_2 \cdot PhH$ (10)	**L29** (200)	N3, N8	dba, Cs_2CO_3, Xylene, 110–125 °C	62–99	10	50
2	I	Cu_2O (0.25–5)	**L30** (0.075–15)	N3, N8	Cs_2CO_3, PEG, *n*-PrCN or NMP or	44–95	17	51
	Br	Cu_2O (5–10)	**L30** (15–30)		MeCN or DMSO, 80–150 °C	60–98	24	52
3	I	CuI (20)	**L29** (20)	N2, N3, N8	KF/Al_2O_3, Xylene, 130–140 °C	74–98	12	53
4	Br	CuI (20)	**L29** (20)	N2, N3, N8	KF/Al_2O_3, Xylene, 130–140 °C	71–92	13	54
5	I	Cu_2O nanoparticles (10)	**L29** (20)	N3, N8	TBAF, 110–115 °C	63–98	6	
	Br				TBAF, 140–145 °C	80–98	11	55
	Cl						7	
6	I	CuI (1–10)	**L24** (10–20)	N1-3, N6-7, N10	K_2CO_3, or Cs_2CO_3 or K_3PO_4, Dioxane, 110 °C	62–99	18	56
	Br					95	1	
7	I	CuI (0.2–10)	**L25** (1–20)	N6	K_3PO_4, Toluene, 80–110 °C	70–98	27	57
	Br					72–93	12	
8	I, Br	CuI (5–10)	**L25, L29** (10–20)	N1-3, N5, N7-9, N12	K_3PO_4, Toluene, 110 °C	67–98	26	58
9	I, Br	CuI (5)	**L48** (10)	N1-3, N5-6, N8, N11	Cs_2CO_3, DMF, 90–110 °C	61–99	25	59
	Cl			N3			3	
10	I	CuI (10)	**L36** (10)	N6	K_2CO_3, MeCN, 82 °C	34–99	17	60
11	I	CuI (5)	**L33** (5)	N6	K_3PO_4, DMF, 110 °C	23–97	15	61
12	I, Br	CuCl (5)	**L44** (10)	N1, N3, N6, N8, N10	NaOH, nBu_4NBr, H_2O, 100 °C	62–97	15	19
	Cl				NaOH, nBu_4NBr, H_2O, KI, 100 °C	66	1	
13	I, Br	CuI (10)	**L1, L3** (20)	N2, N3, N8	K_2CO_3, DMSO, 60–110 °C	22–99	16	7, 62
	I	CuI (10)	**L1** (20)	N1-3, N6, N10	K_2CO_3, DMSO, 75–90 °C	75–93	15	

14	Br	CuI (30)	**L1** (60)	N3, N8	K_2CO_3, [Bmim]BF_4, 105–115 °C	19–87	22	63
15	I, Br	CuI (10)	**L9** (20)	N3, N6	K_2CO_3, DMF, 110 °C	71–95	6	32
	Cl					24–36	2	
16	I, Br, Cl	$Cu(OAc)_2.H_2O$ (20)	**L6** (20)	N1, N3, N6	Cs_2CO_3, DMF, 140 °C	32–99	16	64
17	I, Br	Cu_2O (5)	**L16, L37** (20)	N2	Cs_2CO3, MeCN or DMF, 25–110 °C	12–98	42	65
18	I, Br	Cu_2O (5)	**L37, L16** (20)	N1, N3, N5, N6	Cs_2CO_3, MeCN, 50–82 °C	79–99	17	66
19	Br	CuI (10)	**L13** (10)	N3, N8	$(Et_4N)_2CO_3$, DMF/H_2O, 130 °C	40–90	14	67
20	Br	CuBr (10)	**L15** (20)	N1, N3, N6, N8	Cs_2CO_3, DMSO, 80–90 °C	68–95	14	68
21	I, Br	CuI (5)	**L26, L13** (10)	N1, N3, N5, N6-8, N10, N13	CsF, THF or DMSO, RT–80 °C	65–99	12	69
22	I, Br	CuI (2–10)	**L14** (4–20)	N1-3, N5, N6, N8, N10	K_2CO_3, DMSO, 40–110 °C	30–98	29	70
23	I, Br	CuBr (5)	**L19** (10)	N2, N3, N5, N8	Cs_2CO_3, DMSO, 60–120 °C	65–95	31	71
24	I, Br	CuBr (10)	**L50** (20)	N3, N8	Cs_2CO_3, DMSO, 45–75 °C	23–96	8	72
25	I	CuCl (10)	**L50** (10)	N1	$(CH_3)_4N^+OH^-$, DMSO, 80 °C	74–90	5	14
26	I, Br	CuO/$FeCl_3$ (10/10)	**L58** (20)	N1, N6	Cs_2CO_3, DMF, 90–110 °C	75v89	6	17
27	I, Br	CuI (2.5–10)	**L62, L63, L64** (5–20)	N1, N3, N6, N8	NaOMe, DMSO, 90–110 °C	48–99	22	4
	Cl	CuI (10)	**L63, L64** (20)	N1, N3			7	
28	I, Br	CuI (10)	**L54** (10)	N1-3, N6, N8, N10, N14	K_2CO_3, DMF, 90–110 °C	62–99	27	73
	Cl			N2			3	
29	I, Br	CuI (10)	**L21** (20)	N3, N6	K_3PO_4, DMF, 95–120 °C	85–96	6	34
	Cl			N6		39	1	
30	I	CuBr (10)	**L65** (12)	N2-3, N6, N8	Cs_2CO_3, DMSO, 80 °C	61–98	11	24
31	I, Br	CuBr (5)	**L66** (10)	N2-3, N8	Cs_2CO_3, DMF, 90 °C	65–90	17	74
32	Br	CuBr (10)	**L46** (20)	N3, N8	TBAF, 145–150 °C	30–99	14	75
	Cl						6	
33	I, Br	$CuSO_4$ (10)	**L68** (20)	N3, N6, N8	Cs_2CO_3, H_2O, 120 °C	32–95	21	76

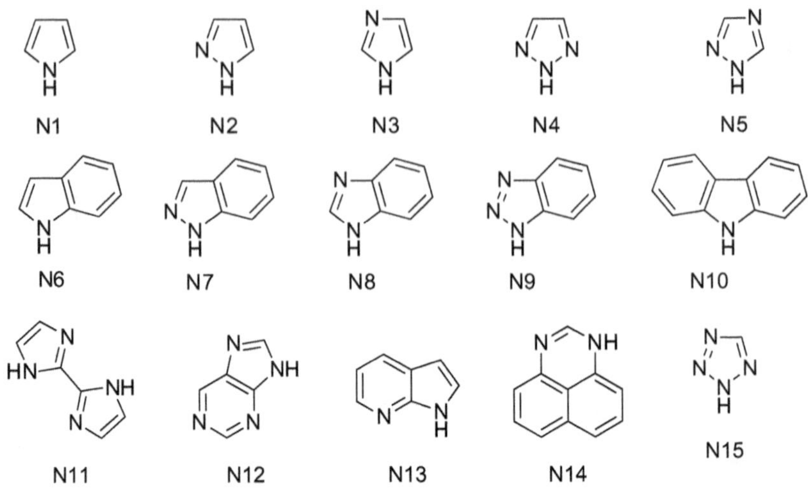

Figure 1.3 *N*-Heterocycles used for the copper-catalyzed coupling reactions.

Copper/ligand catalyzed cross-coupling of aryl halides with *N*-heterocycles has been proven to be one of the most powerful approaches for assembling these compounds. The suitable *N*-heterocycle coupling partners were outlined in Figure 1.3.

Although a stoichiometric quantity of ligand had to be used, 1,10-phenanthroline (**L29**) was the first ligand used for copper-catalyzed arylation of imidazole with aryl iodides (Table 1.5, entry 1).[50] As an extension of this study, Cu_2O/4,7-dimethoxy-1,10-phenanthroline (**L30**) was found to be more effective, leading to the coupling reaction occur at 80–110 °C in the presence of Cs_2CO_3 and using *n*-PrCN as a solvent (entry 2).[51,52] For coupling of iodobenzene and imidazole, complete conversion could be achieved with only 0.05 mol% of copper(I) iodide and 0.075 mol% of ligand. Other improved procedures, including the use of KF/Al_2O_3 or TBAF as the base, have been reported for Cu/1,10-phenanthroline-catalyzed arylation of diazoles, benzoimidazoles and indoles (entries 3–5).[53–55]

Diamine ligands, like *trans*-dicyclohexane-1,2-diamine **L24** and *trans*-*N*,*N*′-dimethyl -dicyclohexane-1,2-diamine **L25**, are also useful promoters for copper catalyzed coupling of aryl halides with *N*-heterocycles (entries 6–8).[56–58] A wide range of heterocycles, including indoles, pyrroles, pyrazoles, imidazoles and triazoles, were successfully employed in this case. Noteworthy is that optimized reaction conditions are slightly different for different *N*-heterocycles (entries 6–8).

Some other *N*,*N*-bidentate ligands have been demonstrated to be able to promote Cu-catalyzed arylation of *N*-heterocycles. Among them, (*S*)-pyrrolidinylimidazole (**L48**) could be used for coupling reaction of activated aryl chlorides with imidazoles (entry 9);[59] *N*,*N*′-dimethyl-1,1′-binaphthyl-2,2′-diamine **L36** and 2-(2′-pyridyl)benzimidazole **L33** were more suitable for the coupling of aryl halides with indoles (entries 10–11);[60,61] and

(1*E*,2*E*)-oxalaldehyde dioxime **L44** was a suitable ligand for coupling in water (entry 12).[19]

For *N*,*O*-bidentate ligands, L-proline (**L1**) has been proven to be a powerful promoter. In this case coupling reaction of aryl iodides with indole, pyrrole, carbazole, imidazole or pyrazole could be carried out at 75–90 °C, and coupling reactions of electron-deficient aryl bromides with imidazole or pyrazole occurs at 60–90 °C (entry 13).[7,62] However, for coupling reaction of electron-rich aryl bromides with imidazole or pyrazole, *N*,*N*-dimethylglycine (**L3**) had to be used and the reaction temperature should be increased to 110 °C. CuI/L-proline also worked well in ionic liquid and thus recycling the catalyst system became possible (entry 14).[63] Pipecolinic acid **L9** and hippuric acid **L6** have been found to be effective (entries 15–16),[32,64] and aryl chlorides with strong electron-withdrawing groups could be used as the coupling partners in this case.

Taillefer and coworkers demonstrated that salicylaldoxime (Salox, **L16**) and Chxn-Py-Al (**L37**) were efficient ligands for Cu_2O-catalyzed *N*-arylation of pyrazoles, imidazoles, pyrroles, indoles and triazoles (entries 17 and 18).[65,66] Both aryl iodides and bromides gave good coupling yields.

8-Hydroxyquinoline **L13** and its derivatives such as 8-hydroquinalidine **L14** and tetrahydro-8-hydroxyquinoline **L15** was another class of effective *N*,*O*-bidentate ligands for Cu-catalyzed *N*-arylation of *N*-heterocycles (entries 20–23),[67–70] while pyridin-2-yl β-ketone **L19** was found to have a similar effect (entry 24).[71]

In case of *O*,*O*-bidentate ligands, ethyl 2-oxocyclohexanecarboxylate (**L19**) and BINOL could make the copper-catalyzed *N*-arylation of imidazoles and benzimidazoles proceed under relatively mild reaction conditions (45–75 °C) (entries 25 and 26).[14,17,72] Additionally, *N*-hydroxyimides (**L62–64**) (entry 27)[4] and 1,3-di(pyridine-2-yl)propane-1,3-dione (**L54**) were also efficient promoters (entry 28).[73]

Several phosphine ligands were applicable for promoting Cu-catalyzed arylation of *N*-heteroarylcycles (entries 29–31).[24,34,74] CuBr/2-aminopyrimidine-4,6-diol (**L46**) catalyzed coupling worked well in solvent-free conditions when TBAF was used as a base. Some aryl chlorides were compatible with these conditions, giving the corresponding coupling products in good yields (entry 32).[75] The arylation of imidazoles, indoles and benzimidazoles was possible in water if copper(II) sulfate and 1,2-bis(2-pyridyl)-ethane-*N*,*N*′-dioxide (**L68**) were employed (entry 33).[76]

Buchwald and coworkers have indicated that there are some interesting trends in regioselectivity for coupling reaction of aryl halides with substituted imidazoles, pyrazoles and indazoles. For unsymmetrical pyrazoles and imidazoles, the *N*-arylation occurred at the less hindered nitrogen (Scheme 1.7).[58] In addition, aryl iodides showed the better *N*-1 selectivity than that of aryl bromides.

1.2.2.2 *Coupling of Vinyl Bromides and* N*-Heterocycles*

Generally, vinyl halides are more reactive coupling partners than aryl halides. Thus, Cu/ligand catalyzed coupling reaction of vinyl bromides with

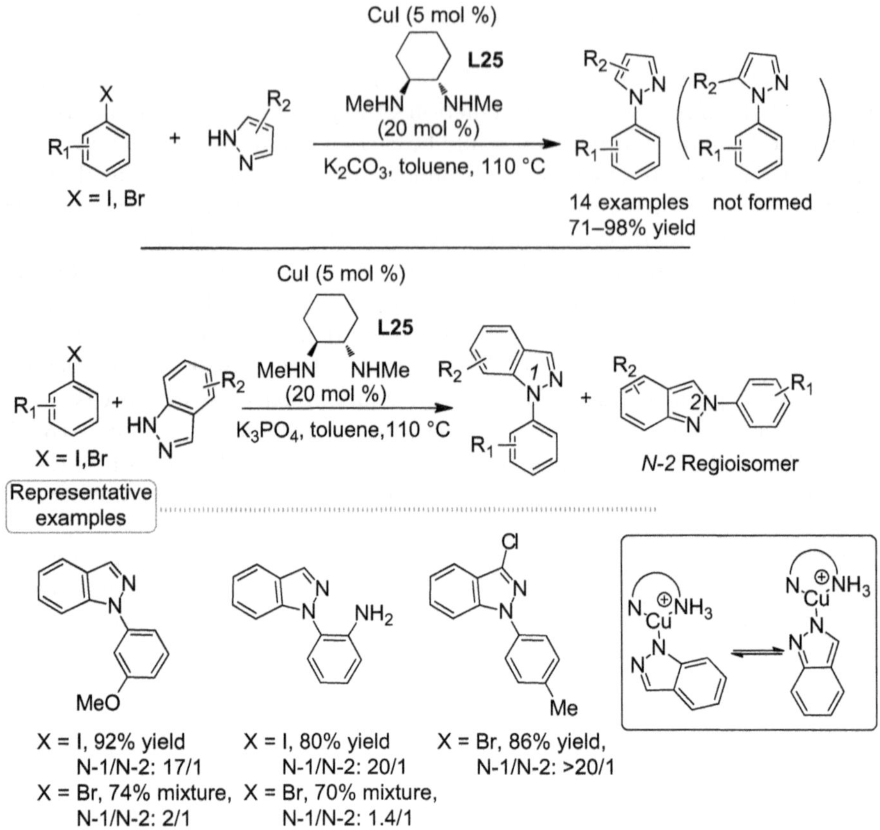

Scheme 1.7 The regioselectivity for coupling of aryl halides and unsymmetrical heterocycles.

N-heterocycles could also take place under mild conditions (Table 1.6, entries 1–6). For example, CuI/L-proline (**L1**) catalyzed coupling reaction of *β*-bromostyrene with imidazoles and benzoimidazoles worked at 110 °C in [BMIM]BF_4 (entry 1).[77] Chxn-Py-Al (**L37**) could make vinylation of azoles complete at relatively low reaction temperatures (entry 4).[78] Notably, CuI/ethylenediamine (EDA, **L27**) has been applied in vinylation of sterically α-bromostyrene and polysubstituted vinyl bromides (entry 6).[79]

1.2.3 Aryl Amidation

The copper-promoted arylation of amides is known as the Goldberg reaction. Similar to the traditional Ullmann reaction, this transformation suffered from high reaction temperature and strictly limited substrates. The introduction of suitable ligands has changed this situation, and the detailed progress is summarized below.

Table 1.6 Copper/ligand catalyzed coupling between vinyl bromides and *N*-heterocycles.

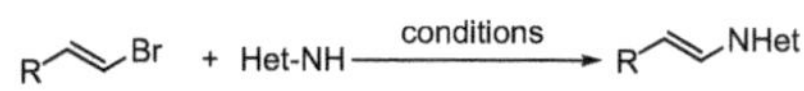

Entry	*X*	*[Cu] (mol%)*	*Ligand (mol%)*	*N-Heterocycle*	*Reaction conditions*	*Yield (%)*	*Number of examples*	*Ref.*
1	Br	CuI (10)	**L1** (20)	N3, N8	[BMIM]BF_4, K_2CO_3, 110 °C	60–93	9	77
2	Br	Cu_2O (5)	**L50** (10)	N3, N8	Cs_2CO_3, CH_3CN, 80–90 °C	53–99	15	80
	Cl			N3	Cs_2CO_3, CH_3CN, 130 °C	45–53	2	
3	Br	CuI (15)	**L50** (30)	N3, N8	Cs_2CO_3, DMSO, 60–80 °C	82–90	8	81
4	Br	CuI (10)	**L37** (5)	N1-3, N5, N6, N7, N9, N15	Cs_2CO_3,CH_3CN, 35–80 °C	50–98	16	78
5	I	CuI (5)	**L33** (5)	N2-3, N5, N6, N7, N15	Cs_2CO_3, DMF, 40–60 °C	81–96	16	82
	Br			N2, N6, N9	Cs_2CO_3, DMF, 80 °C	82–91	3	
6	Br	CuI (10–20)	**L27** (20–40)	N1, N3, N6, N7, N10	K_3PO_4, Dioxane, 110 °C	21–90	17	79

1.2.3.1 Cross-Coupling of Aryl Halides with Amides and Carbamates

1,2-Diamines such as ethylenediamine (EDA, **L27**), *N,N′*-dimethylethylenediamine (DMEDA, **L26**), *trans*-dicyclohexane-1,2-diamine (CyDA, **L24**) and *trans-N,N′*-dimethyl- dicyclohexane-1,2-diamine (DMCyDA, **L25**), were remarkably efficient *N,N*-bidentate ligands for promoting Cu-catalyzed amidation of aryl halides (Table 1.7, entries 1 and 2).[56,83] The catalytic system allowed a wide range of amides (including lactams, open chain alkylamides and arylamides) and carbamates work at about 80–110 °C. Notably, DMCyDA (**L25**) could promote the amidation of deactivated aryl chlorides at higher reaction temperatures (110–130 °C). However, in this case aryl chlorides have to serve as both substrates and solvents.

The Padwa group has extended this methodology to the preparation of 2- and 3-substituted amidofurans and amidothiophenes (entry 3),[84] which are useful building blocks for alkaloid synthesis. Generally, good yields were observed when primary amides, lactams and 2-oxazolidinones were used as coupling partners, while low yields were obtained in the case of secondary amides and carbamates as substrates. Researchers at Novartis reported that CsF could be used to replace the generally used stronger bases like K_2CO_3 and K_3PO_4 in the CuI/DMEDA-catalyzed amidation of aryl iodides. In this case, most aryl iodides gave the amidation products in high yields even at room temperature (entry 4).[69] This improvement will be of benefit for coupling

Table 1.7 Cu/diamine-catalyzed coupling reaction of aryl halides with amides and carbamates.

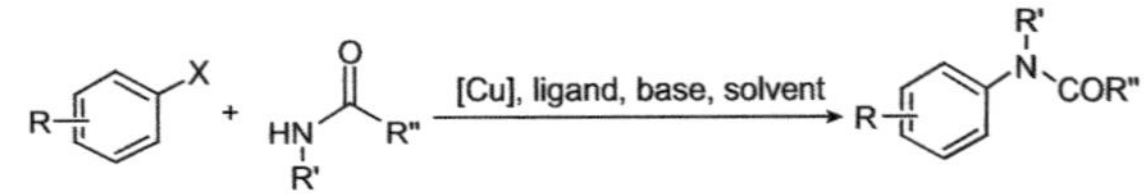

Entry	*X*	*[Cu] (mol%)*	*Ligand (mol%)*	*Amides*	*Reaction conditions*	*Yield (%)*	*Number of examples*	*Ref.*
1	I	CuI (0.2–5)	**L24, L25** (5–10)	Arylamides, alkylamides, lactams, carbamates	K_3PO_4 (or Cs_2CO_3) dioxane (or toluene), 110–130 °C	69–99	16	56
	Br	CuI (1–10)	**L24** (10–20)			81–97	9	
	Cl	CuI (5)	**L25** (10)			51–95	4	
2	I	CuI (0.2–5)	**L24-L27** (5–20)	Arylamides, alkylamides, lactams, carbamates	K_3PO_4 (or Cs_2CO_3) dioxane (or toluene), 80–120 °C	63–99	23	83
	Br	CuI (1–10)				49–99	14	
	Cl	CuI (5)	**L25** (10)	Lactams, arylamides	K_2CO_3, 110–130 °C	51–95	4	
3	Br	CuI (10)	**L26 or L25** (10)	Lactams, arylamides, carbamates, alkylamides, oxzazolidinones	K_2CO_3, Dioxane, 110 °C	11–99	23	84
4	I, Br	CuI (5)	**L26** (10)	Lactams, arylamides, carbamates, alkylamides, oxzazolidinones	CsF, THF or EtOH or MeCN, 25–100 °C	83–99	19	69
5	Br	CuI (3–5)	**L24** (10)	Oxzazolidinones	K_2CO_3, Dioxane, 110–120 °C	20–99	17	85
6	I, Br	CuI (5)	**L26** (10)	Trifluoroacetamides	DMF (or dioxane) K_3PO_4 (or K_2CO_3), 45–75 °C	15–99	19	49
7	I, Br	CuI (5–10)	**L26** (10–20)	Oxindoles	K_2CO_3, MeCN, reflux	14–92	20	86
8	Br	CuI (5)	**L26** (50)	Sulfonamides	K_2CO_3, MeCN, 25–80 °C	62–97	30	87
9	Br	CuI (50)	**L26** (100)	Methyl pyroglutamate	Cs_2CO_3, Dioxane, rt-60 °C	73–95	9	88
10	I	CuI (10)	**L27** (10)	Lactams, arylamides, carbamates, oxzazolidinones	K_3PO_4 or Cs_2CO_3, Dioxane, 110 °C	41–95	13	89
11	I	CuI (10)	**L27** (20)	Lactams, arylamides, alkylamides, oxzazolidinones	K_3PO_4, Dioxane, Reflux	21–95	11	90

reactions using substrates that are incompatible with high temperature and/or strongly basic conditions.

By using *trans*-dicyclohexane-1,2-diamine (CyDA, **L24**) as the ligand, Trehan and coworkers were able to carry out CuI-catalyzed coupling of aryl bromides and 2-oxazolidinone at 110–120 °C (entry 5).[85] Some pharmaceutically important *N*-aryloxazolidinones such as Linezolid and Toloxatone were synthesized conveniently under these conditions. Under the catalysis of CuI/DMEDA (**L26**), trifluoroacetamides, oxindoles, methyl pyroglutamate and sulfonamides could be used for this coupling (entries 6–9).[49,86–88] Additionally, amidation of 2-iodothiophenes and 2-iodoselenophenes could be achieved by using EDA (**L27**) as the promoter (entries 10–11).[89,90]

Notably, Bolm and coworkers found that some aryl amidation reactions could proceed at part-per-million catalyst loading (Scheme 1.8).[91] Two factors appear to be dominant in this case: ligand quantity and the reaction temperature. For complete conversion, they found that the ratio of ligand and copper salt should be much higher and the reaction temperature had to be increased to 135 °C.

Some other efficient ligands for the amidation of aryl halides are summarized in Table 1.8. Normally, higher reaction temperatures are required for this transformation because amides are less reactive than amines. However, two catalytic systems, Cu_2O/Salox (**L16**, or Chxn-Py-Al (**L37**)) and CuBr/ethyl 2-oxocyclohexanecarboxylate (**L50**) could make the coupling reaction complete at relatively lower reaction temperatures (entries 3 and 8).[66,72] Recently, the Taillefer group found that *O*,*O*-bidentate ligand TMHD (**L49**) was suitable for coupling of aryl iodides and secondary aryl amides, affording the tertiary acyclic amides with good yields (entry 11).[100]

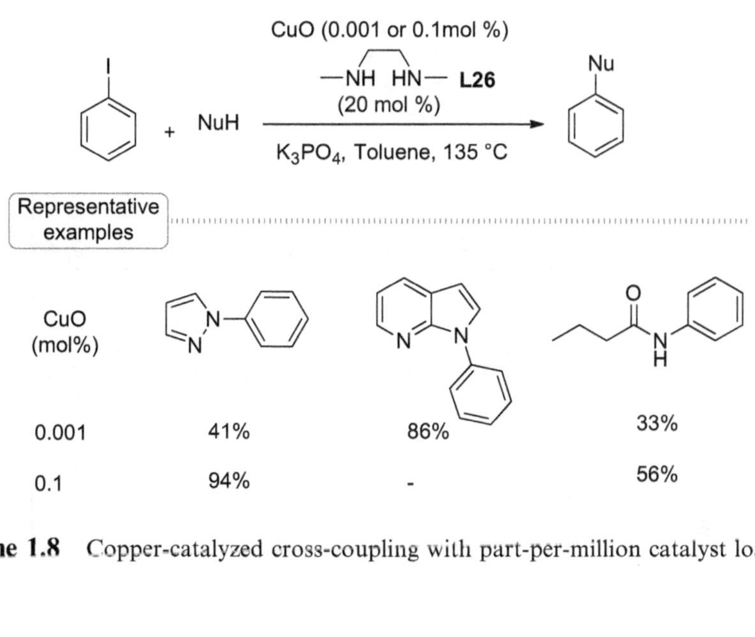

Scheme 1.8 Copper-catalyzed cross-coupling with part-per-million catalyst loading.

Table 1.8 Copper/ligand-catalyzed coupling between aryl halides and amides and carbamates.

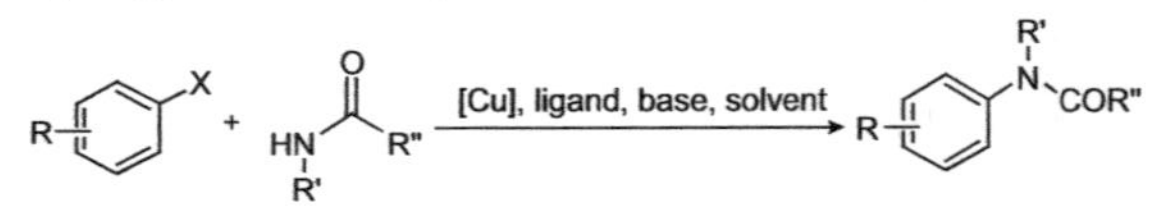

Entry	*X*	*[Cu] (mol%)*	*Ligand (mol%)*	*Amides*	*Reaction conditions*	*Yield (%)*	*Number of examples*	*Ref.*
1	I	CuI (10)	**L29** (10)	Arylamides, alkylamides, lactams	KF/Al_2O_3, Toluene, reflux	67–99	19	92
2	I	CuI (5)	**L43** (10)	Lactams, arylamides, alkylamides, carbamates	K_3PO_4, DMSO, 110–130 °C	25–92	13	93
3	I	Cu_2O (5)	**L16** (20), or **L37** (20)	Sulfonamides, arylamides, alkylamides, lactams, oxzazolidinones 4-hydroxypyridines, pyridazinones	Cs_2CO_3, DMF or MeCN, 50–82 °C	81–97	14	66
4	I	CuI (5)	**L5** (20)	Lactams, arylamides, alkylamides, sulfonamides	K_3PO_4, Dioxane 100 °C	41–98	30	94
5	I	CuI (20)	**L4** (20)	Sulfonamides	K_3PO_4, DMF, 100 °C –reflux	50–98	18	95
	Br		**L3** (20)			68–99	16	
6	I	**C5** (5)		Pyridazinones	K_2CO_3, DMF, 100 °C	70–94	4	96
	Br			140 °C	80–90	3		
7	Br	CuI (5)	**L3** (10)	Oxazolidinone	K_2CO_3, DMF, 120 °C	53–98	19	97
8	I, Br	CuBr (10)	**L50** (20)	Lactams, arylamides, alkylamides 2-hydroxypyridine	Cs_2CO_3, DMSO, RT–75 °C	76–96	12	72
9	I	CuI (10)	**L56 (10)**	Lactams, oxzazolidinones	Cs_2CO_3 or K_3PO_4, DMF/Dioxane, 110 °C	33–98	12	98
10	Br	CuI (20)	**L54 (20)**	Sulfonamides	K_2CO_3, DMF, 110–120 °C	10–99	26	99
11	I	CuI (5)	**L49 (10)**	Secondary aryll acyclic amides	K_3PO_4, Toluene, 130 °C	35–95	24	100

1.2.3.2 Cross-Coupling of Vinyl Halides with Amides or Carbamates

Cu-catalyzed vinylation of amides could also be promoted by some bidentate ligands (Table 1.9). The first successful example was reported by Shen and Porco, in which copper(I) thiophene-2-carboxylate (CuTC) was employed as a catalyst, and reaction took place at 90 °C in NMP to give the corresponding enamides in good yields (entry 1).[101] Soon after that, some diamine ligands were demonstrated to be very effective for this transformation (entries 2–3).[102,103] Interestingly, double amidation of 1,2-dibromo-1-alkenes could be achieved under the assistance of DMEDA (entry 3). Additionally, $Cu(CH_3CN)_4PF_6$/Me_4-Phen (**L31**) and CuI/*N*,*N*-dimethylglycine (**L3**) have been found to be efficient catalytic systems for this transformation (entries 4 and 5).[104,105]

1.2.3.3 Cross-Coupling of Alkynyl Halides with Amides or Carbamates

N,*N*-Bidentate ligands were more suitable for the amidation of alkynyl halides (Table 1.10). Although DMEDA was the first ligand used for this transformation (entry 1),[106] 1,10-phenanthroline seemed more effective (entries 2–5).[107–110] In these cases, cheap and stable $CuSO_4 \cdot 5H_2O$ could be used as the copper source.

Very recently, Evano and coworkers developed a clean and fast oxidative cross-coupling process of an alkynylcopper reagent with amides for the synthesis of ynamides (Scheme 1.9).[111] Chelating ligand TMEDA (tetramethylene-1,2-diamine) as the organic promoter could allow this reaction to be carried out at room temperature and the corresponding ynamides but not diynes were highly selectively obtained. No additional base was required. Although it was a non-catalytic process, this so-called click-alkylnylation is still quite attractive due to its mild reaction conditions.

1.2.4 Cross-Coupling of Aryl Halides or Vinyl Halides with Hydrazide and Hydrazine

Substituted hydrazines are special nucleophiles for Ullmann-type coupling reaction. The key problem for amidation of substituted hydrazine is how to control the regioselectivity of the two nitrogen atoms. Some successful examples were summarized in Table 1.11. *N*-Boc (*tert*-butoxycarbonyl) hydrazine was found to be the best substrate to reach high regioselectivity. In 2001, the Buchwald group reported that CuI/1,10-phenanthroline (**L29**) catalyzed the cross-coupling between aryl iodides, and *N*-Boc hydrazine could provide *N*-arylated *N*-Boc hydrazines exclusively (Table, entry 1).[112] Subsequently, this strategy was extended to prepare bis-Boc biaryl hydrazines and bis-Boc vinyl hydrazines (entries 2–3).[113,114] 2-Picolinic acid (**L10**) and *trans*-4-hydroxy-L-proline (**L2**) are efficient *N*,*O*-bidentate ligands for this transformation (entry

Table 1.9 Copper/ligand catalyzed vinylation of and amides and carbamates.

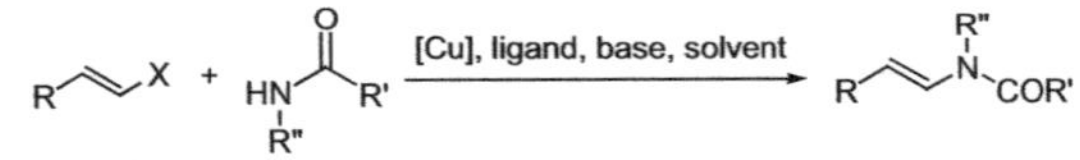

Entry	*X*	*[Cu] (mol%)*	*Ligand (mol%)*	*Amides*	*Reaction conditions*	*Yield (%)*	*Number of examples*	*Ref.*
1	I	CuTC (10)		alkylamides	Cs_2CO_3, NMP, 90 °C	57–75	6	101
2	I, Br	CuI (5–10)	**L26** (20)	Lactams, alkylamides, carbamates, arylamides	Cs_2CO_3, THF, rt–70 °C	76–95	11	102
					K_2CO_3, Toluene 80–115 °C	62–92	6	
3	Br	CuI (12)	**L26** (18)	Lactams	K_3PO_4, Toluene, 80 °C	51–94	8	103
4	I	$Cu(CH_3CN)_4PF_6$ (10–20)	**L31** (20)	Lactams, alkylamides, arylamides	Rb_2CO_3, DMA, 45–60 °C	6–92	13	104
5	I, Br	CuI (10–100)	**L3** (20–100)	Lactams, alkylamides, Arylamides, carbamates	Cs_2CO_3, Dioxane, rt-60 °C	43–87	17	105
					80 °C	62–65	4	

Table 1.10 Copper/ligand catalyzed coupling between alkynyl halides and amides and or carbamates.

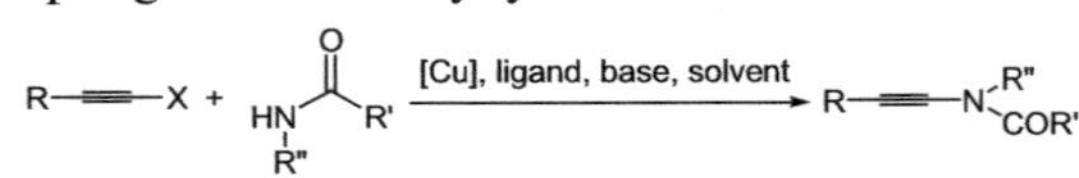

Entry	*X*	*[Cu] (mol%)*	*Ligand (mol%)*	*Amides*	*Reaction conditions*	*Yield (%)*	*Number of examples*	*Ref.*
1	I, Br	CuCN or CuI (1–5)	**L26** (3–10)	Oxazolidinones, lactams, carbamates	K_3PO_4, toluene, 110–150 °C	10–85	23	106
2	Br	$CuSO_4 \cdot 5H_2O$ (5–20)	**L29** (10–40)	Oxazolidinones, sulfonamides, carbamates, lactams	K_3PO_4 or K_2CO_3, toluene, 60–95 °C	37–98	32	107
3	Br	$CuSO_4 \cdot 5H_2O$ (10–20)	**L29** (20–40)	Oxazolidinones, carbamates, alkylamides	K_3PO_4, Toluene, 70–80 °C	52–91	11	108
4	Br	$CuSO_4 \cdot 5H_2O$ (20)	**L29** (40)	Carbamates	K_3PO_4, Toluene 80 °C	22–88	22	109
5	Br	$CuSO_4 \cdot 5H_2O$ (10)	**L29** (20)	Oxazolidinones, carbamates, Sulfonamides	K_2CO_3, Toluene 85 °C	36–95	21	110

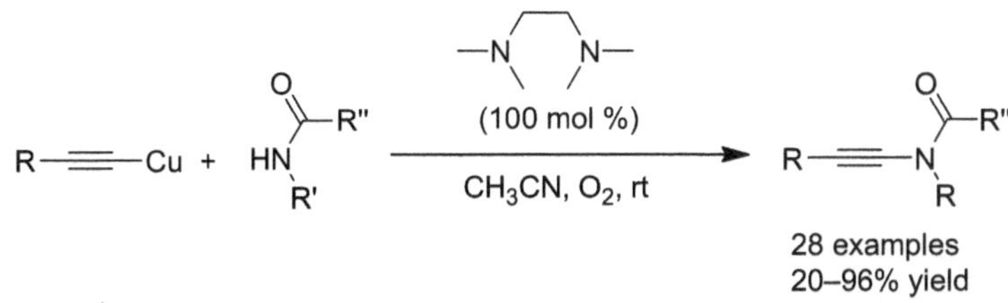

Scheme 1.9 TMEDA-promoted a non-catalytic procedure for the synthesis of ynamides.

4 and 5).[115,38] Noteworthy is that improved results could be obtained for aryl bromide substrates when *trans*-4-hydroxy-L-proline (**L2**) was used. By adding LiCl, CuI/PPAMP (**L21**)-catalyzed *N*-arylation of phenylhydrazine could afford *N*,*N*-biaryl hydrazines with excellent regioselectively (entry 6).[34]

1.2.5 Cross-Coupling of Aryl Halides with Ureas, Guanidines, Hydroxypyridines and *N*, *O*-Substituted Hydroxyamines

The coupling reactions of aryl halides with other related N-nucleophiles are listed in Table 1.12. The diamine ligands, CyDA (**L24**) and *N*,*N*′-dibenzylethylenediamine (**L28**), have been successfully used for arylation of simple urea and phenylurea (entries 1 and 2).[116,117] CuI/*N*,*N*-diethylsalicylamide (**L60**) catalyzed guanidinylation of aryl iodides proceeded smoothly at 80 °C to deliver symmetrical *N*,*N*′-disubstituted guanidines (entry 3).[118] CuOAc/L-proline could be used for catalyzing aryl guanidinylation of a mono-protected guanidine (entry 4).[119] The Ma group discovered that CuI/*N*-methylglycine (**L4**) was a better combination for aryl guanidinylation because both aryl iodides and aryl bromides gave good yields of the coupling products. By tuning addition sequence, they could achieve the assembling of asymmetrical *N*,*N*′-disubstituted guanidines in one-pot from two different aryl iodides (entries 5–6).[120] Phenanthroline ligands were suitable for arylation of *N*-Boc-*O*-functionalized hydroxyamines and hydroxypyridines (entries 7–8).[121,122]

1.2.6 Azidation

With the assistance of L-proline (**L1**), the CuI-catalyzed coupling reaction of aryl halides with sodium azide proceeded smoothly to give aryl azides (Scheme 1.10).[46] Both aryl iodides and bromides worked for aryl azidation, although different reaction conditions were required. For aryl iodides, using DMSO as solvent gave better results, while EtOH/water was more suitable for aryl bromides.

DMCyDA (**L25**) could also promote the CuI-catalyzed formation of aryl azides from aryl halides and sodium azide.[47] Sodium ascorbate was found to have a positive effect on stabilization of the catalytic system (Scheme 1.11). The *in situ* formed aryl azides could further react with terminal alkynes to provide 1,2,3-triazoles,[123,124] or convert into 2-substituted anilines.[125]

Table 1.11 Copper/ligand-catalyzed cross-coupling of aryl halides or vinyl halides with hydrazide and hydrazine.

Entry	*RX*	*[Cu] (mol%)*	*Ligand (mol%)*	*Nucleophile*	*Reaction conditions*	*Product*	*Yield (%)*	*Number of examples*	*Ref.*
1	I	CuI (5–10)	**L29** (10–20)	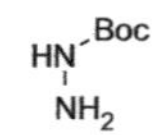	Cs_2CO_3, DMF, 80 °C	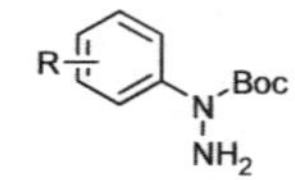	43–97	14	112
2	I, Br	CuI (5)	**L29** (10)	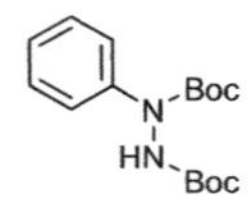	Cs_2CO_3, DMF, 80 °C	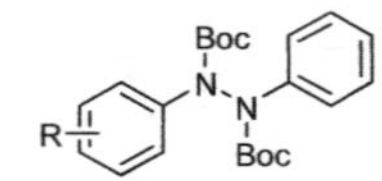	25–99	15	113
3	I	CuI (5)	**L29** (10)	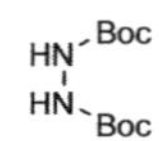	Cs_2CO_3, DMF, 80 °C	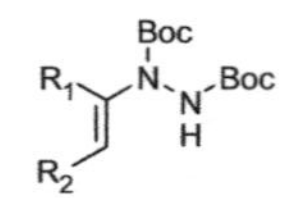	64–94	7	114
4	I	CuI (5)	**L10** (10)	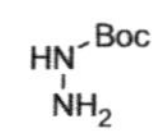	Cs_2CO_3, DMF, 80 °C	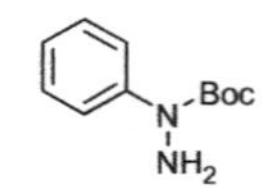	28–73	12	115
5	Br	CuI (10)	**L2** (20)	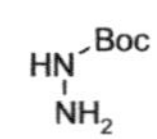	Cs_2CO_3, DMSO, 80 °C	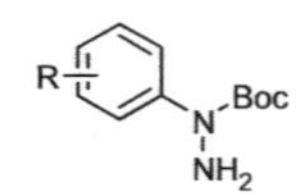	56–91	14	38
6	I, Br	CuI (10)	**L21** (20)	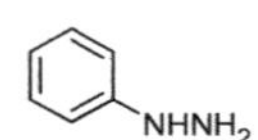	K_3PO_4, LiCl, DMF, 100–110 °C	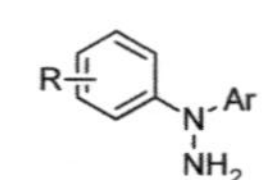	51–84	6	34

Table 1.12 Copper/ligand-catalyzed arylation of other N-nucleophiles.

Entry	*RX*	*[Cu] (mol%)*	*Ligand (mol%)*	*Nucleophile*	*Reaction Conditions*	*Products*	*Yield (%)*	*Number of examples*	*Ref.*
1	I	CuI (10)	**L24 (10)**	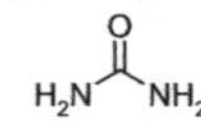	K_3PO_4 DMF 80 °C	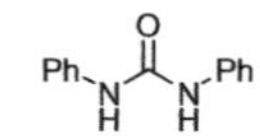	29–65	6	116
2	I	CuI (15)	**L28** (15)	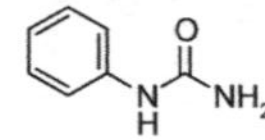	KF/Al_2O_3, THF, 65–70 °C	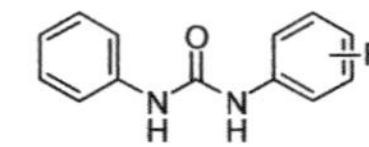	67–88	10	117
3	I	CuI (10)	**L60** (20)	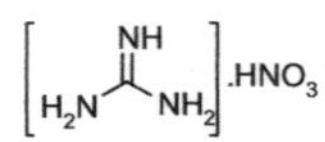	K_3PO_4, MeCN or Dioxane, 80 °C	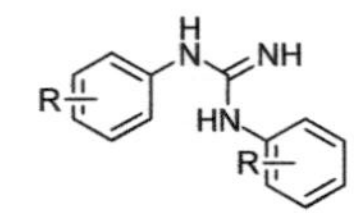	19–92	18	118
4	I, Br	CuOAc (10)	**L1** (20)	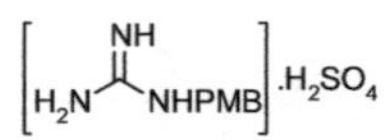	K_3PO_4, MeCN, 100 °C	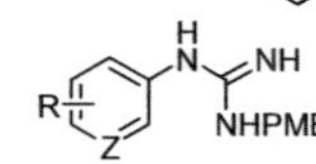	45–96	18	119
	Cl						43	1	
5	I,Br	CuI (5–10)	**L4** (10–20)	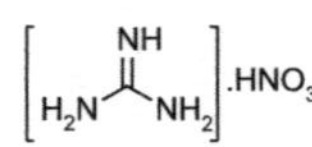	K_3PO_4, MeCN, 70–100 °C	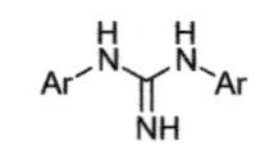	61–92	20	120
6	I	CuI (10)	**L4 (20)**	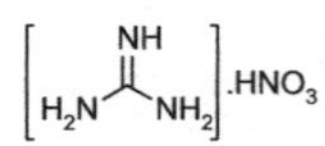	K_3PO_4, MeCN 70 °C	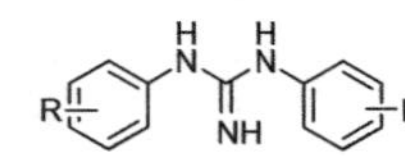	49–70	10	
7	I	CuI (5–10)	**L30** (7.5–20)	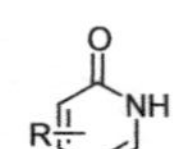	K_2CO_3, DMSO, 110–120 °C	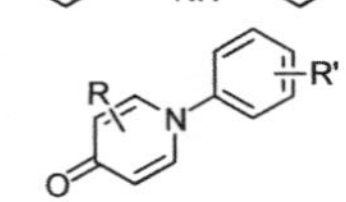	47–95	15	121
			L30 (5–7.5)	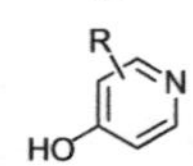					
			L49 (8–40)			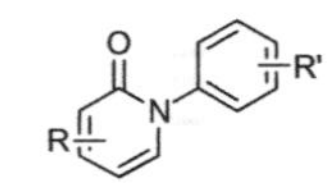			
8	I, Br	CuI (2–10)	**L29** (7.5–40)	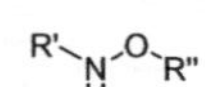	Cs_2CO_3, DMF 80 °C	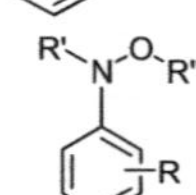	52–89	26	122

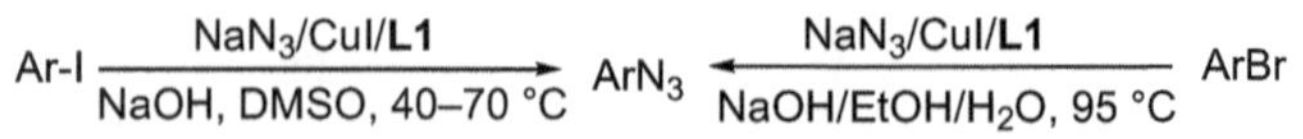

Scheme 1.10 CuI/L-proline catalyzed synthesis of aryl azides.

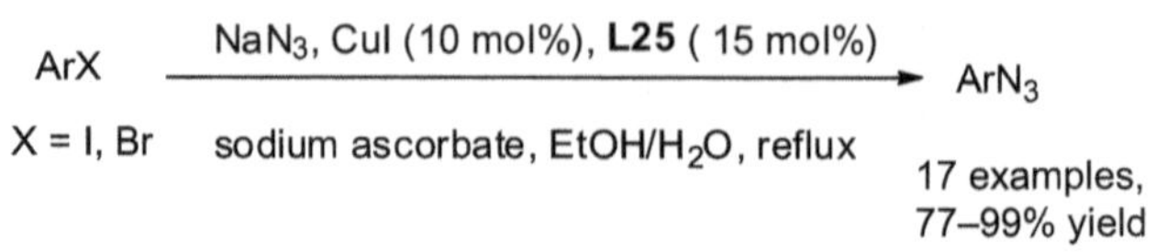

Scheme 1.11 CuI/DMCyDA catalyzed synthesis of aryl azides.

1.3 C–O Bond Formation

1.3.1 Cross-Coupling of Phenols with Aryl Halides and Vinyl Halides

1.3.1.1 Synthesis of Diaryl Ethers

Diaryl ethers are an important class of compounds for biological and material science. The Ullmann-type coupling reaction between phenols and aryl halides provides a direct approach for assembling diaryl ethers. Like other Ullmann-type coupling reactions, harsh reaction conditions such as high temperatures (125–220 °C), stoichiometric quantities of the copper catalyst, and low to moderate yields have greatly limited the utility of the biaryl ether formation reaction. The use of soluble copper salts and some effective ligands made it possible to carry out these reactions under mild reaction conditions (Table 1.13). A number of ligands including *O,O*-bidentate ligands (entries 1–8),[17,72,98,120,126–129] *N,N*-bidentate ligands (entries 9–19),[28,30,130–138] *N,O*-bidentate ligands (entries 20–24)[139–143] as well as phosphine ligands (entry 25)[34] were found effective for promoting this coupling reaction. Usually, aryl halides with electron-withdrawing groups are more reactive for this reaction. In contrast, phenols with electron-donating groups gave better results.

For *O,O*-ligands, TMHD (2,2,6,6-tetramethyl-3,5-heptanedione) (**L49**) has received more attention in promoting Ullmann-type biaryl ether formation (entries 1–4).[121,126–128] Chxn-Py-Al (**L37**) developed by Taillefer and coworkers was a very efficient *N,N*-bidentate ligand for this transformation (entries 12 and 13).[132,133] Some improved results could be obtained if iminopyridine (**L40**) was used (entry 14).[134]

It seems that *N,O*-bidentate ligands have become the primary choice for Ullmann-type coupling reaction between phenols and aryl halides. For example, CuI/*N,N*-dimethylglycine (**L3**) has been used for assembling optically pure diaryl ethers bearing amino ester moieties (entry 20);[139,140] (2-pyridyl)acetone (**L20**) has the ability to promote CuBr-catalyzed coupling of inactive aryl chlorides and phenols (entry 23);[142] while the combination of

Table 1.13 Copper/ligand catalyzed diaryl ether formation from aryl halides and phenols.

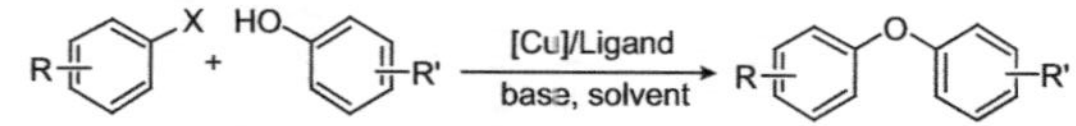

Entry	*X*	*[Cu] (mol%)*	*Ligand (mol%)*	*Reaction conditions*	*Yield (%)*	*Number of examples*	*Ref.*
1	I/Br	CuCl (50)	**L49** (10)	Cs_2CO_3, NMP, 120 °C	51–85	13	126
2	I, Br	CuI (1–10)	**L49** (4–40)	K_3PO_4, DMF or DMSO, 80–130 °C	56–91	7	121
3	Cl	CuBr (10)	**L49** (80)	Cs_2CO_3, DMF, 135 °C	40–99	13	127
4	I, Br	CuI (1)	**L49** (10)	Cs_2CO_3, DMF, 60 °C	12–95	21	128
				80–100 °C	34–86	8	
5	I, Br	CuBr (10)	**L50** (20)	Cs_2CO_3, DMSO, 60–80 °C	72–97	8	72
6	I, Br	CuI (20)	**L59** (20)	Cs_2CO_3, MeCN, 82 °C	54–99	32	129
7	I	CuI (10)	**L56** (10)	Cs_2CO_3, dioxane, 110 °C	81–87	8	98
8	I, Br	CuO/$FeCl_3$(10)	**L58** (20)	Cs_2CO_3, DMF, 110–120 °C	60–89	17	17
9	Br	**C1** (10)		Cs_2CO_3, toluene, 110 °C	31–99	7	28
10	I	CuI (10)	**L29** (20)	KF/Al_2O_3, solvent-free, 100–110 °C	30–98	5	130
11	Br	Cu/C	**L29** (50)	Cs_2CO_3, dioxane, μw, 180–220 °C	69–89	15	131
12	I	Cu_2O (5)	**L37** (20)	Cs_2CO_3, CH_3CN, 82 °C	40–100	12	132
	Br			Cs_2CO_3, DMF, 110 °C	85–100	5	
13	I, Br	CuI (10)	**L37** (5)	K_3PO_4, CH_3CN, 60–80 °C	55–97	13	133
14	Br	CuI (10)	**L40** (10)	K_3PO_4, CH_3CN, 80 °C	68–100	9	134
15	I	CuI (5–10)	**L45** (10–20)	Cs_2CO_3, toluene, reflux	30–99	4	30
16	I, Br	$Cu(OTf)_2$ (20)	**L35** (20)	Cs_2CO_3, dioxane, 110 °C	60–90	22	135
17	Br	CuI (5)	**L41** (5)	K_3PO_4, toluene, 110 °C	69–90	11	136
18	I, Br	CuI (10)	**L47** (10)	Cs_2CO_3, dioxane, 110 °C	78–89	14	137
19	Br	CuI (10)	**L42** (10)	K_3PO_4, MeCN, 60–80 °C	60–92	19	138
20	I, Br	CuI (2)	**L3** (7.5)	Cs_2CO_3, dioxane, 90 °C	33–97	16	139
		(10)	(30)			11	140
21	I	Cu_2O (5)	**L16** (20)	Cs_2CO_3, CH_3CN, 82 °C	45–100	12	132
22	I, Br	CuI (10)	**L12** (20)	K_3PO_4, DMF, 110 °C	30–97	15	141
23	I, Br	CuBr (5–10)	**L20** (10–20)	Cs_2CO_3, DMSO, 80–90 °C	50–98	29	142
	Cl			Cs_2CO_3, DMSO, 120–150 °C	33–97	17	
24	I, Br	CuI (5–10)	**L10** (10–20)	K_3PO_4, DMSO, 80–90 °C	68–96	27	143
25	I, Br	CuI (10)	**L21** (20)	Cs_2CO_3, DMF, 110 °C	60–98	19	34

CuI and picolinic acid (**L10**) is especially useful for synthesis of sterically hindered and heteroaryl diaryl ethers (entry 24).[143]

Room temperature diaryl ether formation for some special substrates has been observed by Ma and coworkers (Scheme 1.12). This may result from the combination of the *ortho*-substituent (directed by an amido group) and ligand effects.[144] These mild reaction conditions showed excellent prospects in the total synthesis of some important natural products.

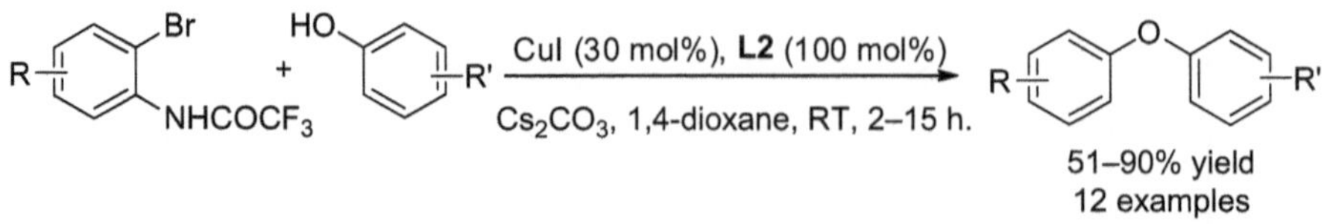

Scheme 1.12 CuI/*N*,*N*-dimethylglycine catalyzed coupling of 2-bromotrifluro-acetanilides with phenols at room temperature.

1.3.1.2 Coupling of Vinyl Halides and Phenols

Aryloxylation of vinyl halides is a convenient route to vinyl aryl ether. Some useful catalytic systems were summarized in Table 1.14. CuI/*N*,*N*-dimethylglycine (**L3**),[145] CuI/Chxn-Py-Al (**L37**)[78] and CuI/ethyl 2-oxocyclohexanecarboxylate (**L50**) have been reported to have a similar ability to promote this transformation (entries 1, 3 and 5).[145,78,81] Eli Lilly researchers found that the combination of CuCl/amino ether (**L8**) could be used for coupling hindered polysubstituted vinyl bromides with phenols (entry 2).[146] The Cook group has indicated that (*Z*)-ethyl 3-iodo-acrylate was a very reactive substrate, whose coupling with phenols could proceed smoothly at room temperature under the catalysis of CuI/2-(2′-pyridyl)benzimidazole (**L33**) (entry 4).[82]

Table 1.14 Copper/ligand catalyzed coupling of vinyl halides and phenols.

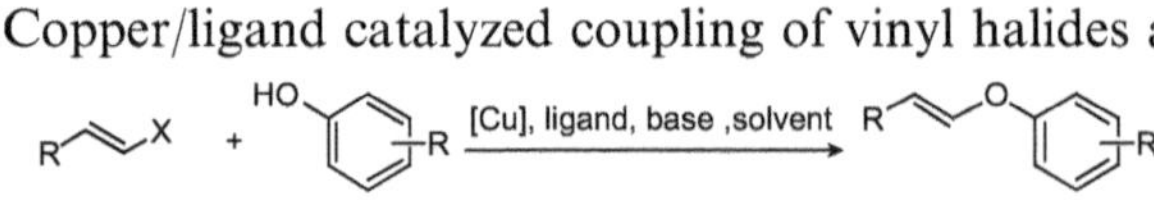

Entry	X	[Cu] (mol%)	Ligand (mol%)	Reaction conditions	Yield (%)	Number of examples	Ref.
1	I, Br	CuI (10)	**L3** (30)	Cs_2CO_3, dioxane, 60–90 °C	65–90	27	145
2	I, Br	CuCl (25)	**L8** (25)	Cs_2CO_3, Toluene, reflux	41–92	8	146
3	Br	CuI (10)	**L37** (5)	Cs_2CO_3, CH_3CN, 50 °C	74–90	5	78
4	I, Br	CuI (5)	**L33** (5)	Cs_2CO_3, DMF, RT–80 °C	81–96	19	82
5	Br	CuI (15)	**L50** (30)	Cs_2CO_3, NMP, 60 °C	79–95	8	81

1.3.2 Cross-Coupling of Aryl Halides with Aliphatic Alcohols

Although aliphatic alcohols were found to be less reactive toward ligand promoted copper-catalyzed arylation, some successful examples have been reported by using special ligands (Table 1.15). In most cases, excess alcohols have to be used as both solvents and nucelphiles to ensure complete conversion of aryl halides. Buchwald and coworkers discovered that if 1,10-phenanthroline (**L29**) was used as a ligand, alkoxylation of aryl iodides with a wide range of aliphatic alcohols (normal and branched, saturated and unsaturated) worked well to give the corresponding products in good yields (entry 1).[147] Switching the base to KF/Al_2O_3 gave similar results (entry 2).[130] Using Me_4-Phen (**L31**) as a ligand afforded milder reaction conditions for this transformation (entry 3).[148] In this case aryl bromides also worked, although only limited cases were described. More importantly, the amount of alcohol could be decreased in their reaction conditions.

CuCl/*N*,*N*-dimethylglycine (**L3**) and CuCl/*trans*-9,10-dihydro-9,10-ethanoanthracene- 11,12-dimethanol (**L59**) were found to be effective for the coupling reaction of aryl iodides with aliphatic alcohols (entry 4 and 7),[149,129] while another two catalytic system, CuI/8-hydroxyquinoline (**L13**) and CuI/ethyl 2-oxocyclohexanecarboxylate (**L50**), were suitable for alkoxylation of aryl bromides (entry 5 and 6).[150,151]

Interestingly, when some amino alcohols was used as the nucleophiles, selective *N*-arylation or *O*-arylation could be achieved by using different ligands (Scheme 1.13).[152] CuI/2-isobutyrylcyclohexanone (**L51**) led to formation of *N*-arylation products at room temperature, while only

Table 1.15 Copper/ligand catalyzed coupling of aryl halides and aliphatic alcohols.

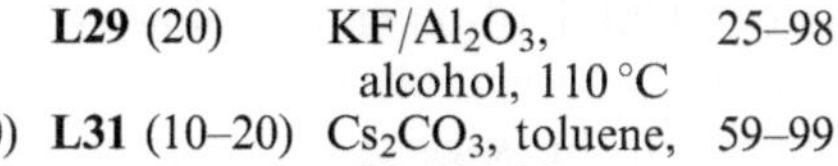

Entry	*X*	*[Cu] (mol%)*	*Ligand (mol%)*	*Reaction conditions*	*Yield (%)*	*Number of examples*	*Ref.*
1	I	CuI (10)	**L29** (20)	Cs_2CO_3, alcohol, 110 °C	40–97	23	147
2	I	CuI (10)	**L29** (20)	KF/Al_2O_3, alcohol, 110 °C	25–98	10	130
3	I, Br	CuI (5–10)	**L31** (10–20)	Cs_2CO_3, toluene, 80–130 °C	59–99	23	148
4	I	CuI (10)	**L3** (20)	Cs_2CO_3, alcohol, 110 °C	25–98	20	149
5	Br	CuI (2–5)	**L13** (4–10)	K_3PO_4, alcohol, 110 °C	78–92(GC)	12	150
6	I, Br	CuI (10)	**L50** (20)	Cs_2CO_3, alcohol, reflux	45–93	17	151
7	I	CuI-**L59** (20)		Cs_2CO_3, alcohol, 82 °C	50–92	10	129

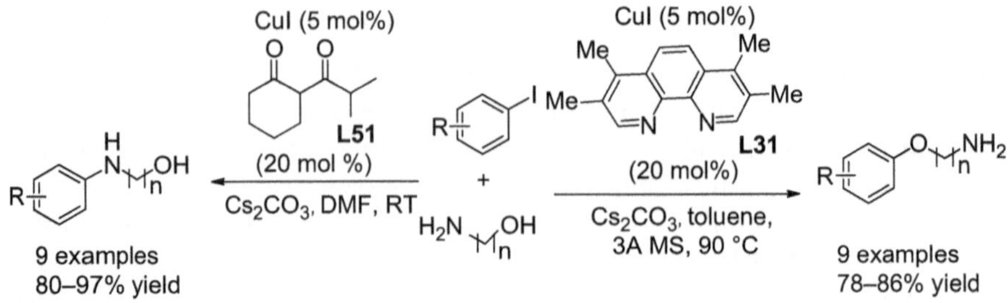

Scheme 1.13 CuI/ligand-catalyzed selective *N*- or *O*-arylation of amino alcohols.

O-arylation products were obtained even at 90 °C under the catalysis of CuI/Me_4-Phen (**L31**).

1.3.3 Direct Hydroxylation of Aryl Halides

Direct copper-catalyzed hydroxylation of aryl halides provide a concise and practical method for preparing substituted phenols. Some special ligands played an important role in this transformation (Table 1.16). You and coworkers reported the first successful case by using 1,10-phenanthroline **L29** as the ligand (entry 1),[153] but the substrates were limited to aryl iodides. With the assistance of a halo-exchange process, CuI/DMEDA (**L26**) could be used for hydroxylation of aryl bromides (entry 2).[154] By adding *n*Bu_4NBr, Cu_2O/pyridine-2-aldoxime (PAO, **L39**)-catalyzed hydoxylation of aryl halides worked in water (entry 3).[155] The Ma group developed an efficient catalytic system that could make the hydroxylation of aryl bromides work well even under low catalyst loadings (entry 4).[156] *O*,*O*-Bidentate ligands, dibenzoylmethane **L53** and D-glucose **L69**, were also useful for Cu-catalyzed hydroxylation of aryl iodides (entries 5–6).[154,157] 8-Hydroxyquinoline and its derivatives were remarkable *N*,*O*-ligands for this transformation (entries 7–9).[158–160] Additionally, lithium pipecolinate (Li salt of **L9**) could be another choice for hydroxylation of aryl iodides (entry 10).[161] Using NaOH as the base and water as the solvent made this catalytic system more practical.

Due to the unstable property of oxime, it is difficult to get *O*-aryloxime by a straight cross-coupling reaction. However, Maitra and coworkers observed that under the catalysis of CuI/1,10-phenanthroline (**L29**), *O*-aryloxime ethers could be isolated in good to acceptable yields from oximes and aryl halides (Scheme 1.14).[162] The key for this success is using Na,K-tartrate as a chelating agent for Cu^{II} (generated during the reaction course) to inhibit deoximation.

1.4 C–S Bond Formation

1.4.1 The Formation of Thioethers

Although thiols are generally stronger nucleophiles than alcohols and amines, the sensitivity of sulfides towards oxidation as well as the formation of disulfide

Table 1.16 Copper/ligand-catalyzed straight hydroxylation for the synthesis of substituted phenols.

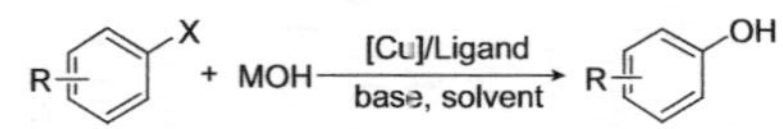

Entry	*X*	*[Cu] (mol%)*	*Ligand (mol%)*	*Nucleophile*	*Reaction conditions*	*Yield (%)*	*Number of examples*	*Ref.*
1	I	CuI (10)	**L29** (20)	KOH	DMSO/H_2O, 100 °C	70–97	16	153
	Br						3	
2	Br	CuI (10)	**L26** (50)	CsOH	NaI, Dioxane/H_2O, 130 °C	70–90	7	154
3	I, Br	Cu_2O (5)	**L39** (10)	CsOH	$(n-Bu)_4NBr$, H_2O, 100–110 °C	52–98	23	155
	Cl						2	
4	Br	CuI (2)	**L34** (4)	KOH or CsOH	*t*-BuOH/DMSO/H_2O, or DMSO/H_2O, 120 °C	73–93	18	156
5	I	CuI (10)	**L53** (50)	CsOH	DMSO/H_2O, 110–130 °C	70–97	20	154
6	I, Br	$Cu(OAc)_2$ (5)	**L69** (5)	KOH	DMSO/H_2O, 120 °C	30–99	20	157
	Cl						2	
7	I, Br	CuI (10)	**L13** (20)	KOH	*t*-BuOH/DMSO/H_2O, 80–120 °C	12–96	22	158
8	I, Br	CuI (10)	**L14** (20)	$(n\text{-}Bu)_4NOH.5H_2O$	DMSO/H_2O, 100–130 °C	85–98	14	159
						74–97	10	
9	I, Br, Cl	CuI (10)	**L67** (40)	CsOH	DMSO/H_2O, 100–130 °C	79–95	20	160
						76–92	17	
						51–86	8	
10	I, Br	CuI (10)	**Li salt of L9** (20)	NaOH	$(n-Bu)_4NF$, H_2O, 130 °C	45–85	25	161

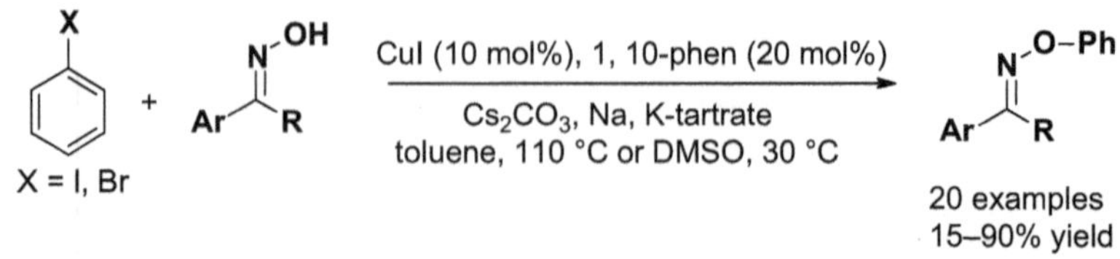

Scheme 1.14 CuI/1,10-phen-catalyzed coupling of aryl halides and aromatic ketoximes and aldoximes.

made the cross-coupling between aryl halides and thiols difficult. In 2002, Venkataraman reported that CuI/neocuproine (**L32**) catalyzed arylation of thiophenols and thiols worked at 110 °C to afford bisaryl- and arylalkylthioethers (Table 1.17, entry 1).[163] In the following years some other *N,N*-bidentate ligands were found to be suitable for this transformation (entries 2–4).[164–166] In particular, CuCl/EDA (**L27**) could catalyze the coupling of thiophenols and activated aryl chlorides (entry 2).[164]

For *O,O*-bidentate ligands, ethylene glycol was the first reported ligand for Cu-catalyzed coupling of aryl iodides with thiols (entry 5).[167] Although excess ligand was required in this case, the wide substrate scope as well as mild reaction conditions make this reaction attractive in organic synthesis. Based on this study, two other diol ligands, 1,1,1-tris(hydroxymethyl)ethane **L56** and *trans*-9,10-dihydro-9,10-ethanoanthracene-11,12-dimethanol **L59**, have been developed. The use of a catalytic amount of these ligands could accelerate this reaction (entries 6–7).[98,168] Additionally, β-keto ester (**L50**) has been found to be effective for this transformation (entries 8–9).[72,169]

For *N,O*-bidentate ligands, *N*-methylglycine (**L4**), L-proline (**L1**) and 1,2,3,4-tetrahydro-8- hydroxyquinoline (**L15**) have been reported to be able to promote Cu-catalyzed coupling of aryl halides and thiols (entries 10–12).[170–172]

1.4.2 The Synthesis of Alkenylsulfides

As shown in Table 1.18, assembly of alkenylsulfides could be achieved through coupling of vinyl halides and thiols. For example, $[Cu(phen)(PPh_3)_2]NO_3$ (**C2**) and CuI/*cis*-1,2-cyclohexanediol (**L55**) catalyzed coupling reaction of vinyl iodides with thiols and thiophenols worked at 30–110 °C, providing the corresponding vinyl sulfides in good to excellent yields (entries 1 and 6).[174,175] Vinyl bromides could serve as the coupling partners if Cu_2O/1,10-phenanthroline (**L29**) and Cu_2O/DMEDA (**L26**) were used as the catalysts (entries 2 and 3).[176,177] Some special reaction conditions are listed in entries 4, 5 and 7.[178,179,81]

1.4.3 Assembly of Aryl Sulfones

The Ma group revealed that CuI-catalyzed coupling of aryl halides and sulfinic acid salts could be promoted by L-proline (**L1**), leading to a reaction that

Table 1.17 Copper/ligand catalyzed coupling between aryl halides and thiols.

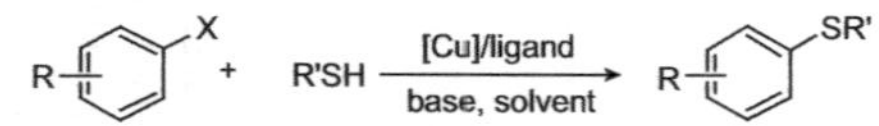

Entry	X	[Cu] (mol%)	Ligand (mol%)	RSH	Reaction conditions	Yield (%)	Number of examples	Ref.
1	I	CuI (10)	**L32** (10)	Thiols, thiophenols	NaO^{t}Bu, toluene, 110 °C	77–98	28	163
2	Cl	CuCl (10)	**L27** (400)	Thiophenols	H_2O, 120 °C	23–97	16	164
3	I, Br, Cl,	Cu(OTf)$_2$ (20)	**L35** (20)	Thiols, thiophenols	Cs_2CO_3, DMF, 110 °C	53–97	14	165
							11	
							3	
4	I, Br	CuI (10)	**L47** (10)	Thiophenols	Cs_2CO_3, Dioxane, 110 °C	65–90	10	166
5	I	CuI (5)	**L57** (200)	Thiols, thiophenols	K_2CO_3, ^{i}PrOH, 80 °C	71–93	27	167
6	I	CuI (10)	**L56** (10)	Thiols, thiophenols	Cs_2CO_3, DMF : dioxane (1 : 9), 110 °C	91–98	8	98
7	I	CuBr (20)	**L59** (20)	Thiols, thiophenols	Cs_2CO_3, CH_3CN, 82 °C	60–98	23	168
8	I	CuBr (10)	**L50** (20)	Thiophenols	Cs_2CO_3, DMSO, 60–75 °C	81–97	7	72
9	I, Br, Cl	Cu_2O (5)	**L50** (10)	Thiols, thiophenols	Cs_2CO_3, DMSO, 80 °C	51–96	33	169
							5	
							5	
10	I	CuI (5)	**L4** (20)	Thiols, thiophenols	KOH, Dioxane, 100 °C	33–98	23	170
11	I	CuI (10)	**L1 (20)**	Thiols, thiophenols	K_2CO_3, DME, 80 °C	33–99	21	171
12	I, Br, Cl	CuBr (10)	**L15** (20)	Thiols, thiophenols	K_2CO_3, DMSO, 80 °C	62–99	17	172
							5	
							1	
13	I	CuI (10)	**L23** (20)	Thiols, thiophenols	Cs_2CO_3, DMF, 90 °C	50–98	21	173

Table 1.18 Copper/ligand catalyzed coupling between vinyl halides and thiols.

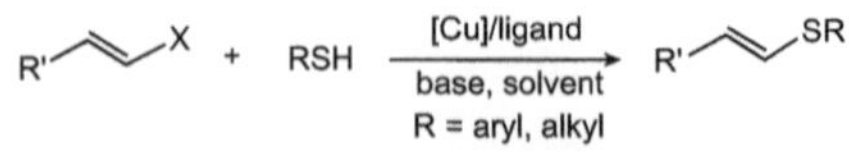

Entry	*X*	*[Cu] (mol%)*	*Ligand (mol%)*	*Reaction conditions*	*Yield (%)*	*Number of examples*	*Ref.*
1	I	**[Cu(phen)(PPh$_3$)$_2$]NO$_3$, C2.** (5)		K_3PO_4 Toluene, 110 °C	80–99	31	174
2	Br	Cu_2O (5)	**L29** (10)	KOH, dioxane, 110 °C	27–99	6	176
3	Br	Cu_2O (10)	**L26** (15)	Cs_2CO_3, DMF, 120 °C	77–97	34	177
4	Br	CuBr (10)	**L1** (20)	K_2CO_3, [BMIM]BF_4, 110–120 °C	75–96	16	178
5	Br	CuI (10)	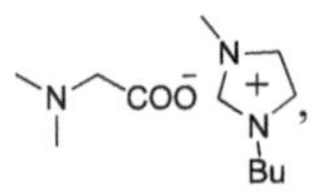	Cs_2CO_3, 90 °C	76–93	15	179
6	I	CuI (10)	**L55** (20)	K_3PO_4, DMF, 30–50 °C	84–98	33	175
7	Br	CuI (15)	**L50** (30)	Cs_2CO_3, NMP, 60 °C	75–92	7	81

Table 1.19 Copper/ligand catalyzed coupling between aryl halides and sulfinic acid salts.

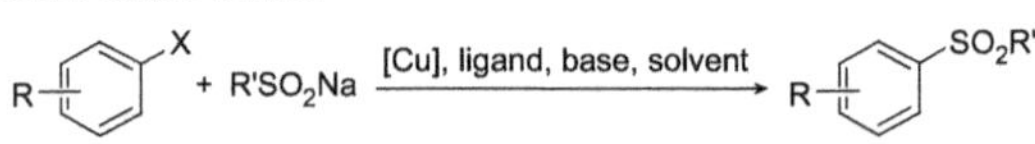

Entry	*X*	*[Cu] (mol%)*	*Ligand (mol%)*	*Reaction conditions*	*Yield (%)*	*Number of examples*	*Ref.*
1	I	CuI	**L1** (as Na$^+$ salt)	DMSO, 80–95 °C	46–93	19	180
	Br	(10)	(20)			8	
2	I	$(CuOTf)_2$·PhH	**L26**	DMSO, 110 °C	24–96	16	181
	Br	(5)	(10)			1	

worked at 80–95 °C to afford aryl sulfones in good yields (Table 1.19, entry 1).[180] Both aryl/heteroaryl iodides and bromides were found as suitable substrates. Prior to the latter, $(CuOTf)_2$·PhH/DMEDA (**L26**) was demonstrated to effect this reaction at 110 °C (entry 2).[181]

1.5 C–P Bond Formation

Venkataraman and coworkers reported that cross-coupling of aryl iodides with Ph_2PH could be catalyzed by CuI, providing the corresponding triarylphophines in good yields (Table 1.20, entry 1).[182] As phosphines themselves

Table 1.20 Copper/ligand catalyzed C-P bond formation.

Entry	*RX*	*P nucleophile*	*[Cu] (mol%)*	*Ligand (mol%)*	*Conditions*	*Yield (%)*	*Number of examples*	*Ref.*
1	Aryl iodides	$HPPh_2$	CuI (5–10)	–	K_2CO_3 or Cs_2CO_3, toluene, 110–115 °C	42–91	12	182
2	aryl iodides	HPR_2	CuI (5)	**L26** (20–35)	Cs_2CO_3, toluene, 110 °C	60–91	17	183
	Vinyl iodide	R: aryl, alkyl						
3	Aryl iodides Vinyl iodides, bromides	$HP(O)(OBu)_2$	CuI (5)	**L26** (20)	Cs_2CO_3, toluene, 70 or 110 °C	64–92	9	183
4	Aryl iodides, bromides	$HP(O)(OR)_2$	CuI (10–15)	**L1** or **L9** (40)	Cs_2CO_3 or DMAP toluene, or DMF, 110 °C	31–91	17	184
		$HP(O)(OR)Ph$ $HP(O)Ph_2$ $H_2P(O)(ONH_4)$						
5	Aryl iodides, bromides	$HP(O)(OR)_2$	CuI (20)	**L21** (20)	Cs_2CO_3 or DMAP Cs_2CO_3 or DMAP toluene, or DMF 110 °C	20–85	18	34
		$HP(O)(OR)Ph$ $HP(O)Ph_2$ $H_2P(O)(ONH_4)$						
6	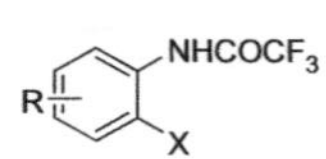 I, Br,	$HP(O)(OR)_2$	CuI (30)	**L12** (100)	Cs_2CO_3, 1,4-dioxane 45–55 °C	50–93	3	185
		$HP(O)Ph_2$					10	
	Cl				110 °C		1	

were good ligands, no extra ligand was required in this case. Some other ligands, DMEDA (**L26**), amino acids (**L1** and **L9**) and pyrrolidine-2-phosphonic acid phenyl monoester (**L21**), have been tested for Cu-catalyzed C-P bond formation. The first one was effective for cross-coupling of aryl iodides with R_2PH and dibutyl phosphite (entries 2 and 3),[183] while the latter two could promote the formation of arylphosphonates, arylphosphinates and arylphosphine oxides (entries 4 and 5).[184,34] With the assistance of the *ortho*-substituent effect, CuI/*N*-methylpyrrolidine-2-carboxamide (**L12**)-catalyzed C–P bond formation of 2-halotrifluoro- acetanilides could proceed at 40–55 °C (entry 6).[185]

1.6 C–X Bond Formation

In 2002, Klapars and Buchwald reported the first copper-catalyzed halogen exchange of aryl bromides. They found that under the catalysis of CuI and DMCyDA (**L25**), aryl bromides could be converted into aryl iodides in the presence of sodium iodide (Scheme 1.15).[186] The method has found applications in copper-catalyzed direct hydroxylation of aryl bromides, in which aryl bromides were *in situ* transferred into aryl iodides and then underwent hydroxylation.[154]

Cu-catalyzed halogen exchange from aryl iodide to aryl fluoride was a challenging task because the strong metal–fluorine bond causes C–F reductive elimination to be slower than competing side reactions. Very recently, the Hartwig group reported the first successful example of this transformation (Scheme 1.16).[187] In this case both a copper-complex and fluoro source were essential for complete conversion. They found that combination of $(t\text{BuCN})_2$CuOTf and AgF could achieve this goal and obtain aryl fluorides from aryl iodides.

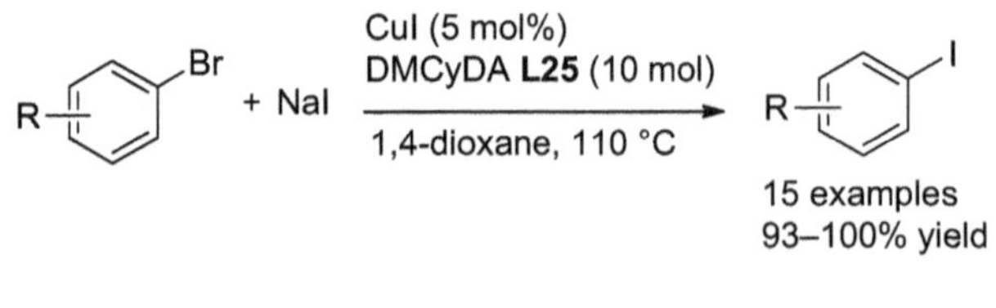

Scheme 1.15 CuI/DMCyDA-catalyzed halogen exchange in aryl halides.

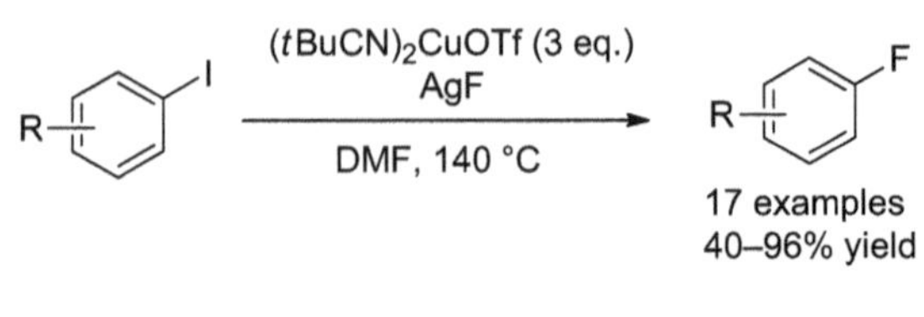

Scheme 1.16 $(t\text{BuCN})_2$CuOTf-catalyzed fluorination of aryl iodides.

1.7 Conclusion

In this chapter we have demonstrated that a number of bidendate ligands such as amino acids (**L1–3, L9**), diamines (**L24–27**), phenanthrolines (**L29–32**), and diketone (**L49, L51**) have remarkable acceleration effects on almost all Cu-catalyzed Ullmann-type coupling reactions. These newly developed conditions have found extensive applications in the synthesis natural products,[1d] designed bioactive compounds and material molecules. However, further investigations in this area are still required to clarify the possible mechanisms for these transformations (see Chapter 2 of this book), extend the reaction scope and improve catalytic efficiency. For examples, in most cases aryl chlorides are not suitable coupling partners. Comparing with palladium-catalyzed coupling reactions, catalytic loadings are still higher in Cu/ligand-catalyzed reactions. Obviously, development of structurally novel ligands will be helpful for solving these problems. These studies will further prompt the discovery of new copper-catalyzed coupling reactions.

References

1. (a) D. S. Surry and S. L. Buchwald, *Chem. Sci.*, 2010, **1**, 13; (b) F. Monnier and M. Taillefer, *Angew. Chem., Int. Ed.*, 2009, **48**, 6954; (c) F. Monnier and M. Taillefer, *Angew. Chem., Int. Ed.*, 2008, **47**, 3096; (d) G. Evano, N. Blanchard and M. Toumi, *Chem. Rev.*, 2008, **108**, 3054; (e) D. Ma and Q. Cai, *Acc. Chem. Res.*, 2008, **41**, 1450; (f) I. P. Beletskaya and A. V. Cheprakov, *Coord. Chem. Rev.*, 2004, **248**, 2337.
2. D. Ma, Y. Zhang, J. Yao, S. Wu and F. Tao, *J. Am. Chem. Soc.*, 1998, **120**, 12459.
3. Z. Lu and R. J. Twieg, *Tetrahedron*, 2005, **61**, 903.
4. H. C. Ma and X. Z. Jiang, *J. Org. Chem.*, 2007, **72**, 8943.
5. A. Ouali, M. Taillefer, J. F. Spindler and A. Jutand, *Organometallics*, 2007, **26**, 65.
6. D. Ma, Q. Cai and H. Zhang, *Org. Lett.*, 2003, **5**, 2453.
7. H. Zhang, Q. Cai and D. Ma, *J. Org. Chem.*, 2005, **70**, 5164.
8. Z. Lu and R. J. Twieg, *Tetrahedron Lett.*, 2005, **46**, 2997.
9. Q. Jiang, D. Jiang, Y. Jiang, H. Fu and Y. Zhao, *Synlett*, 2007, **12**, 1836.
10. D. Wang and K. Ding, *Chem. Commun.*, 2009, 1891.
11. F. Y. Kwong, A. Klapars and S. L. Buchwald, *Org. Lett.*, 2002, **4**, 581.
12. F. Y. Kwong and S. L. Buchwald, *Org. Lett.*, 2003, **5**, 793.
13. A. Shafir and S. L. Buchwald, *J. Am. Chem. Soc.*, 2006, **128**, 8742.
14. Y. S. Feng, Q. S. Man, P. Pan, Z. Q. Pan and H. J. Xu, *Tetrahedron Lett.*, 2009, **50**, 2585.
15. D. Jiang, H. Fu, Y. Jiang and Y. Zhao, *J. Org. Chem.*, 2007, **72**, 672.
16. D. Zhu, R. Wang, J. Mao, L. Xu, F. Wu and B. Wan, *J. Mol. Catal. A: Chem.*, 2006, **256**, 256.
17. Z. Wang, H. Fu, Y. Jiang and Y. Zhao, *Synlett*, 2008, **2008**, 2540.

18. L. Zeng, H. Fu, R. Qiao, Y. Jiang and Y. Zhao, *Adv. Synth. Catal.*, 2009, **351**, 1671.
19. X. Li, D. Yang, Y. Jiang and H. Fu, *Green Chem.*, 2010, **12**, 1097.
20. A. Rolfe and P. R. Hanson, *Tetrahedron Lett.*, 2009, **50**, 6935.
21. L. Xu, D. Zhu, F. Wu, R. Wang and B. Wan, *Tetrahedron*, 2005, **61**, 6553.
22. Z. Zhang, J. Mao, D. Zhu, F. Wu, H. Chen and B. Wan, *Tetrahedron*, 2006, **62**, 4435.
23. Z. Zhang, J. Mao, D. Zhu, F. Wu, H. Chen and B. Wan, *Catal. Commun.*, 2005, **6**, 784.
24. M. Yang and F. Liu, *J. Org. Chem.*, 2007, **72**, 8969.
25. Z. Lu, R. J. Twieg and S. D. Huang, *Tetrahedron Lett.*, 2003, **44**, 6289.
26. Y. Zhang, X. Yang, Q. Yao and D. Ma, *Org. Lett.*, 2012, **14**, 3056.
27. H. B. Goodbrand and N. X. Hu, *J. Org. Chem.*, 1999, **64**, 670.
28. R. K. Gujadhur, C. G. Bates and D. Venkataraman, *Org. Lett.*, 2001, **3**, 4315.
29. Y. H. Liu, C. Chen and L. M. Yang, *Tetrahedron Lett.*, 2006, **47**, 9275.
30. K. T. Wong, S. Y. Ku and F. W. Yen, *Tetrahedron Lett.*, 2007, **48**, 5051.
31. Y. Liu, Y. Bai, J. Zhang, Y. Li, J. Jiao and X. Qi, *Eur. J. Org. Chem.*, 2007, 6084.
32. X. Guo, H. Rao, H. Fu, Y. Jiang and Y. Zhao, *Adv. Synth. Catal.*, 2006, **348**, 2197.
33. R. A. Altman, K. W. Anderson and S. L. Buchwald, *J. Org. Chem.*, 2008, **73**, 5167.
34. H. Rao, Y. Jin, H. Fu, Y. Jiang and Y. Zhao, *Chem. Eur. J.*, 2006, **12**, 3636.
35. H. Rao, H. Fu, Y. Jiang and Y. Zhao, *J. Org. Chem.*, 2005, **70**, 8107.
36. F. Lang, D. Zewge, I. N. Houpis and R. Volante, *Tetrahedron Lett.*, 2001, **42**, 3251.
37. J. Kim and S. Chang, *Chem. Commun.*, 2008, 3052.
38. L. Jiang, X. Lu, H. Zhang, Y. Jiang and D. Ma, *J. Org. Chem.*, 2009, **74**, 4542.
39. N. Xia and M. Taillefer, *Angew. Chem., Int. Ed.*, 2009, **48**, 337.
40. R. Ntaganda, B. Dhudshia, C. L. B. Macdonald and A. N. Thadani, *Chem. Commun.*, 2008, 6200.
41. D. Wang, Q. Cai and K. Ding, *Adv. Synth. Catal.*, 2009, **351**, 1722.
42. Z. Wu, Z. Jiang, D. Wu, H. Xiang and X. Zhou, *Eur. J. Org. Chem.*, 2010, 1854.
43. K. G. Thakur, D. Ganapathy and G. Sekar, *Chem. Commun.*, 2011, **47**, 5067.
44. P. Ji, J. H. Atherton and M. I. Page, *J. Org. Chem.*, 2012. ASAP.
45. X. Gao, H. Fu, R. Qiao, Y. Jiang and Y. Zhao, *J. Org. Chem.*, 2008, **73**, 6864.
46. W. Zhu and D. Ma, *Chem. Commun.*, 2004, 888.
47. J. Andersen, U. Madsen, F. Bjorkling and X. Liang, *Synlett*, 2005, **16**, 2209.
48. J. T. Markiewicz, O. Wiest and P. Helquist, *J. Org. Chem.*, 2010, **75**, 4887.

49. C. Z. Tao, J. Li, Y. Fu, L. Liu and Q. X. Guo, *Tetrahedron Lett.*, 2008, **49**, 70.
50. A. Kiyomori, J. F. Marcoux and S. L. Buchwald, *Tetrahedron Lett.*, 1999, **40**, 2657.
51. R. A. Altman and S. L. Buchwald, *Org. Lett.*, 2006, **8**, 2779.
52. R. A. Altman, E. D. Koval and S. L. Buchwald, *J. Org. Chem.*, 2007, **72**, 6190.
53. R. Hosseinzadeh, M. Tajbakhsh and M. Alikarami, *Synlett*, 2006, **13**, 2124.
54. R. Hosseinzadeh, M. Tajbakhsh and M. Alikarami, *Tetrahedron Lett.*, 2006, **47**, 5203.
55. B. X. Tang, S. M. Guo, M. B. Zhang and J. H. Li, *Synthesis*, 2008, 1707.
56. A. Klapars, J. C. Antilla, X. Huang and S. L. Buchwald, *J. Am. Chem. Soc.*, 2001, **123**, 7727.
57. J. C. Antilla, A. Klapars and S. L. Buchwald, *J. Am. Chem. Soc.*, 2002, **124**, 11684.
58. J. C. Antilla, J. M. Baskin, T. E. Barder and S. L. Buchwald, *J. Org. Chem.*, 2004, **69**, 5578.
59. L. Zhu, L. Cheng, Y. Zhang, R. Xie and J. You, *J. Org. Chem.*, 2007, **72**, 2737.
60. R. K. Rao, A. B. Naidu, E. Jaseer and G. Sekar, *Tetrahedron*, 2009, **65**, 4619.
61. S. Haneda, Y. Adachi and M. Hayashi, *Tetrahedron*, 2009, **65**, 10459.
62. D. Ma and Q. Cai, *Synlett*, 2004, 128.
63. X. Lv, Z. Wang and W. Bao, *Tetrahedron*, 2006, **62**, 4756.
64. J. Mao, J. Guo, H. Song and S. J. Ji, *Tetrahedron*, 2008, **64**, 1383.
65. H. J. Cristau, P. P. Cellier, J. F. Spindler and M. Taillefer, *Eur. J. Org. Chem.*, 2004, 695.
66. H. J. Cristau, P. P. Cellier, J. F. Spindler and M. Taillefer, *Chem.–Eur. J.*, 2004, **10**, 5607.
67. L. Liu, M. Frohn, N. Xi, C. Dominguez, R. Hungate and P. J. Reider, *J. Org. Chem.*, 2005, **70**, 10135.
68. H. Wang, Y. LI, F. Sun, Y. Feng, K. Jin and X. Wang, *J. Org. Chem.*, 2008, **73**, 8639.
69. D. P. Phillips, X. F. Zhu, T. L. Lau, X. He, K. Yang and H. Liu, *Tetrahedron Lett.*, 2009, **50**, 7293.
70. X. Yang, H. Xing, Y. Zhang, Y. Lai, Y. Jiang and D. Ma, *Chin. J. Chem.*, 2012, **30**, 875.
71. H. Chen, D. Wang, X. Wang, W. Huang, Q. Cai and K. Ding, *Synthesis*, 2010, 1505.
72. X. Lv and W. Bao, *J. Org. Chem.*, 2007, **72**, 3863.
73. Z. Xi, F. Liu, Y. Zhou and W. Chen, *Tetrahedron*, 2008, **64**, 4254.
74. J. Kim, S. Y. Lee, J. Lee, Y. Do and S. Chang, *J. Org. Chem.*, 2008, **73**, 9454.
75. Y. X. Xie, S. F. Pi, J. Wang, D. L. Yin and J. H. Li, *J. Org. Chem.*, 2006, **71**, 8324.

76. L. Liang, Z. Li and X. Zhou, *Org. Lett.*, 2009, **11**, 3294.
77. Z. Wang, W. Bao and Y. Jiang, *Chem. Commun.*, 2005, 2849.
78. M. Taillefer, A. Ouali, B. Renard and J. F. Spindler, *Chem.–Eur. J.*, 2006, **12**, 5301.
79. Q. Liao, Y. Wang, L. Zhang and C. Xi, *J. Org. Chem.*, 2009, **74**, 6371.
80. G. Shen, X. Lv, W. Qian and W. Bao, *Tetrahedron Lett.*, 2008, **49**, 4556.
81. W. Bao, Y. Liu and X. Lv, *Synthesis*, 2008, **12**, 1911.
82. M. S. Kabir, M. Lorenz, O. A. Namjoshi and J. M. Cook, *Org. Lett.*, 2009, **12**, 464.
83. A. Klapars, X. Huang and S. L. Buchwald, *J. Am. Chem. Soc.*, 2002, **124**, 7421.
84. K. R. Crawford and A. Padwa, *Tetrahedron Lett.*, 2002, **43**, 7365.
85. B. Mallesham, B. Rajesh, P. R. Reddy, D. Srinivas and S. Trehan, *Org. Lett.*, 2003, **5**, 963.
86. D. P. Phillips, A. R. Hudson, B. Nguyen, T. L. Lau, M. H. McNeill, J. E. Dalgard, J. H. Chen, R. J. Penuliar, T. A. Miller and L. Zhi, *Tetrahedron Lett.*, 2006, **47**, 7137.
87. X. Wang, A. Guram, M. Ronk, J. E. Milne, J. S. Tedraw and M. M. Faul, *Tetrahedron Lett.*, 2012, **53**, 7.
88. A. Ghinet, S. Oudir, J. P. Hénichart, B. Rigo, N. Pommery and P. Gautret, *Tetrahedron*, 2010, **66**, 215.
89. S. K. Kang, D. H. Kim and J. N. Park, *Synlett*, 2002, **2002**, 427.
90. O. S. do Rêgo Barros, C. W. Nogueira, E. C. Stangherlin, P. H. Menezes and G. Zeni, *J. Org. Chem.*, 2006, **71**, 1552.
91. P. F. Larsson, A. Correa, M. Carril, P. O. Norrby and C. Bolm, *Angew. Chem., Int. Ed.*, 2009, **48**, 5691.
92. R. Hosseinzadeh, M. Tajbakhsh, M. Mohadjerani and H. Mehdinejad, *Synlett*, 2004, 1517.
93. T. Mino, Y. Harada, H. Shindo, M. Sakamoto and T. Fujita, *Synlett*, 2008, **19**, 614.
94. W. Deng, Y. F. Wang, Y. Zou, L. Liu and Q. X. Guo, *Tetrahedron Lett.*, 2004, **45**, 2311.
95. W. Deng, L. Liu, C. Zhang, M. Liu and Q. X. Guo, *Tetrahedron Lett.*, 2005, **46**, 7295.
96. Y. M. Pu, Y. Y. Ku, T. Grieme, R. Henry and A. V. Bhatia, *Tetrahedron Lett.*, 2006, **47**, 149.
97. J. Li, Y. Zhang, Y. Jiang and D. Ma, *Tetrahedron Lett.*, 2012, **53**, 3981.
98. Y. J. Chen and H. H. Chen, *Org. Lett.*, 2006, **8**, 5609.
99. X. Han, *Tetrahedron Lett.*, 2010, **51**, 360.
100. E. Racine, F. Monnier, J. P. Vors and M. Taillefer, *Org. Lett.*, 2011, **13**, 2818.
101. R. Shen and J. A. Porco Jr, *Org. Lett.*, 2000, **2**, 1333.
102. L. Jiang, E. Gabriel, A. Klapars and S. L. Buchwald, *Org. Lett.*, 2003, **5**, 3667.
103. A. Coste, F. Couty and G. Evano, *Org. Lett.*, 2009, **11**, 4454.
104. C. Han, R. Shen, S. Su and J. A. Porco Jr, *Org. Lett.*, 2004, **6**, 27.

105. X. Pan, Q. Cai and D. Ma, *Org. Lett.*, 2004, **6**, 1809.
106. M. O. Frederick, J. A. Mulder, M. R. Tracey, R. P. Hsung, J. Huang, K. C. M. Kurtz, L. Shen and C. J. Douglas, *J. Am. Chem. Soc.*, 2003, **125**, 2368.
107. Y. Zhang, R. P. Hsung, M. R. Tracey, K. C. M. Kurtz and E. L. Vera, *Org. Lett.*, 2004, **6**, 1151.
108. K. Dooleweerdt, H. Birkedal, T. Ruhland and T. Skrydstrup, *J. Org. Chem.*, 2008, **73**, 9447.
109. F. M. Istrate, A. K. Buzas, I. D. Jurberg, Y. Odabachian and F. Gagosz, *Org. Lett.*, 2008, **10**, 925.
110. P. Y. Yao, Y. Zhang, R. P. Hsung and K. Zhao, *Org. Lett.*, 2008, **10**, 4275.
111. K. Jouvin, J. Heimburger and G. Evano, *Chem. Sci.*, 2012, **3**, 756.
112. M. Wolter, A. Klapars and S. L. Buchwald, *Org. Lett.*, 2001, **3**, 3803.
113. K. Y. Kim, J. T. Shin, K. S. Lee and C. G. Cho, *Tetrahedron Lett.*, 2004, **45**, 117.
114. M. R. Rivero and S. L. Buchwald, *Org. Lett.*, 2007, **9**, 973.
115. M. S. Lam, H. W. Lee, A. S. C. Chan and F. Y. Kwong, *Tetrahedron Lett.*, 2008, **49**, 6192.
116. M. V. Nandakumar, *Tetrahedron Lett.*, 2004, **45**, 1989.
117. R. Hosseinzadeh, Y. Sarrafi, M. Mohadjerani and F. Mohammadpourmir, *Tetrahedron Lett.*, 2008, **49**, 840.
118. M. Cortes-Salva, B. L. Nguyen, J. Cuevas, K. R. Pennypacker and J. C. Antilla, *Org. Lett.*, 2010, **12**, 1316.
119. H. Hammoud, M. Schmitt, F. Bihel, C. Antheaume and J. J. Bourguignon, *J. Org. Chem.*, 2012, **77**, 417.
120. H. Xing, Y. Zhang, Y. Lai, Y. Jiang and D. Ma, *J. Org. Chem.*, 2012, **77**, 5449.
121. R. A. Altman and S. L. Buchwald, *Org. Lett.*, 2007, **9**, 643.
122. K. L. Jones, A. Porzelle, A. Hall, M. D. Woodrow and N. C. O. Tomkinson, *Org. Lett.*, 2008, **10**, 797.
123. J. Andersen, U. Madsen, F. Bjorkling and X. Liang, *Synlett*, 2005, **16**, 2941.
124. L. Ackermann, H. K. Potukuchi, D. Landsberg and R. Vicente, *Org. Lett.*, 2008, **10**, 3081.
125. H. zhao, H. Fu and R. Qiao, *J. Org. Chem.*, 2010, **75**, 3311.
126. E. Buck, Z. J. Song, D. Tschaen, P. G. Dormer, R. Volante and P. J. Reider, *Org. Lett.*, 2002, **4**, 1623.
127. N. Xia and M. Taillefer, *Chem.–Eur. J.*, 2008, **14**, 6037.
128. F. F. Yong, Y. C. Teo, Y. K. Yan and G. L. Chua, *Synlett*, 2012, **2012**, 101.
129. A. B. Naidu, E. Jaseer and G. Sekar, *J. Org. Chem.*, 2009, **74**, 3675.
130. R. Hosseinzadeh, M. Tajbakhsh, M. Mohadjerani and M. Alikarami, *Synlett*, 2005, **7**, 1101.
131. B. H. Lipshutz, J. B. Unger and B. R. Taft, *Org. Lett.*, 2007, **9**, 1089.

132. H. J. Cristau, P. P. Cellier, S. Hamada, J. F. Spindler and M. Taillefer, *Org. Lett.*, 2004, **6**, 913.
133. A. Ouali, J. F. Spindler, H. J. Cristau and M. Taillefer, *Adv. Synth. Catal.*, 2006, **348**, 499.
134. A. Ouali, J. F. Spindler, A. Jutand and M. Taillefer, *Adv. Synth. Catal.*, 2007, **349**, 1906.
135. A. B. Naidu, O. Raghunath, D. Prasad and G. Sekar, *Tetrahedron Lett.*, 2008, **49**, 1057.
136. T. Mino, F. Yagishita, M. Shibuya, K. Kajiwara, H. Shindo, M. Sakamoto and T. Fujita, *Synlett*, 2009, 2457.
137. N. R. Jogdand, B. B. Shingate and M. S. Shingare, *Tetrahedron Lett.*, 2009, **50**, 4019.
138. Y. H. Liu, G. Li and L. M. Yang, *Tetrahedron Lett.*, 2009, **50**, 343.
139. D. Ma and Q. Cai, *Org. Lett.*, 2003, **5**, 3799.
140. Q. Cai, G. He and D. Ma, *J. Org. Chem.*, 2006, **71**, 5268.
141. X. Liu, H. Fu, Y. Jiang and Y. Zhao, *Synlett*, 2008, **19**, 221.
142. Q. Zhang, D. Wang, X. Wang and K. Ding, *J. Org. Chem.*, 2009, **74**, 7187.
143. D. Maiti and S. L. Buchwald, *J. Org. Chem.*, 2010, **75**, 1791.
144. Q. Cai, B. Zou and D. Ma, *Angew. Chem., Int. Ed.*, 2006, **45**, 1276.
145. D. Ma, Q. Cai and X. Xie, *Synlett*, 2005, **16**, 1767.
146. Z. Wan, C. D. Jones, T. M. Koenig, Y. J. Pu and D. Mitchell, *Tetrahedron Lett.*, 2003, **44**, 8257.
147. M. Wolter, G. Nordmann, E. Gabriel and S. L. Buchwald, *Org. Lett.*, 2002, **4**, 973.
148. R. A. Altman, A. Shafir, A. Choi, P. A. Lichtor and S. L. Buchwald, *J. Org. Chem.*, 2008, **73**, 284.
149. H. Zhang, D. Ma and W. Cao, *Synlett*, 2007, **2**, 243.
150. J. Niu, P. Guo, J. Kang, Z. Li, J. Xu and S. Hu, *J. Org. Chem.*, 2009, **74**, 5075.
151. D. Vuluga, J. Legros, B. Crousse and D. Bonnet-Delpon, *Eur. J. Org. Chem.*, 2009, **2009**, 3513.
152. A. Shafir, P. A. Lichtor and S. L. Buchwald, *J. Am. Chem. Soc.*, 2007, **129**, 3490.
153. D. Zhao, N. Wu, S. Zhang, P. Xi, X. Su, J. Lan and J. You, *Angew. Chem., Int. Ed.*, 2009, **48**, 8729.
154. A. Tlili, N. Xia, F. Monnier and M. Taillefer, *Angew. Chem., Int. Ed.*, 2009, **48**, 8725.
155. D. Yang and H. Fu, *Chem.–Eur. J.*, 2010, **16**, 2366.
156. J. Jia, C. Jiang, X. Zhang, Y. Jiang and D. Ma, *Tetrahedron Lett.*, 2011, **52**, 5593.
157. K. G. Thakur and G. Sekar, *Chem. Commun.*, 2011, **47**, 6692.
158. S. Maurer, W. Liu, X. Zhang, Y. Jiang and D. Ma, *Synlett*, 2010, 976.
159. R. Paul, M. A. Ali and T. Punniyamurthy, *Synthesis*, 2010, **24**, 4296.
160. K. Yang, Z. Li, Z. Wang, Z. Yao and S. Jiang, *Org. Lett.*, 2011, **13**, 4340.
161. L. Jing, J. Wei, L. Zhou, Z. Huang, Z. Li and X. Zhou, *Chem. Commun.*, 2010, **46**, 4767.

162. P. De, K. Nonappa, U. Pandurangan, Maitra and S. Wailes, *Org. Lett.*, 2007, **9**, 2767.
163. C. G. Bates, R. K. Gujadhur and D. Venkataraman, *Org. Lett.*, 2002, **4**, 2803.
164. M. T. Herrero, R. SanMartin and E. Domínguez, *Tetrahedron*, 2009, **65**, 1500.
165. D. Prasad, A. B. Naidu and G. Sekar, *Tetrahedron Lett.*, 2009, **50**, 1411.
166. N. R. Jogdand, B. B. Shingate and M. S. Shingare, *Tetrahedron Lett.*, 2009, **50**, 6092.
167. F. Y. Kwong and S. L. Buchwald, *Org. Lett.*, 2002, **4**, 3517.
168. D. Prasad and G. Sekar, *Synthesis*, 2010, 79.
169. H. J. Xu, X. Y. Zhao, J. Deng, Y. Fu and Y. S. Feng, *Tetrahedron Lett.*, 2009, **50**, 434.
170. W. Deng, Y. Zou, Y. F. Wang, L. Liu and Q. X. Guo, *Synlett*, 2004, **15**, 1254.
171. H. Zhang, W. Cao and D. Ma, *Synth. Commun.*, 2007, **37**, 25.
172. Y. Feng, H. Wang, F. Sun, Y. Li, X. Fu and K. Jin, *Tetrahedron*, 2009, **65**, 9737.
173. D. Zhu, L. Xu, F. Wu and B. Wan, *Tetrahedron Lett.*, 2006, **47**, 5781.
174. C. G. Bates, P. Saejueng, M. Q. Doherty and D. Venkataraman, *Org. Lett.*, 2004, **6**, 5005.
175. M. S. Kabir, M. L. Van Linn, A. Monte and J. M. Cook, *Org. Lett.*, 2008, **10**, 3363.
176. H. L. Kao and C. F. Lee, *Org. Lett.*, 2011, **13**, 5204.
177. H. Xu, W. Chen, J. Chen and H. Wu, *Synthesis*, 2012, **44**, 380.
178. Y. Zheng, X. Du and W. Bao, *Tetrahedron Lett.*, 2006, **47**, 1217.
179. Z. Wang, H. Mo and W. Bao, *Synlett*, 2007, **18**, 91.
180. W. Zhu and D. Ma, *J. Org. Chem.*, 2005, **70**, 2696.
181. J. M. Baskin and Z. Wang, *Org. Lett.*, 2002, **4**, 4423.
182. D. Van Allen and D. Venkataraman, *J. Org. Chem.*, 2003, **68**, 4590.
183. D. Gelman, L. Jiang and S. L. Buchwald, *Org. Lett.*, 2003, **5**, 2315.
184. C. Huang, X. Tang, H. Fu, Y. Jiang and Y. Zhao, *J. Org. Chem.*, 2006, **71**, 5020.
185. D. Jiang, Q. Jiang, H. Fu, Y. Jiang and Y. Zhao, *Synthesis*, 2008, 3473.
186. A. Klapars and S. L. Buchwald, *J. Am. Chem. Soc.*, 2002, **124**, 14844.
187. P. S. Fier and J. F. Hartwig, *J. Am. Chem. Soc.*, 2012, **134**, 5524.

CHAPTER 2

Mechanistic Understanding of Copper-Catalyzed Aryl–Heteroatom Bond Formation: Dependence on Ancillary Ligands

ALICIA CASITAS

Max-Planck-Institut für Kohlenforschung, Kaiser-Wilhelm-Platz 1, D-45470 Mülheim an der Ruhr, Germany
Email: casitas@mpi-muelheim.mpg.de

2.1 Introduction

The copper-catalyzed arylation of heteroatom nucleophiles, nowadays known as Ullmann condensation reactions, have emerged as a practical method for the construction of C_{aryl}–O, C_{aryl}–N, or C_{aryl}–S bonds, not only in simple but also in complex aryl halide substrates (Scheme 2.1 a). These reactions originate from the pioneering work of Fritz Ullmann and Irma Goldberg in the early 1900s but have not been used to their full potential until 2000.[1–3] The requirement of harsh reaction conditions (such high reaction temperatures (>200 °C) and long reaction times in highly polar solvents), the limited synthetic scope and the moderate yields obtained, reduced the applicability of these reactions for a long time.[4,5] However, earlier contributions from Ma, Hauptman and others in the late nineties, showed that some organic

RSC Catalysis Series No. 11
C–H and C–X Bond Functionalization: Transition Metal Mediation
Edited by Xavi Ribas

Published by the Royal Society of Chemistry, www.rsc.org

additives enhanced the reaction rate, presumably increasing the solubility and stability of the copper catalyst.[6–9] A breakthrough in the field appeared in 2001 by the research groups of Buchwald[10–12] and Taillefer[13,14] that reported the use of simple auxiliary ligands in combination with bases for promoting the cross-coupling reactions at lower reaction temperatures and lower copper catalyst loadings. Since then a variety of new copper/ligand systems have been developed for obtaining the cross-coupling reactions under mild reaction conditions. Usually, nitrogen and oxygen bidentate ligands have been used in these reactions, for instance phenanthrolines,[8] 1,2-diamines,[15] 1,3-diketones,[16,17] imino-pyridines[18,19] and α-amino acids[7,20,21] (Scheme 2.1 b). The attractiveness of copper, in comparison to more toxic and expensive palladium, has led to a renaissance of Ullmann-type coupling reactions. Many efforts have focused on developing even more efficient synthetic procedures in order to compete with the palladium-catalyzed methods not only in academic laboratories but also in large-scale industrial production.[22–25]

Mechanistic understanding is key for developing more efficient methodologies using a more rational criteria and not a trial an error approach for each system. The experimental data collected is not enough to elucidate the reaction mechanism and, therefore, many proposals are mainly supported by extensive computational studies. Although several efforts have been made to obtaining relevant experimental data, the ability of copper to participate in coordination equilibria as well as in redox processes, makes it difficult to obtain and interpret kinetic data. In addition, most copper-catalyzed cross-coupling reactions are carried out under heterogeneous conditions owing to the

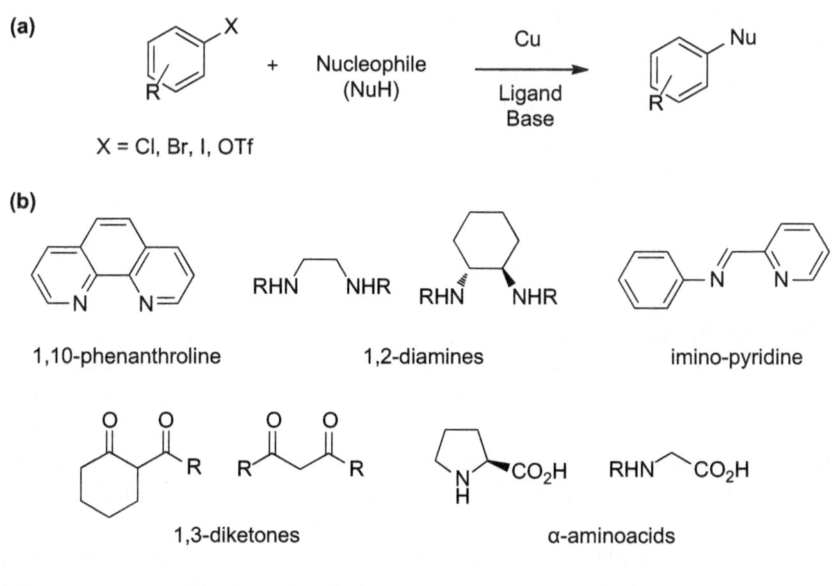

Scheme 2.1 (a) Ullmann condensation reaction. (b) Representative collection of bidentate ligands used for promoting copper-catalyzed cross-coupling reactions.

insolubility of the inorganic base, the copper halide salt or the nucleophile at the high concentrations applied. The heterogeneity of the system increases the complexity of mechanistic studies under reaction conditions.

In this context, different mechanistic pathways have been proposed for Ullmann condensation reactions. Hereafter, the aim of this chapter is to disclose the main mechanistic proposals found in the literature in order to achieve a general comprehension in this topic.

2.2 Oxidation State of the Active Catalyst

Initial reports concerning the mechanism of copper-mediated nucleophilic aromatic substitutions in aryl halides focused on unraveling the oxidation state of the active copper catalyst. Copper sources in three different oxidation states (0, +1 and +2), for instance, $CuBr_2$, $CuCl_2$, $Cu(OAc)_2$, CuI, CuBr, CuCl or even elemental copper, were found to be effective in Ullmann-type reactions.[5,6,26] Owing to the high ability of copper to participate in redox processes, copper may be present in solution in more than one oxidation level, with variable proportions during the reaction. On the other hand, all experimental data collected supported that copper(I) species is the active catalyst and, whether starting from copper(II) or copper(0) sources, copper(I) complexes are formed *in situ* by chemical reduction or oxidation processes, respectively.

By means of electron paramagnetic resonance (EPR) spectroscopy, Weingarten showed that copper(II) bromide was reduced *in situ* in the presence of phenoxide anions to cuprous salts.[6] Similarly, Kondratov and Shein observed EPR signal decay over time when copper(II) species react with several amines, which was attributed to ligand oxidation to amide and the corresponding formation of copper(I) species.[27] Later on, Paine studied the coupling reaction of diphenylamine (Ph_2NH) and aryl halides to form triphenylamines in the presence of copper in three different oxidation states.[26] The formation of tetraphenylhydrazine (Ph_2NNPh_2) in excess of copper(II) salts suggested *in situ* reduction of cupric salts to copper(I) by the nucleophile before catalysis occurred. In addition, when copper metal was used as catalyst, the formation of a layer of Cu_2O was detected by means of scanning electron microscopy (SEM). The Cu_2O formed on the surface of the metallic copper was responsible of the coupling reaction, since it was dissolved upon coordination with the nucleophile Ph_2NLi affording the active copper(I) species in solution.

Since modern copper-catalyzed nucleophilic substitutions are carried out in the presence of auxiliary ligands, it is also important to determine the oxidation state of the active ligand-copper catalyst. In this context, only very recently, Jutand and coworkers reported that the electrogenerated [Cu^0(1,10-phenanthroline)] complex in acetonitrile is oxidized *in situ* by the aryl iodide or aryl bromide substrate *via* an inner sphere electron transfer mechanism.[28] This redox process takes place at the very beginning of the reaction resulting in the formation of the active catalyst $[Cu^I(1,10\text{-phenanthroline})]^+$ complex with the concomitant aryl anion that is protonated to the arene. This example shows again the feasibility of copper complexes to participate in redox processes in the presence of bidentate ligands before the coupling reaction, also supporting copper(I) as the active catalyst.

2.3 Studies Related to the Active Catalyst Structure

Unravelling the structure of the active catalyst in a given copper/ligand system is very challenging due to presence of different equilibria of copper with ligand molecules as well as with coordinating solvents, anions and nucleophile molecules presents in the reaction media. The high variability of coordination possibilities, in addition to the versatility of copper in adopting different geometries, increases the difficulty of determining the actual structure of the active catalyst prior the activation of the aryl halide, which is usually considered to be the rate-determining step. Although most synthetic procedures for Ullmann condensation reactions consist of adding different ratios of copper and auxiliary ligand to generate the copper catalyst *in situ*, several authors showed also good activities using well-defined copper(I) complexes that serve as efficient precatalysts.[29,30]

In this context, Whitesides, Weingarten, Yamamoto and others authors showed in the seventies that copper(I) complexes bearing nitrogen and oxygen nucleophiles (alkoxide, phenoxide and amidate groups) were able to participate in stoichiometric nucleophilic substitutions with alkyl and aryl halides.[6,31–34] The kinetic studies carried out by Paine in the catalytic reaction of diphenylamine with aryl halides to form triphenylamines prompted him to propose nucleophile coordination to the copper(I) at the beginning of the reaction.[26] The reaction was zero order in the amine nucleophile, indicating that copper(I) species in solution coordinate to the nucleophile in a fast and irreversible step to afford a cuprous nucleophile species [Ph_2NCu^I], which reacts with iodobenzene at the rate determining step.

Based on this previous work, Buchwald studied mechanistically the copper-catalyzed amidation of 3,5-dimethyliodobenzene promoted by *N*,*N*-cyclohexyldiamine ligand.[35,36] High ligand concentrations favor the formation of a copper(I)-amidate complex, which is a chemically and kinetically competent intermediate in the N-arylation reaction of the iodobenzene. On the other hand, low ligand concentrations favor the formation of inactive copper species that resulted from multiple coordination of amide ligands (Scheme 2.2). These results were supported computationally by Guo and coworkers who studied a similar reaction: the coupling reaction between bromobenzene and acetamide using the catalytic copper(I)/ethylenediamine system.[37] They calculated the concentration of multiple ligated species in solution and predicted that the copper(I)-amidate complex was the major species in the reaction mixture.

Similarly, Hartwig and coworkers reported the synthesis and characterization of several copper(I)-imidate and copper(I)-amidate complexes as well as copper(I)-phenoxide complexes based on bidentate N,N and P,P donor auxiliary ligands such as 1,10-phenanthroline, bipyridine, diamines and diphosphines.[38–40] By NMR spectroscopy and conductivity experiments, it was found that complexes in solution exist as dimeric ionic species

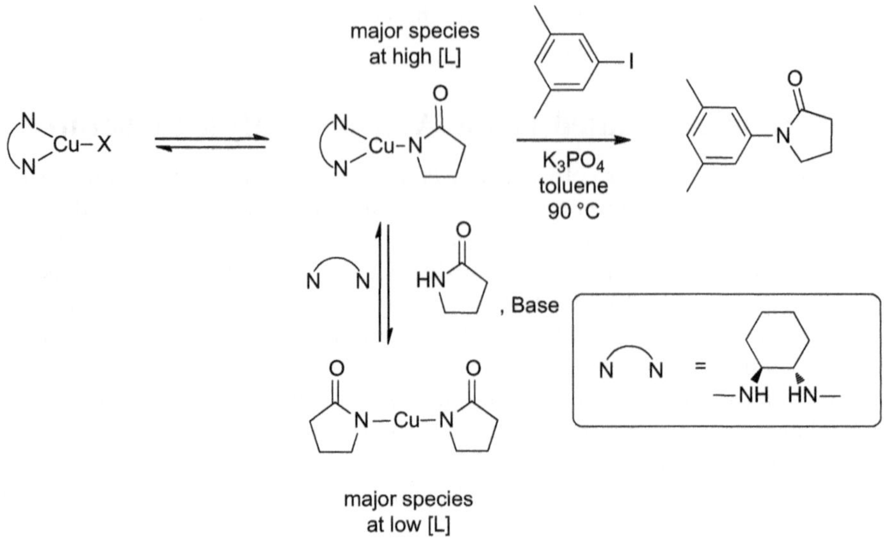

Scheme 2.2 Proposed role of the diamine ligand in the copper(I) catalyzed N-arylation reaction between 2-pyrrolidinone and 3,5-dimethyliodobenzene.

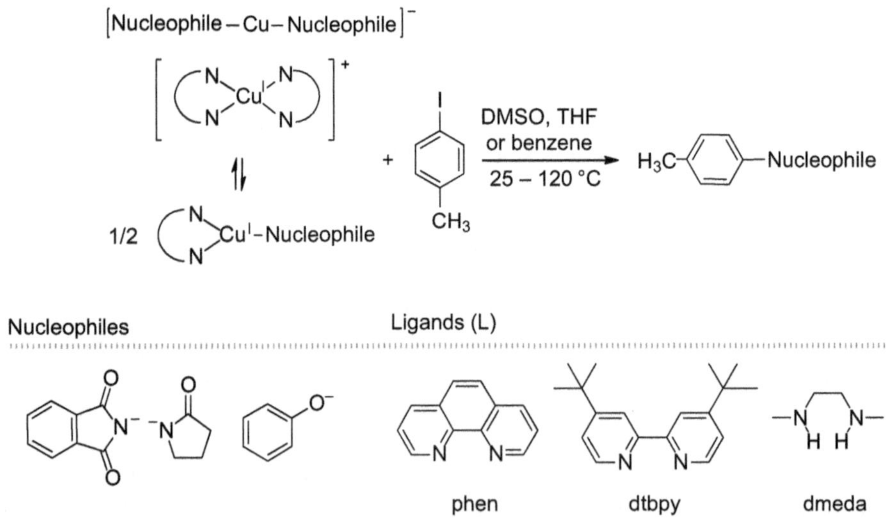

Scheme 2.3 Reaction of several Cu^I-imidate, -amidate and -phenoxide complexes with aryl iodide, as reported by Hartwig.

[L_2Cu^I][Cu^I(nucleophile)$_2$] in equilibrium with neutral [LCu^I(nucleophile)] species (Scheme 2.3). The three-coordinated neutral species were proposed to be the active intermediates for the formation of both N- or O-arylation coupling products. The authors rejected the idea that the anionic complex

$[Cu^I(nucleophile)_2]^-$ was capable of activating the haloarene, since the independently synthesized complex $[Cu^I(phth)_2][Bu_4N]$ was found to be inactive in the amidation reaction.

However, Shyu and coworkers reported some controversial data in their studies in the N-arylation of diphenylamine with the catalytic system CuI/1,10-phenanthroline/*t*BuONa by means of electrospray ionization mass spectrometry (ESI-MS).[41] The observation that the isolated complex $[Cu^I(NPh_2)_2][(phen)_3Na]$ reacted with 4-iodotoluene prompted the authors to propose that the anionic $[Cu^I(NPh_2)_2]^-$ species are intermediates in the C–N coupling reaction. By ESI-MS the authors observed ligand redistribution in solution at high temperatures favoring the formation of ionic $[Cu^I(phen)_2]^+$ and $[Cu^I(NPh_2)_2]^-$ under reaction conditions that suggest different equilibrium coordination at copper. Nevertheless, it is worth to mention that ESI-MS experiments are not valid for discarding the presence of neutral species such as $[Cu^I(phen)(NPh_2)]$ complex.

Most experimental data is in agreement with coordination of the nucleophile to copper(I) before the activation of the aryl halide occurs in the presence of auxiliary ligands. The addition of auxiliary ligands to copper salts in the catalytic reactions prevents the formation of less reactive, multiply ligated cuprate structures.[35–37,42] Therefore, the nature and concentration of the auxiliary ligand used in Ullmann-type condensation reactions have a big impact in the equilibrium between different Cu^I complexes present in solution, and in the formation of the active catalysts as well.

Finally, the solubility of the active copper complex strongly influences the C-heteroatom coupling reaction. Despite the fact that the auxiliary ligand is considered to increase the solubility of the copper salt, Taillefer and coworkers showed that in certain systems the ligand does not help in copper(I) solubilization but causes the formation of insoluble copper(I) reservoir complexes preventing any degradation process in solution.[43] The coordination of copper(I) iodide to the tetradentate bis(imino-pyridine) ligand led to the formation of highly insoluble dimeric complex $[L_2Cu_2]I_2$ (Scheme 2.4). A very small fraction of the latter is soluble in acetonitrile, but its dimeric structure is lost to release a monomeric copper(I) complex [LCu]I, which is the active precatalyst in the biaryl ether formation reaction.

2.4 Activation Mode of the Aryl Halide

Since the activation of the aryl halide is usually the turnover-limiting step of the catalytic cycle of Ullmann condensation reactions, the detection of intermediate species after this step has been very scarce, thus most mechanistic proposals are drawn from kinetic and computational studies. However, the experimental data collected is not enough for relying only on one proposal for the activation step of the aryl halide.

Initial proposals considered that the reaction took place without changing the oxidation state of the metal center that remained +1. In this context, the activation of the C–halogen bond can be achieved by (a) σ-bond methatesis

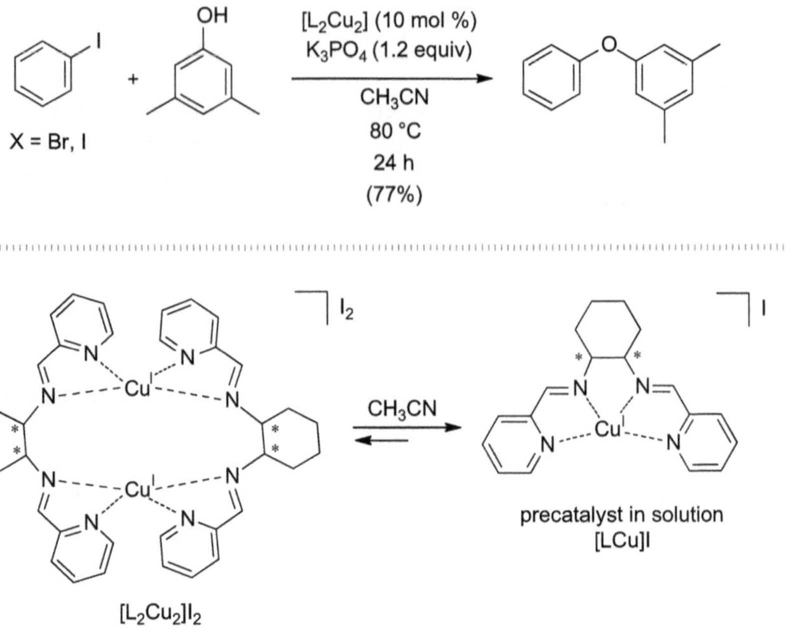

Scheme 2.4 Insoluble dimeric Cu^{I} complex acting as Cu^{I}-reservoirs in bis(imino-pyridine) ligand-based system.

through a four center intermediate or (b) by π-complexation between the copper(I) on the aryl halide. Alternatively, several proposals considered that the activation of the aryl halide occurs with the concomitant oxidation of the metal center *via* one or two electron redox processes. Several authors have proposed single electron transfer or halide atom transfer mechanisms involving oxidation to copper(II) species with the formation of aryl radical intermediates, which corresponds to a Cu^{I}/Cu^{II} catalytic cycle. On the other hand, oxidative addition/reductive elimination pathway has often been invoked for Ullmann condensation reactions. By a Cu^{I}/Cu^{III} catalytic cycle, arylcopper(III) intermediates are proposed to be formed *via* two-electron oxidation of the copper(I) complex with the aryl halide. The different mechanistic postulates will be reviewed hereafter (Scheme 2.5).

2.4.1 Mechanism Involving σ-Bond Metathesis

In 1964 Bacon and Hill studied stoichiometric substitution reactions between aryl halides and cuprous salts in polar organic solvents at 110–140 °C.[44–46] They proposed a mechanistic pathway that involves the activation of the aryl halide *via* σ-bond metathesis with the copper, which remains with at oxidation state +1. The copper(I) nucleophile activates the aryl halide *via* a four-centered intermediate in which the metal center forms a σ-complex with the lone electron pair of the halogen atom. Then, the C–X bond is polarized causing a

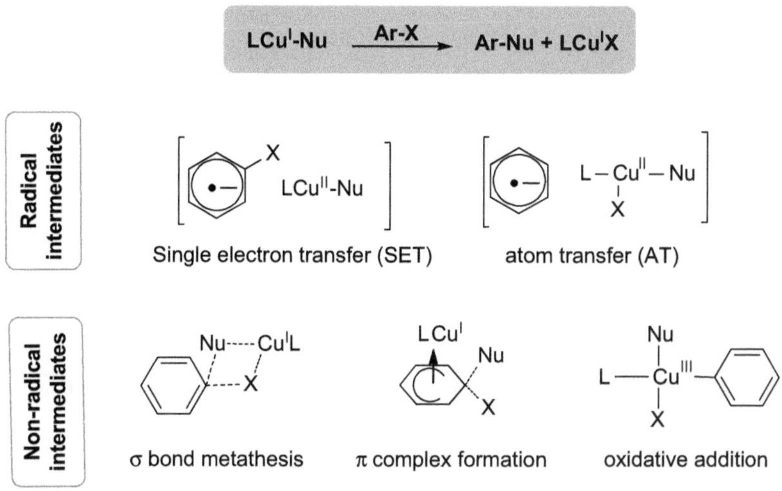

Scheme 2.5 Radical and non-radical intermediates proposed for Ullmann condensation reactions.

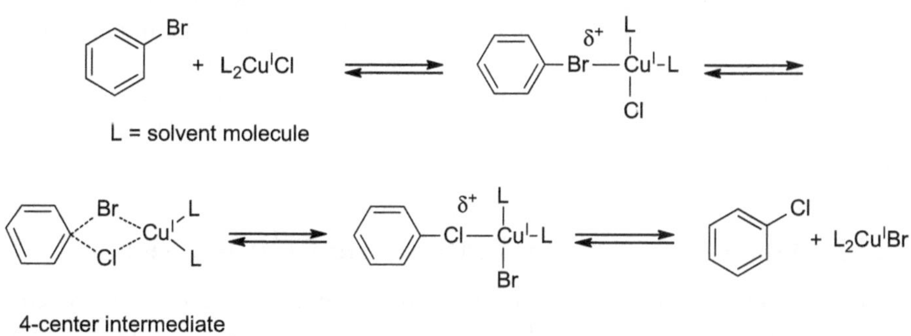

Scheme 2.6 Bacon and Hill's proposal based on the σ-bond metathesis pathway.

partial positive charge on the *ipso*-carbon that facilitates the attack by the nucleophile to give the corresponding coupling product and the copper complex (Scheme 2.6).

2.4.2 Mechanism Involving π-Complexation

A mechanism involving nucleophilic aromatic substitution *via* a π-bound organocuprate species was first reported by Weingarten in 1964.[6] In this mechanistic proposal the polarization of the C-halogen bond is caused by the interaction of the Cu^I-nucleophile complex with the π electrons of the aromatic ring instead of those of the leaving halogen atom. The formation of a Wheland intermediate facilitates the substitution of the halide by the nucleophile, while the copper species maintains its +1 oxidation state (Scheme 2.7). Weingarten

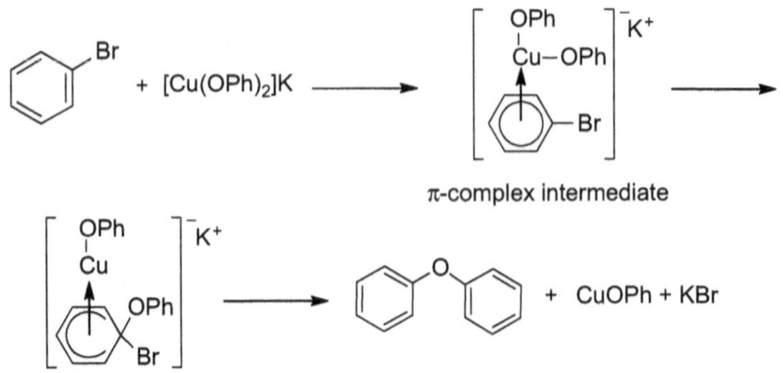

Scheme 2.7 Proposed mechanism for the coupling reaction of bromobenzene with potassium phenoxide in the presence of copper salt.

proposed this mechanism for the coupling reaction of bromobenzene with potassium phenoxide in the presence of copper salts.[47] Competition experiments between several aryl halides showed a reactivity pattern that is in agreement with the C-halogen bond strength, being faster for ArI > ArBr > ArCl, indicating that C-halogen bond cleavage is involved in the rate determining step, as in nucleophilic aromatic substitutions. Ma and coworkers also proposed this mechanism for the reaction of α-amino acids with aryl halides in the presence of catalytic amount of copper(I), although in a recent report they suggested an alternative reaction pathway based on oxidative addition/reductive elimination steps.[7,20]

2.4.3 One Electron Redox Processes *via* Cu^I/Cu^{II}: SET and AT

The intermediacy of aryl radicals formed *via* sequential one-electron redox processes have been proposed for Ullmann condensation reactions. Two plausible mechanistic pathways have been considered for a Cu^I/Cu^{II} catalytic cycle: (a) single electron transfer (SET) and (b) atom transfer (AT) (Scheme 2.8). In the SET mechanism, the copper(I) complex is oxidized by inner-sphere one-electron process by the aryl halide, yielding to the formation of a haloarene radical anion intermediate that in the subsequent step couples with the nucleophile *via* a second SET, and regenerates the catalyst. Some authors have invoked an atom transfer pathway in which the first electron transfer is associated to the transference of the aromatic halide to the copper complex.

These mechanistic proposals originate from the work of Kochi and Hida reported more than 50 years ago. Kochi supported an atom transfer pathway through their studies in the reactivity of free aryl radicals with copper(II) halide salts.[48,49] Aryl radicals, formed by decomposition of peroxides and iodonium salts in the presence of copper, are able to react with copper(II) halide salts yielding the corresponding aryl halide and copper(I) species through atom transfer mechanism.

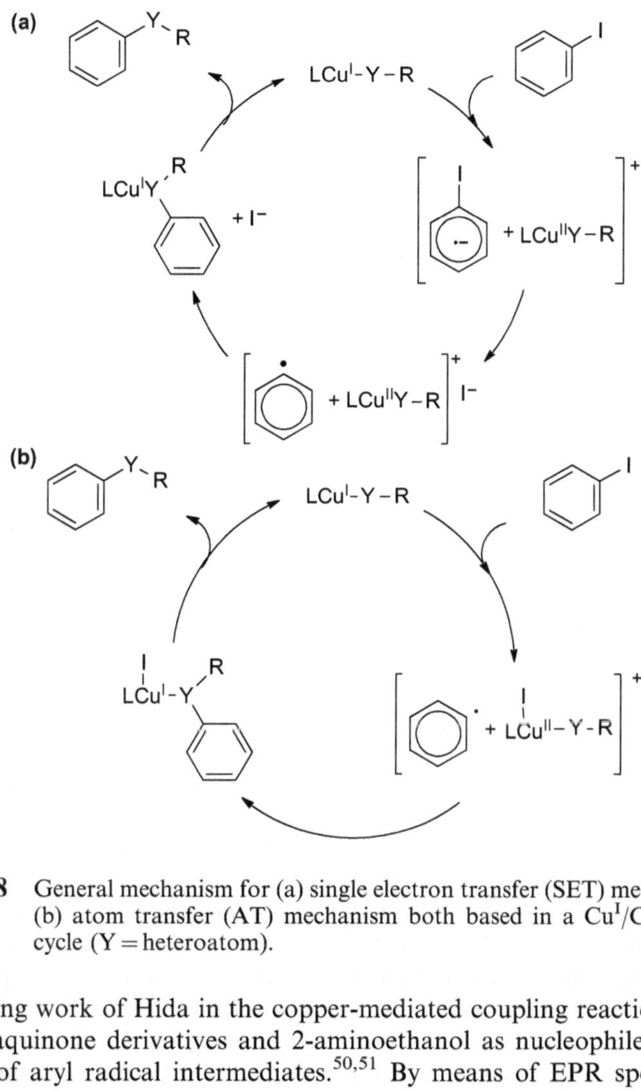

Scheme 2.8 General mechanism for (a) single electron transfer (SET) mechanism and (b) atom transfer (AT) mechanism both based in a Cu^I/Cu^{II} catalytic cycle (Y = heteroatom).

Pioneering work of Hida in the copper-mediated coupling reaction between haloanthraquinone derivatives and 2-aminoethanol as nucleophile led to the detection of aryl radical intermediates.[50,51] By means of EPR spectroscopy, Hida observed the formation of paramagnetic 1-bromoanthraquinone radical anion and copper(II) species in solution, which are formed by SET from copper(I) species to 1-bromoanthraquinone. Thus, the presence of the reduced product anthraquinone together with the C–N coupled product, prompted Hida to propose SET pathway as a general mechanism for Ullmann-type reactions even though haloanthraquinones are very specific substrates.

Classical radical scavengers have been used for proving the presence of aryl radical intermediates in Ullmann condensation reactions. In most catalytic systems, the reaction rate is not inhibited by radical scavengers, which has been used for supporting non-radical mechanisms. However, a negative result in radical trapping experiments cannot exclude the involvement of very short-lived solvent-cage radicals in the reaction pathway. Therefore, several groups

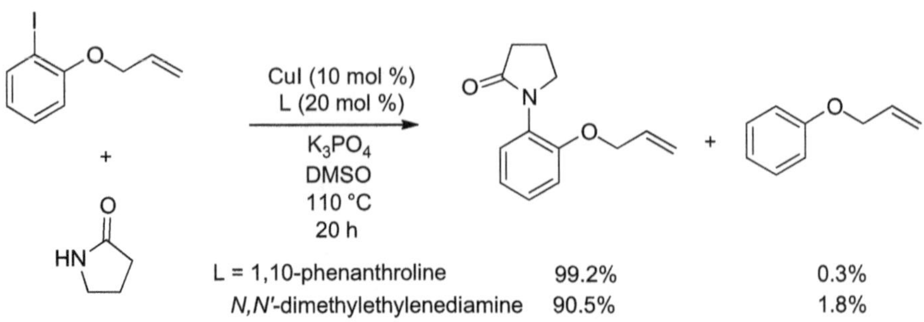

Scheme 2.9 Radical clock tests in C–O bond-forming reactions performed by Hartwig.

have performed radical clock tests as a more reliable proof for discarding radical intermediates. Following initial reports by Bowman and coworkers,[52] Hartwig and coworkers have reported radical clock tests in copper-catalyzed C–N and C–O coupling reactions in the presence of bidentate auxiliary ligands.[38–40] The reaction between *o*-(allyloxy)iodobenzene and 2-pyrrolidone in the presence of catalytic amounts of copper and 1,10-phenanthroline or *N*,*N*′-dimethylethylenediamine afforded high yields of the C–N coupled product with minor amounts of phenyl allyl ether obtained by hydrodeiodination (Scheme 2.9).[38] It is known that the aryl radical generated from *o*-(allyloxy)iodobenzene undergoes extremely fast cyclization to yield 2,3-dihydrobenzofuranyl-methyl radical, which can abstract hydrogen from the solvent, dimerize or combine with the nucleophile ligand to form the C-heteroatom coupled product. The lack of cyclization products derived from *o*-(allyloxy)iodobenzene substrate was interpreted as a proof against the involvement of solvent cage radicals in the mechanism. However, Van Koten and coworkers claimed that these radical trapping experiments are not a reliable proof against radicals. The assumption is that intramolecular rearrangement reactions are faster than bimolecular reactions but it has not been proven that this concept is the same in the presence of transition metals.[53]

In this context, Van Koten and co-workers explored a family of aminoarenethiolato-copper(I) complexes as pre-catalysts for the copper-catalyzed C–N and C–O coupling reaction of aryl bromide (Scheme 2.10).[54,55] They observed that the reaction does not match with the common reactivity order that follows the leaving group ability of the halide (ArI > ArBr > ArCl) since only aryl bromides were able to react. In addition, the use of radical traps slowed down or even stopped the reaction.[53] These observations prompted the authors to propose a sequential single electron transfer (SET) mechanism. In the first step, one electron transfer from the copper(I) complex to the aryl bromide causes the formation of an aryl radical and copper(II) intermediates. Then, the aryl radical couples with the amine moiety through a second SET that regenerates the copper(I) catalyst and releases the C–N coupling product. The lack of reactivity of aryl iodides was explained by the overstabilization of copper(I) species by the soft iodide anion that prevents the formation of copper(II) in the first SET step.

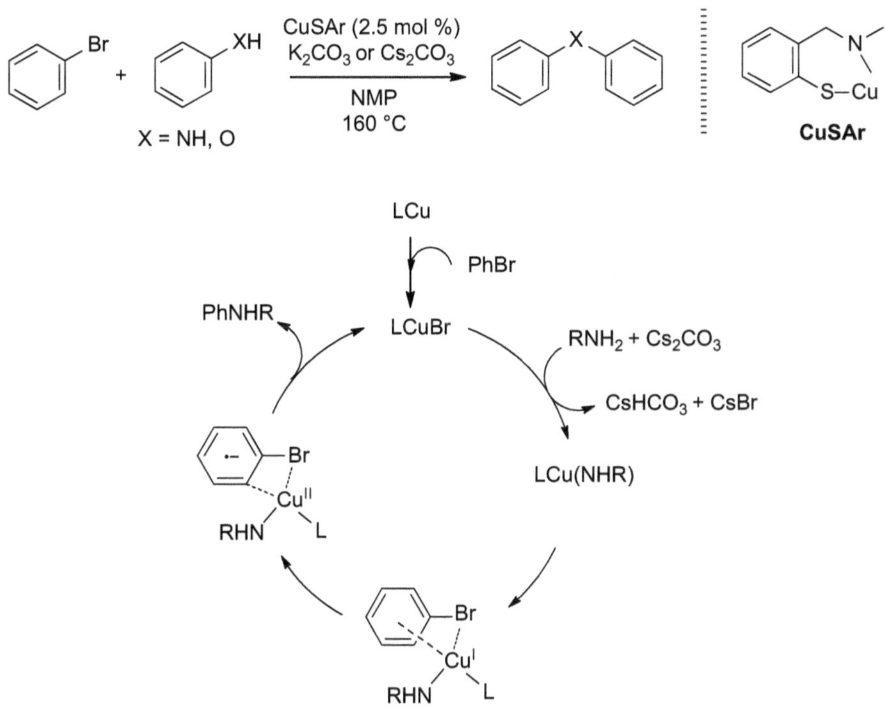

Scheme 2.10 Reaction mechanism for the coupling reaction of bromobenzene with nitrogen nucleophiles reported by Van Koten and coworkers.

Remarkably, very recently Peters, Fu and coworkers reported the first copper/ligand system that operates by a single electron transfer mechanism, thus demonstrating the viability of the radical pathway.[56] They used a photoluminescent copper-carbazolide complex bearing monophospine ligands for promoting the photoinduced C–N bond forming reaction. Using a suitable irradiation lamp, not only aryl iodides but also aryl bromides and chlorides could be coupled under mild reaction conditions. Upon photoexcitation of the copper-carbazolide complex, a copper containing radical is formed, as detected by EPR spectroscopy. This radical intermediate reacts with aryl halides *via* single electron transfer to afford the corresponding C–N coupling product. Interestingly, a radical clock test with deuterated *o*-(allyloxy)iodobenze afforded exclusive formation of cyclized compounds as a 1 : 1 mixture of diastereoisomers, which is consistent with a radical pathway (Scheme 2.11 a). Moreover, competition experiments between 1-bromonapthalene and 1-chlorobenzonitrile are also in agreement with a SET pathway since the latter, which has a more favorable reduction potential, reacts predominantly (Scheme 2.11). This important contribution demonstrated that C–X cleavage as well as the C–N bond forming steps can proceed *via* single electron processes. Although the copper-catalyzed photoinduced C–N reaction is demonstrated only for a single nucleophile, the work of Fu, Peters and coworkers opens the

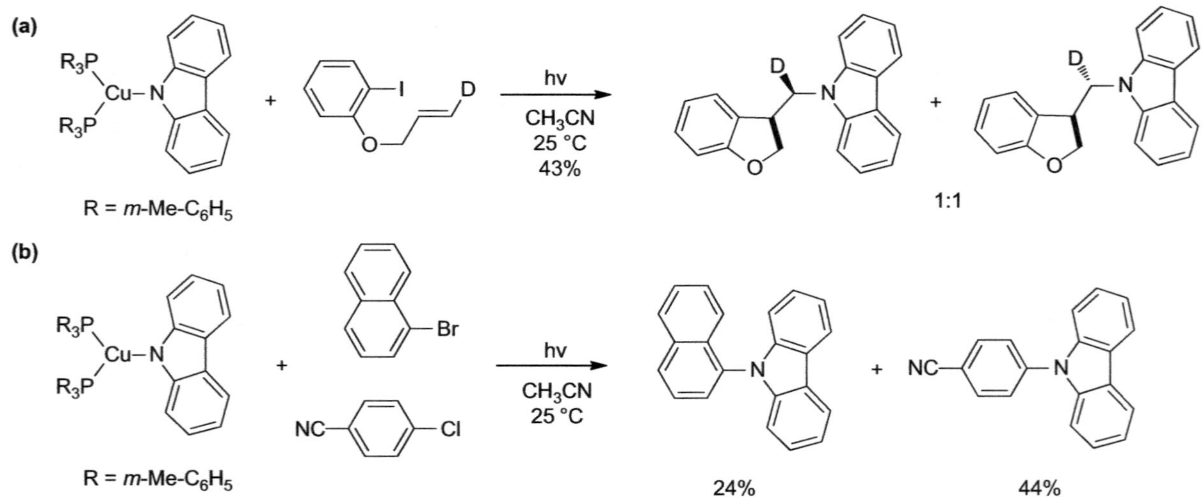

Scheme 2.11 Radical Clock test (a) and competition experiments (b) as mechanistic probes for supporting a single electron transfer mechanism in copper-carbazolide complex.

door to explore other photoinduced C-heteroatom bond forming reactions under mild reaction conditions.

2.4.4 Oxidative Addition/Reductive Elimination Pathway

A catalytic cycle Cu^{I}/Cu^{III} based on two-electron redox processes has often been invoked for Ullmann condensation reactions. In this proposal the aryl halide is activated by oxidative addition of the copper(I)-nucleophile complex to the C-halogen bond forming an arylcopper(III) intermediate. The copper(III) complex undergoes C-heteroatom reductive elimination that affords the coupling product between the nucleophile and the aryl moiety and regenerates the copper(I) (Scheme 2.12).

Despite the inherent instability of copper(III) complexes, early reports already proposed copper(III) species as key intermediates in Ullmann chemistry. In this context, Cohen proposed the formation of arylcopper(III) complexes in the homocoupling reaction of *o*-bromonitrobenzene mediated by copper.[57,58] The formation of biaryl 2,2′-dinitrobiphenyl and small amounts of the reduced product nitrobenzene, which increases in the presence of proton sources, was rationalized by the intermediacy of organocopper complexes that were susceptible to protonation. Aryl radical intermediates were discarded by the lack of rate inhibition in the presence of radical scavengers. In addition, in Ullmann coupling reaction of vinyl halides, the homocoupled products showed retention of the configuration that discards again the intermediacy of radical species.[58,59] With these mechanistic considerations in hand, Cohen proposed that organocopper(III) species, obtained by oxidative addition from aryl or vinyl halides, were key intermediate species in Ullmann reactions.

Since the work of Cohen on Ullmann C–C coupling reactions, several authors have proposed copper(III) intermediates in C–heteroatom bond

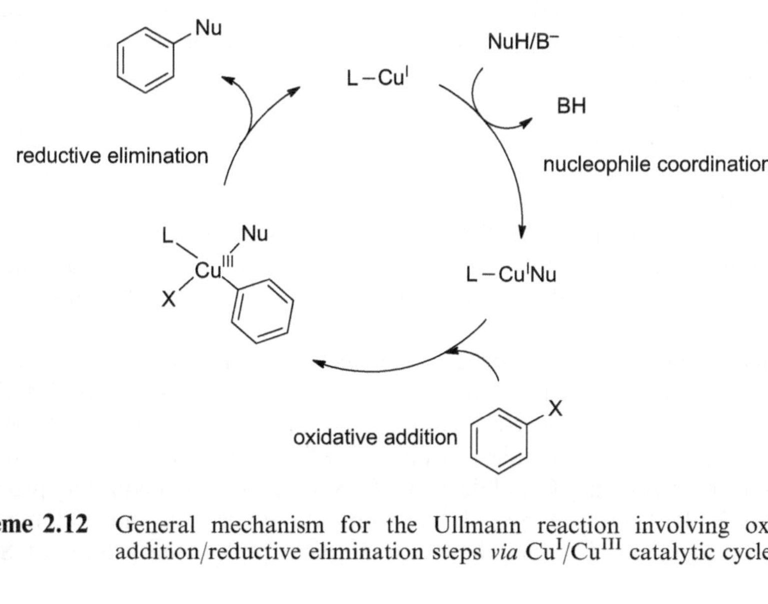

Scheme 2.12 General mechanism for the Ullmann reaction involving oxidative addition/reductive elimination steps *via* Cu^{I}/Cu^{III} catalytic cycle.

forming reactions catalyzed by copper with minor or no supportive mechanistic data. Recently, Taillefer and coworkers proposed oxidative addition/reductive elimination pathway for the arylation of phenols based on the relationship between the structure of the auxiliary ligand and their catalytic activity.[19] Among the *N*,*N*-bidentate ligands explored, the most efficient ligands contained one imine and one pyridine binding site acting in a synergistic manner. More electron-donating groups in the *para* position in the pyridine ring improved the catalytic activity owing to a more favorable oxidative addition, as indicated by the lower oxidation potential. The catalytic activity was also improved with more electron-withdrawing *para*-substituents in the imine binding site, which was explained by a more favorable ligand exchange or reductive elimination steps. Taillefer showed that the reaction pathway is strongly influenced by little electronic differences in the auxiliary ligand. However, these studies did not identify Cu^{III} intermediates in the catalytic cycle.

The elusive short-lived arylcopper(III) intermediates that are formed after the rate-determining step have not been detected under reactions conditions. Thus, the proposal of the Cu^{I}/Cu^{III} cycle is mainly supported by the lack of rate inhibition with radical scavenging and by computational studies (Section 2.5). On the other hand, fundamental studies in organometallic high valent copper complexes may provide greater understanding of Ullmann condensation mechanism. However, the difficulty to stabilize well-defined copper(III) complexes have precluded study of their reactivity until recently.

A remarkable contribution to the characterization of well-defined Cu^{III} species was the synthesis of the first isolated arylcopper(III) complex in 2002 reported by Ribas, Stack and Llobet.[60] The copper(III) was stabilized owing to the σ-donor properties of amines in the macrocyclic ligand. In 2008, Huffman and Stahl demonstrated that the arylcopper(III) complex can react stoichiometrically with different amide-type nitrogen nucleophiles, such as pyridone, phthalimide and acetanilide to form the corresponding C–N coupling products (Scheme 2.13).[61] The groups of Ribas and Stahl also explored the involvement of arylcopper(III) complex in C–O bond formation with several carboxylic acids, phenols and acidic aliphatic alcohols.[62] The authors proposed a mechanism that consists of nucleophile coordination to the arylcopper(III) complex, deprotonation assisted by an external base and reductive elimination. Kinetic data was rationalized by a change in the rate-determining step depending on the pK_A of the nucleophile. Whereas proton loss of the nucleophile was the RDS for the less acidic amides and phenols, reductive elimination determined the reaction rate for the more acidic carboxylic acids.

Similarly Wang and coworkers reported the synthesis and reactivity of another monoaryl copper(III) complex based in an azacalix[1]arene[3]pyridine ligand.[63,64] The arylcopper(III) complex reacts with a number of anionic nucleophiles, including halides, cyanide, isothiocyanate and carboxylates, to yield the corresponding C–halide, C–C, C–S, C–C and C–O coupling products (Scheme 2.14). Therefore, by isolating copper(III) complexes and studying their reactivity in front of several heteroatom nucleophiles, the groups of Stahl,

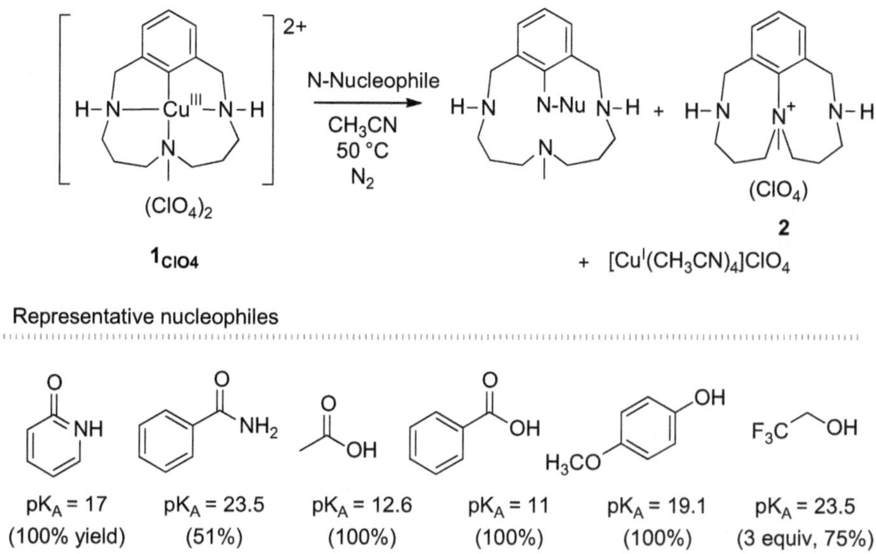

Scheme 2.13 C–N and C–O bond forming reaction from well-defined aryl-copper(III) complexes reported by Ribas and Stahl.

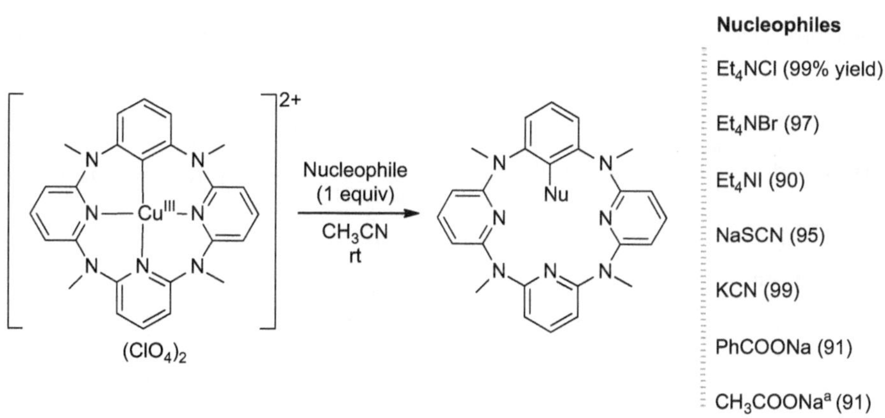

Scheme 2.14 Stoichiometric C-heteroatom coupling reactions between aryl-Cu^{III} complex and anionic nucleophiles reported by Wang. [a]Two equivalents were used.

Ribas and Wang have demonstrated the feasibility of arylcopper(III) complexes to participate in C–heteroatom bond forming reactions.

Remarkably, the synthesis and characterization of a family of arylCu^{III}-halide complexes enabled Ribas, Stahl and coworkers to observe the fundamental Cu^I/Cu^{III} redox steps relevant the Ullmann condensation reactions, the C(aryl)-halogen reductive elimination as well as the reversal oxidative addition (Scheme 2.15).[65] The C-halogen reductive elimination was triggered by the addition of proton sources, such as triflic acid. Indeed, once reductive elimination of arylCu^{III}-Br complex upon addition of acid is finished, and full formation of protonated coupling products $[L\text{-}Br\text{-}H]^+$ and Cu^I salts is achieved, the addition of Proton-sponge® as a non-coordinating base to deprotonate the amine causes the instantaneous reversal oxidative addition reaction. Therefore, the intramolecular C–halogen bond formation by reductive elimination within these triazamacrocyclic ligands is a reversible, pH dependent reaction.

In addition Ribas and Stahl demonstrated that arylcopper(III)-halide complexes are relevant in copper-catalyzed C–heteroatom bond forming reactions.[65] They reported C–O and C–N bond forming reactions catalysed by copper in macrocyclic aryl halide substrates (Scheme 2.16). By means of

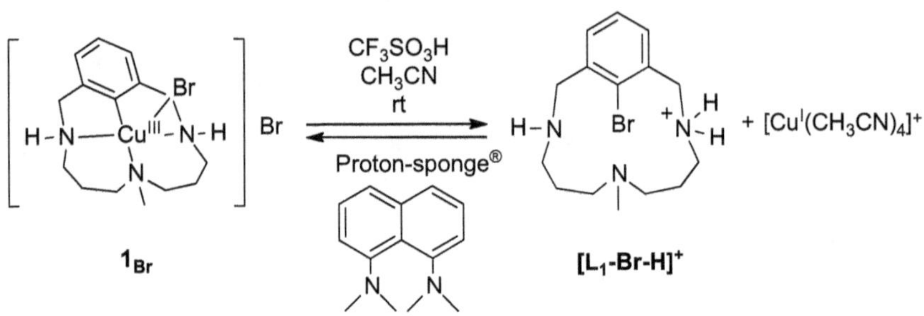

Scheme 2.15 Reversible reductive elimination/oxidative addition under pH control.

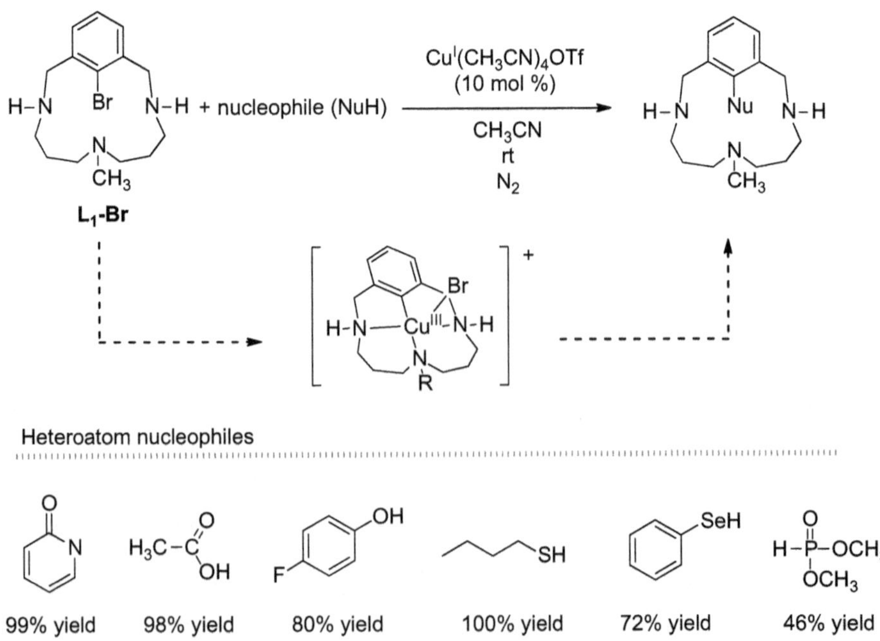

Scheme 2.16 Catalytic C–heteroatom bond forming reactions in macrocyclic aryl halide substrate through arylcopper(III) intermediates.

UV-Vis and NMR spectroscopy, a steady-state concentration of copper(III) intermediate was detected, which indicates the involvement of the arylcopper(III)-halide complex in the turnover-limiting step of the catalytic reaction. This observation was proof of principle for the experimental feasibility of an oxidative addition/reductive elimination Cu^I/Cu^{III} catalytic cycle. Moreover, they also developed copper-catalyzed nucleophilic fluorination of aryl chloride and bromide substrates using silver fluoride at room temperature through arylcopper(III) intermediates.[66] The mechanistic pathway was expanded to C–S, C–P and C–Se bond forming reactions, which supports the idea that the oxidative addition/reductive elimination pathway can be generalized to obtain very versatile C–heteroatom bonds.[67] These studies demonstrate experimentally that copper(III) complexes can be catalytically relevant in these transformations even though they do not resolve the controversy concerning the mechanism of conventional Ullmann condensation reactions.

2.5 Computational Studies for Unravelling the Mechanism

In recent years, an increasing number of computational studies for evaluating the mechanism of copper-catalyzed C-heteoratom bond forming reactions have been reported. Due to the high success of C–N bond formation protocols with bidentate ligands, amidation and amination reactions have been systematically investigated from a computational point of view.[37,68,69] In agreement to the experimental data, calculations of the concentration of the different copper species in equilibrium usually showed that the nucleophile is coordinated to the active copper(I) catalyst (see section 2.3). With regard to the activation of the aryl halide, most theoretical investigations support an oxidative addition/reductive elimination pathway *via* arylcopper(III) intermediates.[38–40,70] Nevertheless, a reliable theoretical study should take into account all mechanistic possibilities for a given copper/ligand system, since theoretical data gains value when meaningful comparisons can be done not only with experimental results but also with other computed mechanisms.

A detailed theoretical study of the copper-catalyzed acetamide arylation with bromobenzene was reported by Guo and coworkers in 2007 (Scheme 2.17).[37] The copper catalyst bearing ethylenediamine as a chelating ligand undergoes oxidative addition to the aryl bromide yielding a square-pyramidal arylcopper(III) intermediate. The pentacoordinated copper(III) complex contains the two coupling partners situated in a *trans* position. By a pseudo-rotation step the phenyl and amide group are placed *cis* to each other before the reductive elimination step, which occurs with a low energetic barrier. The oxidative addition was the rate limiting step of the aryl amidation reaction catalyzed by copper(I) with an overall free energy barrier of 26 kcal mol^{-1}. Guo and coworkers also evaluated ligand effects by calculating the mechanistic pathway using other chelating diamines. Theoretical data indicated that *trans-N,N*-dimethylcyclohexane-1,2-diamine was the most reactive amine

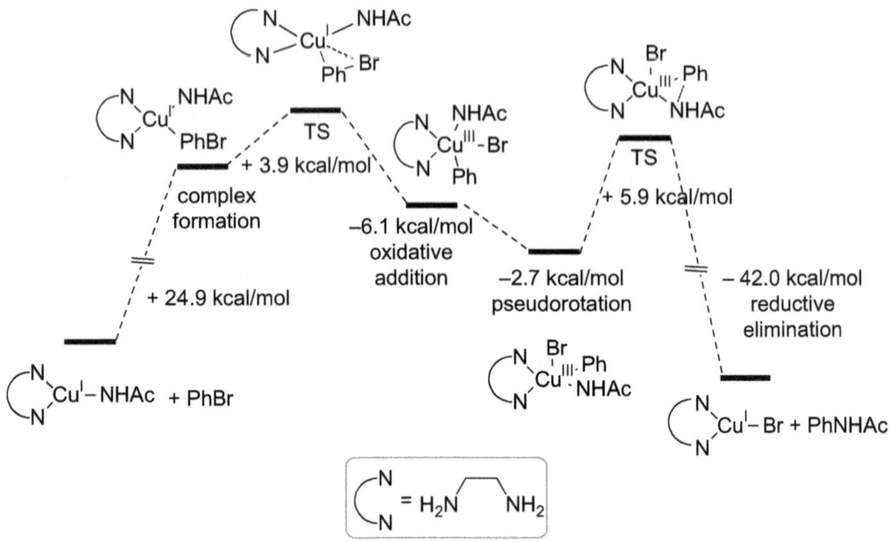

Scheme 2.17 Energetic profile calculated for the reaction of bromobenzene and acetamide catalyzed by Cu^{I}/ethylenediamine.

whereas *N,N*-tetramethylethylenediamine was the less reactive, in agreement with the experimental results reported previously by Buchwald.[35]

The oxidative addition/reductive elimination pathway for the coupling of bromobenzene with methylamine in the presence of copper(I)/acetylacetonate has been evaluated by Ding and coworkers.[68] Analogous to copper-amidation reactions, they found that the active species is a neutral copper(I) complex containing both ligand and nucleophilic substrate, and oxidative addition of bromobenzene to this complex is the rate-determining step. Similarly, Jutand and coworkers proposed a Cu^{I}/Cu^{III} catalytic cycle for the amination reaction of iodobenzene with *N*-cyclohexylamine catalyzed by copper/2-acetylcyclohexanone system.[71] In this case, an initial activation of the aryl iodide by the nucleophilic amido ligand coordinated to copper(I) was computationally found to occur before the oxidative addition step. This unusual adduct formation leads to an intramolecular oxidative addition that has a transition state of 22.9 kcal mol^{-1}, which was a more favored value than the one computed for intermolecular oxidative addition with a similar *β*-diketone ligand.[72]

Cundari and coworkers reported an extensive computational study of a monomeric copper(I)-anilido complex bearing a N-heterocyclic carbene (NHC) ligand.[69] They performed DFT calculations for four possible pathways including oxidative addition, nucleophilic substitutions, π-complex formation as well as a SET mechanism. The oxidative addition/reductive elimination pathway was calculated to have the lowest energetic barrier of 25.6 kcal mol^{-1}. On the other hand, the nucleophilic substitution, where the iodide of the haloarene reacts with the anilido ligand without coordinating to the copper(I), was only 4.3 kcal mol^{-1} higher in energy than the oxidative addition pathway. The authors discarded the nucleophilic substitution mechanism with experimental results, since the [(NHC)Cu-NHPh] complex does not react with phenyl *p*-toluenesulfonate, which is a good leaving group for nucleophilic substitutions.

The ancillary ligand plays a role in the selectivity of copper-catalyzed C–heteroatom bond forming reactions. Buchwald demonstrated ligand-directed selectivities in N- *versus* O-arylation of 5-amino-1-pentanol as bifunctional nucleophile and iodobenzene in the presence of copper(I) and bidentate ligand.[73] Whereas the *β*-diketone ligand causes selective formation of the N-arylation product in DMF at room temperature, the 1,10-phenanthroline ligand promotes selective O-arylation in toluene at 90 °C (Scheme 2.18 a). In order to explain these selectivity patterns, Buchwald and Houk conducted DFT calculations on the simplified model reactions of iodobenzene with methanol and methylamine promoted by copper(I) complexes containing either *β*-diketone or phenanthroline type ligands.[72] In their study, four different pathways for the activation of the iodobenzene were evaluated: (a) oxidative addition/reductive elimination, (b) SET, (c) IAT and (d) σ-bond metathesis. The pathways showing a lower computed energy barrier were the SET and IAT mechanisms, which involve aryl radical intermediates (Scheme 2.18 b). Moreover, they proposed that the selectivity derived from the step involving aryl halide activation, which is the rate determining step. When a *β*-diketone is used, SET is the most favorable pathway for both N-arylation and O-arylation. Although the LCu^{I}-NHMe complex is less stable than LCu^{I}-OMe, the N-bound pathway has a lower activation barrier than the O-bound pathway. A different scenario is observed with the 1,10-phenanthroline ligand, since SET and IAT have similar barriers and either may occur depending on the nucleophile. The copper-catalyzed O-arylation reaction proceeds *via* IAT and it is lower in energy than the N-arylation that proceeds *via* SET.

On the other hand, Fu and coworkers have proposed an alternative computational pathway for explaining the selectivities observed experimentally in Buchwald's system,[73] using the real nucleophile 5-amino-1-pentanol bifunctional substrate.[74] Fu's DFT results indicated that the most plausible mechanism is oxidative addition/reductive elimination pathway *via* arylcopper(III) complexes instead of radical mechanisms such as SET and IAT (Scheme 2.19). Therefore, calculations performed with 1,10-phenanthroline ligand indicated that nucleophile coordination is prior to the oxidative addition that is the rate-determining step. The most favorable transition state consisted in a pentacoordinated copper(III) complex bound to the deprotonated alcohol of the substrate that lead to the C–O coupling product by reductive elimination. Otherwise, DFT calculations with the *β*-diketone ligand showed that activation of the aryl halide occurs before coordination of the nucleophile. Therefore, the most favorable transition state for the oxidative addition does not involve the amino alcohol nucleophile. Oxidative addition produced a tetracoordinated copper(III) intermediate that binds to the nucleophile at the RDS. In this case, amine coordination to form a pentacoordinated copper(III) complex is more

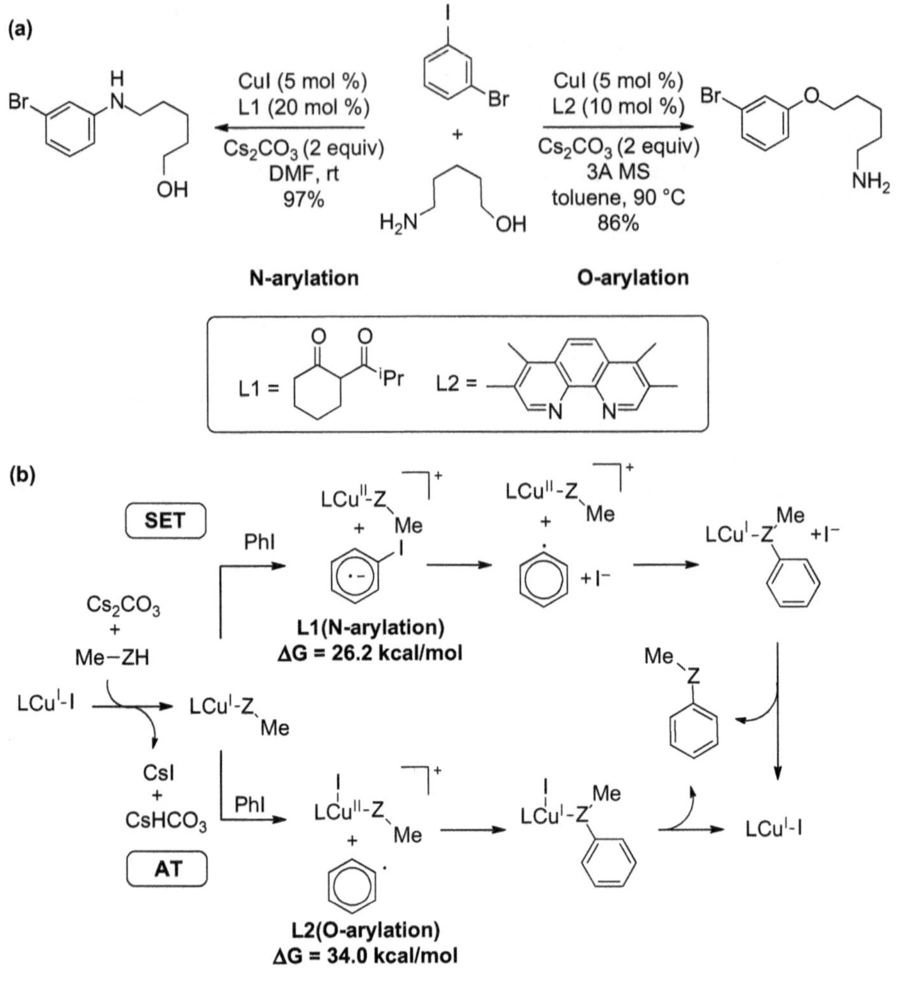

Scheme 2.18 (a) Chemoselective N-arylation and O-arylation of 5-amino-1-pentanol with 3-bromo-3-iodobenzene catalyzed by copper and bidentate ligands. (b) DFT calculations with MeOH and $MeNH_2$ as model nucleophiles, involving rate-limiting SET and/or IAT mechanistic pathways depending on the ancillary ligand used.

favorable than the coordination by the alcohol group of the nucleophile. The C–N coupling product is finally obtained after deprotonation of the amino group and reductive elimination.

Theoretical studies are a useful and complementary tool when experimental data is scarce for understanding the reaction mechanism. Despite the several efforts that have been made for elucidating the mechanism of Ullmann condensation reactions by computational studies, controversial results have been obtained even for similar copper/ligand catalyst systems.

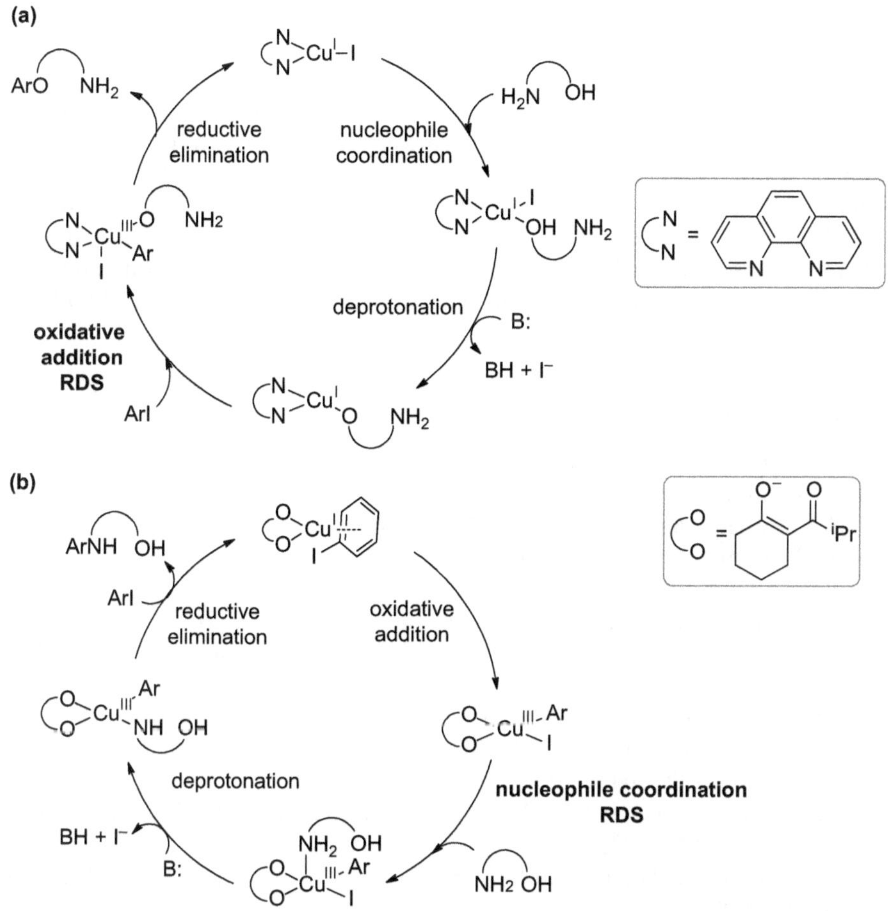

Scheme 2.19 Computed mechanism for the cross-coupling reaction between aryl iodide and 5-amino-1-pentanol, involving oxidative addition and reductive elimination steps using (a) β-diketone and (b) 1,10-phenanthroline ligand.

2.6 Conclusions and Perspectives

During the last few years much progress has been made in order to gain mechanistic understanding of copper-catalyzed arylation of heteroatom nucleophiles. There is a general consensus that copper in oxidation state +1 is the active catalyst and that the nucleophile coordination precedes the activation of the aryl halide. The cleavage mode of the C–halogen bond in the haloarene by the copper complex is the most discussed mechanistic aspect. The experimental and computational data do not converge to a single unified mechanism, sometimes even for the same reaction. On the other hand, the great diversity of catalytic systems and nucleophiles makes it difficult to include all Ullmann condensation reactions in one single mechanism. It is plausible that depending on the system, the reaction could proceed through one mechanism or another. Many mechanistic questions remain to be uncovered and their answers will help to design more efficient copper catalysts for the formation of C–heteroatom bonds. Besides, mechanistic comprehension of these reactions will have an impact in other related copper-catalyzed cross-coupling reactions, for instance Chan-Lam reactions,[75] aromatic Finkelstein reactions[76] and cyanation reactions,[77,78] that can be accommodated by analogous mechanisms. Moreover, the chemical understanding of metals in high oxidation states is a key topic to unravel mechanistic proposals in catalysis (see Chapters 4 and 5 of this book).[79]

References

1. F. Ullmann and J. Bielecki, *Chem. Ber.*, 1901, **34**, 2174.
2. F. Ullmann, *Ber. Dtsch. Chem. Ges.*, 1903, **36**, 2389.
3. I. Goldberg, *Ber. Dtsch. Chem. Ges.*, 1906, **39**, 1691–1696.
4. P. E. Fanta, *Synthesis*, 1974, 9–21.
5. J. Lindley, *Tetrahedron*, 1984, **40**, 1433–1455.
6. H. Weingarten, *J. Am. Chem. Soc.*, 1964, **29**, 3624–3626.
7. D. Ma, Y. Zhang, J. Yao, S. Wu and F. Tao, *J. Am. Chem. Soc.*, 1998, **120**, 12459–12467.
8. H. B. Goodbrand and N.-X. Hu, *J. Org. Chem.*, 1999, **64**, 670–674.
9. P. J. Fagan, E. Hauptman, R. Shapiro and A. Casalnuovo, *J. Am. Chem. Soc.*, 2000, **122**, 5043–5051.
10. A. Kaplars, J. C. Antilla, X. Huang and S. L. Buchwald, *J. Am. Chem. Soc.*, 2001, **123**, 7727–7729.
11. A. Klapars, X. Huang and S. L. Buchwald, *J. Am. Chem. Soc.*, 2002, **124**, 7421–7428.
12. S. L. Buchwald, A. Klapars, J. C. Antilla, G. E. Job, M. Wolter, F. Y. Kwong, G. Nordmann and E. J. Hennessy, *Patent WO 02/085838 A1*, 2001.
13. M. Taillefer, H.-J. Cristau, P. P. Cellier, J.-F. Spindler and A. Ouali, *Patent Fr 2840303-WO 03101966 (Pr. Nb. Fr 2002 06717)*, 2002.
14. M. Taillefer, H.-J. Cristau, P. P. Cellier and J.-F. Spindler, *Patent Fr 2833947-WO 0353225 (Pr. Nb. Fr. 01 16547)*, 2001.
15. D. S. Surry and S. L. Buchwald, *Chem. Sci.*, 2010, **1**, 13–31.
16. A. Shafir and S. L. Buchwald, *J. Am. Chem. Soc.*, 2006, **128**, 8742–8743.
17. B. de Lange, M. H. Lambers-Verstappen, L. Schmieder-van de Vondervoort, N. Sereinig, R. de Rijk, A. H. M. de Vries and J. G. de Vries, *Synlett*, 2006, 3105–3109.
18. H.-J. Cristau, P. P. Cellier, J.-F. Spindler and M. Taillefer, *Chem.–Eur. J.*, 2004, **10**, 5607–5622.
19. A. Ouali, J.-F. Spindler, A. Jutand and M. Taillefer, *Adv. Synth. Catal.*, 2007, **349**, 1906–1916.
20. D. Ma and Q. Cai, *Acc. Chem. Res.*, 2008, **41**, 1450–1460.

21. D. Ma, Q. Cai and H. Zhang, *Org. Lett.*, 2003, **5**, 2453–2455.
22. S. V. Ley and A. W. Thomas, *Angew. Chem., Int. Ed.*, 2003, **42**, 5400–5449.
23. I. P. Beletskaya and A. V. Cheprakov, *Coord. Chem. Rev.*, 2004, **248**, 2337–2364.
24. G. Evano, N. Blanchard and M. Toumi, *Chem. Rev.*, 2008, **108**, 3054–3131.
25. F. Monnier and M. Taillefer, *Angew. Chem., Int. Ed.*, 2009, **48**, 6954–6971.
26. A. J. Paine, *J. Am. Chem. Soc.*, 1987, **109**, 1496–1502.
27. S. A. Kondratov and S. M. Shein, *Zh. Org. Khim.*, 1979, **15**, 2160.
28. M. Mansour, R. Giacovazzi, A. Ouali, M. Taillefer and A. Jutand, *Chem. Commun.*, 2008, 6051–6053.
29. R. K. Gujadhur, C. G. Bates and D. Venkataraman, *Org. Lett.*, 2001, **3**, 4315–4317.
30. E. Haldón, E. Alvarez, M. C. Nicasio and P. J. Pérez, *Organometallics*, 2009, **28**, 3815–3821.
31. T. Kawaki and H. Hashimoto, *Bull. Chem. Soc. Jpn.*, 1972, **45**, 1499–1500.
32. G. M. Whitesides, J. S. Sadowski and J. Lilburn, *J. Am. Chem. Soc.*, 1974, **96**, 2829–2835.
33. M. Kubota and A. Yamamoto, *Bull. Chem. Soc. Jpn.*, 1978, **51**, 2909–2915.
34. T. Yamamoto, Y. Ehara, M. Kubota and A. Yamamoto, *Bull. Chem. Soc. Jpn.*, 1980, **53**, 1299–1302.
35. E. R. Strieter, D. G. Blackmond and S. L. Buchwald, *J. Am. Chem. Soc.*, 2005, **127**, 4120–4121.
36. E. R. Strieter, B. Bhayana and S. L. Buchwald, *J. Am. Chem. Soc.*, 2009, **131**, 78–88.
37. S.-L. Zhang, L. Liu, Y. Fu and Q.-X. Guo, *Organometallics*, 2007, **26**, 4546–4554.
38. J. W. Tye, Z. Weng, A. M. Johns, C. D. Incarvito and J. F. Hartwig, *J. Am. Chem. Soc.*, 2008, **130**, 9971–9983.
39. R. Giri and J. F. Hartwig, *J. Am. Chem. Soc.*, 2010, **132**, 15860–15863.
40. J. W. Tye, Z. Weng, G. Giri and J. F. Hartwig, *Angew Chem., Int. Ed.*, 2010, **49**, 2185–2189.
41. C.-K. Tseng, C.-R. Lee, C.-C. Han and S.-G. Shyu, *Chem.–Eur. J.*, 2011, **17**, 2716–2723.
42. E. Nakamura and S. Mori, *Angew. Chem., Int. Ed. Engl.*, 2000, **39**, 3750–3771.
43. A. Ouali, M. Taillefer, J.-F. Spindler and A. Jutand, *Organometallics*, 2007, **26**, 65–74.
44. R. G. R. Bacon and H. A. O. Hill, *J. Chem. Soc.*, 1964, 1097–1107.
45. R. G. R. Bacon and H. A. O. Hill, *J. Chem. Soc.*, 1964, 1108–1112.
46. R. G. R. Bacon and H. A. O. Hill, *J. Chem. Soc.*, 1964, 1112–1119.
47. H. Weingarten, *J. Org. Chem.*, 1964, **29**, 977–978.
48. J. K. Kochi, *J. Am. Chem. Soc.*, 1957, **79**, 2942–2948.

49. J. K. Kochi and R. V. Subramanian, *J. Am. Chem. Soc.*, 1965, **87**, 1508–1514.
50. S. Arai, T. Yamagishi, S. Ototake and M. Hida, *Bull. Chem. Soc. Jpn.*, 1977, **50**, 547–548.
51. S. Arai, M. Hida and T. Yamagishi, *Bull. Chem. Soc. Jpn.*, 1978, **51**, 277–282.
52. W. R. Bowman, H. Heaney and P. H. G. Smith, *Tetrahedron Lett.*, 1984, **25**, 5821.
53. E. Sperotto, G. P. M. v. Klink, G. v. Koten and J. G. d. Vries, *Dalton Trans.*, 2010, **39**, 10338–10351.
54. E. Sperotto, G. P. M. van Klink, J. G. de Vries and G. van Koten, *Tetrahedron*, 2010, **66**(2010), 9009–9020.
55. E. Sperotto, G. P. M. van Klink, J. G. de Vries and G. van Koten, *Tetrahedron*, 2010, **66**, 3478–3484.
56. S. E. Creutz, K. J. Lotito, G. C. Fu and J. C. Peters, *Science*, 2012, **338**, 647–651.
57. T. Cohen and I. Cristea, *J. Am. Chem. Soc.*, 1976, **98**, 748–753.
58. T. Cohen and I. Cristea, *J. Org. Chem.*, 1975, **40**.
59. T. Cohen and T. Poeth, *J. Am. Chem. Soc.*, 1972, **94**, 4363–4364.
60. X. Ribas, D. A. Jackson, B. Donnadieu, J. Mahía, T. Parella, R. Xifra, B. Hedman, K. O. Hodgson, A. Llobet and T. D. P. Stack, *Angew.Chem., Int. Ed.*, 2002, **41**, 2991–2994.
61. L. M. Huffman and S. S. Stahl, *J. Am. Chem. Soc.*, 2008, **130**, 9196–9197.
62. L. M. Huffman, A. Casitas, M. Font, M. Canta, M. Costas, X. Ribas and S. S. Stahl, *Chem.–Eur. J.*, 2011, **17**, 10643–10650.
63. B. Yao, D.-X. Wang, Z.-T. Huang and M.-X. Wang, *Chem. Commun.*, 2009, 2899–2901.
64. Z.-L. Wang, L. Zhao and M.-X. Wang, *Org. Lett.*, 2011, **13**, 6560–6563.
65. A. Casitas, A. E. King, T. Parella, M. Costas, S. S. Stahl and X. Ribas, *Chem. Sci.*, 2010, **1**, 326–330.
66. A. Casitas, M. Canta, M. Costas, M. Solà and X. Ribas, *J. Am. Chem. Soc.*, 2011, **133**, 19386–19392.
67. M. Font, T. Parella, M. Costas and X. Ribas, *Organometallics*, 2012, **31**, 7976–7982.
68. Y. Zhang and Y. Ding, *Organometallics*, 2011, **30**, 633–641.
69. S. A. Delp, L. A. Goj, M. J. Pouy, C. Munro-Leighton, J. P. Lee, T. B. Gunnoe, T. R. Cundari and J. L. Petersen, *Organometallics*, 2011, **30**, 55–57.
70. Z. Huang and J. F. Hartwig, *Angew. Chem. Int. Ed. Engl.*, 2012, **51**, 1028–1032.
71. G. Franc, Q. Cacciuttolo, G. Lefèvre, C. Adamo, I. Ciofini and A. Jutand, *ChemCatChem*, 2011, **3**, 304–309.
72. G. O. Jones, P. Liu, K. N. Houk and S. L. Buchwald, *J. Am. Chem. Soc.*, 2010, **132**, 6205–6213.
73. A. Shafir, P. A. Lichtor and S. L. Buchwald, *J. Am. Chem. Soc.*, 2007, **129**, 3490–3491.

74. K. M. Engle, T.-S. Mei, X. Wang and J.-Q. Yu, *Angew. Chem., Int. Ed.*, 2011, **50**, 1478–1491.
75. J. X. Qiao and P. Y. S. Lam, *Synthesis*, 2011, 829–856.
76. A. Klapars and S. L. Buchwald, *J. Am. Chem. Soc.*, 2002, **124**, 14844–14845.
77. J. Zanon, A. Klapars and S. L. Buchwald, *J. Am. Chem. Soc.*, 2003, **125**, 2890–2891.
78. H.-J. Cristau, A. Ouali, J.-F. Spindler and M. Taillefer, *Chem–Eur. J.*, 2005, **11**, 2483–2492.
79. A. J. Hickman and M. S. Sanford, *Nature*, 2012, **484**, 177–1185.

CHAPTER 3

Fundamental Pd^0/Pd^{II} Redox Steps in Cross-coupling Reactions: Homogeneous, Hybrid Homogeneous–Heterogeneous to Heterogeneous Mechanistic Pathways for C–C Couplings

IAN J. S. FAIRLAMB*[a] AND ADAM F. LEE[b]

[a] Department of Chemistry, University of York, Heslington, York, YO10 5DD, UK; [b] Cardiff Catalysis Institute, School of Chemistry, Cardiff University, Main Building, Park Place, Cardiff, CF10 3AT, UK, LeeAF@Cardiff.ac.uk
*Email: ian.fairlamb@york.ac.uk

3.1 Introduction

C–C bond-forming, cross-coupling reactions of organohalides with nucleophilic compounds (*e.g.* organometallics, olefins or alkynes) catalysed by palladium, often in combination with an activating ligand and/or co-additives/base, are amongst the most important chemical reactions available to the synthetic chemist. Nobel prize (2010) winning Heck, Negishi and Suzuki reactions highlight this point. Interest in 'evolved' and related transformations have received widespread recent attention, *e.g.* direct functionalization of one C–H bond (leading to loss of HX, see Chapter 9 of this book), or oxidative

RSC Catalysis Series No. 11
C–H and C–X Bond Functionalization: Transition Metal Mediation
Edited by Xavi Ribas

Published by the Royal Society of Chemistry, www.rsc.org

transformations involving the activation of two C–H bonds (leading to the formal loss of H_2, see Chapter 10 of this book) (Scheme 3.1). The intimate mechanisms of these types of reactions have received an incredible amount of attention from all areas of chemistry. A raft of experimental techniques has been used to study these reactions under a variety of different conditions. Whilst it was thought that cross-coupling reactions in the main involve homogeneous palladium species (including neutral and anionic pathways), clear evidence implicating the involvement of higher order Pd clusters in these reactions has emerged, particularly in classic cross-coupling processes. The field has been able to identify reactions where there is hybrid homogeneous–heterogeneous behaviour (*i.e.* surface chemistry and metal atom leaching). Indeed, catalytic manifolds have been identified that are truly heterogeneous. A simplified mechanistic picture is depicted in Scheme 3.2, which illustrates the interplay of mononuclear Pd species with higher order species; clearly this is more complex than the current mechanism presented to undergraduate students studying chemistry degrees across the world (*i.e.* by the *textbook mechanism*).

In this chapter, the background to the mechanisms of homogeneous Pd-catalysed cross-coupling reactions will be presented. There are many reviews[2] and books[3] that have dealt with this subject in significant detail (*e.g.* Heck, Kumada, Negishi, Stille, Suzuki–Miyaura), therefore only exemplar cases will be presented, with a particular emphasis on Suzuki–Miyaura (hereafter Suzuki) cross-couplings, where recent studies have highlighted the importance of Pd^{II}-hydroxo species. It is pertinent to mention that much less is known about Sonogashira cross-coupling reactions. Also mentioned are recent impurities in common Pd precursor catalysts (pre-catalysts), from nitrite anions to Pd nanoparticles (PdNPs) – both can be viewed in a positive or negative light depending on the reaction under study. Mechanistic work on Pd^{III} and Pd^{IV} catalytic manifolds is presented elsewhere[4] and does not form a detailed part of the following discussion, although it is equally important, especially in oxidative transformations (see Chapters 4 and 5 of this book).

A focus is then placed on hybrid homogeneous–heterogeneous catalytic cycles,[5,6] especially the propagation of Pd clusters (nanoparticles/colloids) derived from common Pd precursor complexes {*e.g.* $Pd(OAc)_2$}, and their involvement as catalytically relevant species in cross-coupling cycles. Many research groups have intentionally prepared stabilised PdNPs ($\sim$1.5–20 nm in size) and employed these as catalysts for a raft of cross-coupling reactions. Several mechanistic studies are highlighted as leading examples, wherein Pd leaching is seen as central to the success of the catalytic process. Pd leaching can be either in the form of single Pd atoms, ions or small clusters, but clearly this will depend on the reaction conditions, identity of the substrates and additives used, in addition to the type of PdNPs. Where appropriate the activating ligand at Pd will be discussed.

Also presented in this chapter are heterogeneous cross-coupling processes and their mechanisms. New methods and techniques for investigating the heterogeneous character of certain cross-coupling reactions have emerged in recent years, and been embraced by a few research groups, especially EXAFS.[7]

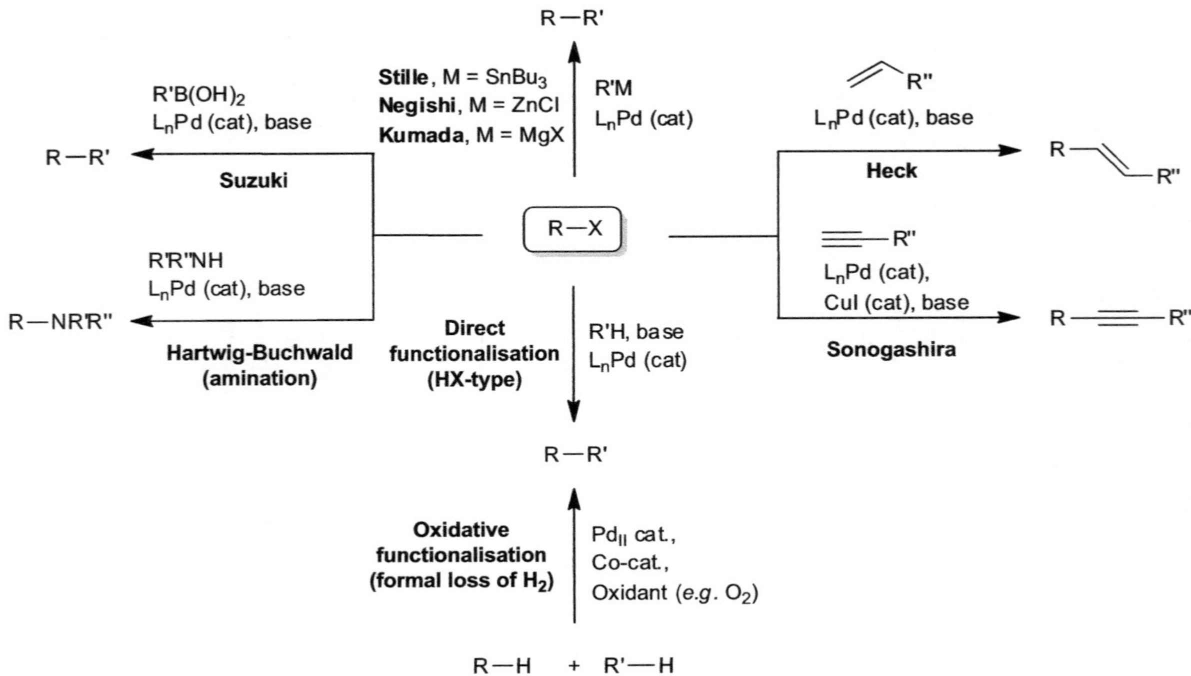

Scheme 3.1 Common Pd-mediated cross-coupling processes, from classical to C–H bond functionalization processes. Adapted from Ref. 1.

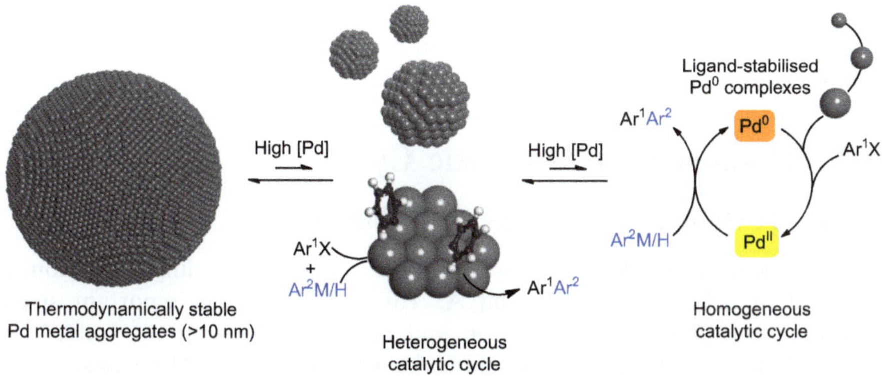

Scheme 3.2 A mechanistic picture illustrating the involvement of both homogeneous and heterogeneous Pd species in a generic cross-coupling catalytic cycle (shown with aromatic coupling partners, but not limited to).

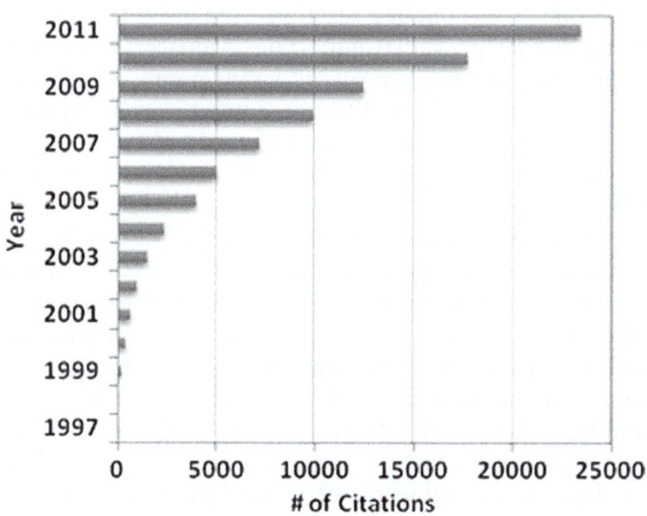

Figure 3.1 Statistics showing the number of paper citations reporting "PdNPs" from 1997 to 2011 (data obtained using Web of Knowledge).

This topic has been reviewed in 2012.[8] With interest in PdNPs (and other late metal NPs) increasing over the last 15 years, as illustrated in Figure 3.1, clearly the question of whether reactions can be truly heterogeneous has come up many times, yet has rarely been comprehensively investigated. Truly heterogeneous systems offer many advantages, including catalyst longevity, protection from trace metal contamination of the organic products, and ease of catalyst separation/re-use, hence one could consider this as the 'holy-grail of cross-coupling catalysis'. Preventing Pd leaching is crucial, as rules on acceptable levels of trace Pd in therapeutic compounds (and other commercial products) continue to tighten {*e.g.* European Medicines Agency (2008), <0.5 ppm Pd}. Therefore the last section of this chapter focuses on promising PdNP catalyst systems.

3.2 Homogeneous Catalytic Cycles

3.2.1 Textbook Mechanisms (Including Anionic Variants)

The classical textbook mechanism for the a "cross-coupling" reaction is depicted in Scheme 3, mechanism A. This involves three important steps, oxidative addition, transmetallation and reductive elimination.[9] Indeed, in simple form, this adequately accounts for the changes in Pd species during Kumada, Negishi, Stille and related Hiyama, Suzuki and Sonogashira reactions. The mechanism of the Heck reaction is different, as it involves carbopalladation and a β-hydrogen elimination step (instead of transmetallation). There is a substantial body of mechanistic work dedicated to this area, and it is well established and therefore not elaborated upon here (with the exception of Suzuki cross-coupling reaction – see below).

Amatore and Jutand presented important evidence (using electrochemical methods) indicating the involvement of halide/pseudohalide anions in Pd-catalysed cross-coupling.[10,11] A different catalytic cycle involving anionic complexes such as $L_2Pd^0Cl^-$ and $L_2Pd^0OAc^-$ ($L = PPh_3$, for example) was found to be operative; the anionic Pd^0/Pd^{II} catalytic cycle is depicted in mechanism B, Scheme 3.3.[12]

Halides serve different functions in Pd-catalysed cross-coupling processes, as they do in many other transition metal-catalysed processes.[13] With Pd, halides have the propensity to affect intrinsic reactivity, *via* all the common oxidative states Pd^0 to Pd^{IV}. The reactivity of the halide in oxidative addition and transmetallation steps is quite different, that is a C–Cl bond is harder to activate than a C–Br bond at Pd^0 (it being a stronger bond), yet the resultant oxidative addition species, "C–Pd–X", will react faster in transmetallation where $X = Cl$ with organometallic reagents (*e.g.* an organozinc or organostannane). The various influences of halide ligands need to be considered, *e.g.* (i) steric effects; (ii) electronic properties (either σ- and π-bonding with the Pd centre); (iii) polarizability; (iv) nucleophilicity; and (v) the *trans*-effect.[12]

The organohalide, metal salt additive or by-product provides an adequate source of halide. Given that most researchers employ a Pd^{II} precatalyst, *e.g.* $Pd(Ph_3P)_2Cl_2$ or $Pd(OAc)_2$, then a source of chloride and acetate is available to influence the chemistry and preferred reaction mechanism. It is also pertinent to mention that reduction of $Pd(Ph_3P)_2Cl_2$ gives anionic complexes, $Pd^0(PPh_3)_2Cl^-$, $Pd^0(PPh_3)_2Cl_2^{2-}$ and $Pd^0{}_2(PPh_3)_4(\mu\text{-}Cl)_2^{2-}$.[14] Studies on three $Pd(Ph_3P)_2{\times}_2$ complexes (where $X = I$, Br or Cl) indicate that iodide shows the greatest stabilisation (in order of stabilisation, $I > Br > Cl$).[15]

Quantum-chemical DFT studies support the anionic Pd^0/Pd^{II} catalytic cycle.[16] The most energetically favorable process for oxidative addition of aryl halides to $Pd^0(PPh_3)_2OAc^-$ was that involving a stable four-coordinate anionic

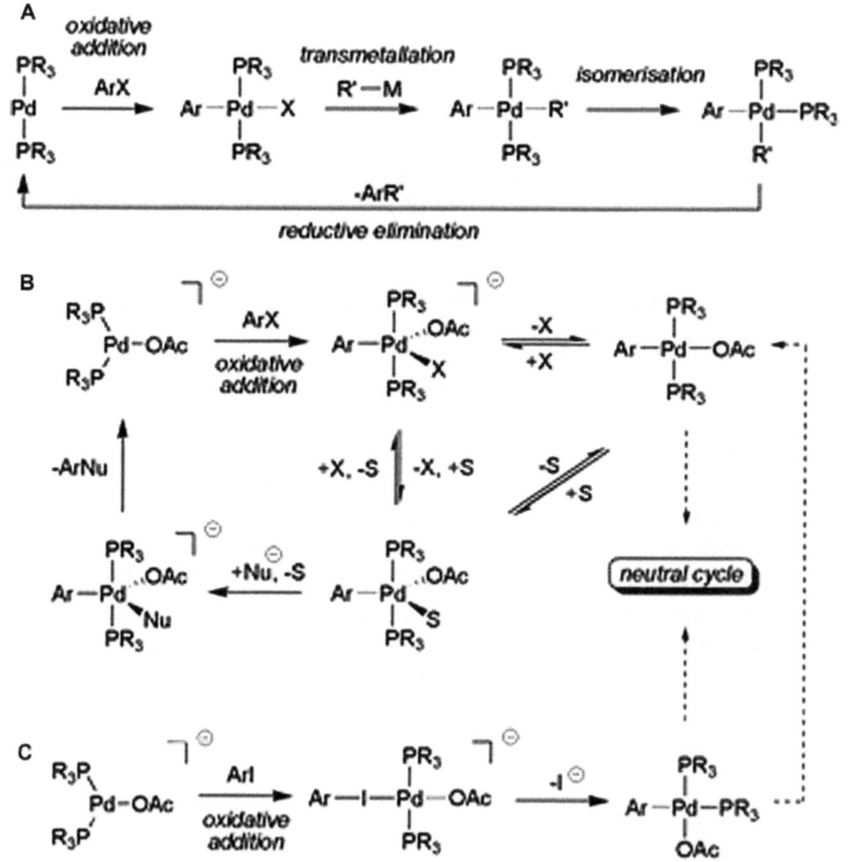

Scheme 3.3 A: Neutral Pd^0/Pd^{II} catalytic cycle (textbook mechanism); B: anionic Pd^0/Pd^{II} catalytic cycle *via* five-coordinate intermediates; C: anionic Pd^0/Pd^{II} catalytic cycle *via* four-coordinate intermediates. Reproduced from Ref. 12.

intermediate (Mechanism C, Scheme 3). However, it is difficult to know how one would detect such a species by current experimention, unless it could be trapped in the solid-state.

Other anionic Pd^0 species containing acetate and trifluoroacetate anions have also been assessed, particularly from $Pd(OAc)_2$ and $Pd(OCOCF_3)_2$.[17] Acetate and triflate anions are known to influence the degradation rates of palladacycles (by altering their reduction potential). Palladacycles have a rich history in organometallic chemistry, and their application in cross-coupling processes is widespread.[18] Crucially, both catalyst activity and lifetime are dependent on the identity of the pseudohalide, in that the latter can play a stabilising role.[19,20] It is proposed that neutral Pd^0 species undergo faster aggregation to give Pd clusters, which form large metallic Pd precipitates *via* Ostwald ripening. These Pd precipitates are thought to be inactive species – in an organic chemistry laboratory their observation often hints that a reaction is complete, or catalytically dead. Therefore halide/pseudohalide binding serves to stabilise lower order L_nPd^0 species, or soluble Pd clusters/colloids, which can serve as an important Pd reservoir.[21] This is the case particularly at high reaction temperatures (see later), wherein such reservoirs can release lower order Pd 'on-demand'.[22] This could be potentially exploited as a concept in catalyst recycling, although we are not aware of such a system having been developed from a common precursor like $Pd(OAc)_2$, for example.

3.2.2 A Key Role for Pd^{II} Hydroxide Species in Suzuki Cross-Coupling

Given the role played by halide and pseudohalides in cross-coupling cycles, it is perhaps unsurprising to find that hydroxides play different roles in Suzuki cross-coupling reactions (particularly at Pd^{II}). Separate studies by Jutand and Hartwig (using complementary experimental techniques) showed that "Pd^{II}-hydroxos", *e.g. trans*-$[ArPd(OH)(PPh_3)_2]$, are important in activating organoboronic acids in the transmetallation step.[23,24] Independent reactions of *trans*-$[ArPd(OH)(PPh_3)_2]$ and *trans*-$[ArPd(X)(PPh_3)_2]$ with arylboronic acids showed that the former complex reacted quicker than the latter. $ArB(OH)_3^-$ was found to be a poorly reactive species toward *trans*-$[ArPd(X)(PPh_3)_2]$. These observations can be explained by the high oxophilicity of boron (route a in Scheme 3.4). It is also of interest that the hydroxide anion was shown

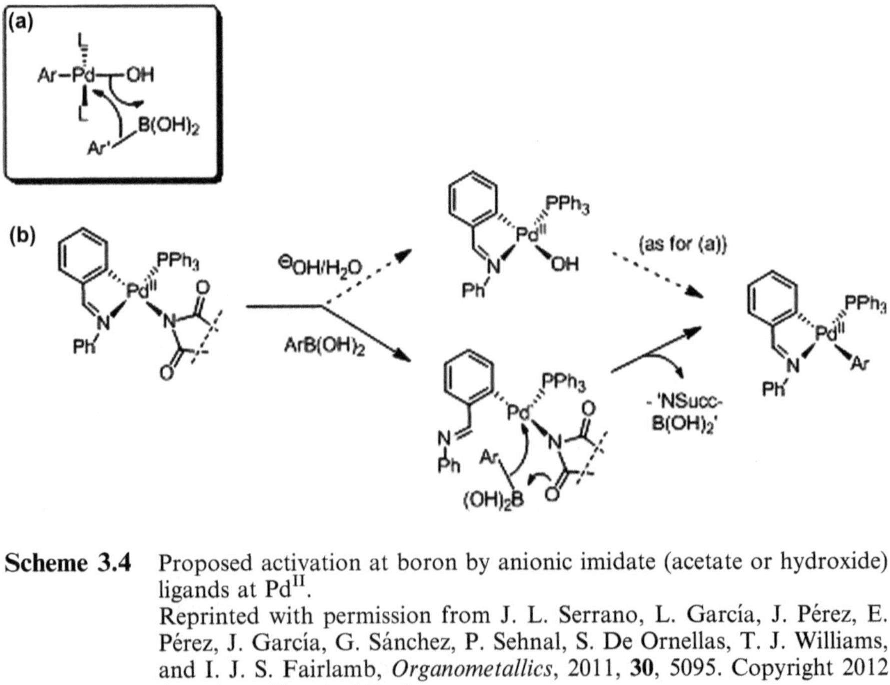

Scheme 3.4 Proposed activation at boron by anionic imidate (acetate or hydroxide) ligands at Pd^{II}.
Reprinted with permission from J. L. Serrano, L. García, J. Pérez, E. Pérez, J. García, G. Sánchez, P. Sehnal, S. De Ornellas, T. J. Williams, and I. J. S. Fairlamb, *Organometallics*, 2011, **30**, 5095. Copyright 2012 American Chemical Society.

to enhance reductive elimination from the intermediate complex *trans*-[$ArPdAr'(PPh_3)_2$] *via* coordination to Pd^{II}.[22] This work also links to that reported on palladacycles by Fairlamb and Serrano, where palladacycles (containing acetate or imidate anions) were found to degrade in the presence of arylboronic acids (use of an exogenous base was unnecessary for both transmetallation and reductive elimination steps).[25] During catalysis however, *i.e.* in the presence of Na_2CO_3 in a biphasic THF–H_2O solvent system, hydrolysis to a similar "Pd-hydroxo" species could easily be invoked. It was proposed that imidates[26] could activate boron in these palladacyclic systems (route b, Scheme 4), with the caveat that trace water may also play a key role (in the stoichiometric examples tested). Further work on palladacycles degradation under Suzuki cross-coupling conditions is needed.

Amatore and Jutand have shown that there are antagonistic effects operating when bases such as carbonate and hydroxide are used in Suzuki cross-couplings. These anions are also modulated by the identity of their counter-ions (*e.g.* Na^+, K^+ or Cs^+).[27] Interestingly, these cations slow the transmetallation step *via* their complexation to the hydroxide ligand in the reactive complex '$ArPd(OH)L_2$'. A decreasing reactivity order was observed when the cation was associated with hydroxide, that is $K^+ > Cs^+ > Na^+$ (note n-Bu_4N^+ was markedly more reactive than these three metal cations). Worthy of note is the finding that *trans*-$ArPd(OAc)L_2$ does not undergo transmetallation with $ArB(OH)_2$, which explains why acetate cannot be used as an independent base in Suzuki cross-couplings. Reactions performed with Cs_2CO_3 were slower than those conducted with n-Bu_4NOH at the same concentration. Even on water addition to Cs_2CO_3, causing in situ generation of low concentrations of hydroxide, a similar outcome was noted. It was proposed that the overall reactivity (in transmetallation and reductive elimination) is finely tuned by the concentration of hydroxide and the ratio $[OH^-]/[ArB(OH)_2]$. Thus hydroxide, however generated or used, and its counterion (Na^+, K^+ or Cs^+), play four different kinetic roles: (a) accelerating transmetallation by formation of '$ArPd(OH)L_2$', (b) accelerating reductive elimination, (c) retarding transmetallation by formation of unreactive $ArB(OH)_3^-$, (d) complexation of '$ArPd(OH)L_2$' by the counter cation.

The importance of hydrolysis in the context of organotrifluoroborates (RBF_3K) reagents (developed by Molander[28] for use in Suzuki cross-couplings) for 'trickle-feeding' low concentrations of $RB(OH)_2$ has been studied by Lloyd-Jones and Lennox.[29] In a reaction system employing THF/H_2O and Cs_2CO_3 at 55 °C it was found that gearing the rate of hydrolysis to the rates of the cross-coupling catalytic cycle was key (Scheme 3.5).

Preventing accumulation of $RB(OH)_2$ minimises side reactions such as oxidative homocoupling and protodeboronation. Somewhat surprisingly, it was found that there is an 'acid-base paradox' in Suzuki cross-couplings employing certain types of RBF_3K reagents (*e.g.* p-F-phenyl, naphthyl, furyl, and benzyl moieties). In these reactions, the reactor shape, material, size and stirring rate have a profound effect on the hydrolysis profiles of these RBF_3K reagents. Moreover, the Cs_2CO_3 causes partial phase-splitting of the THF/H_2O solvent

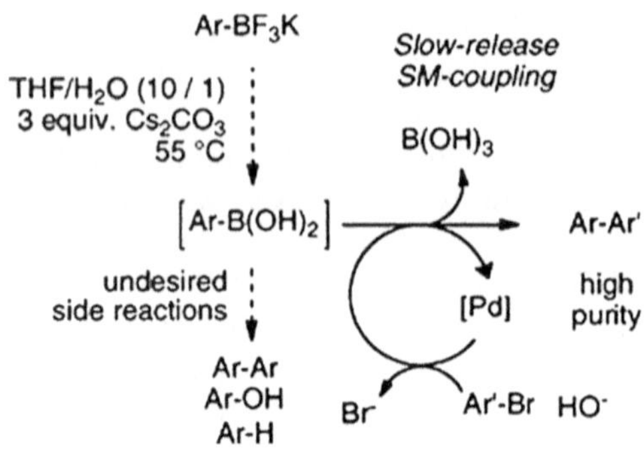

Scheme 3.5 Suzuki cross-coupling of $ArBF_3K$ reagents, *via* $ArB(OH)_2$ under Molander's conditions (SM = SuzukiMiyaura).
Reprinted with permission from A. J. J. Lennox, and G. C. Lloyd-Jones, *J. Am. Chem. Soc.*, 2012, **134**, 7431. Copyright 2012 American Chemical Society.

system, resulting in a lower pH in the bulk reaction medium. Crucially, acidic sites in the reaction glassware affect the hydrolysis rates. For other RBF_3K reagents bearing isopropyl, beta-styryl and anisyl moieties, fast hydrolysis occurs, whereas reagents bearing alkynyl or nitrophenyl moieties hydrolyse very slowly.

3.2.3 Issues with the Purity of Pd Precatalysts

In 2012, the purity of common Pd pre-catalysts has been highlighted, particularly in the case of $Pd(OAc)_2$ and $Pd_2(dba)_3$, which is an issue for mechanistic studies which routinely employ both .

The non-trivial behaviour of $Pd(OAc)_2$,[30] which is formally the trinuclear complex, $Pd_3(OAc)_6$, has been reported by Cotton and Murillo (Figure 3.2). Reaction with trace H_2O in $CDCl_3$ (*e.g.* laboratory grade $CDCl_3$) affords a new complex, namely $Pd_3(OAc)_6(OH_2)$. However, a more serious issue is the presence of nitrite that often accompanies this material. This relates to how the $Pd_3(OAc)_6$ is prepared, *i.e.* by the oxidation of metallic Pd with nitric acid and acetic acid.[31] The nitrogen oxides formed are removed by a flow of N_2 gas. However, when this is incomplete or inefficient $Pd_3(OAc)_5NO_2$ may also form, which can impact on the synthesis of complexes derived from it, and potentially catalytic applications.[32] While no catalytic reactions directly benefit from this impurity (at the time of writing), nitrite is a redox active ligand and therefore could play an important role in catalytic oxidation chemistry at palladium, and perhaps other related couplings.[33]

It has been known for some time that $Pd_2(dba)_3$ (and its various solvated adducts) can be used to prepare PdNPs, usually under reducing conditions with heating. A study by Ananikov and Zalesskiy[34] has shown that $Pd_2(dba)_3$ can contain large Pd particles (note: degradation of this complex in solution is well established in the absence of a donor ligand like phosphine, indeed it is the premise for its use as a starting precursor complex for PdNPs[35]). It was shown

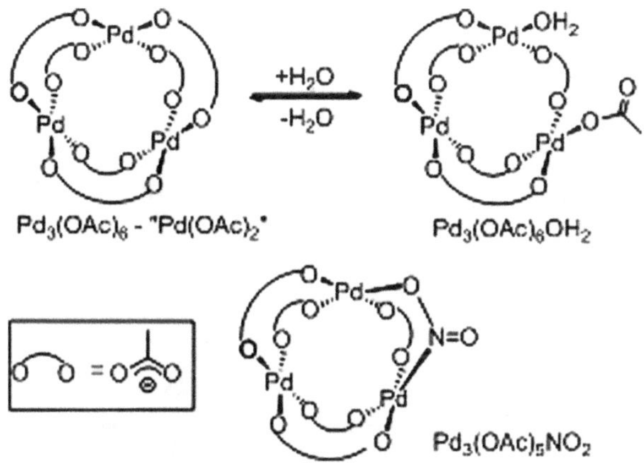

Figure 3.2 Non-trivial behaviour of $Pd(OAc)_2$.
Reproduced from Ref. 32.

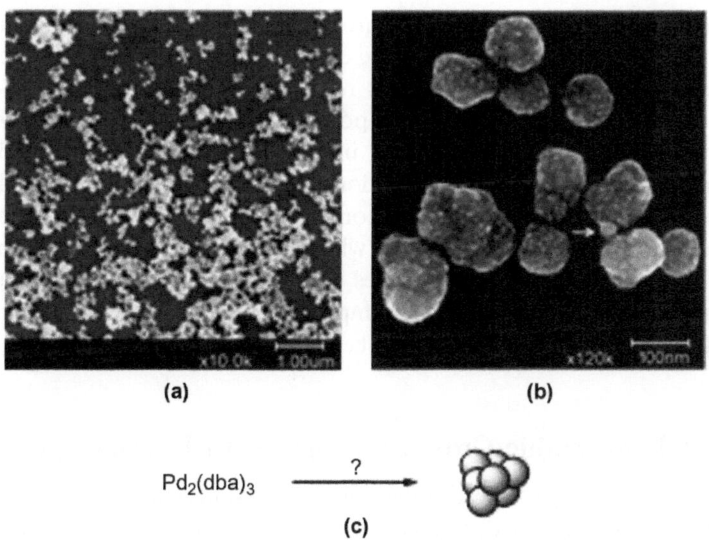

Figure 3.3 PdNPs formed by degradation of $Pd_2(dba)_3$: (a) low magnification image; (b) higher magnification image; (c) cartoon schematic for $Pd_2(dba)_3 \rightarrow$ large PdNPs.
Reprinted with permission from S. S. Zalesskiy, and V. P. Ananikov, *Organometallics*, 2012, **31**, 2302. Copyright 2012 American Chemical Society.

that commercial samples can contain up to 40% PdNPs (from 10–200 nm in size) (Figure 3.3), which could account for some variability seen in catalysis.

It should be also borne in mind that $Pd_2(dba)_3$ is sensitive to photo-oxidation.[36] The issue is whether these large Pd particles exhibit similar activity to $Pd_2(dba)_3$, and whether the mechanistic information being abstracted is

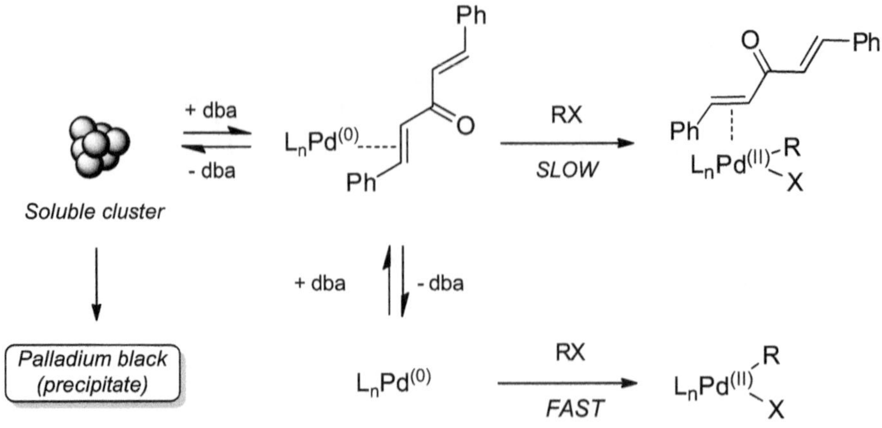

Scheme 3.6 Dba effects in oxidative addition, and playing a role in stabilising Pd^0.

meaningful. Further studies are required to clarify these issues, specifically the impact of aged $Pd_2(dba)_3$ on catalytic cross-coupling reactions. It is clear that the dba ligand plays a non-innocent role[37] in the majority of cross-coupling reactions, which has been exploited positively in the design of dba-Z ligands (where Z = aryl substituent), which influence oxidative addition, β-hydride elimination and reductive elimination steps.[38] An example is shown in Scheme 3.6. Here the dba ligand controls the concentration of the highly reactive L_nPd^0 species (L = 2 or 1), which reacts quickly with organohalides (RX). The dba effectively equilibrates the rates of the individual steps of the catalytic cycle. This is particularly important where transmetallation is rate-limiting and oxidative addition is fast.

3.2.4 'Homeopathic Cross-couplings' – An Unusual Case?

For the Heck coupling of bromobenzene with *n*-butyl acrylate, in 2003, A. H. M. de Vries[20] showed the optimal catalyst loading was 0.08 mol% (Scheme 3.7). Higher loadings suppressed catalyst activity, attributed to Pd aggregation (Pd black precipitation). A key question is whether the active species are mononuclear, Pd clusters, or PdNPs? A similar observation was made in 2004 by Fairlamb and co-workers[39] in the Sonogashira coupling of 4-bromoacetophenone with phenylacetylene; efficient catalysis occurs at 0.01 mol% Pd {note: lower loadings give higher turnover frequencies (TOFs) but lower conversions}. Despite differing conditions (solvent, catalyst type, additives *etc.*), these Heck and Sonogashira couplings are highly active at ~40 ppm Pd.

Leadbeater and co-workers reported transition-metal free Sonogashira-[10] and Suzuki-type cross-couplings (Scheme 3.8).[11] At the time there was debate whether a radical-based mechanism[12] or indeed trace 'contaminant' Pd

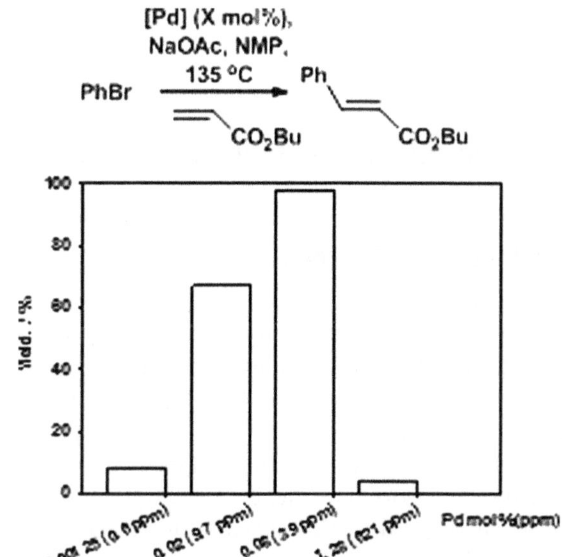

Scheme 3.7 Observations from A. H. M. de Vries *et al.* on a Heck cross-coupling reaction.

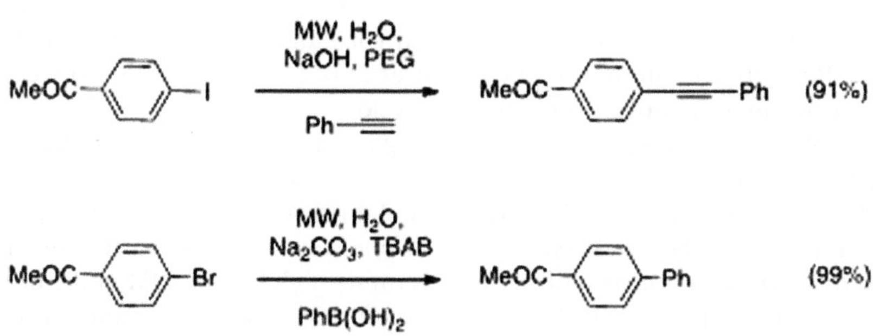

Scheme 3.8 Homeopathic catalysis with ppb levels of Pd.

(*i.e.* ppm levels) could explain these exciting observations. In a rigorous reassessment, the latter was determined to the case, with the surprising finding that Na_2CO_3 carries enough Pd (~20–50 ppb) to effect catalysis under microwave conditions in water.[13] Interestingly, K_2CO_3 (Pd ~0.09 ppb) did not work in the original studies. Reaction of activated 4-bromoacetophenone with phenylboronic acid in the presence of $Pd(OAc)_2$ (100 ppb) using K_2CO_3 gave the coupled product in good yield. It is interesting to contrast these findings with the Heck reactions reported by de Vries (see above), where some catalytic turnover was also observed at sub-ppm levels of Pd. A 'take home' message is that activated aryl halide substrates, such as those shown in Scheme 3.8, are mediated by homeopathic levels of Pd, and are poor candidates for the development of new Pd catalyst systems.

3.3 Hybrid Homogeneous–Heterogeneous Catalytic Cycles

3.3.1 Pd Leaching and Hybrid Behaviour

The central premise for a hybrid heterogeneous catalytic cycle is that a reservoir of Pd (whether supported or stabilised by a polymer or a suitable stabilising solvent, *e.g.* DMF, DMA, NMP or ionic liquid) is able to release Pd {assuming this species is mononuclear, although it could be a Pd_n cluster ($n = 5$–100 atoms, *i.e.* at the detection limit of conventional TEM analysis)}. Most likely the Pd species released are from defect sites in the surfaces of such nanostructures.

Arai and co-workers reported evidence for Pd leaching,[40] working with supported Pd catalysts (*e.g.* Pd/C, Pd/SiO_2 and others) in Heck cross-coupling reactions. It was found that Pd leaching into solution was apparent; indeed it was this 'dissolved' Pd that catalysed the Heck cross-coupling. Moreover, the dissolved Pd species were able to recombine with their supports following the reaction. This finding led the authors to recycle 10 wt% Pd/C in these coupling reactions with good success. A clear question is how is the Pd leached into solution – what is the mechanism for leaching? Work on Pd/Al_2O_3 revealed that ArX was necessary[41] for leaching to occur (presumably *via* oxidative addition to the Pd surface), inferred by the appearance of soluble 'ArPdX' species that enter a 'normal' homogeneous catalytic cycle.

Halide and pseudohalides can both play a key role in stabilising Pd reservoirs, but on fully supported surfaces, acetate ions can enhance metal leaching. A crucial point is that Pd accumulation *via* leaching is a slow process from supported Pd catalyst systems. So, if one stops a reaction after only a few catalytic turnovers, one is unlikely to determine high levels of Pd in the filtrate. How one does this, and the tests that exist for determining the amount of Pd leached in such reactions, has been extensively covered elsewhere (particularly the review by Pagliaro[42]), and new researchers to this field are warmly encouraged to read a recent review by Finke.[43] In brief, ICP chemical analysis of the reaction filtrate, and comparing catalytic activity prior to and following a hot-filtration of the active Pd catalyst, will show whether Pd is leached or not. It is worth pointing out that Pd leaching seems apparent in most supported Pd catalysts.[41]

Reetz and de Vries[44] recognised the importance of leaching and the formation of soluble Pd nanoclusters (identified in solution), pointing to the importance of the catalyst/substrate ratio in addition to Pd concentration (*i.e.* in Heck reactions mediated by $Pd(OAc)_2$ and palladacycles, and Suzuki cross-couplings).

Later Rothenberg and co-workers[45] reported that leaching of Pd species occurs from PdNPs (*ca.* 15 nm in size) for both Heck and Suzuki cross-coupling reactions. They showed this by constructing a specialised membrane reactor, which allowed the transfer of <5 nm Pd atoms/ions. Under non-oxidising conditions it was shown that Pd leaching occurred (note: classical cross-coupling reactions can be considered reducing environments, although clearly reactive organohalides are oxidants). Using mass spectrometry, UV-visible spectroscopy and ^{13}C NMR spectroscopy, oxidative addition products were characterised, which were able to diffuse through the membrane. It was suggested that PdNPs are not the true catalysts in cross-coupling reactions, *i.e.* catalysis occurs at Pd^0 atoms or Pd^{II} ions that leach into solution. The simple fact, however, that leaching occurs at all, highlights that the larger PdNPs act at the very least as a reservoir of soluble Pd. However, it should be pointed out that <4 nm Pd clusters are catalytically feasible species for a raft of cross-coupling reactions (see sections 3.2, 3.4, 4.2, 4.3 and 5.2). Indeed, when one purposely characterises PdNPs formed in cross-coupling reactions, the size of such particles are typically 2–4 nm.

3.3.2 Trapping *In Situ* Generated Pd Nanoparticles

Work by Fairlamb and co-workers[46] demonstrated that it is possible to trap and characterise the metal-containing nanoparticles formed in a direct arylation reaction of adenosine (Scheme 3.9). However, a Hg-drop test[47] proved inconclusive. Such chemistry was previously believed to operate by a homogeneous catalytic cycle.[48–50] TEM, used widely to characterise metal-containing nanoparticles,[22,39,51] allowed characterisation of the particles isolated from two separate samples (2 × 1 mL reaction samples removed after 1.5 h, which corresponds to *ca.* 45% conversion by HPLC analysis). One of the major concerns with TEM analysis is that the analysis sample is concentrated under high vacuum, which would seemingly favour aggregation (*i.e.* DMF is a potential metal colloid stabiliser – see later). The first sample was concentrated under high vacuum at 40 °C to remove the DMF, which was then re-suspended in EtOH and analysed directly by TEM. To address the aggregation issue *N*-polyvinylpyrrolidone (PVP) ($M_w = 29\,000$) was added to stabilise *in situ* generated metal nanoparticles that were formed (concentration *in vacuo* should not affect the size of the particles). The TEM micrographs showed similar sized metal-containing nanoparticles, although it is difficult to know whether all the surface DMF molecules were removed. XPS analysis of unstabilised metal nanoparticles showed the presence of Cu^I and both Pd^0 and Pd^{II}. Quantification revealed that the Cu:Pd atomic ratio was ~19 : 1 (theoretical ratio = 60 : 1). It was proposed that this high Pd content be attributable to Pd segregation to form a capped Pd-Cu particle, although further work is necessary to confirm this.

Independently characterised[52] Pd-PVP nanoparticles (mean diameter = 2 nm) mediate the direct arylation reaction of adenosine with iodobenzene in DMF (giving the product in 51% isolated yield, comparing to 65% using $Pd(OAc)_2$. Larger Pd-PVP particles (3.4 nm) performed less well (18% yield). Indeed, these transformations are some of the lowest temperatures used for this type of

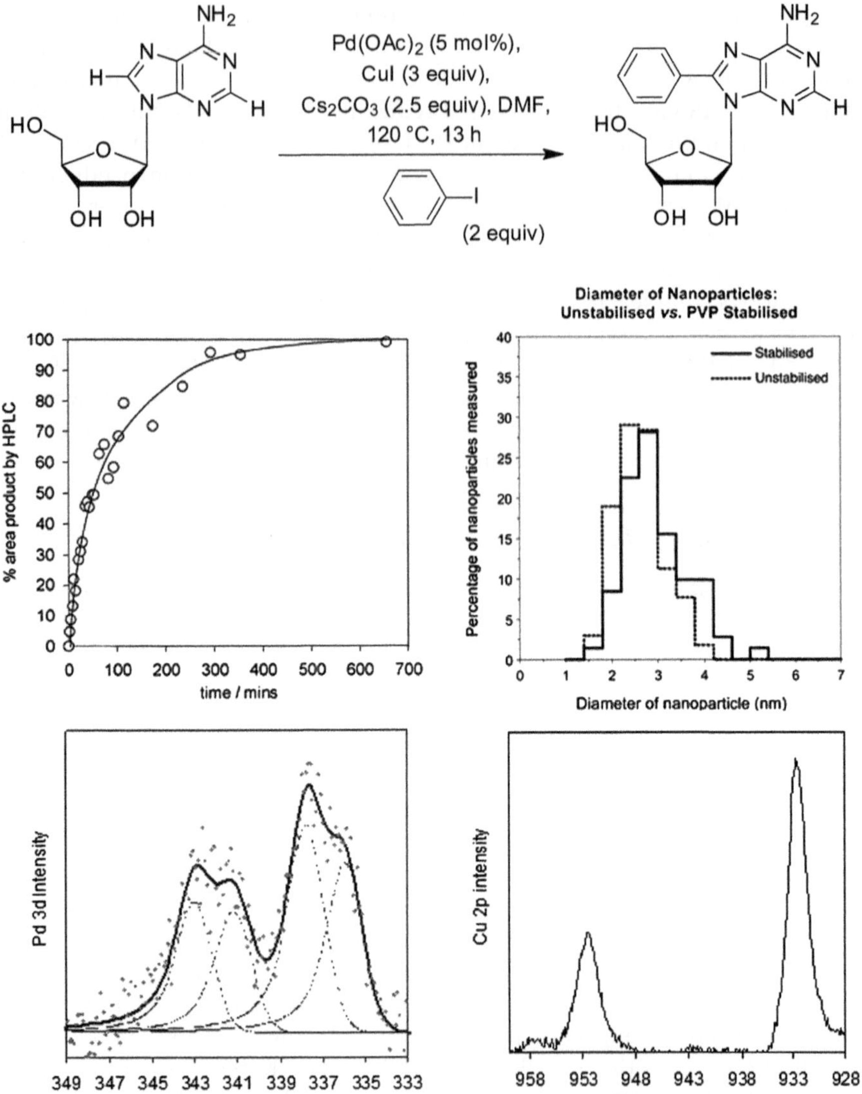

Scheme 3.9 Product formation in the arylation of adenosine (by HPLC); distribution of metal-containing nanoparticles by TEM under stabilised and non-stabilised conditions; Pd 3d region and Cu 2p peaks identified by X-ray photoelectron spectroscopy (XPS).

coupling chemistry (140–160 °C is typical). Further work on 2′-deoxyadenosine has demonstrated that reactions with Pd-PVP occur at temperatures of 60–80 °C, where $Pd(OAc)_2$ is ineffective (in the absence of an enhancing piperidine ligand).[53] It is known that Pd-PVP particles can increase in size during recycling experiments in related Heck cross-couplings leading to lower product yields.[54] The O_2 adsorbed on the surface of these PdNPs plays a pivotal role in promoting the ripening process, which explains the observed loss in catalytic activity. It is interesting to note that the adenosinyl direct arylation reactions are very sensitive to air, which may also ripen PdNPs in DMF (and oxidise Cu^I to Cu^{II}).

It is interesting to note that similar-sized metal nanoparticles are formed in the Sonogashira cross-coupling reaction of 8-bromoguanosine with phenylacetylene on an Amberlite-base resin.[55] It is most likely that these supports release catalytically active Pd into solution (either mononuclear or small clusters), allowing partial or full homogeneous reactions to take place, akin to the behaviour of other heterogenised sources that have been reported.[56]

3.3.3 Encapsulated Pd Nanoparticles Possessing Useful Catalytic Activity

In 2002, Ley and co-workers reported polyurea-microencapsulated Pd catalysts {derived from $Pd(OAc)_2$}, which was given the tradename 'PdEnCat'.[57,58] Two catalyst systems were commercialised, Pd^{II} EnCat and Pd^0 EnCat, the latter containing PdNPs. Such systems also allow the use of phosphines (*e.g.* Pd EnCat TPP30 and TOTP30, the former contains PPh_3 whereas the latter contains P(*o*-tolyl)$_3$ as catalyst modifiers). The Pd EnCat catalysts were found to be extremely useful for both Heck and Suzuki cross-couplings, where the support provided a 'stable' reservoir of catalytically active Pd.

McQuade and co-workers[59] investigated the issue of leaching from three PdEnCats, two of which contained phosphine, using the three-phase test.[60,61] For the three-phase test an immobilized catalyst is exposed to both soluble and solid-supported reagents. Any active Pd species leached from the immobilised media (PdEnCat) will result in conversion of the solid-supported reagent. For this purpose McQuade employed a base-stable Tenta-Gel Macrobead-supported iodoarene reagent (Figure 3.4). It is also necessary to run a homogeneous reaction and the solid-phase reaction simultaneously in the same vessel, as the leached species could be a Pd^{II} oxidative intermediate (note: the leached species could also be Pd^0).

4-Bromoacetophenone and *n*-butyl acrylate were identified as suitable soluble reagents (Scheme 3.10). McQuade applied the three-phase test to this reaction, recording high conversion of the supported iodoarene reagent to the corresponding cinnamate derivative for all three PdEnCat catalysts (in addition to full conversion of the 4-bromoacetophenone). This indicated that the active Pd catalyst leached out of the polyurea matrix during the Heck reaction. The outcome of the three-phase test showed that the same Pd species might catalyse

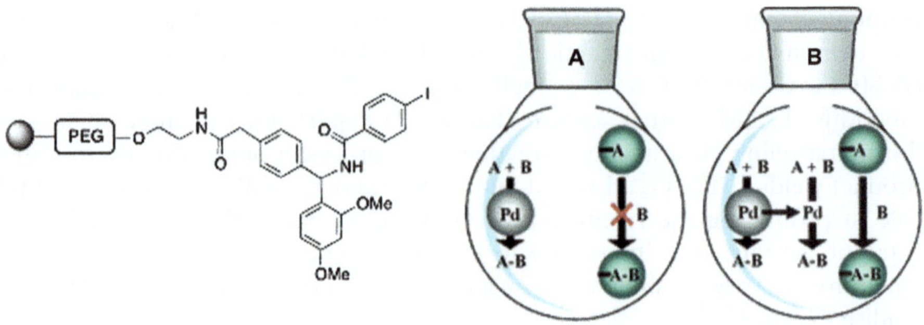

Figure 3.4 Supported iodoarene required for the three-phase test (left). Representation of the three-phase test (right): A: The Pd catalyst remains on the support, allowing only solution-phase reagents to react, *i.e.* the resin-bound iodoarene reagent does not react. B: The palladium catalyst leaches out of the PdEnCat support, allowing both solution-phase and resin-bound reagents to react.
Reprinted with permission from S. J. Broadwater and D. Tyler McQuade, *J. Org. Chem.*, 2006, **71**, 2131. Copyright 2012 American Chemical Society.

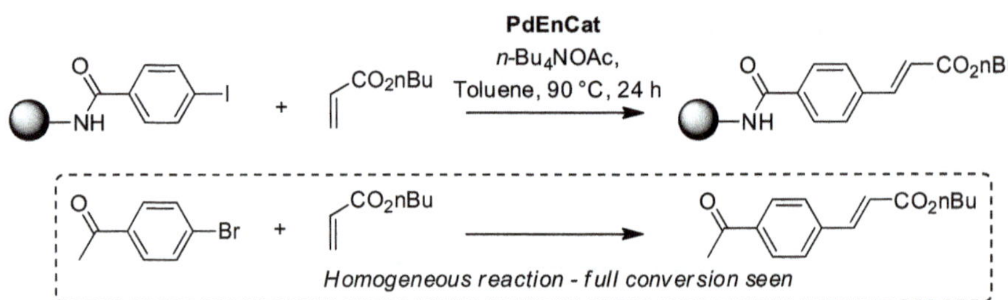

Catalyst	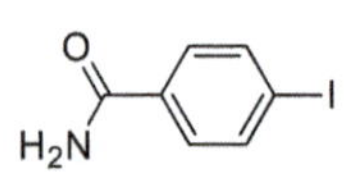	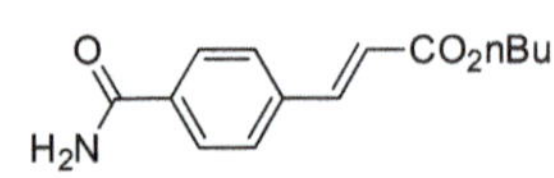
PdEnCat 30	**8%**	**92%**
PdEnCat TPP30	**<1%**	**>99%**
PdEnCat TOTP30	**<1%**	**>99%**

Scheme 3.10 Three-phase test results from a Heck cross-coupling reaction.

the reaction both on and off the solid support. Leaching from these supports was noted when using DMF or toluene in Heck cross-coupling, although *iso*-propyl alcohol proved to be a compatible solvent where < 10 % Pd leaching was recorded. The overall take home message from this study was that there is a solution-phase contribution to the catalysis, but one cannot rule out surface catalytic chemistry taking place as a competing process.

3.3.4 Questions About the Active Catalyst Phase

Fagnou reported intermolecular and intramolecular direct arylation reactions mediated by Pearlman's catalyst, $Pd(OH)_2/C$ (Scheme 3.11).[62] As with the preceding McQuade study, Fagnou questioned whether the catalyst was active in the homogeneous or heterogeneous phase. The three-phase test was used (without a homogenous reaction in the same reaction vessel). Two separate reactions were conducted. An aryl halide anchored to a Wang resin was fully converted to the cyclic product in the presence of $Pd(OH)_2/C$, indicating Pd leaching and a homogeneous process (if the reaction was purely heterogeneous then interaction between the supported substrate and catalyst should be negligible). In a separate reaction, a silica-supported, thiol-based resin was used to scavenge any homogeneous catalyst species. It was proposed that the heterogeneous metal scavenger should not inhibit a heterogeneous Pd species. The result was no reaction, *i.e.* in the absence of the thiolated-silica resin full product conversion was recorded.

Pd leaching was therefore found necessary for the reaction – Fagnou's tests are fairly convincing – however the identity of the true catalyst species was not determined.[63] The experiments do not confirm whether it is mononuclear Pd species or Pd clusters that are responsible for the catalysis. This issue reflects the pore size of the Wang resin, which can effectively act as a micro-reactor, and is large enough to encapsulate Pd clusters (and provide sites allowing the seeding of Pd clusters). For example, related thiolated-resins (Quadrasil-MP) are perfectly capable of sequestering and encapsulating PdNPs (Figure 3.5).

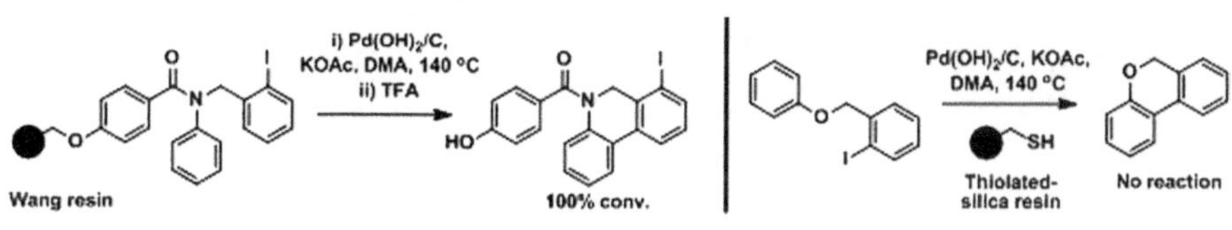

Scheme 3.11 Three-phase tests: left, heterogeneous aryl halide and $Pd(OH)_2/C$; right, a reaction in the presence of a thiolated silica-resin.

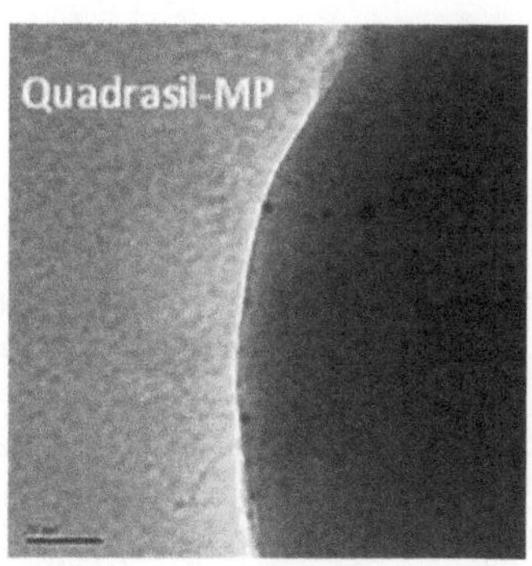

Figure 3.5 Encapsulation (trapping) of palladium nanoparticles in Quadrasil-MP (imaged by TEM spectroscopy).
Reproduced from Ref. 81.

The answer to this question thus remains – are mononuclear Pd species or Pd clusters/PdNPs responsible for the catalysis? It is pertinent to mention that high $Pd(OH)_2/C$ catalyst loadings are required for efficient catalysis (<10 mol% Pd led to substantially less catalytic activity), which would be consistent with a leaching process.

Since 2005, researchers working on 'direct arylation' reactions have broadly assumed that catalytic cycles involve mononuclear Pd species (as exemplified by the best reviews in the field).[64–66] In light of the discoveries that Pd clusters and nanoparticles play an important role in classical cross-coupling reactions, this viewpoint could be questioned in the future.

Studies by Dupont and co-workers[67] showed that PdNPs (~2nm, formed by degradation of a NC-palladacycle) disperse well in ionic liquids (1-*n*-butyl-3-methylimidazolium hexafluorophosphate) and act as a catalytic Pd reservoir in Heck cross-couplings (*e.g.* reaction of *n*-butylacrylate with iodobenzene) (Scheme 3.12). It was proposed that these types of reactions proceed *via* a sequence of oxidative additions to the PdNP surface, resulting in the Pd^{II} oxidative addition intermediates detaching from the surface (being labile) and entering into a 'homogeneous' catalytic cycle. *In situ* TEM analysis of the post-catalytic ionic liquid solution shows the formation of larger PdNPs (~6 nm). The Pd content in the organic phase showed high levels of Pd leaching (~34%)

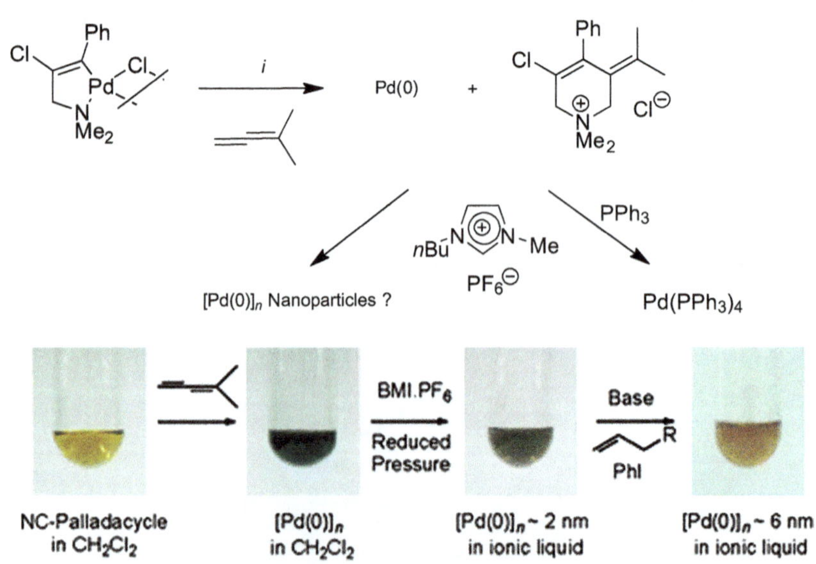

Scheme 3.12 Image showing the changes in solution, following degradation/activation of a NC-palladacycle to give well-dispersed Pd nanoparticles, and then the final composition of the PdNPs following a Heck cross-coupling reaction.
Reprinted with permission from C. C. Cassol, A. P. Umpierre, G. Machado, S. I. Wolke and J. Dupont, *J. Am. Chem. Soc.*, 2005, **127**, 3298. Copyright 2012 American Chemical Society.

at low substrate conversions, which dropped to 5–8% leaching at higher substrate conversions. The proposal is in line with that reported by de Vries and Reetz on Heck couplings[43] using ligand-free Pd sources.

3.4 Heterogeneous Cross-Coupling Reactions

3.4.1 Early Evidence for Heterogeneous Behaviour

As highlighted above there is a rich literature associated with the use of heterogeneous Pd catalysts to effect classic cross-couplings, in which for example conventional Pd/C,[68] colloidal Pd,[69,70] or immobilised complexes,[71,72] are proposed as the active species. However, as questions regarding *in situ* leaching and the potential role of soluble homogeneous pathways have come to the fore over the past decade,[73] much of this earlier work requires careful re-evaluation. In this section, we therefore focus on more detailed mechanistic investigations into heterogeneous cross-couplings.

Evidence for a true heterogeneous pathway in Heck couplings was first produced by Blackmond and co-workers in 1999,[74] who observed a strong structure-sensitivity in the reaction of *p*-bromobenzaldehyde with butyl acrylate over PVP-stabilised (PVP) palladium colloids, implicating surface defects as the active sites, as previously alluded to by Augustine and co-workers for oxide supported PdNPs.[75] Control experiments, in which the $Pd(dba)_2$ {formally $Pd_2(dba)_3$.dba} precursor was used without the PVP stabiliser, afforded much slower rates than attainable from the pre-formed (PVP)PdNPs, suggesting that Heck coupling did not occur *via* solution-phase catalysis by leached Pd species. The case of Suzuki coupling over (PVP)PdNPs is more complex. Initial studies by El Sayed and co-workers favoured a surface mechanism,[76] driven by coordinately unsaturated vertex and edge atoms akin to Heck coupling,[68] however subsequent TEM investigations by the same authors uncovered significant changes in palladium nanoparticle shape/size during reaction (Figure 3.6),[77] that likely reflect extensive Pd dissolution. Subsequently, El Sayed reported a strong shape-sensitivity for the same Suzuki coupling of phenylboronic acid and iodobenzene over PVP-stabilised *platinum* NPs, wherein tetrahedral NPs show moderate activity whereas spherical NPs are inert, suggesting a heterogeneous pathway.[78]

3.4.2 Comparison of a Suzuki Cross-Coupling Mediated by $Pd(OAc)_2$ and (PVP)PdNPs

In an effort to rationalise the preceding observations, Lee and Fairlamb conducted kinetic studies on the Suzuki cross-coupling of *p*-bromoanisole with phenylboronic acid, mediated by (PVP)PdNPs (*ca.* 1.8 nm, ~236 Pd atoms), which showed that the Hg-poisoning test worked well, as confirmed by X-ray photoelectron spectroscopy (XPS) measurements of the Hg-coated PdNPs (Figure 3.7).[79–81] Deliberate spiking with 0.03 mol% of soluble $Pd(OAc)_2$, equivalent to the maximum theoretical amount of Pd that can leach from such

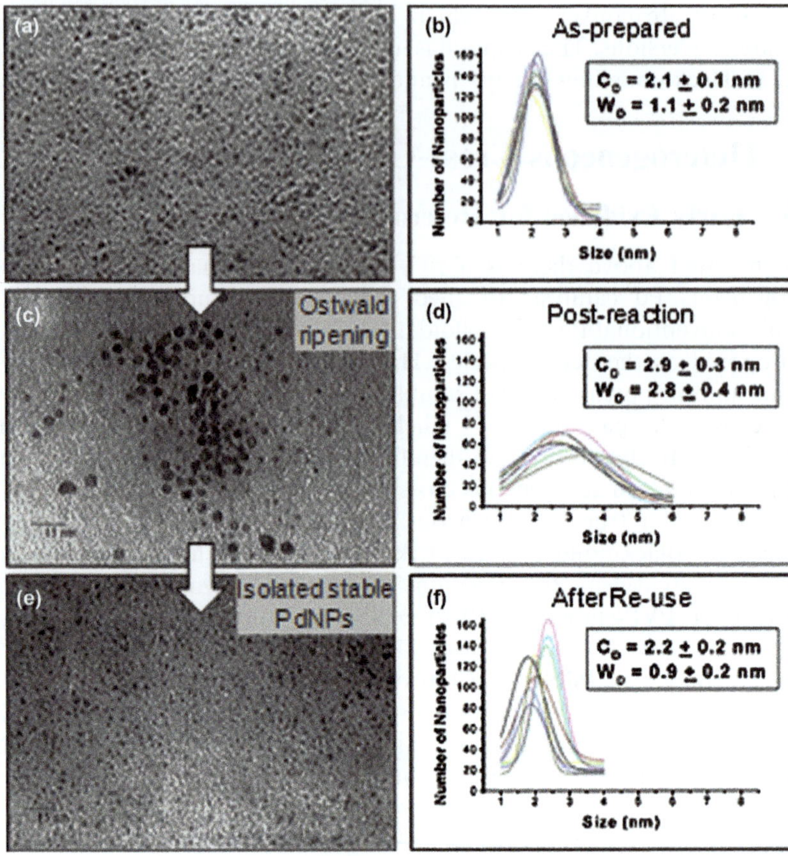

Figure 3.6 TEM images and Gaussian fits of the associated particle size distributions of (PVP)PdNPs: (a,b) before Suzuki coupling; after one reaction cycle; and (e,f) after two reaction cycles. Particles grow *via* dissolution and Ostwald ripening during the first cycle, and are lost from solution upon product separation, leaving only a lower concentration of smaller nanoparticles to catalyse the second cycle.
Reprinted with permission from R. Narayanan and M. A. El-Sayed, *J. Am. Chem. Soc.*, 2003, **125**, 8340. Copyright 2012 American Chemical Society.

particles, did not perturb the reaction kinetics, evidencing the active catalyst phase under these relatively mild conditions as the surface of PdNP directly, not leached Pd. The (PVP)PdNPs exhibit first-order kinetics with respect to Pd, precisely as expected if coupling rates were proportional to the total surface area of the added PdNPs. In contrast, the same reaction mediated by $Pd(OAc)_2$ exhibited a strong negative order dependence on [Pd], in accord with the expected behaviour of molecular Pd catalysts which tend to aggregate forming very large (>100 nm) and inert Pd black in the absence of surface stabilisers.

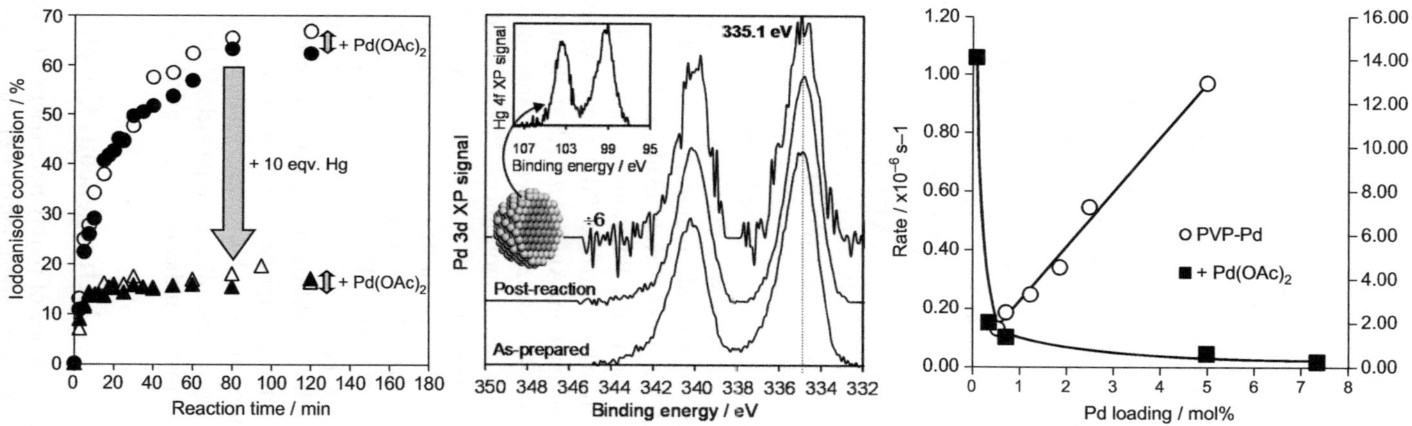

Figure 3.7 A: Hg poisoning of Suzuki coupling; B: XPS of Hg-coated PdNPs; C: differential Suzuki coupling order dependence on [Pd]. Reproduced from Ref. 81.

3.4.3 X-Ray Absorption Spectroscopy (XAS) as a Spectroscopic Tool to Probe Catalytic Species

Several groups have turned to X-ray absorption spectroscopy (XAS) as an excellent spectroscopic tool to probe the local chemical environment and oxidation state of PdNPs, during cross-couplings, as first demonstrated in an *operando* investigation of Suzuki coupling over size-selected (PVP)PdNPs (Figure 3.8).[79–81] A key advantage of this approach is that the penetrating power of hard X-rays permits the study of both PdNPs (< 5 nm) and Pd clusters during continuous catalytic turnover within liquid-phase reactors. In tandem with kinetic analyses, this study provided strong evidence that (PVP)PdNPs can function as heterogeneous catalysts, as long as mild reaction conditions (low temperature and halide concentration) are employed to minimise leaching. Considering the latter, it is important to note that even with current 3rd generation synchrotrons and ultrafast X-ray detectors, the sensitivity and time-scales available for such *operando* XAS measurements cannot yet eliminate the possibility of trace Pd atoms reversibly leaching from small PdNPs and playing (albeit a very small) role in the overall catalytic cycle.

In contrast, a related *in situ* QEXAFS study of the Heck coupling of bromobenzene and styrene over a Pd/Al_2O_3 catalyst placed within a fixed-bed reaction cell, identified $[PdBr_4]^{2-}$ and $[Pd_2Br_6]^{2-}$ within the solution phase at high temperature (150 °C),[82] coincident with *trans*-stilbene product formation, suggesting minimal surface coupling occurred. Reduction of surface PdO present in the initial commercial catalyst was postulated the rate-limiting step, since oxidative addition of the aryl halide requires the presence of Pd^0. The latter transformation has itself been extensively explored by *ex situ*,[83] *in situ*[84] and *operando*[85] XAS with respect to the Pd-catalysed, liquid phase aerobic selective oxidation of alcohols, in which surface PdO reduction to metal conversely poisons the heterogeneous reaction. An *in situ* XAS study of the thermo-chemical behaviour of a Pd/USY-zeolite under different organic solvents, has suggested an intermediate scenario, wherein monomeric Pd species dispersed within the USY framework under *o*-xylene, are responsible for the (heterogeneous) Suzuki–Miyaura coupling reaction of bromobenzene and phenylboronic acid.

These XAS studies confirm the existence of subtle equilibria between active ligated palladium and multi-nuclear Pd clusters/small nanoparticles, and larger, stable and essentially inert, PdNPs (Scheme 3.13), as originally hypothesised by Reetz and de Vries. Similar equilibria likely operate in the Sonagashira cross-coupling of iodobenzene and phenylacetylene, over silica- and titania-supported Au_{55} clusters and nanoparticles.[86] However, solubilised gold atoms appear completely inert, with larger metallic nanoparticles[87] both more active and selective than the smaller clusters, indicative of an entirely surface-driven process.

3.4.4 Other Evidence for Heterogeneous Coupling Pathways

Definitive proof for the existence of purely heterogeneous coupling pathways is provided by ultra-high vacuum experiments over metal single-crystal surfaces,

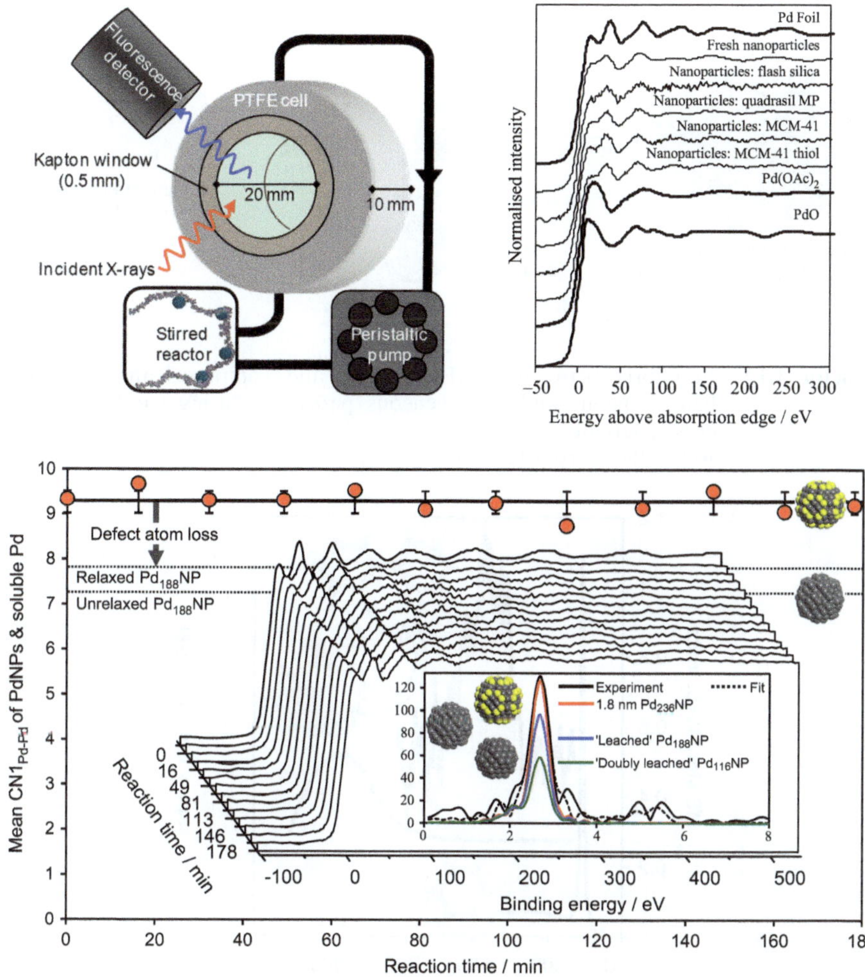

Figure 3.8 Top: Recirculating reactor setup for operando fluorescence XAS. A solution of PVP-stabilised PdNPs (50 ppm metal) is flowed through the cell during spectral acquisition while the reactor is simultaneously sampled for off-line GC analysis. Bottom: The Suzuki coupling of iodoanisole with phenylboronic acid was monitored over a family of PVP-stabilised nanoparticles while acquiring fluorescence Pd K-edge XAS spectra using the purpose-built cell shown above.
Reprinted by permission from John Wiley and Sons: P. J. Ellis, I. J. S. Fairlamb, S. F. J. Hackett, K. Wilson and A. F. Lee, *Angew. Chem., Int. Ed.*, 2010, 49, 1820, copyright 2012.

in which there is no possibility of homogeneous contributions. Ullmann coupling of iodobenzene over Cu(111),[88] and analogous homo-coupling to biphenyl over Ag(111)[89] and Au(111)[90] model catalysts has been well-documented in the surface science literature. Pd(111) is likewise able to activate

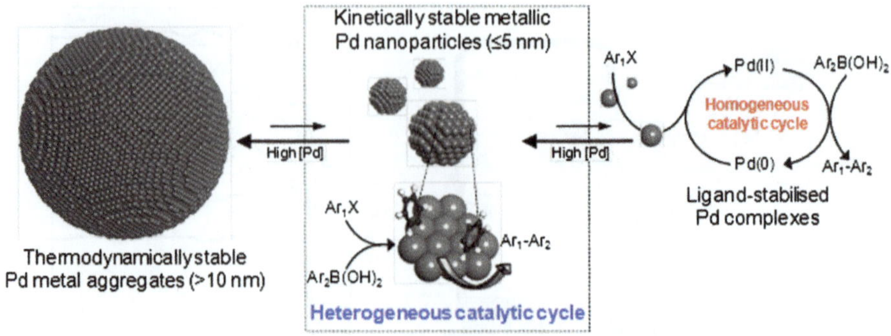

Scheme 3.13 Equilibria between PdNPs, clusters and complexes and associated heterogeneous and homogeneous pathways during catalytic Suzuki cross-coupling.
Reproduced from Ref. 81.

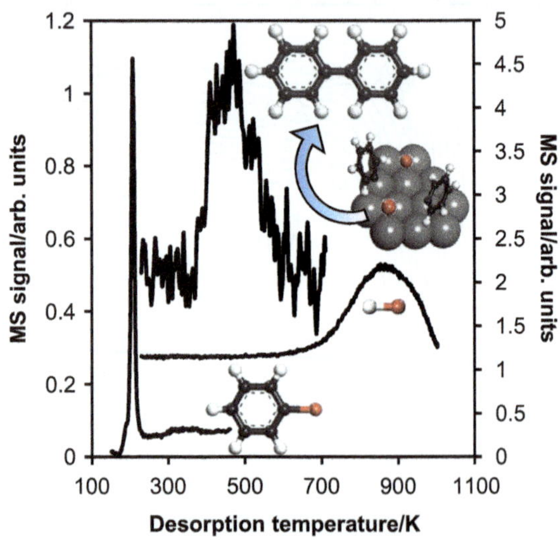

Figure 3.9 Reactively-formed biphenyl desorbing from the homo-coupling of bromobenzene over a Pd(111) single crystal (temperature-programmed desorption mass spectra from a reacting bromobenzene adlayer adsorbed at 120 K over a Pd(111) single crystal).
Reproduced from Ref. 81.

aryl halides, however the focus of C–C bond formation over this surface is dominated by alkyne coupling and associated cyclisation chemistry.[91] A synchrotron XPS study by von Schenck and co-workers produced strong evidence for iodobenzene homo-coupling over Pd(111) and Pd(110), wherein biphenyl production must occur at the Pd(111) solid surface.[92] Similar observations were recently made during the temperature-programmed reaction of bromobenzene over Pd(111), in which the homocoupled biphenyl product evolved from the palladium surface above 65 °C (Figure 3.9). Although cross-coupled products have yet to be reported over palladium single crystals, these latter observations are the strongest to date that such heterogeneous mechanisms exist.

In contrast to palladium, Sonagashira cross-coupling of iodobenzene and phenylacetylene was recently demonstrated over Au(111) by Lambert and co-workers.[93] Using time and temperature-dependent mass spectrometry (Figure 3.10) and high-resolution XPS, they unequivocally demonstrated diphenylacetylene production over a smooth, gold surface, from flat-lying reactants (as identified by NEXAFS).

A final, elegant attempt to demonstrate palladium catalysed heterogeneous Suzuki and Heck couplings in the solution phase is that utilising a Pd-coated AFM tip to induce spatially-resolved cross-coupling over a surface-assembled, aryl bromide self-assembled monolayer (SAM), in boronic acid solution.[94] Catalytically functional AFM probes were first generated by incubating silicon nitride probes in an aqueous solution of (PVP)PdNPs, and these were subsequently scanned over an aryl bromide dialkyl sulfide-modified gold surface while both probe and substrate were incubated in a methanolic solution of phenylboronic acid and sodium acetate (Figure 3.11). Reaction-induced changes in surface free energy were mapped, offering convincing evidence for spatially-resolved chemical coupling of surface-bound and solution phase reagents initiated by the Pd-modified AFM tip.

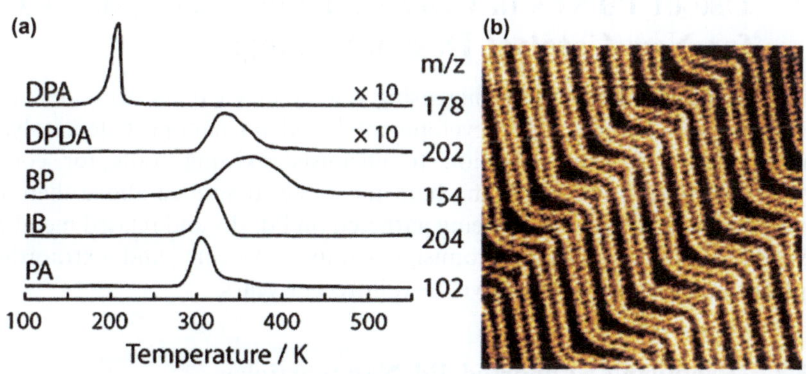

Figure 3.10 (a) Temperature-programmed reaction spectra of phenylacetylene (PA), iodobenzene (IB) and reactively-formed diphenylacetylene (DPA), diphenyldiacetylene (DPDA), and biphenyl (BP) products following adsorption of 0.55 ML IB + 0.25 ML PA on Au(111) at 90 K. (b) STM image of clean Au(111) surface showing the herringbone reconstruction prior to reactant adsorption (65 nm×65 nm, $V_{gap} = 2.41$ V, $I = 0.21$ nA). Reprinted with permission from V. K. Kanuru, G. Kyriakou, S. K. Beaumont, A. C. Papageorgiou, D. J. Watson and R. M. Lambert, *J. Am. Chem. Soc.*, 2010, **132**, 8081. Copyright 2012 American Chemical Society.

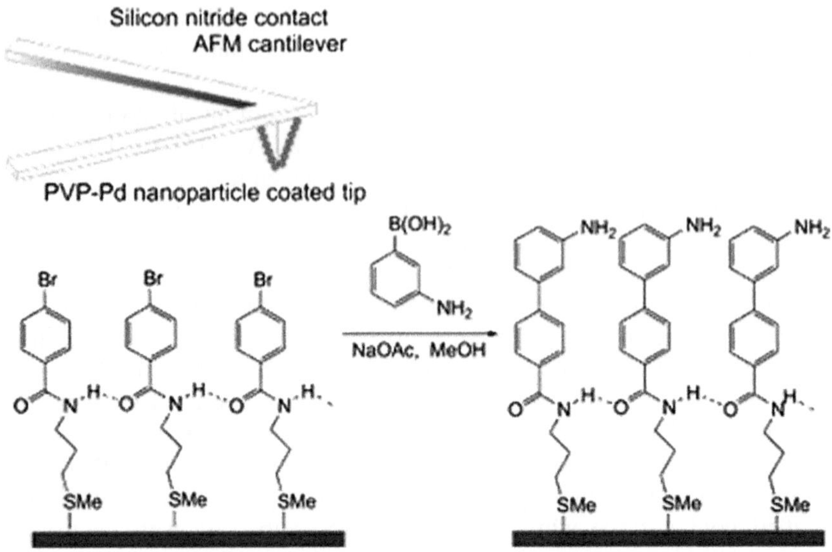

Figure 3.11 Schematic of heterogeneously-catalysed Suzuki reaction initiated on an aryl bromide monolayer by a (PVP)PdNP-functionalized AFM probe in phenylboronic acid solution.
Reprinted with permission from J. J. Davis, C. B. Bagshaw, K. L. Busuttil, Y. Hanyu and K. S. Coleman, *J. Am. Chem. Soc.*, 2006, **128**, 14135. Copyright 2012 American Chemical Society.

3.5 Use of PdNPs in Cross-Coupling Catalysis: A Case for New Catalyst Design Strategies?

Porco recently[95] stated "With increasing research endeavors being devoted to this field, new methodology development based on nanoparticle catalysis is emerging as a valuable addition to chemists' existing tools for complex molecule synthesis." With the mechanistic studies described above there is an increasing bank of knowledge being gathered on PdNPs and related metal NPs. Described below are some promising catalyst systems, and extraordinary applications, that have been developed in recent years.

3.5.1 Promising Supported Pd Nanoparticles

Many of the metal NPs described within the literature are based on solution-phase colloidal chemistry, and are stabilised by various agents, *e.g.* dendrimers, inorganic supports, polymers or surfactants. In order to make a stable and robust catalyst system, the stabilising agent should have a strong affinity for the metal surface. This is an issue for higher catalytic activities (enhanced TOFs, catalyst longevity *etc.*) as the stabilising agent takes up coordination sites, which can also serve as active catalytic sites (*i.e.* edge/defect sites). The key is to maintain a healthy balance between stability and activity. Binding Pd clusters

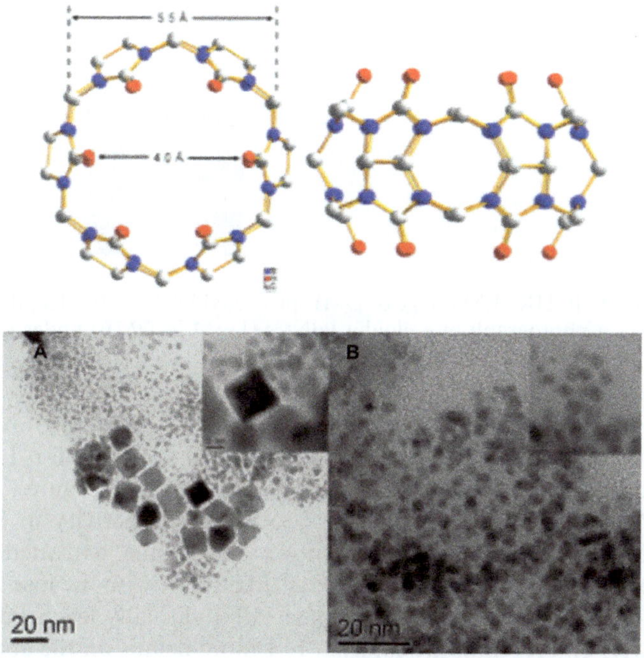

Figure 3.12 Top (left) and side (right) views of the pumpkin-shaped CB[6]. Bottom: TEM images of CB[6]-Pd NPs illustrate the variation in morphology when the ratio of $PdCl_2$ to CB[6] is equal to 2 : 1 (A) and 1 : 2 (B). Insets show corresponding HRTEM images, scale bar: 5 nm.
Reproduced from Ref. 95.

by means of a solid-support represents an interesting alternative approach. One such example was reported by Cao and co-workers based on cucurbit[6]uril,[96] abbreviated as CB[6], which can be utilized as a protecting agent for the formation of stable and well-dispersed PdNPs, which are active catalysts for Suzuki cross-coupling (Figure 3.12).

It is proposed that the surface atoms of PdNPs bind to the 'carbonyl portals' of CB[6] through electrostatic interactions. As CB[6] is non-polymeric, it bridges neighbouring PdNPs with the two identical carbonyl fringes, which thereby prevent further aggregation (even at elevated temperatures). The CB[6]/PdNPs are stable in the solid-state for long periods, and CB[6] has been described as a 'solid solvent' for dispersing metal NPs. Remarkably CB[6]/PdNP catalyses the reaction of chlorobenzene with phenyl boronic acid at 80 °C to give biphenyl in 87 % yield in 1 h! This is competitive with the performance of well-defined, electron-rich mononuclear Pd catalyst systems,[97] and an equivalent Suzuki cross-coupling catalysed by naked PdNPs.[98]

3.5.2 *In Situ* Generated Naked Pd Nanoparticles

The design and application of naked PdNPs in cross-coupling reactions has been reported by Obora and co-workers.[99] Relatively small photoluminescent

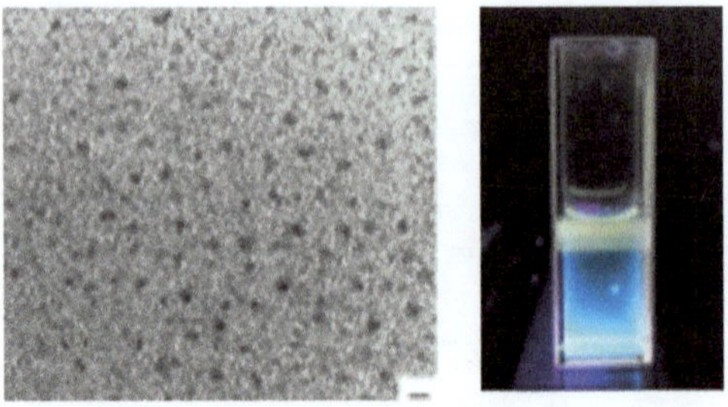

Figure 3.13 Left: HRTEM image of DMF-protected Pd NCs. Bar length: 2 nm; right: a photograph of colloidal PdNCs (1 mM in DMF) under UV light. Reproduced from Ref. 98.

Pd nanoclusters, abbreviated PdNCs, with an average size of 1–1.5 nm, were synthesised by refluxing $PdCl_2$ in refluxing DMF for 6 h (Figure 3.13).

These Pd nanoclusters exhibit very good catalytic activity in Suzuki and Heck reactions (up to 6.0×10^8 turnovers for some activated iodoarene substrates), and could be recycled at least five times without loss of catalytic activity. The latter is of interest as naked PdNPs usually leach Pd and deactivate (slowly) on recycling the catalyst. For more details on the application of PdNPs in Suzuki cross-couplings see the authoritative review.[100]

3.5.3 Heterogeneous Pd Catalysts for Use in Biological Applications

One of the most extraordinary reports in the recent literature is the intracellular Pd chemistry reported by Bradley and co-workers.[101] The study shows the first Pd^0-based heterogeneous catalyst (Pd^0 microspheres) with the ability to cross cell membranes and conduct cross-coupling *in cellulo*. The heterogeneous catalyst localises innocently within the cytoplasm (remaining for 48 h), which then mediates Pd-catalysed cross-coupling chemistry. The approach allows one to carry out artificial chemistry within a HeLa type cell. Analogies can be drawn here with how metalloproteins function intracellularly.

Bradley prepared PdNPs (*ca.* 5 nm) trapped within polystyrene microspheres, containing a suitable dye that could be imaged by confocal microscopy (Scheme 3.14). The Powder XRD pattern would suggest that the dye-tagged Pd microspheres are predominantly in zero oxidation state. An established fluorescent dye (anthofluorescein) was identified as being an ideal candidate for *in cellulo* Pd-catalysed Suzuki reaction of two non-fluorescent cross-coupling precursors, namely an aryl triflate and alkylaminophenylboronate, which affords the fluorescent product on exposure to the Pd microspheres (Scheme 3.15). The fluorescent product is able to localise at the mitochondria due to the presence of a targeting alkyltriphenylphosphonium moiety, which is shown by the images at the bottom of Scheme 3.15. The unique approach offers

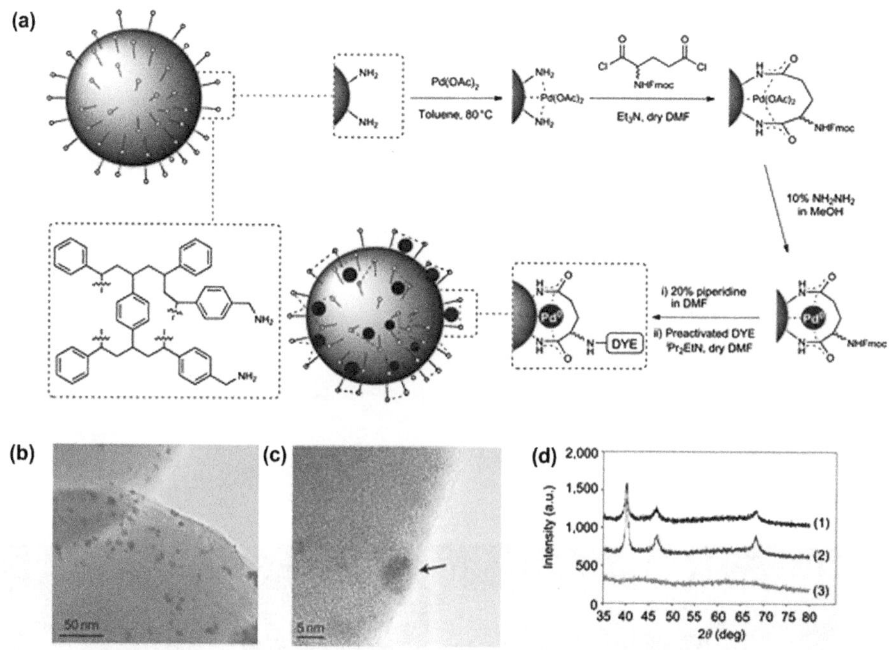

Scheme 3.14 (a) Synthetic route to fluorescently labelled Pd^0 microspheres; (b) and (c) TEM images of the microsphere entrapped PdNPs at different magnifications (PdNPs are *ca.* 5 nm size); (d) powder XRD patterns of (1) Pd^0 microspheres, (2) commercial Pd^0 powder, and (3) naked microspheres.
Reprinted by permission from Macmillan Publishers Ltd: Nature Chemistry, R. M. Yusop, A. Unciti-Broceta, E. M. V. Johansson, R. M. Sánchez-Martín and M. Bradley, *Nature Chem.* 2011, **3**, 239, copyright 2012.

the opportunity to customise heterogeneous PdNP catalysts for creative use in chemical biology, pharmacology and potentially medicine.

3.6 Future Perspectives and Conclusions

The interplay of homogeneous and heterogeneous catalytic cycles in cross-coupling chemistry is undoubtedly complex. However, by understanding hybrid homogeneous heterogeneous behaviour new catalyst systems have been identified. Indeed under the right conditions one can employ PdNPs as heterogeneous catalysts. Mechanistic studies will continue to inform the design of cross-coupling catalyst systems, which should not be limited to Pd. The remarkable chemistry that is possible using suitable PdNPs in cellular systems (*e.g.* the work reported by Bradley and co-workers) exemplifies what is possible and the opportunities that lie ahead in the future.

One should bear in mind that there is only a finite source of Pd on Earth, yet it is used in so many different processes and applications. This is seen in the

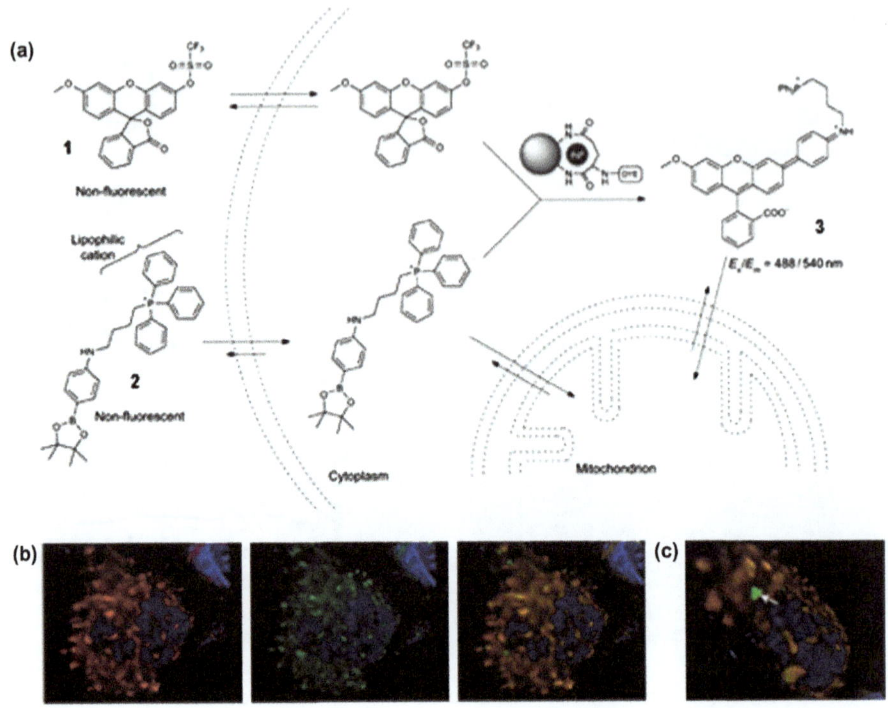

Scheme 3.15 (a) Cartoon depiction showing the intracellular cross-coupling of aryl triflate **1** with aryl boronate ester **2** which forms a mitochondria-localized fluorescent compound **3**. (b) and (c) show HeLa cells containing intracellular Pd^0 microspheres, incubated for 48 h with **1** and **2**. Cells were incubated for 30 min with a mitochondrial stain (MitoTracker Deep Red), fixed, incubated with a nuclei stain and imaged by confocal microscopy. Left image shows the cell nucleus (blue) and mitochondria (red). The centre image shows the cell nucleus (blue) and the *in cellulo* synthesised **3** (green). The right image is a merged image, with orange indicating co-localization. (c) shows an image of the same cell observed from a different angle, and the white arrow indicates the presence of a Texas Red-labelled Pd^0 microsphere (non-localized), which was imaged with a 550/20 nm emission filter together with compound **5**.
Reprinted by permission from Macmillan Publishers Ltd: Nature Chemistry, R. M. Yusop, A. Unciti-Broceta, E. M. V. Johansson, R. M. Sánchez-Martín and M. Bradley, *Nature Chem.* 2011, **3**, 239, copyright 2012.

value of palladium over the last 10 years or so. At present palladium is at a high price ($700 per Troy ounce, *ca.* 31.1 g), and this is unlikely to change in the foreseeable future {in 2000 it reached a high of $1100, and in 2008 a low of $200 per Troy ounce). Scavenging it from waste steams and our local environment, *e.g.* using organisms or plants to scavenge palladium, will be necessary for sustainable cross-coupling chemistry.

In light of this scarcity/cost issue, it is imperative that we learn to recycle and reuse spent palladium catalysts. Moreover, it is of paramount importance to develop heterogeneous catalyst systems, drawing on fundamental knowledge gathered from comprehensive mechanistic studies, in order to identify appropriate stabilised PdNPs able to function as heterogeneous catalysts under certain reaction conditions. Whether it is possible to do this for an array of reaction conditions and using diverse substrate systems remains an open question, but progress has been made in recent years, and the future for PdNP catalyst design is ripe for exploitation.

Porco eloquently stated that the "*Future development of nanoparticle catalysis may also significantly benefit from further understanding of reaction mechanisms that will ultimately require interdisciplinary collaborations; for example, between organic chemists and material scientists. Such mechanistic studies may include experiments on reaction kinetics, theoretical modeling, catalyst surface characterization, and validation of reaction intermediates*".[90] That is an apt way in bringing a close to this chapter, with the final remark that collaboration with biologists, pharmacologists and clinicians can now be anticipated.

References

1. I. J. S. Fairlamb, *Chem. Soc. Rev.*, 2007, **36**, 1036.
2. C. C. C. Johansson Seechurn, M. O. Kitching, T. J. Colacot and V. Snieckus, *Angew. Chem., Int. Ed.*, 2012, **51**, 5062. Also, see special issue dedicated to this topic: M. Beller, *Chem. Soc. Rev.*, 2011, **40**, 4891.
3. Transition Metals for Organic Synthesis, ed. M. Beller, and C. Bolm, vol. 1 & 2 , 2nd edn, Wiley-VCH GmbH & Co. KGaA, 2004.
4. P. E. Sehnal, R. J. K. Taylor and I. J. S. Fairlamb, *Chem. Rev.*, 2010, **110**, 824.
5. K. Kölhler, W. Kleist and S. S. Pröckl, *Inorg. Chem.*, 2007, **46**, 1876.
6. D. Astruc, *Inorg. Chem.*, 2007, **46**, 1884.
7. Modern Surface Organometallic Chemistry, ed. J.-M. Bassset, R. Psaro, D. Roberto and R. Ugo, Wiley-VCH GmbH & Co. KGaA, 2008.
8. S. K. Beaumont, *J. Chem. Technol. Biotechnol.*, 2012, **87**, 595.
9. J. F. Hartwig, in *Organotransition Metal Chemistry From Bonding to Catalysis*, University Science Books, 2010.
10. C. Amatore and A. Jutand, *Acc. Chem. Res.*, 2000, **33**, 314.
11. C. Amatore and A. Jutand, *J. Organomet. Chem.*, 1999, **576**, 254.
12. I. J. S. Fairlamb, R. J. K. Taylor, J. L. Serrano and G. Sanchez, *New. J. Chem.*, 2006, **30**, 1695.
13. K. Fagnou and M. Lautens, *Angew. Chem., Int. Ed.*, 2002, **41**, 26, and references cited therein.
14. C. Amatore, M. Azzabi and A. Jutand, *J. Organomet. Chem.*, 1989, **363**, C41.
15. C. Amatore, M. Azzabi and A. Jutand, *J. Am. Chem. Soc.*, 1991, **113**, 8375.

16. L. J. Gooßen, D. Koley, H. Hermann and W. Thiel, *Chem. Commun.*, 2004, 2141.
17. C. Amatore, A. Jutand, F. Lemaître, J. L. Ricard, S. Kozuch and S. Shaik, *J. Organomet. Chem.*, 2004, **689**, 3728.
18. Palladacycles: Synthesis, Characterization and Applications, ed. J. Dupont and M. Pfeffer, Wiley-VCH GmbH & Co. KGaA, 2006.
19. R. B. Bedford, C. S. J. Cazin, S. J. Coles, T. Gelbrich, P. N. Horton, M. B. Hursthouse and M. E. Light, *Organometallics*, 2003, **22**, 987.
20. R. B. Bedford, C. S. J. Cazin, S. J. Coles, T. Gelbrich, M. B. Hursthouse and V. J. M. Scordia, *J. Chem. Soc., Dalton Trans.*, 2003, 3350.
21. (a) A. H. M. de Vries, J. M. C. A. Mulders, J. H. M. Mommers, H. J. W. Henderickx and J. G. de Vries, *Org. Lett.*, 2003, **5**, 3285. (b) For the involvement of PdNPs in Heck coupling, see: C. Cassol, A. P. Umpierre, G. Machado, S. I. Wolke and J. Dupont, *J. Am. Chem. Soc.*, 2005, **127**, 3298; (c) C. S. Consorti, F. R. Flores and J. Dupont, *J. Am. Chem. Soc.*, 2005, **127**, 12054.
22. J. G. de Vries, *Dalton Trans.*, 2006, 421.
23. C. Amatore, A. Jutand and G. Le Duc, *Chem. Eur. J.*, 2011, **17**, 2492.
24. B. P. Carrow and J. F. Hartwig, *J. Am. Chem. Soc.*, 2011, **133**, 2116.
25. J. L. Serrano, L. García, J. Pérez, E. Pérez, J. García, G. Sánchez, P. Sehnal, S. De Ornellas, T. J. Williams and I. J. S. Fairlamb, *Organometallics*, 2011, **30**, 5095.
26. Imidate ligand effects (*e.g.* succinimide) are well established in cross-coupling reactions, see: S. Burling, C. M. Crawforth, I. J. S. Fairlamb, A. R. Kapdi, R. J. K. Taylor and A. C. Whitwood, *Tetrahedron*, 2005, **61**, 9736.
27. C. Amatore, A. Jutand and G. Le Duc, *Chem.–Eur. J.*, 2012, **18**, 6616.
28. G. A. Molander, P. E. Gormisky and D. L. Sandrock, *J. Org. Chem.*, 2008, **73**, 2052, and references cited therein.
29. A. J. J. Lennox and G. C. Lloyd-Jones, *J. Am. Chem. Soc.*, 2012, **134**, 7431.
30. V. I. Bakhmutov, J. F. Berry, F. A. Cotton, S. Ibragimov and C. A. Murillo, *Dalton Trans.*, 2005, 1989.
31. The fully balanced equation is: $3Pd^0 + 6HNO_3 + 6HOAc \rightarrow Pd_3(OAc)_6 + 6NO_2 + 6H_2O$.
32. S. E. Bajwa, T. E. Storr, L. E. Hatcher, T. J. Williams, C. G. Baumann, A. C. Whitwood, D. R. Allan, S. J. Teat, P. R. Raithby and I. J. S. Fairlamb, *Chem. Sci.*, 2012, **3**, 1656–1661.
33. K. J. Stowers, A. Kubota and M. S. Sanford, *Chem. Sci.*, 2012, **3**, 3192–3195.
34. S. S. Zalesskiy and V. P. Ananikov, *Organometallics*, 2012, **31**, 2302.
35. E. Ramirez, S. Jansat, K. Philippot, P. Lecante, M. Gomez, A. M. Masdeu-Bultó and B. Chaudret, *J. Organomet. Chem.*, 2004, **689**, 4601.
36. M. S. Hubig, M. A. Drouin, A. Michel and P. D. Harvey, *Inorg. Chem.*, 1992, **31**, 5375.

37. C. Amatore, A. Jutand, F. Khalil, M. A. M'Barki and L. Mottier, *Organometallics*, 1993, **12**, 3168.
38. I. J. S. Fairlamb, *Org. Biomol. Chem.*, 2008, **6**, 3645.
39. I. J. S. Fairlamb, A. R. Kapdi, A. F. Lee, G. Sánchez, G. López, J. L. Serrano, L. García and E. Pérez, *Dalton Trans.*, 2004, 3970.
40. F. Zhao, B. M. Bhanage, M. Shirai and M. Arai, *Chem.–Eur. J.*, 2000, **6**, 843.
41. A. Biffis, M. Zecca and M. Basato, *Eur. J. Inorg. Chem.*, 2001, 1131.
42. M. Pagliaro, V. Pandarus, R. Ciriminna, F. Béland and P. D. Carà, *ChemCatChem*, 2012, **4**, 432.
43. (a) W. M. Alley, I. K. Hamdemir, K. A. Johnson and R. G. Finke, *J. Mol. Catal. A: Chem.*, 2010, **315**, 1; (b) E. Bayram, J. C. Linehan, J. L. Fulton, J. A. S. Roberts, N. K. Szymczak, T. D. Smurthwaite, S. Ozkar, M. Balasubramanian and R. G. Finke, *J. Am. Chem. Soc.*, 2011, **133**, 18889.
44. M. T. Reetz and J. G. de Vries, *Chem. Commun.*, 2004, 1559.
45. A. V. Gaikwad, A. Holuigue, M. B. Thathagar, J. E. ten Elshof and G. Rothenberg, *Chem.–Eur. J.*, 2007, **13**, 6908.
46. T. E. Storr, A. G. Firth, K. Wilson, K. Darley, C. G. Baumann and I. J. S. Fairlamb, *Tetrahedron*, 2008, **64**, 6125.
47. J. A. Widegren and R. G. Finke, *J. Mol. Catal. A: Chem.*, 2003, **198**, 317.
48. F. Bellina, S. Cauteruccio and R. Rossi, *Eur. J. Org. Chem.*, 2006, 1379.
49. F. Bellina, S. Cauteruccio, L. Mannina, R. Rossi and S. Viel, *Eur. J. Org. Chem.*, 2006, 693.
50. F. Bellina, C. Calandri, S. Cauteruccio and R. Rossi, *Tetrahedron*, 2007, **63**, 1970.
51. M. T. Reetz and J. G. de Vries, *Chem. Commun.*, 2004, 1559.
52. R. Narayanan and M. A. El Sayed, *J. Am. Chem. Soc.*, 2003, **125**, 8340. For catalytic applications of PVP-stabilised palladium nanoparticles, see: (a) A. Gniewek, A. M. Trzeciak, J. J. Ziółkowski, L. Kępiński, J. Wrzyszcz and W. Tylus, *J. Catal.*, 2005, **229**, 332; (b) A. M. Trzeciak, W. Wojtków, J. J. Ziółkowski, J. Wrzyszcz and M. Zawadzki, *New J. Chem.*, 2004, **28**, 859.
53. (a) T. E. Storr, C. G. Baumann, R. J. Thatcher, S. De Ornellas, A. C. Whitwood and I. J. S. Fairlamb, *J. Org. Chem.*, 2009, **74**, 5810; (b) T. E. Storr, PhD thesis, University of York, UK, 2010; (c) S. de. Ornellas, PhD thesis, University of York, UK, 2013.
54. J. Zou, S. G. Stewart, C. L. Raston and K. S. Iyer, *Chem. Commun.*, 2011, **47**, 1803.
55. A. G. Firth, K. Wilson, C. G. Baumann and I. J. S. Fairlamb, *Nucleosides, Nucleotides and Nucleic acids*, 2011, **30**, 168.
56. K. Köhler, W. Kleist and S. S. Pröckl, *Inorg. Chem.*, 2007, **46**, 1876.
57. S. V. Ley, C. Ramarao, R. S. Gordon, A. B. Holmes, A. J. Morrison, I. F. McConvey, I. M. Shirley, S. C. Smith and M. D. Smith, *Chem. Commun.*, 2002, 1134.

58. S. V. Ley, C. Ramarao, I. M. Shirley, S. C. Smith and D. J. Tapolczay, *Microencapsulated Catalyst Methods of Preparation and Methods of Use Thereof*, U.S. Patent Application, US20040254066, 2004.
59. S. J. Broadwater and D. Tyler McQuade, *J. Org. Chem.*, 2006, **71**, 2131.
60. J. Rebek and F. Gavina, *J. Am. Chem. Soc.*, 1974, **96**, 7112.
61. I. W. Davies, L. Matty, D. L. Hughes and P. J. Reider, *J. Am. Chem. Soc.*, 2001, **123**, 10139.
62. M. Parisien, D. Valette and K. Fagnou, *J. Org. Chem.*, 2005, **70**, 7578.
63. Following discussions between IJSF and the late Prof. Fagnou (in early 2009) on the hybrid homogeneous–heterogeneous catalyst behaviour in direct arylation chemistry, it became evident that the problem in proving such behaviour could come down to the availability of (new) techniques to unequivocally detect catalytically relevant Pd_n clusters under working conditions.
64. L. Ackermann, R. Vicente and A. R. Kapdi, *Angew. Chem., Int. Ed.*, 2009, **48**, 9792.
65. D. Lapointe and K. Fagnou, *Chem. Lett.*, 2010, **39**, 1118.
66. L. Ackermann, A. R. Kapdi, H. K. Potukuchi and S. I. Kozhushkov, Syntheses *via* C-H Bond Functionalizations, in *Handbook of Green Chemistry*, ed. C.-J. Li, Wiley-VCH, 2012.
67. C. C. Cassol, A. P. Umpierre, G. Machado, S. I. Wolke and J. Dupont, *J. Am. Chem. Soc.*, 2005, **127**, 3298.
68. G. Marck, A. Villiger and R. Buchecker, *Tetrahedron Lett.*, 1994, **35**, 3277.
69. S.-W. Kim, M. Kim, W. Young Lee and T. Hyeon, *J. Am. Chem. Soc.*, 2002, **124**, 7642.
70. Y. Li, E. Boone and M. A. El-Sayed, *Langmuir*, 2002, **18**, 4921.
71. Q. Du, W. Zhang, H. Ma, J. Zheng, B. Zhou and Y. Li, *Tetrahedron*, 2012, **68**, 3577.
72. A. Modak, J. Mondal and A. Bhaumik, *Green Chem.*, 2012, **14**, 2840.
73. (a) N. T. S. Phan, M. Van Der Sluys and C. W. Jones, *Adv. Synth. Catal.*, 2006, **348**, 609; (b) V. Farina, *Adv. Synth. Catal.*, 2004, **346**, 1553; (c) M. Perez-Lorenzo, *J. Phys. Chem. Lett.*, 2012, **3**, 167; (d) L. D. Pachon and G. Rothenberg, *App. Organomet. Chem.*, 2008, **22**, 288.
74. J. Le Bars, U. Specht, J. S. Bradley and D. G. Blackmond, *Langmuir*, 1999, **15**, 7621.
75. R. L. Augustine and S. T. O'Leary, *J. Mol. Catal. A: Chem.*, 1995, **95**, 277.
76. Y. Li, X. M. Hong, D. M. Collard and M. A. El-Sayed, *Org. Lett.*, 2000, **2**, 2385.
77. (a) R. Narayanan and M. A. El-Sayed, *J. Am. Chem. Soc.*, 2003, **125**, 8340; (b) R. Narayanan and M. A. El-Sayed, *J. Phys. Chem. B*, 2004, **108**, 8572; (c) R. Narayanan, C. Tabor and M. A. El-Sayed, *Top. Catal.*, 2008, **48**, 60.
78. R. Narayanan and M. A. El-Sayed, *Langmuir*, 2005, **21**, 2027.
79. A. F. Lee, *Abstr. Pap. Am. Chem. Soc.*, 2010, 240.

80. P. J. Ellis, I. J. S. Fairlamb, S. F. J. Hackett, K. Wilson and A. F. Lee, *Angew. Chem., Int. Ed.*, 2010, **49**, 1820.
81. A. F. Lee, P. J. Ellis, I. J. S. Fairlamb and K. Wilson, *Dalton Trans.*, 2010, **39**, 10473.
82. S. Reimann, J. Stötzel, R. Frahm, W. Kleist, J. -D. Grunwaldt and A. Baiker, *J. Am. Chem. Soc.*, 2011, **133**, 3921.
83. S. E. J. Hackett, R. M. Brydson, M. H. Gass, I. Harvey, A. D. Newman, K. Wilson and A. F. Lee, *Angew. Chem., Int. Ed.*, 2007, **46**, 8593.
84. A. F. Lee, C. V. Ellis, J. N. Naughton, M. A. Newton, C. M. A. Parlett and K. Wilson, *J. Am. Chem. Soc.*, 2011, **133**, 5724.
85. A. F. Lee and K. Wilson, *Green Chem.*, 2004, **6**, 37.
86. G. Kyriakou, S. K. Beaumont, S. M. Humphrey, C. Antonetti and R. M. Lambert, *ChemCatChem*, 2010, **2**, 1444.
87. S. K. Beaumont, G. Kyriakou and R. M. Lambert, *J. Am. Chem. Soc.*, 2010, **132**, 12246.
88. M. Xi and B.E. Bent, *J. Am. Chem. Soc.*, 1993, **115**, 7426.
89. G. J. Szulczewski and J. M. White, *Surf. Sci.*, 1998, **399**, 305.
90. D. Syomin and B. E. Koel, *Surf. Sci.*, 2001, **490**, 265.
91. (a) W. T. Tysoe, G. L . Nyberg and R. M. Lambert, *J. Chem. Soc., Chem. Commun.*, 1983, 623; (b) R. M. Ormerod and R. M. Lambert, *J. Chem. Soc., Chem. Commun.*, 1990, 1421; (c) A. F. Lee, K. Wilson, R. L. Middleton, A. Baraldi, A. Goldoni, G. Paolucci and R. M. Lambert., *J. Am. Chem. Soc.*, 1999, **121**, 7969.
92. H. von Schenck, J. Weissenrieder, S. Helldén, B. Åkermark and M. Göthelid, *App. Surf. Sci.*, 2003, **212**, 508.
93. V. K. Kanuru, G. Kyriakou, S. K. Beaumont, A. C. Papageorgiou, D. J. Watson and R. M. Lambert, *J. Am. Chem. Soc.*, 2010, **132**, 8081.
94. J. J. Davis, C. B. Bagshaw, K. L. Busuttil, Y. Hanyu and K. S. Coleman, *J. Am. Chem. Soc.*, 2006, **128**, 14135.
95. H. Cong and J. A. Porco, Jr., *ACS Catal.*, 2012, **2**, 65.
96. M. Cao, J. Lin, H. Yang and R. Cao, *Chem. Commun.*, 2010, **46**, 5088.
97. For examples, see: I. J. S. Fairlamb, Symposium in Print 113, *Tetrahedron*, 2005, **61**, 9661.
98. V. Calò, A. Nacci, A. Monopoli and F. Montingelli, *J. Org. Chem.*, 2005, **70**, 6040.
99. M. Hyotanishi, Y. Isomura, H. Yamamoto, H. Kawasaki and Y. Obora, *Chem. Commun.*, 2011, **47**, 5750.
100. A. Fihri, M. Bouhrara, B. Nekoueishahraki, J. -M. Basset and V. Polshettiwar, *Chem. Soc. Rev.*, 2011, **40**, 5181.
101. R. M. Yusop, A. Unciti-Broceta, E. M. V. Johansson, R. M. Sánchez-Martín and M. Bradley, *Nature Chem.*, 2011, **3**, 239.

CHAPTER 4

Pd^{II}/Pd^{IV} *Redox Couple Mediated C–X Bond Formation*

ANDREI N. VEDERNIKOV

Department of Chemistry and Biochemistry, University of Maryland, College Park, MD 20742 USA
Email: avederni@umd.edu

4.1 Introduction

The last decade has seen an unprecedented growth of interest to organometallic chemistry of palladium in its high oxidation states, both from synthetic organic and organometallic chemists. While the most common oxidation states of palladium in its organometallic derivatives are 0 and +2, high-valent organopalladium complexes contain palladium atoms in oxidation states +3 and +4 (Chart 4.1).[1] High-valent palladium complexes can be very versatile including one, two or three hydrocarbyl ligands such as alkyl, aryl, benzyl, allyl, acyl, propargyl and others.

Most of the known organopalladium(III) complexes are dinuclear. They feature a Pd–Pd bond and contain bridging ligands such as carboxylates (Chart 4.1, compound **3**) or amidinates. Only recently, the first mononuclear organopalladium(III) complexes were prepared and characterized (see, e.g., Chart 4.1, **4**).[2] An important key feature of most high-valent organopalladium compounds is their ability to undergo a thermodynamically favorable and often facile C–X (X = F, Cl, Br, I, C, O, S, N *etc.*) bond elimination to form organic products along with a stable low-valent palladium(II) compound. One of the first examples of C–C reductive elimination from a well-characterized

RSC Catalysis Series No. 11
C–H and C–X Bond Functionalization: Transition Metal Mediation
Edited by Xavi Ribas

Published by the Royal Society of Chemistry, www.rsc.org

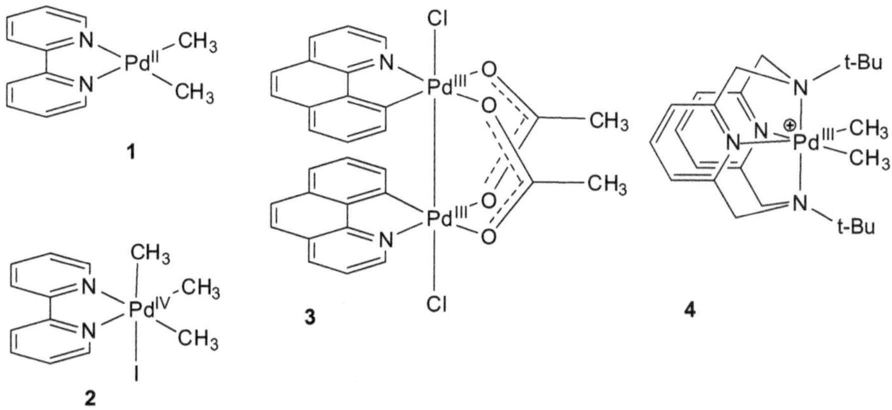

Chart 4.1 Some representative organopalladium compounds: mononuclear palladium(II), **1**, mononuclear palladium(IV), **2**, dinuclear palladium(III), **3** and mononuclear palladium(III), **4**.

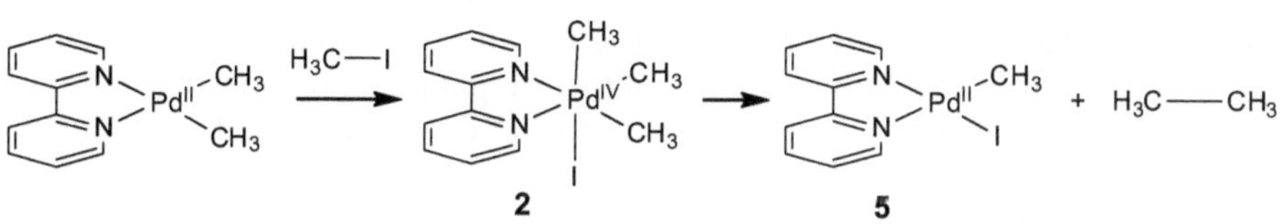

Scheme 4.1 Preparation and C–C reductive elimination of a trimethylpalladium(IV) complex.

palladium(IV) species was ethane elimination of trimethylpalladium(IV) complex **2**. Palladium(II) complex **5** is another reaction product (Scheme 4.1).[3]

The ability of high-valent organopalladium species to reductively eliminate products with new C–X bonds is of the key importance for application of such complexes in palladium-mediated oxidation reactions and synthesis of various functionalized organic compounds.

High-valent organometallic derivatives of palladium can be generated from appropriate palladium(II) precursors by action of strong oxidants such as peroxosulfates, a number of other peroxocompounds, iodine(III)-containing reagents, N-halosuccinimides, free halogens and others.[4,5] Recently, it was found that oxidation of organopalladium(II) to organopalladium(IV) complexes can also be achieved with dilute aqueous hydrogen peroxide[6] and even with oxygen.[7] Correct choice of a facially chelating ligand at the metal is critical in the two latter cases.[8] A correct choice of an oxidant for a particular metal-catalyzed oxidative functionalization reaction also implies that more electron-rich organometallic derivatives will be oxidized only, so that an "inorganic" palladium(II) precursor will be available for palladation of organic substrate present in reaction mixture.

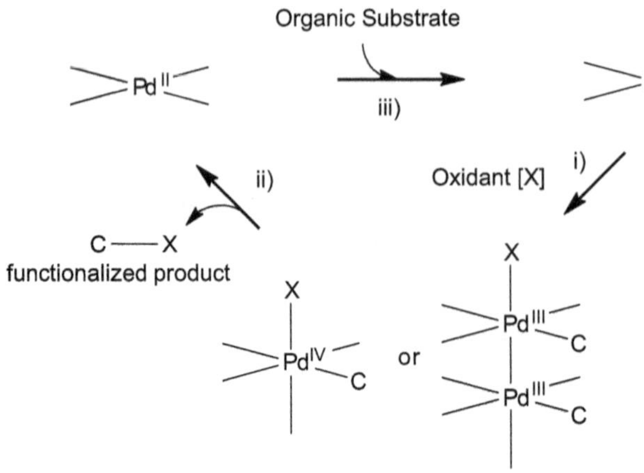

Scheme 4.2 High valent palladium mediated oxidative functionalization.

A simplified catalytic cycle showing a Pd^{II}/Pd^{IV} or Pd^{II}/Pd^{III} redox couple-mediated oxidative transformation of an organic substrate is given in Scheme 4.2. The reaction sequence includes (i) oxidation of an organopalladium(II) precursor to form a high-valent organopalladium(III) or palladium(IV) intermediate, (ii) reductive elimination of a functionalized product featuring a new C–X bond and (iii) transformation of the palladium(II) complex resulting from the C–X bond reductive elimination back to the organopalladium(II) precursor.

The chemistry of both high oxidations states of palladium, +3 and +4, can be tightly connected, and both can be involved in a particular oxidative functionalization reaction sequence. For instance, it was shown that the mononuclear palladium(IV) complex **8** in Scheme 4.3 is formed *via* an oxidation–fragmentation sequence involving palladium(III) intermediates. Here, oxidation of a palladium(II) precursor **6** leads initially to a dinuclear acetoxy-bridged palladium(III) species **7** that undergoes subsequent disproportionation to give the palladium(II) complex **6** and the palladium(IV) product **8**.[9]

Hence, the mechanism of oxidation of palladium(II) compounds may be complex and may include interconversion of palladium(III) and palladium(IV) species. Determining the identity of high-valent palladium species involved in catalytic oxidative functionalization reaction often constitutes a significant problem. This is the case when C–X bond elimination of a high-valent intermediate is not the rate determining step of the whole catalytic cycle and, therefore, concentration of the high-valent intermediate in the reaction mixture is extremely low. For instance, in oxidative C–H acetoxylation of arenes with $PhI(OAc)_2$[10,11] the C–H bond cleavage and formation of an organopalladium(II) complex is the rate determining step (Scheme 4.2iii). Faster rates of subsequent oxidation (Scheme 4.2i) and reductive elimination reactions (Scheme 4.2ii) complicate experimental characterization of high-valent palladium intermediates leaving without answer questions about the palladium oxidation state in these species.

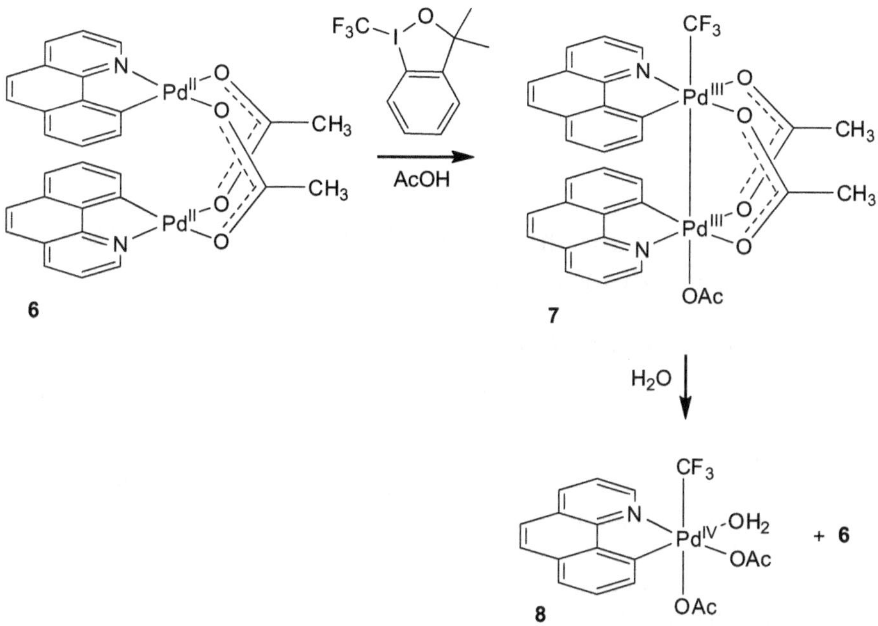

Scheme 4.3 Intermediate palladium(III) complex **7** in the oxidation of palladium(II) precursor **6** to form palladium(IV) compound **8**.

There are only few well-characterized catalytic systems involving high-valent palladium complexes where the nature of the catalytically competent organopalladium intermediates is known. One of these reactions is aromatic C–H chlorination with *N*-chlorosuccinimide mediated by catalytically competent dinuclear palladium(III) species.[12] Another example is catalytic aromatic trifluoromethylation involving palladium(IV) intermediates.[13] As a result, in the subsequent review of catalytic high-valent palladium chemistry presented in this chapter the identity of most of the catalytically competent metal intermediates is proposed or postulated.

Analysis of catalytic reaction mechanisms is usually complemented by studies of structure and reactivity of relevant model high-valent organopalladium complexes that are sufficiently stable for such studies. Two reactions that are of great interest here are oxidation of organopalladium(II) precursors to produce the model high-valent palladium complexes and C–X bond elimination reactivity of the latter. Poor knowledge of such reactivity might be one of the major reasons why, in spite of numerous proposals of intermediacy of organopalladium(IV) species in palladium-catalyzed oxidation reaction and a relatively early discovery of some high-valent organopalladium-catalyzed transformations,[10] their synthetic potential was not revealed until recently. In spite of the limitations characterized above, the number of new oxidation reactions catalyzed by organometallic derivatives of palladium(III) and palladium(IV) species is rapidly increasing, which is reflected in the growing number of reviews in the field.[4,5,14,15]

Chapter 5 of this book will introduce the reader into the field of organopalladium(III) chemistry and its applications in organic synthesis. The main goal of the Chapter 4 is to characterize some reactivity of organopalladium(IV) compounds serving as intermediates in organic synthesis and catalysis leading to the formation of various C–X bonds.

4.2 C–X Bond Formation by Reductive Elimination of Organopalladium(IV) Complexes

4.2.1 Formation of C–C Bonds

Formation of C–C bonds was one of the first well-characterized reactions of isolated organopalladium(IV) complexes produced by oxidative addition of methyl iodide and dihydrocarbyl palladium(II) species.[1] Palladium(IV) complexes with various combinations of hydrocarbyl ligands can be prepared in a similar way.[16] An example of a C(sp^3)–C(sp^3) oxidative coupling leading to formation of ethane is given in Scheme 4.1. This reaction is facile in polar solvents such as acetone and acetonitrile. The mechanism of the C–C reductive elimination includes dissociation of Pd^{IV}–I bond to form a labile cationic palladium(IV) solvento complex which undergoes concerted three-center C–C oxidative coupling. A concerted three-center elimination mechanism is most typical for palladium(IV) complexes.[17] Aryl dimethyl complexes with structures similar to that of complex **2** can form several products, those with a new C(sp^3)–C(sp^2) bond and those with a new C(sp^3)–C(sp^3) bond. In fact, elimination of both arylalkane and ethane occurs concurrently.[1]

Selective catalytic C(sp^3)–C(sp^2) oxidative coupling reactions may be of high practical interest. One of the recent examples of such reactions involving arene substrates and S-trifluoromethyldibenzothiophenium as the second coupling partner is shown in Scheme 4.4. In this case it was shown that the palladium(IV) complex **8** is the catalytically competent species.[13]

Catalytic formation of C(sp^2)–C(sp^2) bonds to form both symmetric and unsymmetrical diaryls is another synthetically useful transformation. This reaction can be used, in particular, in palladium-catalyzed oxidative coupling of heteroarenes. Unsymmetrical diaryls can be produced using diaryliodonium tetrafluoroborates, *e.g.*, $[Ph_2I]BF_4$, as a source of one of the aryl groups,[18] whereas symmetrical heterodiaryls can be prepared using Oxone as the oxidant.[19]

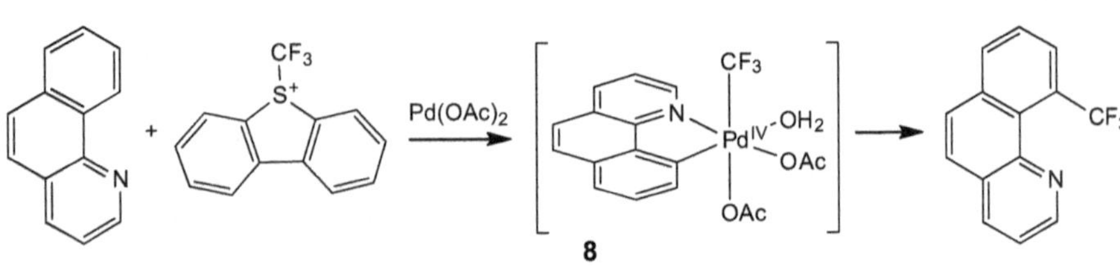

Scheme 4.4 Catalytic trifluoromethylation of a C(sp^2)–H bond donor.

The catalytic synthesis of unsymmetrical diaryls is an example of a reaction where the identity of catalytically competent species could not be revealed readily. Oxidation of organopalladium(II) complexes with diaryliodonium cations was thought initially to lead to mononuclear palladium(IV) diaryl complexes[19] but subsequent studies pointed to the most likely involvement of dinuclear Pd(III) compounds.[4]

4.2.2 Formation of C–O Bonds

Aromatic carbon–oxygen bond oxidative coupling leading to formation of arylcarboxylates **11** has been studied in detail using some well characterized isolable model diaryl dicarboxylatopalladium(IV) complexes **9** with variably substituted arylcarboxylate ligands (Scheme 4.5).[20]

According to the studies performed, the mechanism of reaction in Scheme 4.5 involves dissociation of the carboxylate ligand in complex **9** that is *trans* to the aryl ligand to produce a five-coordinate palladium(IV) intermediate **10** that undergoes subsequent concerted three-center C–O bond elimination. This type of reductive elimination reactions is most likely involved in oxidative acetoxylation of aromatic substrates bearing heteroatom-containing directing groups. Oxidative acetoxylation of C(sp^3)–H bond donors has also been demonstrated in these systems but model studies of palladium(IV) alkyls capable of C(sp^3)–O elimination are yet to be performed. High-valent organopalladium species in the reactions mentioned above are produced by the action of various oxidants such as $PhI(OAc)_2$, Oxone, $K_2S_2O_8$, IOAc,[21] or, in some cases, *tert*-butylhydroperoxyacetate.[22]

Another type of products with aromatic carbon–oxygen bonds that can result from C(sp^2)–O reductive elimination of palladium(IV) species are phenols. Well characterized palladium(IV) hydroxo and 1-hydroxo-1,1-diarylmethoxo complexes **13** and **16** produce phenols **14** or palladium(II) phenoxide **17** under mild conditions (Scheme 4.6).[6] The latter reaction occurs already at 20 °C and is virtually quantitative. Based on mechanistic tests and DFT calculations, the C(sp^2)–O bond formation in this reaction was proposed to proceed directly from six-coordinate palladium(IV) species **13** and **15**.

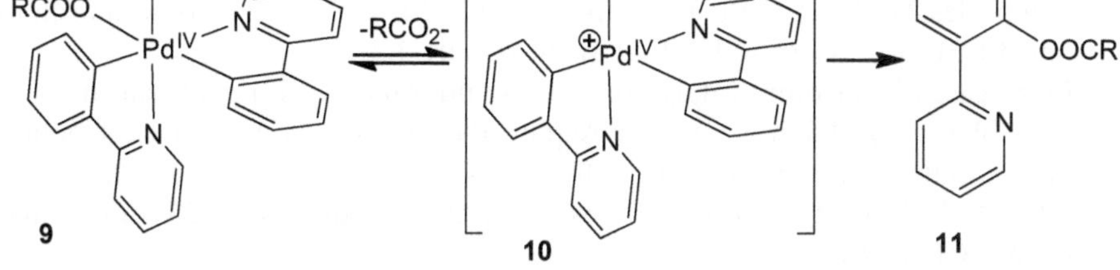

Scheme 4.5 C(sp^2)–O reductive elimination of arylcarboxylates.

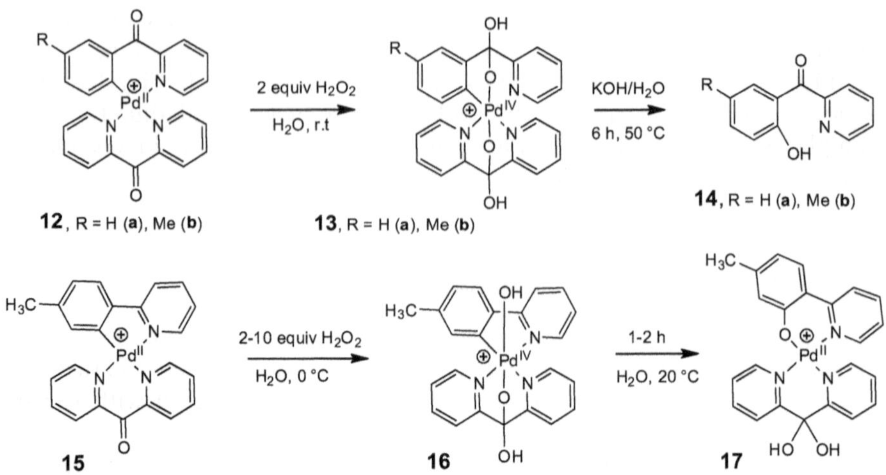

Scheme 4.6 Elimination of phenols from 1-hydroxomethoxopalladium(IV) aryl complexes.

While high-valent palladium hydrocarbyls are typically generated in reaction mixtures using various waste-producing oxidants such as peroxosulfates, iodine(III) compounds or *N*-halosuccinimides, palladium(IV) complexes **13** and **16** can be prepared readily by oxidation of palladium(II) precursors **12** and **15** with environmentally benign dilute aqueous H_2O_2.

In this respect, it is worth mentioning that oxidative C(sp^3)–H acetoxylation of various substituted 8-methylquinolines **18** to produce corresponding 8-acetoxymethylquinolines **19** (R = H, F, Cl, Br, I, NO_2, Me, OMe) can be performed with O_2 as the only oxidant (Scheme 4.7).[23] Potential intermediacy of palladium(IV) complexes such as **20** was analyzed in this work using DFT calculations.

4.2.3 Formation of C–Halogen Bonds

Reductive elimination of C–halogen bonds (halogen = F, Cl, Br, I) is involved in a number of palladium-catalyzed halogenation reactions of C(sp^2)–H and C(sp^3)–H bond donors. Typical oxidants used in these reactions are *N*-halosuccinimides **21** (X = Cl, Br, I), $ArIX_2$ (X = Cl, **22**; F, **25**), or a combination of I_2 and $PhI(OAc)_2$ as an iodinating agent (Chart 4.2).[4]

In the case of palladium-catalyzed oxidative fluorination of organic substrates,[24] donors of C(sp^2)–H bond typically require the use of a source of electrophilic fluorine atoms such as Selectfluor, *N*-fluoro-2,4,6-trimethylpyridinium tetrafluoroborate **23**, *N*-fluorodisulfonylimide **24**, *p*-iodotoluenedifluoride **25** and similar reagents (Chart 4.2).

Importantly, utilization of these expensive electrophilic fluorinating agents can be avoided in some cases. For instance, fluorination of benzylic C(sp^3)–H

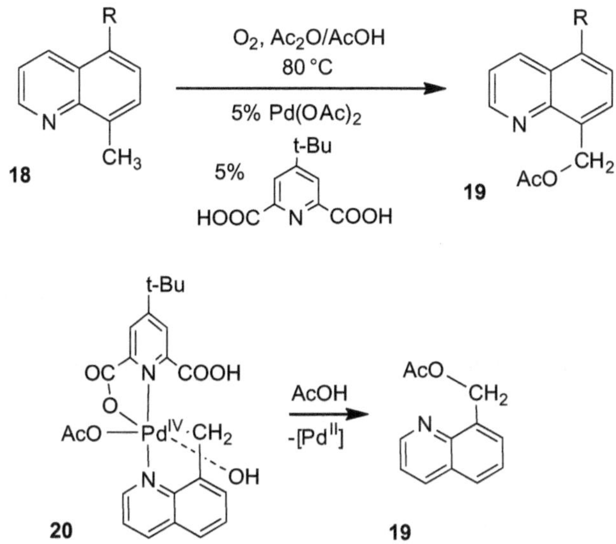

Scheme 4.7 Oxidative C(sp^3)H acetoxylation of 8-methylquinolines **18** with O_2 as the oxidant.

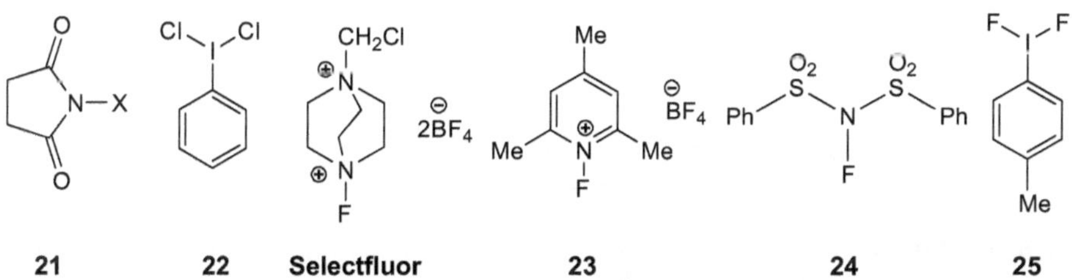

Chart 4.2 Some reagents used for oxidative halogenation of organic substrates.

bond in various substituted 8-methylquinolines **26** (R = H, F, Br, I, NO_2, CN, Ph, Me, OMe) to form 8-fluoromethylquinolines **27** can be performed in good yield with $PhI(OPiv)_2$ as an oxidant and AgF as a source of nucleophilic fluorine (Scheme 4.8).[25] The minor reaction products are the corresponding 8-pivaloylmethylquinolines **28**.

Reductive elimination of C(sp^2)–halogen (halogen = F, Cl, Br)[26,27] and C(sp^3)–F bonds[28] from individual palladium(IV) species has been studied using various model palladium(IV) complexes (Schemes 4.9–4.12). Selective elimination of a C(sp^3)–F bond from a palladium(IV) center was first demonstrated using complex **30** (Scheme 4.9). This moderately stable dihydrocarbyl palladium(IV) difluoride was prepared by oxidative fluorination of pallada(II)indane **29** with *N*-fluoro-2,4,6-trimethylpyridinium triflate. This mixed dihydrocarbyl palladium(IV) complex eliminates C(sp^3)–F bond to form selectively a fluoroalkylphenyl palladium(II) compound **31**. Another potential

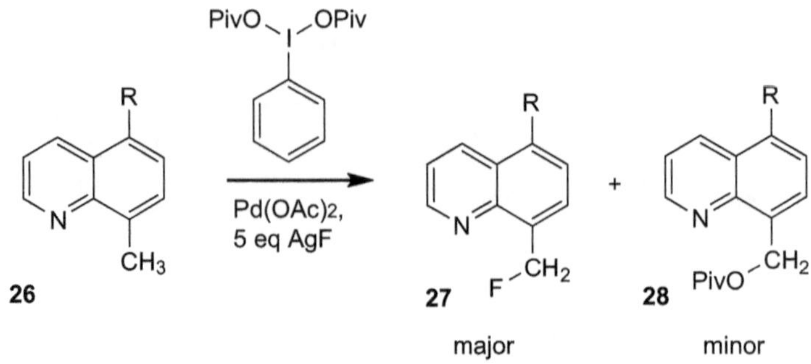

Scheme 4.8 Fluorination of C(sp^3)–H bond donors using nucleophilic source of fluorine.

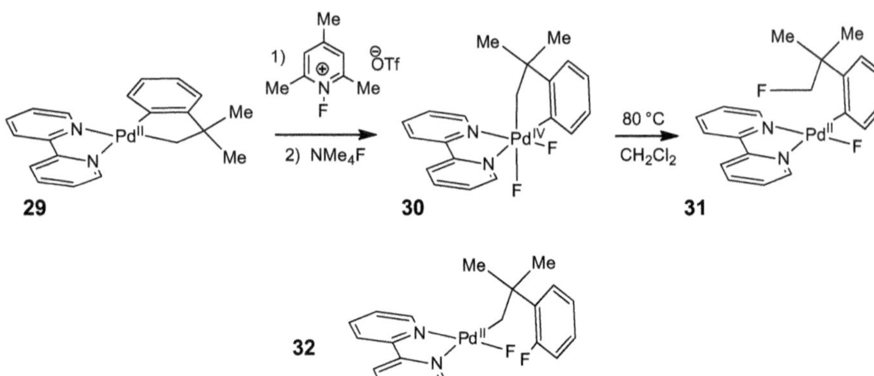

Scheme 4.9 Reductive elimination of C(sp^3)–F bond from model fluoro alkylpalladium(IV) complex **30**.

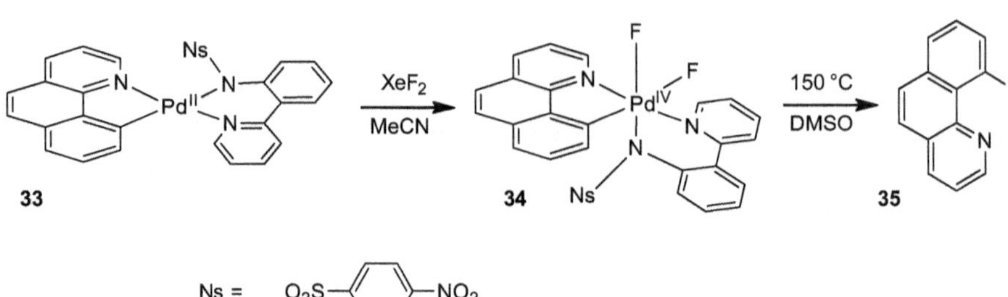

Scheme 4.10 C(sp^2)–F bond elimination of a palladium(IV) aryl fluoro complex **34**.

reaction product, an isomeric fluoroarylalkyl complex **32**, was not detected in reaction mixtures.

Clean elimination of a C(sp^2)–F bond from a palladium(IV) center was demonstrated using a model aryldifluoropalladium(IV) complex **34** (Scheme 4.10).[27]

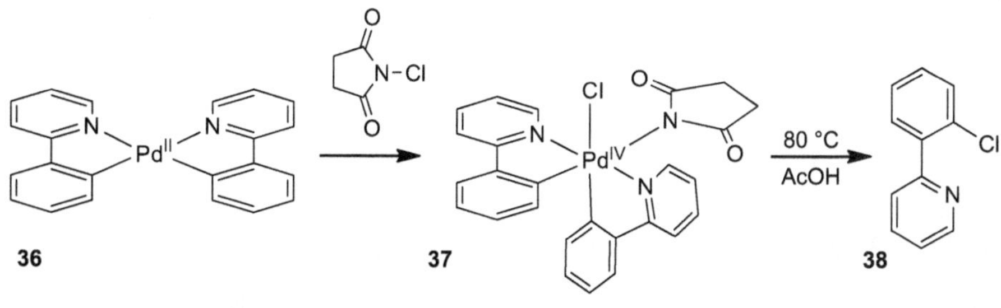

Scheme 4.11 Reductive elimination of C(sp^2)–Cl bond of diarylchloropalladium(IV) complex **37**.

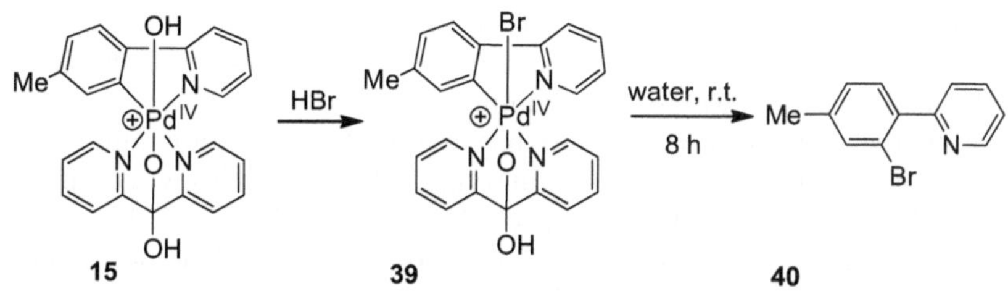

Scheme 4.12 Elimination of C(sp^2)–Br bond from an arylpalladium(IV) bromide **39**.

The latter was produced by oxidation of arylpalladium(II) precursor **33** with a strong electrophilic fluorinating agent, xenon difluoride. The elimination reaction of **34** leading to arylfluoride **35** is high-yielding.

Palladium-catalyzed chlorination of C–H bonds was one of the reactions where intermediacy of organopalladium(IV) complexes was studied in detail[29] and selective elimination of C(sp^2)–Cl bond from model palladium(IV) species is well characterized.

One of the early model studies used diarylchloroimido complex **37**.[26] Arylchloride **38** was solely produced in this elimination reaction (Scheme 4.11) while products of potential C–N bond elimination were not detected. Oxidation of diarylpalladium(II) precursor **36** to form palladium(IV) species **37** was performed using *N*-chlorosuccinimide.

Catalytic bromination of C–H bond donors mediated by palladium complexes was also developed as a convenient practical tool.[29] At the same time, a limited number of model studies of C(sp^2)–Br elimination from palladium(IV) center exist. An example of C(sp^2)–Br bond elimination from an arylpalladium(IV) bromide is given in Scheme 4.12.[6] High-yielding elimination of corresponding arylbromide **40** from palladium(IV) species **39** is facile in aqueous solutions at 20 °C. The bromopalladium(IV) complex **39** could be produced in two steps from palladium(II) precursor **15** (Scheme 4.6) by its oxidation with dilute aqueous hydrogen peroxide with subsequent treatment of the resulting hydroxopalladium(IV) derivative **16** with hydrogen bromide.

Though palladium-catalyzed C(sp^2)–H and C(sp^3)–H iodination reactions have been demonstrated previously using such iodinating reagents as *N*-iodosuccinimide[29] and $I_2 + PhI(OAc)_2$,[30] the nature of the catalytically active species was not revealed and no model iodopalladium(IV) complexes have been studied in C–I reductive elimination reactions.

4.2.4 Formation of C–S and C–Se Bonds

Some early examples of dimethylpalladium(IV) complexes bearing two phenylsulfido or two diphenylselenido ligands and supported by 2,2′-bipyridine and 1,10-phenathroline were reported in the end of 1990s.[31] These reactive compounds exhibited complex reaction behavior including unselective elimination of several organic products with C–C and C(sp^3)–S or C(sp^3)–Se bonds.

Oxidative sulfonylation of C(sp^2–H) bond donors with arylsulfonyl chlorides and palladium(II) complexes as catalysts is one of the rare examples of synthetically useful oxidative palladium-catalyzed C–S bond formation.[32] Recently the same research group has reported preparation and reductive elimination reactivity of a series of diarylpalladium(IV) sulfonyl complexes **42** (Scheme 4.13; R = CF_3, $PhCH_2$, *p*-MeC_6H_4, *p*-$MeOC_6H_4$, *p*-FC_6H_4).[33]

The C–S elimination reaction of **41** can produce arylsulfones **43** in 50–70% yield with a major by-product resulting from C–C coupling of two aryl ligands. Mechanistic tests performed here that involved the use of solvents of various polarities and additives of silver(I) salts showed that the C–S elimination reaction of **42** proceeds most likely *via* an intermediate five-coordinate palladium(IV) transient resulting from dissociation of the Pd^{IV}–Cl bond.

4.2.5 Formation of C–N Bonds

Palladium-catalyzed oxidative amination of C–H bond donors leading to formation of a new C(sp^2)–N bond is another relatively rare type of oxidative palladium-catalyzed reactions. Some of these reactions were proposed to occur *via* intermediate amidopalladium(IV) species. Both intramolecular[34] (Scheme 4.14) and intermolecular[35] versions are known. In the example shown in Scheme 4.14, *ortho*-amino-substituted biphenyl **44** is converted to a

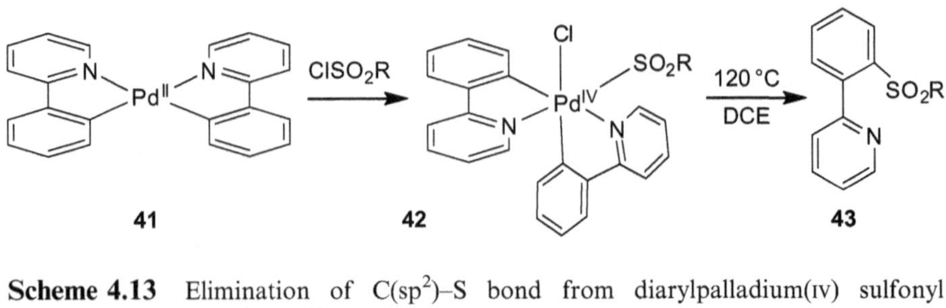

Scheme 4.13 Elimination of C(sp^2)–S bond from diarylpalladium(IV) sulfonyl complexes **42**.

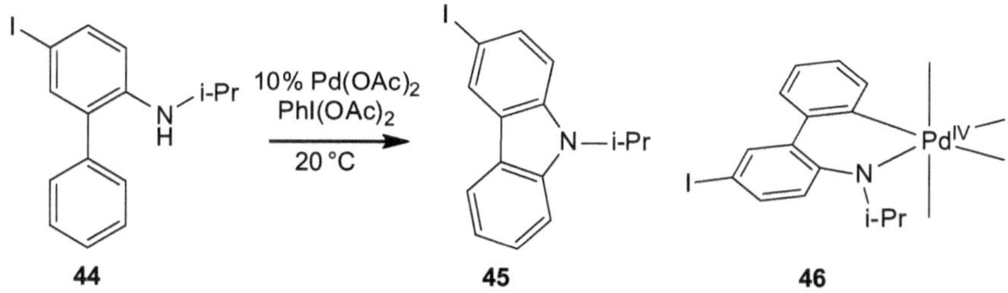

Scheme 4.14 Intramolecular oxidative C–H amination presumably mediated by palladium(IV) species **46**.

carbazole **45** *via* a presumed amidoarylpalladium(IV) intermediate **46**.[36] Similarly, using *N*-fluoro-2,4,6-trimethylpyridinium triflate as oxidant, indolines can be prepared.[36]

Currently there are no studies detailing C–N bond elimination reactivity of organopalladium(IV) complexes.

4.3 Summary and Outlook

The high-valent organopalladium(IV) chemistry described above is an already mature field that has found many practical applications in catalytic oxidative functionalization of various organic substrates. A number of important organic functional groups can be introduced into an organic molecule using C–X reductive elimination from a palladium(IV) center (Scheme 4.2ii). A great variety of oxidants can be used to convert low-valent organopalladium species into their high-valent derivatives (Scheme 4.2i). Though our analysis of oxidative functionalization reactions was limited mostly to the cases where step iii in Scheme 4.2 is a C–H activation at a palladium(II) center, many other methods exist that allow creation of a new Pd–C bond and introduce a hydrocarbyl ligand into the palladium(II) coordination sphere. This fact makes the importance and utility of the oxidative palladium-catalyzed reactions described in this chapter even greater. Appropriate information discussing the use of olefins, alkynes, hydrocarbyl halides and other organic substrates in palladium-catalyzed oxidative functionalization can be found in some more current reviews.[4,5,14,15] One can imagine that in some near future an even greater variety of C–X elimination reactions from palladium(IV) center will be disclosed and, no less importantly, less expensive and environmentally more benign reagents could be used to achieve conversion of organopalladium(II) species to their high-valent organopalladium derivatives (step i, Scheme 4.2). The latter achievement may become possible as we improve our currently very limited understanding of mechanisms of oxidation of organopalladium(II) complexes. Last but not least, the renaissance of high-valent organopalladium chemistry introduced first about forty years ago, which was boosted by new demanding practical applications of synthetic organic chemistry, might lead to exploration and extensive growth of high-valent organometallic chemistry of some other transition metals such as copper, silver and gold, in particular.[14]

References

1. A. J. Canty, *Acc. Chem. Res.*, 1992, **25**, 83.
2. J. R. Khusnutdinova, N. P. Rath and L. M. Mirica, *J. Am. Chem. Soc.*, 2010, **133**, 7303.
3. P. K. Byers, A. J. Canty, B. W. Skelton and A. H. White, *J. Chem. Soc., Chem. Commun.*, 1986, 1722.
4. T. W. Lyons and M. S. Sanford, *Chem. Rev.*, 2010, **110**, 1147.
5. P. Sehnal, R. J. K. Taylor and I. J. S. Fairlamb, *Chem. Rev.*, 2010, **110**, 824.
6. W. N. Oloo, P. Y. Zavalij, J. Zhang, E. Khaskin and A. N. Vedernikov, *J. Am. Chem. Soc.*, 2010, **133**, 14400.
7. J. R. Khusnutdinova, F. Qu, Y. Zhang, N. P. Rath and L. M. Mirica, *Organometallics*, 2012, **31**, 4627.
8. A. N. Vedernikov, *Acc. Chem. Res.*, 2012, **45**, 803.
9. D. C. Powers, E. Lee, A. Ariafard, M. S. Sanford, B. F. Yates, A. J. Canty and T. Ritter, *J. Am. Chem. Soc.*, 2012, **134**, 12002.
10. T. Yoneyama and R. H. Crabtree, *J. Mol. Catal., A*, 1996, **108**, 35.
11. K. J. Stowers and M. S. Sanford, *Org. Lett.*, 2009, **11**, 4584.
12. D. C. Powers, D. Y. Xiao, M. A. L. Geibel and T. Ritter, *J. Am. Chem. Soc.*, 2010, **133**, 14530.
13. (a) N. D. Ball, J. W. Kampf and M. S. Sanford, *J. Am. Chem. Soc.*, 2010, **133**, 2878; (b) Y. Ye, N. D. Ball, J. W. Kampf and M. S. Sanford, *J. Am. Chem. Soc.*, 2010, **133**, 14682.
14. A. J. Hickman and M. S. Sanford, *Nature*, 2012, **484**, 177.
15. K. Muniz, *Angew. Chem., Int. Ed.*, 2009, **48**, 9412.
16. A. J. Canty in *Handbook of Organopalladium Chemistry for Organic Synthesis*, ed. E.-i. Negishi, Wiley, New York, 2002, vol. 1, p. 189.
17. A. N. Vedernikov, in *Topics in Organometallic Chemistry*, ed. A. Vigalok, Springer, 2010, vol. 31, p. 101.
18. (a) D. Kalyani, N. R. Deprez, L. V. Desai and M. S. Sanford, *J. Am. Chem. Soc.*, 2005, **127**, 7330; (b) N. R. Deprez and M. S. Sanford, *Inorg. Chem.*, 2007, **46**, 1924.
19. K. L. Hull, E. L. Lanni and M. S. Sanford, *J. Am. Chem. Soc.*, 2006, **128**, 14047.
20. (a) J. M. Racowski, A. R. Dick and M. S. Sanford, *J. Am. Chem. Soc.*, 2009, **131**, 10974; (b) A. R. Dick, J. W. Kampf and M. S. Sanford, *J. Am. Chem. Soc.*, 2005, **127**, 12790.
21. (a) A. R. Dick, K. L. Hull and M. S. Sanford, *J. Am. Chem. Soc.*, 2004, **126**, 2300; (b) L. V. Desai, H. A. Malik and M. S. Sanford, *Org. Lett.*, 2006, **8**, 1141; (c) L. V. Desai, K. L. Hull and M. S. Sanford, *J. Am. Chem. Soc.*, 2004, **126**, 9542.

22. R. Giri, J. Liang, J. Q. Lei, J. J. Li, D. H. Wang, X. Chen, I. C. Naggar, C. Guo, B. M. Foxman and J. Q. Yu, *Angew. Chem., Int. Ed.*, 2005, **44**, 7420.
23. J. Zhang, E. Khaskin, N. P. Anderson and A. N. Vedernikov, *Chem. Commun.*, 2008, 3625.
24. (a) K. L. Hull, W. Q. Anani and M. S. Sanford, *J. Am. Chem. Soc.*, 2006, **128**, 7134; (b) X. Wang, T. S. Mei and J. Q. Yu, *J. Am. Chem. Soc.*, 2009, **131**, 7520.
25. K. B. McMurtrey, J. M. Racowski and M. S. Sanford, *Org. Lett.*, 2012, **14**, 4094.
26. S. R. Whitfield and M. S. Sanford, *J. Am. Chem. Soc.*, 2007, **129**, 15142.
27. T. Furuya and T. Ritter, *J. Am. Chem. Soc.*, 2008, **130**, 10060.
28. J. M. Racowski, J. B. Gary and M. S. Sanford, *Angew. Chem., Int. Ed.*, 2012, **51**, 3414.
29. D. Kalyani, A. R. Dick, W. Q. Anani and M. S. Sanford, *Tetrahedron*, 2006, **62**, 11483.
30. (a) R. Giri, X. Chen and J. Q. Yu, *Angew. Chem., Int. Ed.*, 2005, **44**, 2112; (b) R. Giri, X. Chen, X. S. Hao, J. J. Li, J. Liang, Z. P. Fan and J. Q. Yu, *Tetrahedron: Asymmetry*, 2005, **16**, 3502; (c) T. S. Mei, R. Giri, N. Maugel and J. Q. Yu, *Angew. Chem., Int. Ed.*, 2008, **47**, 5215.
31. A. J. Canty, H. Jin, B. W. Skelton and A. H. White, *Inorg. Chem.*, 1998, **37**, 3975.
32. X. Zhao, E. Dimitrijevic and V. M. Dong, *J. Am. Chem. Soc.*, 2009, **131**, 3466.
33. X. Zhao and V. M. Dong, *Angew. Chem., Int. Ed.*, 2011, **50**, 932.
34. J. A. Jordon-Hore, C. C. C. Johansson, M. Gulia, E. M. Beck and M. J. Gaunt, *J. Am. Chem. Soc.*, 2008, **130**, 16184.
35. H. Y. Thu, W. Y. Yu and C. M. Che, *J. Am. Chem. Soc.*, 2006, **128**, 9048.
36. T. S. Mei, X. Wang and J. Q. Yu, *J. Am. Chem. Soc.*, 2009, **131**, 10806.

CHAPTER 5

Organometallic Pd^{III} Complexes in C–C and C–Heteroatom Bond Formation Reactions

JULIA R. KHUSNUTDINOVA AND LIVIU M. MIRICA*

Department of Chemistry, Washington University, CB 1134, One Brookings Drive, St. Louis, MO 63130, USA
*Email: mirica@wustl.edu

5.1 Introduction

Palladium complexes are extensively employed as efficient catalysts for a wide range of synthetically useful organic transformations such as C–H functionalization, C–C coupling, and hydrocarbon oxidation reactions.[1–3] Whereas catalytic processes involving Pd complexes in oxidation states +2 and 0 have been recognized and studied for several decades,[2,4–7] more recent studies have shown that high-valent Pd species can also act as catalytically active intermediates in various organic transformations.[8–13] For example, the reactivity of isolated organometallic Pd^{IV} complexes has been studied for the past two decades,[14–23] and Pd^{IV} intermediates have been shown to be involved in a variety of functional group-directed C–H functionalization reactions (see Chapter 4 of this book).[8–13] By contrast, much less is known about the organometallic chemistry of Pd^{III} complexes and only a limited number of isolated mononuclear and dinuclear organometallic Pd^{III} complexes have been reported to date.[24,25] Moreover, dinuclear Pd^{III} complexes have been recently proposed as reactive intermediates in oxidative C–H functionalization reactions,[24,26–29]

RSC Catalysis Series No. 11
C–H and C–X Bond Functionalization: Transition Metal Mediation
Edited by Xavi Ribas

Published by the Royal Society of Chemistry, www.rsc.org

while mononuclear Pd^{III} intermediates have been proposed in oxidatively induced C–C bond formation, O_2 activation or insertion, radical Kumada coupling and other reactions.[24,25]

Herein, we overview the synthesis and reactivity of isolated mononuclear and dinuclear Pd^{III} complexes and describe the various C–C and C–heteroatom bond formation reactions in which the intermediacy of Pd^{III} species has been demonstrated or invoked.

5.2 Mononuclear Pd^{III} Complexes

5.2.1 Electronic and Structural Properties of Pd^{III} Coordination Complexes

The electronic configuration of Pd in oxidation states +2, +3, and +4 is d^8, d^7, and d^6, respectively (Figure 5.1). Pd^{II} complexes are typically square planar, while the paramagnetic Pd^{III} d^7 centers prefer a Jahn–Teller distorted octahedral geometry and the diamagnetic Pd^{IV} d^6 centers exhibit an octahedral geometry.

While the first confirmed Pd^{III} species was the ionic compound $NaPdF_4$ obtained in 1982 in the solid state,[30] the first structurally characterized molecular Pd^{III} coordination compounds were the mononuclear homoleptic complexes of type $[L_2Pd^{III}]^{3+}$ supported by the tridentate macrocyclic ligands 1,4,7-trithiacyclononane and 1,4,7-triazacyclononane (Scheme 5.1).[31,32] These complexes were obtained by electrochemical or chemical oxidation with $HClO_4$ and their reactivity in one-electron redox processes was investigated.[31–33] EPR characterization suggests that the unpaired electron resides predominantly at the metal center,[32,33] although it was shown recently that the sulfur atoms of the macrocyclic ligands may contribute significantly to the singly occupied molecular orbital of the Pd^{III} complex.[34] In addition, the X-ray structures reveal a tetragonally distorted octahedral geometry, consistent with a Jahn–Teller distorted Pd^{III} d^7 center.[31,32]

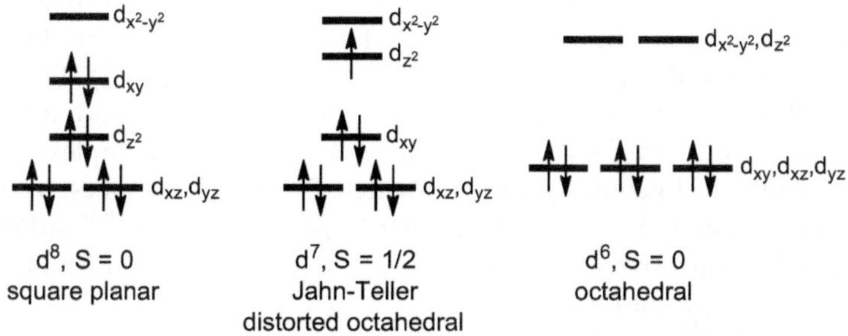

Figure 5.1 Representative d electron configuration of Pd^{II}, Pd^{III}, and Pd^{IV} oxidation states.

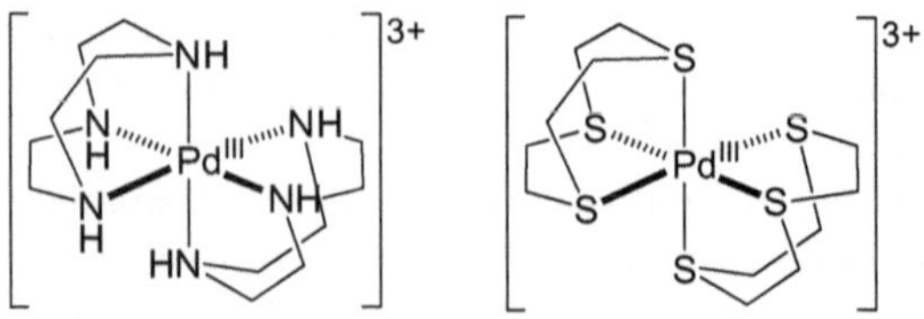

1 2

Scheme 5.1 Pd^{III} coordination compounds supported by N- or S-donor tridentate macrocyclic ligands.

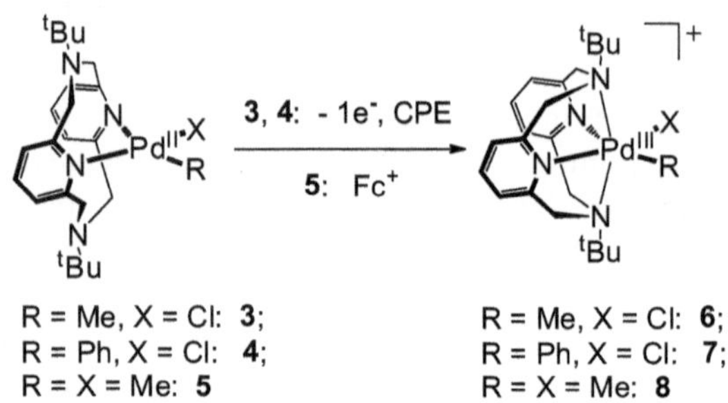

Scheme 5.2 Synthesis of organometallic (^{tBu}N4)Pd^{III} complexes (CPE – controlled potential electrolysis; Fc^{+}-ferrocenium).

5.2.2 Reactivity of Isolated Organometallic Pd^{III} Complexes

Although the intermediacy of mononuclear Pd^{III} organometallic species was proposed in a number of C–C and C–heteroatom bond formation reactions, the first isolated mononuclear organometallic Pd^{III} complexes were reported by Mirica *et al.* in 2010.[35] The tetradentate macrocyclic ligand *N*,*N′*-di-*tert*-butyl-2,11-diaza[3.3](2,6)pyridinophane (^{tBu}N4) was employed in order to stabilize the distorted octahedral geometry of a Pd^{III} d^{7} center. The complexes [(^{tBu}N4)Pd^{III}RX]$^{+}$ (**6**: R = Me, X = Cl; **7**: R = Ph, X = Cl; **8**: R = X = Me) were obtained by controlled potential electrolysis (CPE) or chemical oxidation with ferrocenium (Fc^{+}) of the corresponding Pd^{II} precursors (^{tBu}N4)Pd^{II}RX (**3**: R = Me, X = Cl; **4**: R = Ph, X = Cl; **5**: R = X = Me) (Scheme 5.2). The resulting Pd^{III} complexes were stable in the absence of light in the solid state and in solution for several weeks. The structures of these Pd^{III} complexes reveal a distorted octahedral geometry with elongated Pd–N$_{axial}$ distances, as expected for a d^{7} Pd^{III} center. In addition, the EPR spectra of the [(^{tBu}N4)Pd^{III}RX]$^{+}$ species exhibit anisotropic signals with g_{ave} = 2.12–2.13, confirming a Pd^{III} center with a d$_{z2}$ ground state.

The isolated Pd^{III} complexes are stable in the dark, however they exhibit C–C (and C–Cl) bond formation reactivity in the presence of visible light.[35] For example, when a solution of [(^{tBu}N4)Pd^{III}MeCl]BF$_4$ in MeCN is exposed to visible light, formation of ethane (25% yield), MeCl (8%), and methane (9%) occurs, along with the Pd^{II} product $[(^{tBu}N4)Pd^{II}Cl(MeCN)]^+$ (80%, Scheme 5.3a). Reactivity studies suggest a radical mechanism for the observed light-induced reactions. For example, the formation of ethane from $[(^{tBu}N4)Pd^{III}MeCl]BF_4$ is completely suppressed in the presence of the radical trap TEMPO (TEMPO = (2,2,6,6-tetramethylpiperidin-1-yl)oxyl), and the TEMPO–Me adduct forms as the major product (Scheme 5.3a). Notably, O_2 was also reported to act as an efficient Me radical trap and irradiation of $[(^{tBu}N4)Pd^{III}MeCl]BF_4$ in the presence of O_2 generates a number of oxygenated products such as methanol, formaldehyde, and formic acid (Scheme 5.3a). Similar C–C bond formation photoreactivity was observed for $[(^{tBu}N4)Pd^{III}PhCl]^+$ to generate biphenyl in 21% yield along with benzene (13%) and $[(^{tBu}N4)Pd^{II}Cl(MeCN)]^+$ (82%) (Scheme 5.3b). Interestingly, while biaryl elimination from monoaryl Pd compexes is known (*vide infra*), the elimination of ethane from a monomethyl Pd complex has not been reported previously and is currently of great interest due to its relevance to catalytic methane oligomerization.[36]

In addition, irradiation of a solution of $[(^{tBu}N4)Pd^{III}Me_2]^+$ generated ethane and the monomethyl complex $[(^{tBu}N4)Pd^{II}Me(MeCN)]^+$ in 41% and 83% yield, respectively (Scheme 5.3b). Such a reactivity resembles the Me group transfer and ethane elimination reactions induced by chemical oxidation of $PdMe_2$ complexes in which Pd^{III} intermediates were proposed but not detected (*vide infra*),[37,38] thus confirming that Pd^{III} complexes are viable intermediates in such processes. Interestingly, the Pd^{II} dimethyl complex $(^{tBu}N4)Pd^{II}Me_2$ (**5**) also exhibits a non-radical ethane elimination induced by oxidation with O_2 or peroxides through the intermediate formation of a $[(^{tBu}N4)Pd^{III}Me_2]^+$ complex **8**.[39] The oxidation of $(^{tBu}N4)PdMe_2$ with O_2 in C_6H_6-MeOH leads

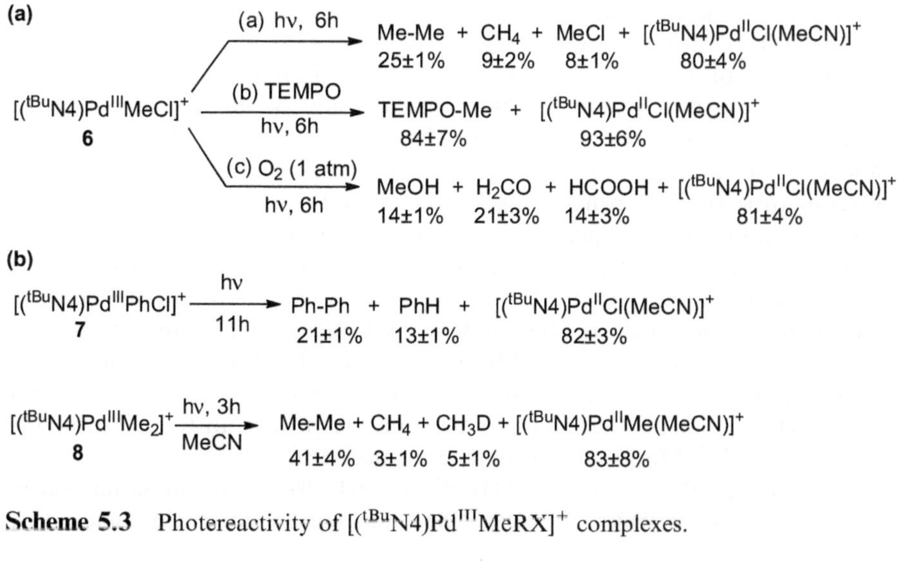

Scheme 5.3 Photereactivity of $[(^{tBu}N4)Pd^{III}MeRX]^+$ complexes.

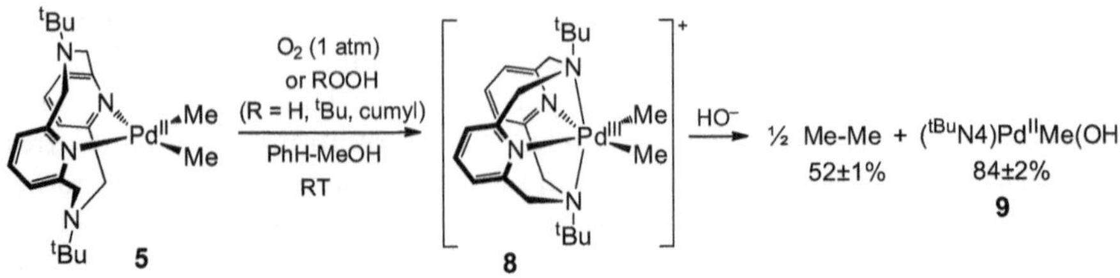

Scheme 5.4 Aerobic reactivity of (^{tBu}N4)PdMe$_2$ (**5**).

to the initial formation of [(^{tBu}N4)Pd^{III}Me$_2$]$^+$, as detected by UV-vis and EPR, followed by elimination of ethane and (^{tBu}N4)Pd^{II}Me(OH) (**9**) in 52% and 84% yield, respectively (Scheme 5.4). The presence of protic solvents such as alcohols or water was required for oxidation, as no reaction was observed in anhydrous aprotic solvents. Similar reactivity was also observed when (^{tBu}N4)Pd^{II}Me$_2$ was oxidized with H_2O_2 or ROOH (R = ^{t}Bu, cumyl).[39]

Use of the spin trap 5,5-dimethyl-1-pyrroline-*N*-oxide (DMPO) and cyclic voltammetry (CV) studies suggest an inner-sphere mechanism for the aerobic oxidation and formation of a Pd^{III}-superoxide transient intermediate (Scheme 5.5a). Therefore, one-electron oxidation and formation of a Pd^{III} center is likely a pre-requisite for the aerobic oxidation of the Pd^{II} complex. Similar to the mechanism proposed for the aerobic oxidation of Pt^{II} dimethyl complexes,[40–42] protonation of the Pd^{III}-superoxide intermediate generates a transient Pd^{IV}-hydroperoxo and/or Pd^{IV}-hydroxo species, and a subsequent comproportionation in the presence of the Pd^{II}Me$_2$ complex yields a Pd^{III}Me$_2$ species. In addition, mechanistic crossover reactivity studies (Scheme 5.6) are consistent with a mechanism involving a Me group transfer and formation of a [(^{tBu}N4)Pd^{IV}Me$_3$]$^+$ species as the key intermediate responsible for the C–C bond formation step (Scheme 5.5b), while a radical mechanism involving formation of Me radicals is not supported by radical trapping studies. The proposed mechanism is strongly supported by the use of a less sterically hindered N–Me substituted analog of the ^{tBu}N4 ligand, *N*,*N*′-dimethyl-2,11-diaza[3.3](2,6)pyridinophane (^{Me}N4), which leads to the detection of the proposed intermediates [(^{Me}N4)Pd^{IV}Me$_2$(OOH)]$^+$, [(^{Me}N4)Pd^{IV}Me$_2$(OH)]$^+$, and [(^{Me}N4)Pd^{IV}Me$_3$]$^+$ by ESI mass spectrometry or NMR spectroscopy.[43] Overall, these studies strongly suggest that both Pd^{III} and Pd^{IV} intermediates are involved in the observed aerobic C–C bond formation reactivity.

The isolated [(^{tBu}N4)Pd^{III}Me$_2$]ClO$_4$ complex **8**[ClO$_4$] also undergoes C–C bond formation reactivity in the absence of light upon addition of 1 equiv. of coordinating ions such as OH$^-$ or CN$^-$ (Scheme 5.7). This reactivity is not affected by the radical trap TEMPO, suggesting that the formation of a Me radical is not involved in ethane elimination under these conditions, while crossover experiments (Scheme 5.6) suggest a Me group transfer and formation of a [(^{tBu}N4)Pd^{IV}Me$_3$]$^+$ intermediate (Scheme 5.5b).[39]

Interestingly, the complex (^{tBu}N4)Pd^{II}Me(OH) (**9**), formed upon the aerobic ethane elimination from (^{tBu}N4)Pd^{II}Me$_2$, can react with weakly acidic C–H

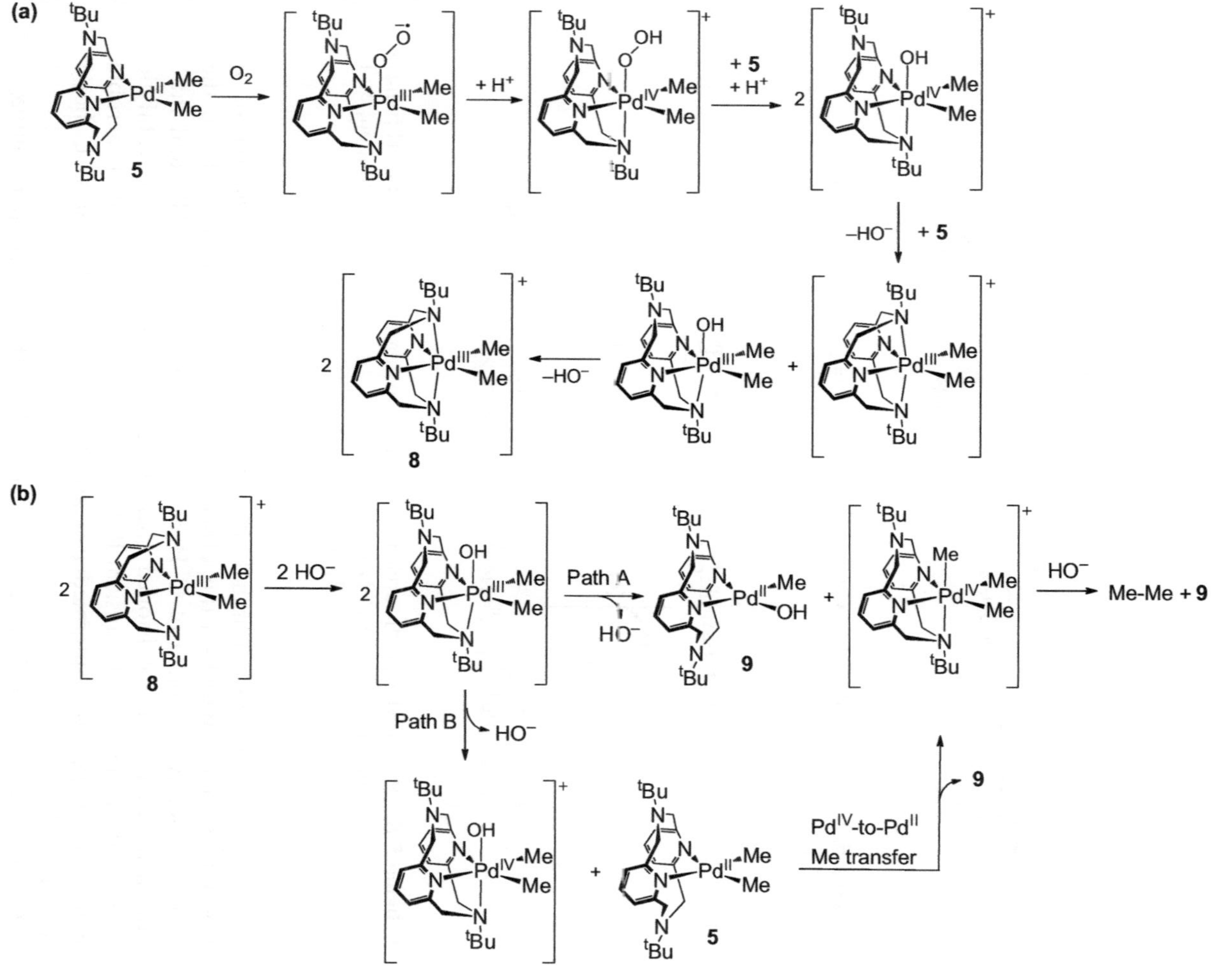

Scheme 5.5 Proposed mechanisms of (a) aerobic oxidation of $(^{tBu}N4)PdMe_2$ and (b) subsequent C–C bond formation.

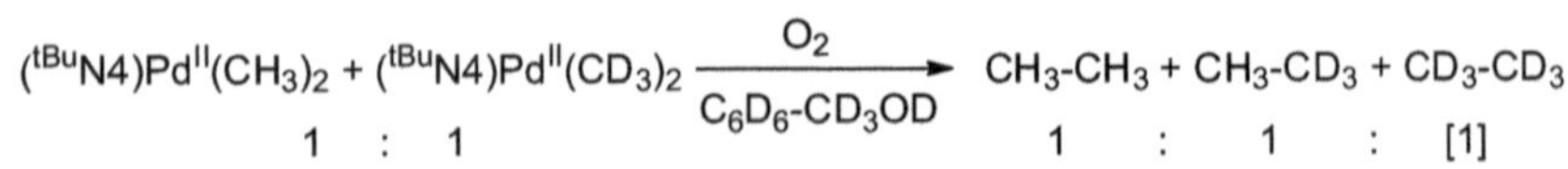

Scheme 5.6 Crossover reactivity of $(^{tBu}N4)Pd^{II}Me_2$ and $[(^{tBu}N4)Pd^{III}Me_2]^+$.

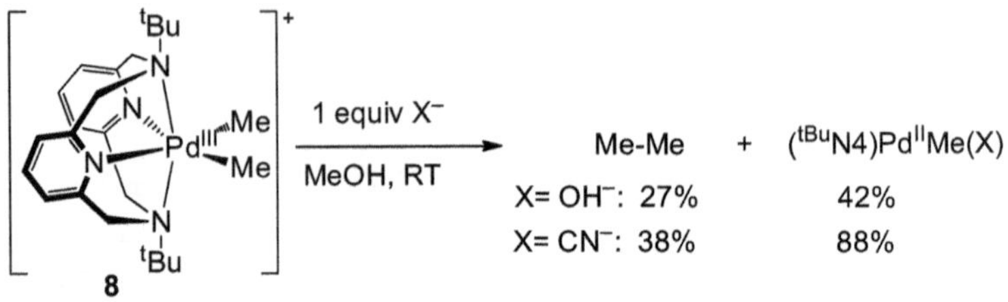

Scheme 5.7 Anion-induced C–C bond formation from $[(^{tBu}N4)Pd^{III}Me_2]^+$.

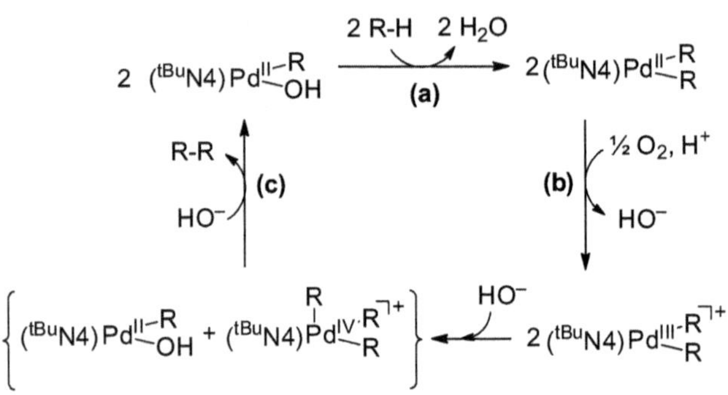

Scheme 5.8 Proposed catalytic cycle for the aerobic oxidative coupling of C–H bonds that involves (a) C–H bond activation, (b) aerobic Pd^{II} oxidation, and (c) C–C bond formation.

bonds of acetone and terminal alkynes leading to the formation of a new Pd^{II}–C bond. Based on this observed reactivity, a catalytic cycle for the aerobic oxidative coupling of C–H bonds that involves both Pd^{III} and Pd^{IV} intermediates can be envisioned (Scheme 5.8). In this proposed catalytic process, the formation of transient Pd^{III} species is expected to promote aerobic oxidation, while Pd^{IV} intermediates are necessary for efficient C–C bond formation.[39]

Overall, these results show that isolated Pd^{III} methyl and phenyl complexes exhibit C–C bond formation reactivity, and the mechanisms of these transformations may involve either radical or non-radical pathways for the C–C bond elimination step. While the radical pathway for the C–C bond formation proceeds through an initial photoinduced Pd^{III}–C bond homolysis, the non-radical C–C elimination most likely involves the formation of Pd^{IV} intermediates.

5.2.3 Proposed Pd^{III} Intermediates in C–C Bond Formation Reactions

5.2.3.1 Proposed Pd^{III} Intermediates in C_{sp3}–C_{sp3} Coupling

The oxidatively induced C–C bond formation from dialkyl Pd^{II} complexes with phosphine ligands has been studied by Trogler and Seligson.[37] Oxidation of (dmpe)Pd(CH_2SiMe_3)$_2$ (dmpe = 1,2-bis(dimethylphosphino)ethane) complex **10** with Fc^+ in the presence of 1 equiv. PPh_3 gave a mixture of Si(CH_3)$_4$, $Me_3SiCH_2CH_2SiMe_3$, and the Pd^{II} monoalkyl product [(dmpe)Pd^{II}(CH_2SiMe_3)(PPh_3)]$^+$ (Scheme 5.9). A mechanism involving a Me_3SiCH_2 radical was proposed, as the oxidation of $MgCH_2SiMe_3$ with Fc^+ produced Si(CH_3)$_4$ and $Me_3SiCH_2CH_2SiMe_3$ in a ~1 : 1 ratio. Interestingly, a more selective C–C bond formation was observed in the absence of PPh_3, to give $Me_3SiCH_2CH_2SiMe_3$ as the major product accompanied by complete decomposition of a metal complex. The oxidation of the analogous dimethyl complex (dmpe)PdMe_2 was reported to produce ethane.[37]

The selective ethane formation from the (^{t}Bu$_2$bipy)Pd^{II}Me$_2$ complex **11** induced by one-electron oxidation has been recently reported by Sanford, Mayer, *et al.*[38] Oxidation of (^{t}Bu$_2$bipy)Pd^{II}Me$_2$ with one-electron oxidants such as Fc^+, the thianthrenyl radical cation, or Ag^+ was shown to produce ethane in ~50% yield and the monomethyl product [(^{t}Bu$_2$bipy)Pd^{II}Me(acetone)]$^+$ quantitatively (Scheme 5.10a). Several mechanisms were considered for the C–C bond formation (Scheme 5.10b), all involving the initial one-electron oxidation of a Pd^{II}Me$_2$ complex to form an undetected Pd^{III}Me$_2$ intermediate. In path (i), subsequent Pd–Me bond homolysis generates a Me radical and ethane is produced by combination of two Me radicals. In path (ii), a Me group transfer occurs upon reaction of the Pd^{III}Me$_2$ species with **11** and 1 equiv. of Fc^+ to generate [(^{t}Bu$_2$bipy)Pd^{IV}Me$_3$]$^+$ and [(^{t}Bu$_2$bipy)Pd^{II}Me(solv)]$^+$, followed by reductive elimination from the former species. Path (iii) involves direct C–C reductive elimination from a Pd^{III} center to afford ethane and a Pd^I product. Path (i) was ruled out since the yield of ethane remains almost unchanged in the presence of Me radical traps 1,4-cyclohexadiene or styrene. Moreover, low temperature NMR studies and crossover experiments suggest that path (ii) is operative. Oxidation of (^{t}Bu$_2$bipy)Pd^{II}Me$_2$ with Fc^+ at –80 °C generates a new species assigned as the [(^{t}Bu$_2$bipy)Pd^{IV}Me$_3$(solv)]$^+$ species along

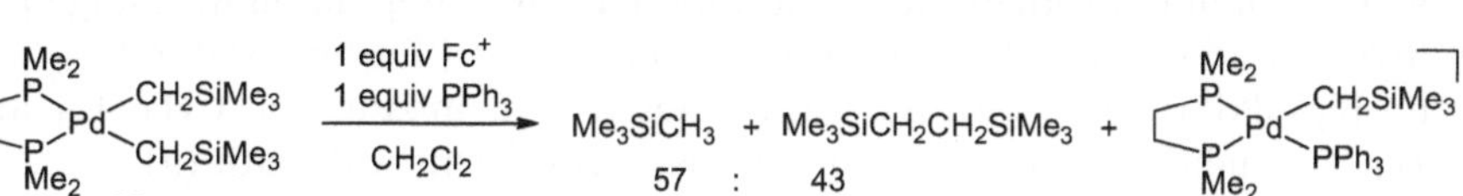

Scheme 5.9 Oxidatively induced C–C bond formation from (dmpe)Pd^{II}(CH_2SiMe_3)$_2$.

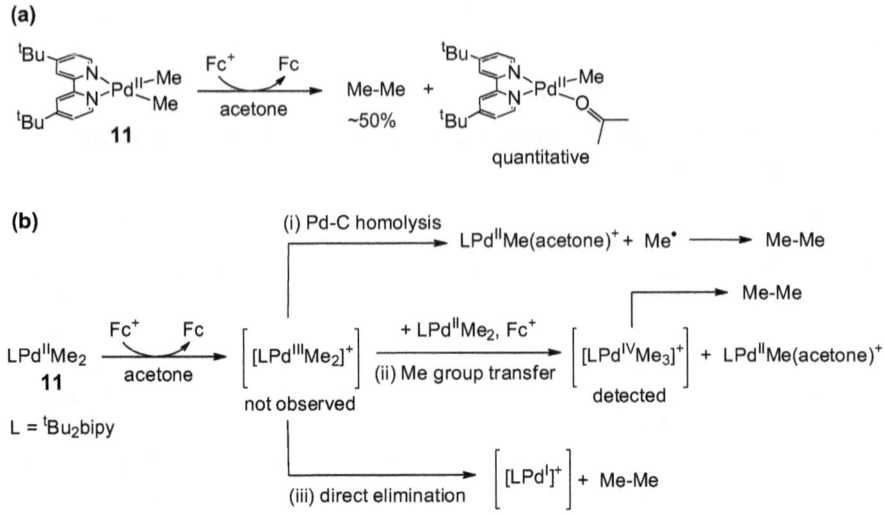

Scheme 5.10 (a) Oxidatively induced ethane elimination from (^{t}Bu$_2$bipy)PdMe$_2$ and (b) mechanisms considered for ethane formation.

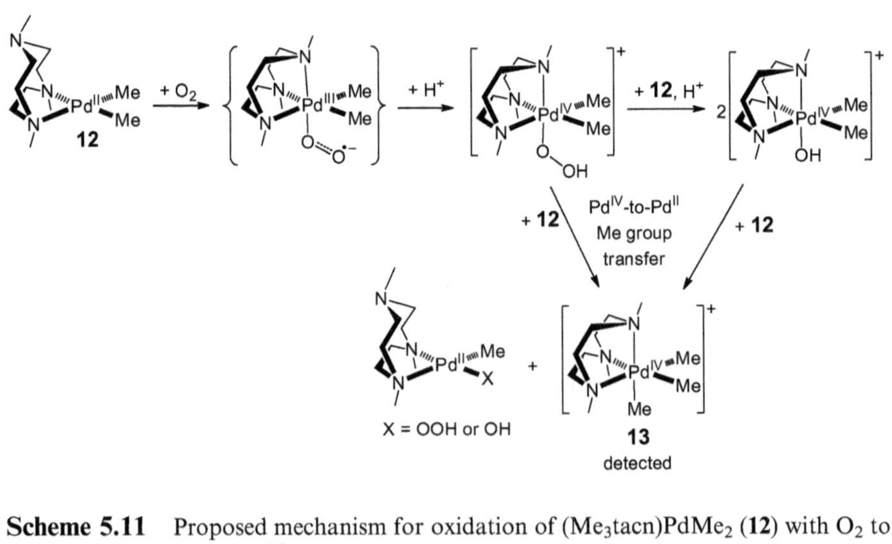

Scheme 5.11 Proposed mechanism for oxidation of (Me$_3$tacn)PdMe$_2$ (**12**) with O$_2$ to generate **13**.

with [(^{t}Bu$_2$bipy)Pd^{II}Me(solv)]$^+$. While [(^{t}Bu$_2$bipy)Pd^{IV}Me$_3$(solv)]$^+$ is stable at –80 °C, it eliminates ethane and [(^{t}Bu$_2$bipy)Pd^{II}Me(solv)]$^+$ upon warming up to –30 °C. In addition, oxidation of a 1 : 1 mixture of (^{t}Bu$_2$bipy)Pd^{II}(CH$_3$)$_2$ and (^{t}Bu$_2$bipy)Pd^{II}(CD$_3$)$_2$ with Fc$^+$ generates a 1 : 2 mixture of CH$_3$CH$_3$ and CH$_3$CD$_3$, indicative of a Me group transfer consistent with path (ii).

The transient formation of a Pd^{III} intermediate was also recently proposed by Mirica *et al.* during the aerobic oxidation of (Me$_3$tacn)Pd^{II}Me$_2$, **12** (Me$_3$tacn = *N*,*N*′,*N*″-trimethyl-1,4,7-triazacyclononane, Scheme 5.11).[44] The inner-sphere oxidation of the Pd^{II} center by O_2 is proposed to form a Pd^{III}–$O_2^{\bullet-}$ species – as suggested by detection of an EPR-active radical adduct trapped by DMPO, followed by further oxidation of the metal center to form detected Pd^{IV}–OOH and Pd^{IV}–OH intermediates and ultimately the $[(Me_3tacn)Pd^{IV}Me_3]^+$ species **13**, which eliminates ethane at elevated temperatures.[44]

5.2.3.2 Pd^{III} Intermediates in Oxidative Transformations of Pd^{II} Aryl Complexes

The mechanism of aryl–aryl coupling of Pd aryl complexes in presence of aryl iodides or one-electron oxidants has been studied by Milstein *et al.*[45] The reaction of (PCP)Pd^{II}(Ph) (PCP = 2,6-($^iPr_2PCH_2)_2C_6H_3$) (**14**) with aryl iodides results in the formation of (PCP)Pd^{II}(I) and a biaryl product. The use of substituted aryl iodides results in the formation of a mixed biaryl and homo-coupled products in a nearly statistical ratio (Scheme 5.12a), suggesting a radical mechanism involving an initial one-electron transfer from the electron-rich Pd phenyl complex **14** to the aryl iodide to form $[(PCP)PdPh]^{\bullet+}$ (**15**), which undergoes Pd–C bond homolysis to generate Ph radicals (Scheme 5.12b). Fragmentation of aryl iodide radical anion produces aryl radical and iodide,

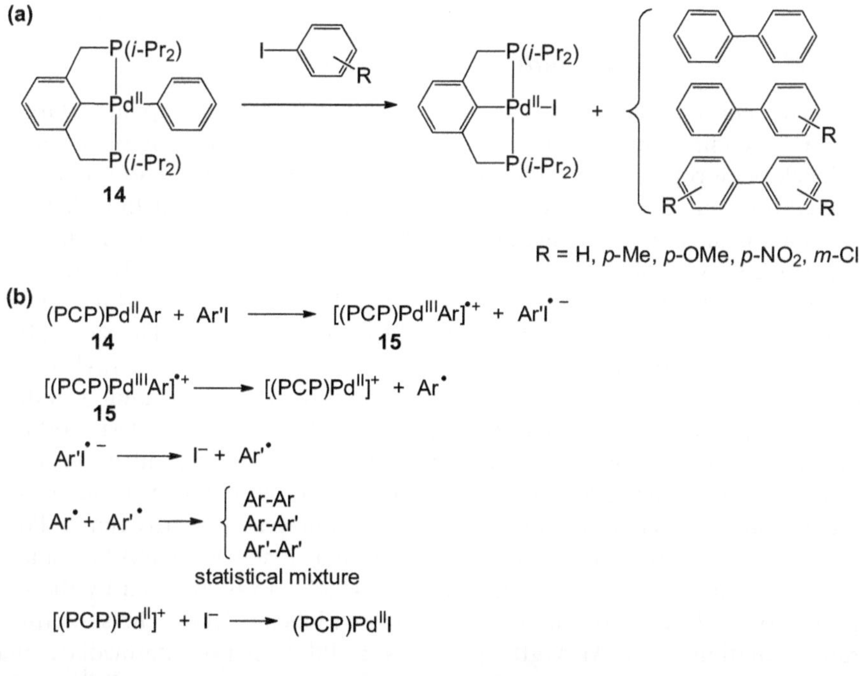

Scheme 5.12 (a) Biaryl formation upon reaction of (PCP)PdPh with aryl iodides and (b) proposed radical mechanism.

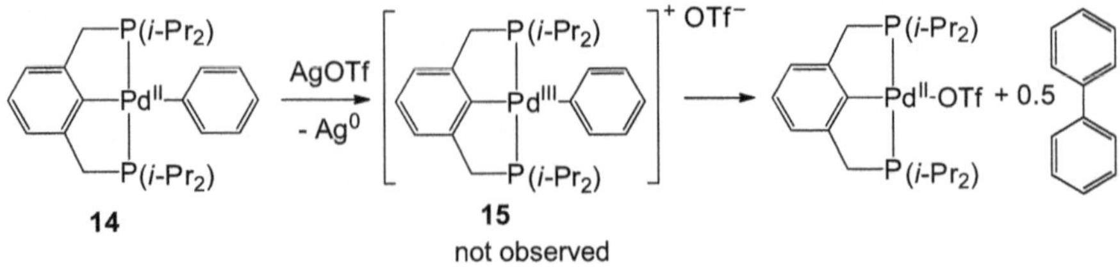

Scheme 5.13 One-electron oxidation of (PCP)PdPh with Ag^+ and Ph–Ph bond formation.

and the combination of Ph and aryl radicals then furnishes a statistical mixture of biaryls. A similar mechanism has been proposed for Ni-catalyzed aryl halide coupling.[46,47]

Complex **14** also reacts with one-electron oxidants such as galvinoxyl and AgOTf, and formation of a transient Pd^{III} species $[(PCP)Pd^{III}Ph]^+$ was proposed (Scheme 5.13).[45] As no products resulting from the reaction of a phenyl radical with the deuterated solvent was observed, the authors suggested the formation of aryl-bridged intermediates prior to C–C bond formation. The possibility of such bridged intermediates was also considered for the reactions involving aryl iodide as an oxidant (*vide supra*).

5.2.3.3 Pd^{III} Intermediates in Kumada Coupling

A recent report by Knochel *et al.* indicated that Kumada coupling catalyzed by Pd compounds such as 1,3-bis(2,6-diisopropylphenyl)imidazol-2-ylidene](3-chloropyridyl)palladium(II) dichloride (**16**) shows a dramatic rate acceleration in the presence of ^{i}PrI.[48] While a slow reaction between *p*-bromoanisole and phenylmagnesium chloride that was prepared from PhCl and Mg/LiCl was observed, significantly faster reactions were obtained with PhMgCl prepared by I/Mg exchange from PhI and ^{i}PrMgCl·LiCl, which generates ^{i}PrI as a stoichiometric byproduct (Scheme 5.14a). The same effect was observed in the presence of other alkyl iodides such as methyl iodide, neopentyl iodide, 1-iodoadamantane, or cyclohexyl iodide, and a radical mechanism was proposed. Indeed, use of a radical clock reaction test produced a mixture of the coupled and the cyclized product, the latter likely resulting from the ring closure between the aryl radical and an olefin (Scheme 5.14b). The proposed mechanism involves an initiation reaction between a Pd(0) complex and ^{i}PrI to form a Pd^I intermediate that reacts with aryl bromide to generate a Pd^{II} species and an aryl radical, which can be trapped by the Pd^{II} species to produce a Pd^{III} monoaryl complex (Scheme 5.15). The subsequent transmetalation with Ar_2MgBr produces a Pd^{III} biaryl intermediate that reductively eliminates the coupling product and regenerates the Pd^I species. However, it remains to be seen whether Pd^{III} species can undergo direct

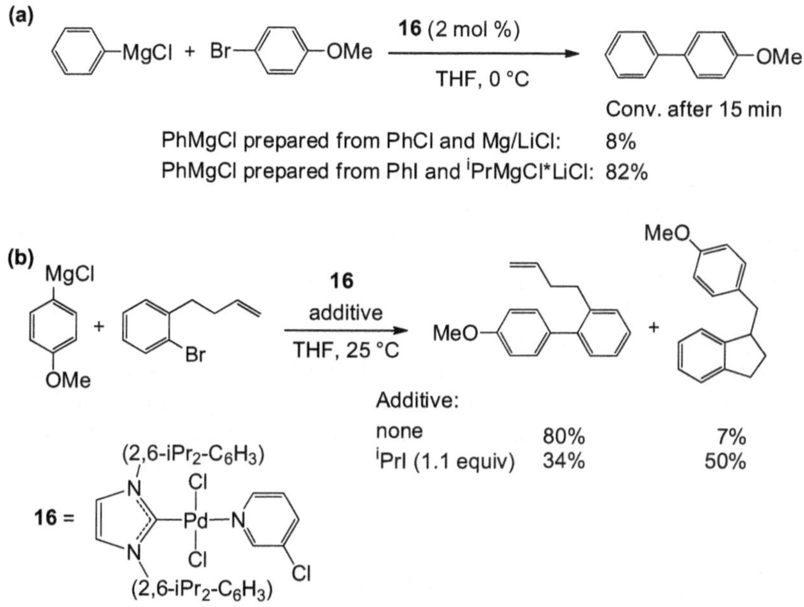

Scheme 5.14 (a) Effect of ^{i}PrI on Pd-catalyzed Kumada coupling and (b) radical clock test for the presence of aryl radicals.

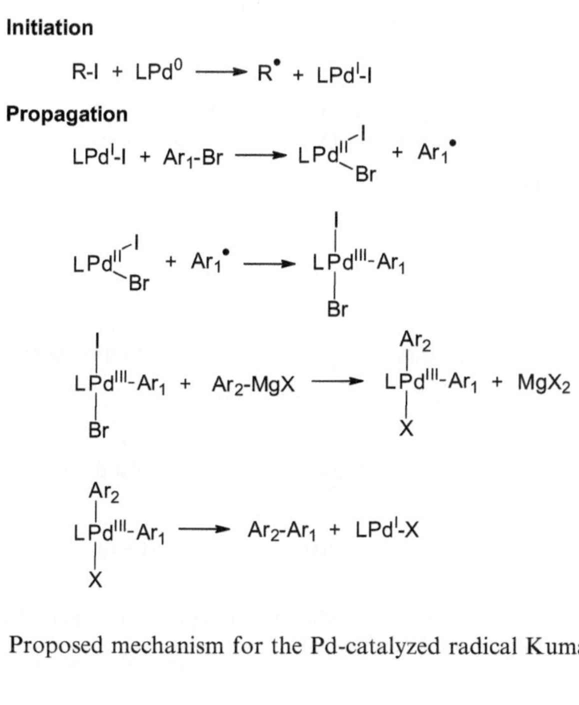

Scheme 5.15 Proposed mechanism for the Pd-catalyzed radical Kumada coupling.

reductive elimination to form C–C or C–heteroatom bond, as to date no isolated mononuclear Pd^{III} species were shown to exhibit such reactivity.

5.2.3.4 *Pd^{III} Intermediates in Homocoupling of Aryl Triflates*

The Pd^{III} intermediates were proposed in the electrosynthesis of binaphthyls by reductive coupling of naphthyl triflates catalyzed by Pd complexes with bidentate phosphine ligands (P-P)$PdCl_2$ (P-P = dppe, dppp; Scheme 5.16a).[49] Electrochemical studies suggest a mechanism that involves an initial activation of the catalyst by two-electron reduction to form a Pd(0) species, which reacts with aryl triflate to generate a Pd^{II} monoaryl complex (Scheme 5.16b). CV studies indicate that the reduction of the (dppe)Pd^{II} monoaryl complex is a one-electron process,[50] and a mechanism similar to the (dppe)$NiCl_2$-catalzyed aryl halide coupling was proposed[51] that involves the formation of Pd^{I} and Pd^{III} species, followed by the C–C reductive elimination of biaryl from a $Pd^{III}Ar_2$ intermediate (Scheme 5.16b).[49]

5.2.3.5 *Pd^{III} Intermediates in Other Arylation Reactions*

Pd^{III} intermediates were proposed by Sanford *et al.* in a photocatalyzed C–H arylation reaction using diazonium salts in the presence of $Pd(OAc)_2$ and a $Ru(bpy)_3Cl_2$ photocatalyst (Scheme 5.17a).[52] The proposed mechanism

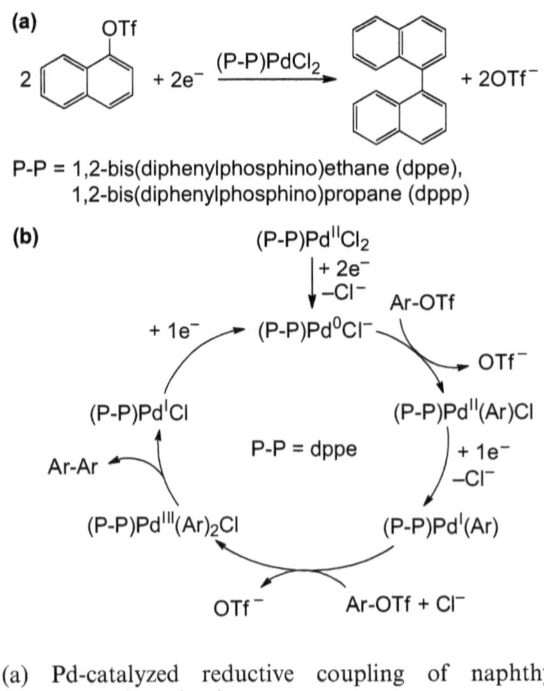

Scheme 5.16 (a) Pd-catalyzed reductive coupling of naphthyl triflates and (b) proposed mechanism.

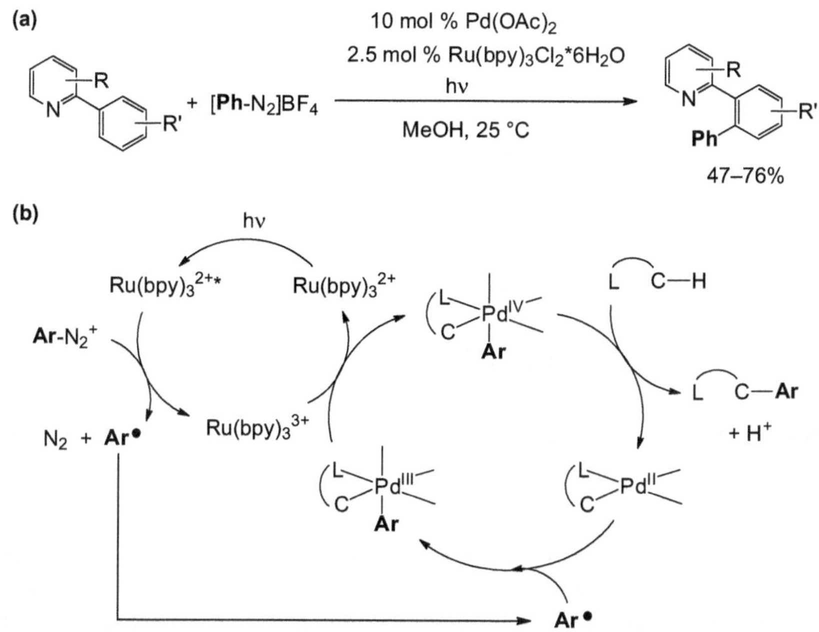

Scheme 5.17 (a) Pd/Ru-catalyzed C-H arylation with aryldiazonium salts and (b) proposed mechanism.

involves the reduction of diazonium salt with the photoexcited $Ru(bpy)_3^{2+*}$ species to generate an aryl radical, which reacts with the palladacycle to produce a Pd^{III} intermediate (Scheme 5.17b). A subsequent one-electron oxidation of the Pd^{III} species is proposed to form a Pd^{IV} intermediate that is responsible for the C–C bond reductive elimination.[52]

Yu *et al.* reported Pd-catalyzed C–H bond arylation and ethoxycarbonylation reactions that are proposed to involve the generation of aryl or ethoxyacyl radicals and formation of high-valent Pd intermediates, although it remains unclear whether Pd^{III} or Pd^{IV} species form under these conditions.[53,54]

Interestingly, formation of a Pd^{III} intermediate by reaction of a Pd^{II} precursor with a phenyl radical source, phenylazotriphenylmethane (PAT), was also proposed by Baird *et al.*[55] Thermal decomposition of PAT forms both Ph and trityl radicals that react with the Pd^{II}-allyl complex **17** to give a Pd^{II}Ph species and 4,4,4-triphenyl-1-butene (Scheme 5.18). The proposed mechanism involves an initial attack of the Ph radical at the Pd^{II} center to generate a Pd^{III}Ph(allyl) species **18**, which reacts with the trityl radical to give the observed products (Scheme 5.18).

Overall, the formation of Pd^{III} intermediates was proposed in a number of C_{sp3}–C_{sp3} and C_{sp2}–C_{sp2} coupling reactions and such intermediates can be formed by either oxidation of Pd^{II} precursors with one-electron oxidants, oxidative addition of ArX (X = halide or triflate) to Pd^{I} species, or reaction of

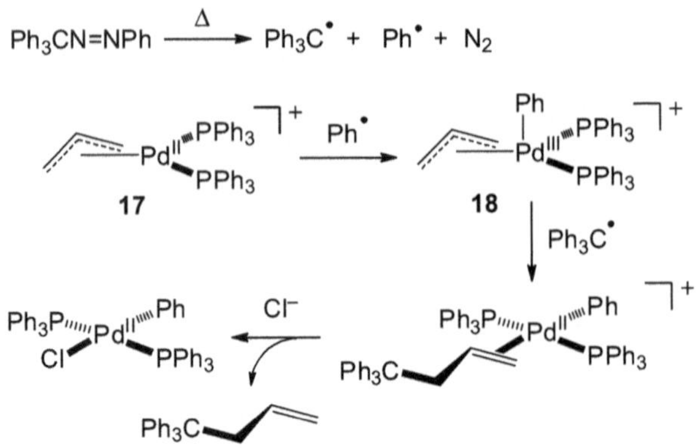

Scheme 5.18 Proposed formation of a Pd^{III} intermediate in presence of phenyl radical.

Pd^{II} complexes with carbon-centered radicals. The formation of the Pd^{III} intermediates in a number of C–C coupling reactions can thus be considered as a viable alternative to the commonly accepted Pd^0/Pd^{II} mechanisms.[56] It is important to note that the Ag^+ salt additives that are commonly employed in Pd-catalyzed C–H functionalization reactions can also act as one-electron oxidants and promote formation of Pd^{III} species.

5.2.4 Proposed Pd^{III} Intermediates in C–Heteroatom Bond Formation Reactions

5.2.4.1 Pd^{III} Intermediates in the Radical O_2 Insertion into a Pd–Me Bond

The radical insertion of dioxygen into the Pd–C bond of a dimethyl Pd^{II} complex was reported by Goldberg *et al.*[57] When a solution of (bipy)$Pd^{II}Me_2$ (**19**) in benzene was exposed to O_2 (1–10 atm) at RT, formation of the Pd^{II} methylperoxo complex (bipy)Pd^{II}Me(OOMe) (**20**) was observed in up to ~70% yield (Scheme 5.19a). The reaction was accelerated in the presence of light or radical initiator additives, and reproducible reaction rates were obtained in the presence of excess AIBN radical initiator. Detailed kinetic studies suggest a radical chain mechanism in which a chain propagation step involves a stepwise associative homolytic substitution at the Pd center (Scheme 5.19b). Addition of the MeOO radical to a square planar Pd^{II} center generates a five-coordinate Pd^{III} intermediate **21**, which then extrudes a Me radical with concomitant formation of (bipy)Pd^{II}Me(OOMe) (**20**). Analogous reactions of alkylperoxo radicals with first row transition complexes leading to a one-electron oxidation of a metal center have been reported previously,[58–60] and also proposed in the radical-chain aerobic oxidation of a Pt^{II} complex in the presence of ^{i}PrI.[61]

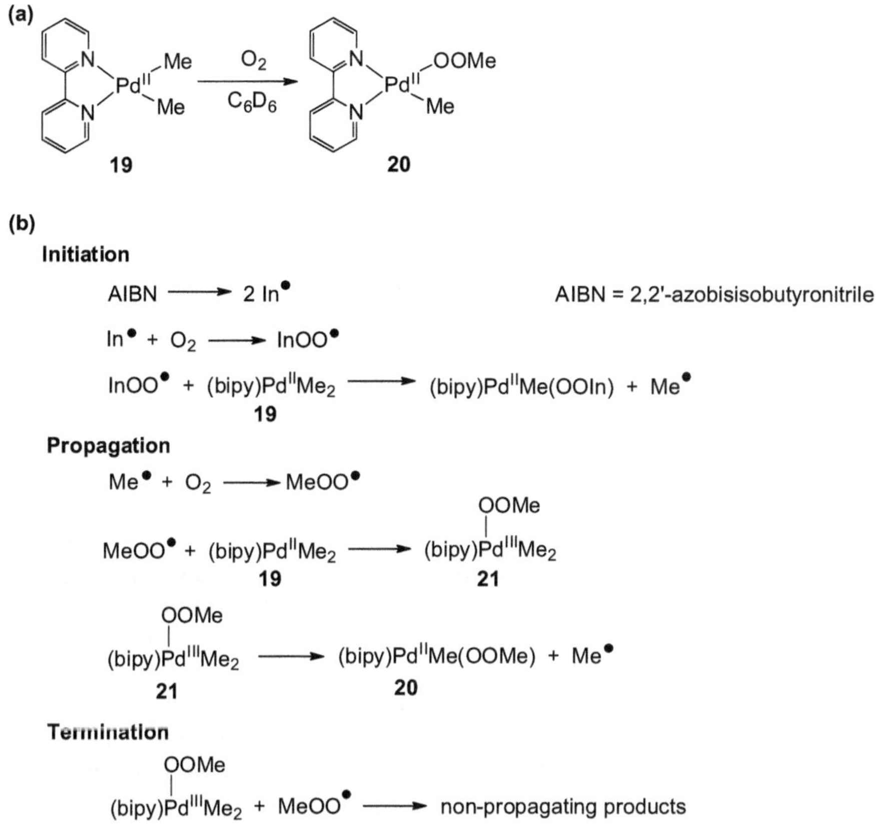

Scheme 5.19 (a) Reaction of (bipy)$PdMe_2$ (**19**) with O_2 and (b) proposed mechanism for the insertion of O_2 into the Pd–Me bond.

5.2.4.2 Pd^{III} Intermediates in Catalytic C–H Functionalization Reactions

Yu *et al.* reported intramolecular C–H amination reactions catalyzed by $Pd(OAc)_2$ in the presence of the one-electron oxidant $Ce(SO_4)_2$ (Scheme 5.20).[62] The formation of a Pd^{III} intermediate was proposed, although it remains unclear whether its further oxidation to Pd^{IV} occurs before the C–N reductive elimination step.[62]

The involvement of high-valent Pd intermediates as an alternative to the more common Pd^0/Pd^{II} catalytic cycles was also discussed by Glorius in the C–H amination in the presence of AgOAc, which can act as a one-electron oxidant (Scheme 5.21).[63]

In general, the oxidation of Pd^{II} complexes with one-electron oxidants may yield either Pd^{III} or Pd^{IV} species, depending on the relative stability of the two oxidation states and the ligand environment. For example, stable Pd^{IV}

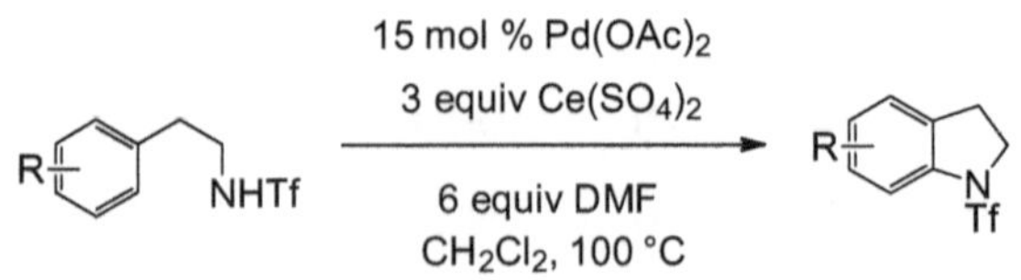

Scheme 5.20 Pd-catalyzed intramolecular C–H amination in presence of $Ce(SO_4)_2$.

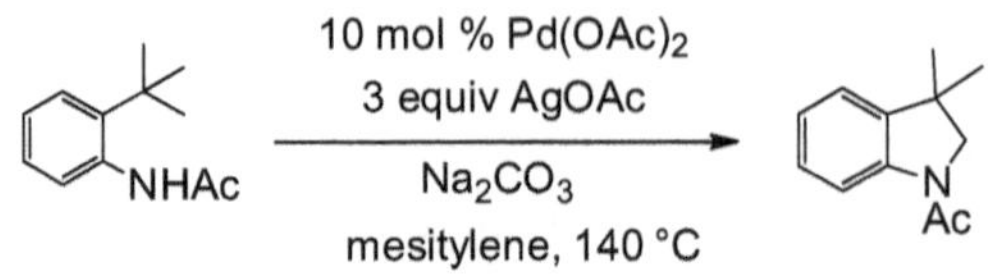

Scheme 5.21 Pd-catalyzed intramolecular C–H amination in presence of AgOAc.

complexes were obtained by oxidation with Fc^+ of palladacycles $LPd^{II}(CH_2CMe_2C_6H_4)$ (L = hydrotris(pyrazolyl)borate anion, Tp^-), since the Tp^- ligand is known to stabilize Pd^{IV} complexes.[64] In addition, examples of disproportionation of Pd^{III} to Pd^{IV} and Pd^{II} species,[65–67] as well as the comproportionation of Pd^{IV} and Pd^{II} to form Pd^{III} species are known.[33] Therefore, it may not be possible to unambiguously assign the oxidation state of high-valent Pd intermediates, and both Pd^{III} and Pd^{IV} species may be involved in oxidatively-induced C–C or C–heteroatom bond formation reactions.

5.2.4.3 Pd^{III} Intermediates in C–Cl Bond Formation Reactions

The transient formation of a Pd^{III} intermediate was proposed in a number of reactions involving CCl_4 or CCl_3Br as the oxidant, such as the C–H chlorination of saturated hydrocarbons with CCl_4 catalyzed by $(L)_2PdX_2$ (L = PPh_3, *p*-Tol_3P, MeCN; X = Cl, Br, I; Scheme 5.22a).[68,69] Based on the reaction selectivity, a mechanism was proposed in which the Pd^{II} complex reacts with CCl_4 to produce a transient $Pd^{III}Cl$ species and a CCl_3 radical, which reacts with the C–H bond of an alkane to generate chloroform and an alkyl radical that then reacts with the $Pd^{III}Cl$ species to give alkyl chloride and regenerate the Pd^{II} complex (Scheme 5.22b).

A similar mechanism was proposed for the Kharasch addition of CCl_3Br to alkenes catalyzed by Pd^{II} complexes.[70,71] For example, the Pd^{II} complexes $(Me_3tacn)PdCl_2$ and $(Me_3tacn)PdBr_2$ (Me_3tacn = *N,N′N′*-trimethyl-1,4,7-triazacyclononane) were shown to be good catalysts for the radical addition of CCl_3Br to alkenes, most likely due to their low $Pd^{II/III}$ oxidation potentials.[71] While the proposed catalytic mechanism involves a mononuclear trihalide Pd^{III} intermediate that was not detected, a dinuclear Pd^{III} species was observed in the reaction mixture (*vide infra*).

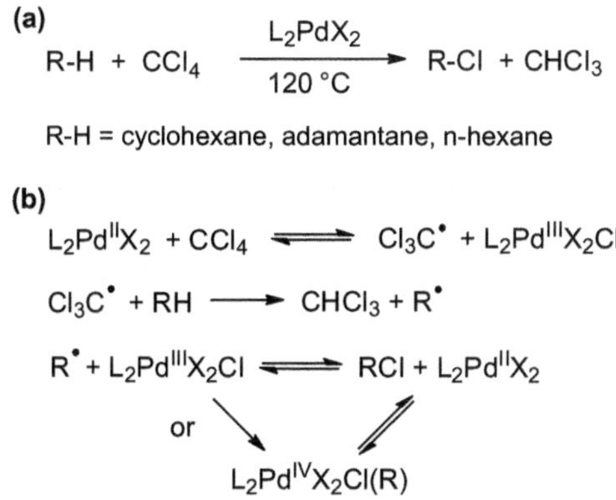

Scheme 5.22 (a) Pd-catalyzed C–Cl bond formation and (b) proposed mechanism.

5.3 Dinuclear Pd^{III} Complexes

5.3.1 Electronic Properties of Dinuclear Pd^{III} Complexes

The interaction of two square planar Pd^{II} d^8 centers along the z-axis leads to mixing of the d orbitals and a molecular orbital diagram in which both the bonding and antibonding molecular orbitals are filled (Figure 5.2), and thus no metal–metal bonding interaction is expected for such complexes. However, due to symmetry-allowed mixing of Pd $5p_z$ and 5s orbitals into the $4d_{z2}$ orbital, weak metal–metal bonding interactions may be present in such complexes.[72] One-electron or two-electron oxidation of Pd^{II} dinuclear species removes electrons from antibonding orbitals and generates mixed-valent Pd^{II}–Pd^{III} or Pd^{III}–Pd^{III} complexes with Pd–Pd bond orders of 0.5 or 1, respectively.[24,25]

5.3.2 Dinuclear Pd^{II} Complexes

Cyclometalated Pd^{II} complexes are common precursors in Pd-mediated oxidative C–H bond functionalization reactions. In this context, the geometry of bridged dinuclear Pd complexes depends on the nature of the bridging ligands and directly affects their redox reactivity. For example, while the dichloro-bridged palladacycle **22** is essentially planar and exhibits no metal–metal interactions, the acetato- (**23**) or trifluoroacetate-bridged complexes adopt an open "clamshell" geometry with two Pd centers in close proximity (2.862 Å Pd···Pd distance for complex **23**) and exhibit weak attractive d^8–d^8 interactions between two Pd centers and π-stacking interactions between two phenylpyridine ligands (Scheme 5.23).[72]

Interestingly, weak d^8–d^8 interactions between two cationic Pd^{II} centers were also observed for Pd^{II} complexes supported with the macrocyclic tetradentate ligand 2,11-dithia[3.3](2,6)pyridinophane (N2S2), **24** and **25** (Scheme 5.24).[73]

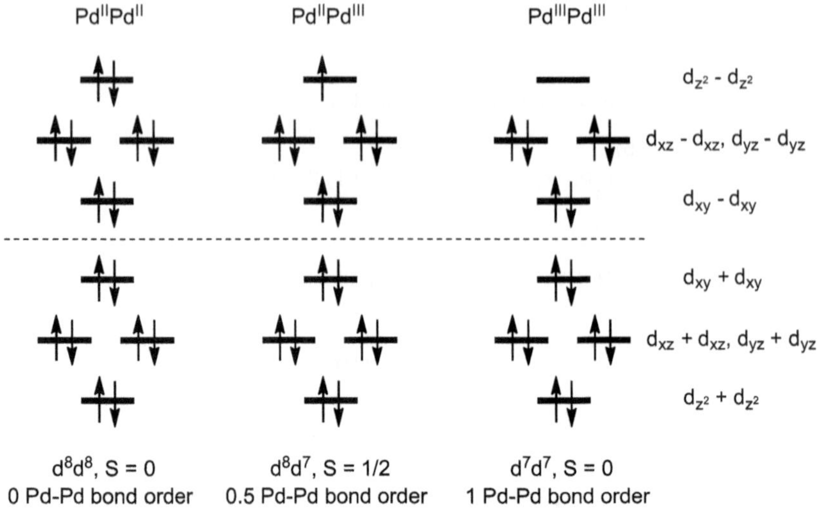

Figure 5.2 Molecular orbital diagram for Pd^{II}Pd^{II}, Pd^{II}Pd^{III}, and Pd^{III}Pd^{III} dinuclear complexes with interacting metal centers. The *z* axis is along the Pd–Pd vector and the dashed line represents the separation between bonding and antibonding molecular orbitals.

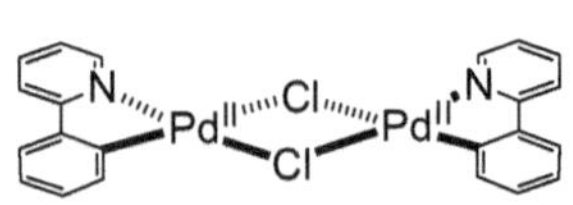

22

closest Pd^{II}...Pd^{II} contact: 3.691 Å

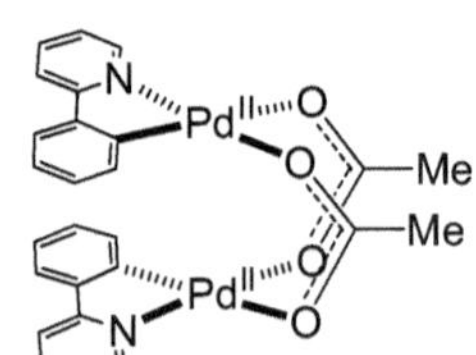

23

Pd^{II}...Pd^{II} 2.862 Å

Scheme 5.23 Examples of dinuclear bridged Pd^{II} complexes with planar (**22**) or "clamshell" (**23**) geometry.

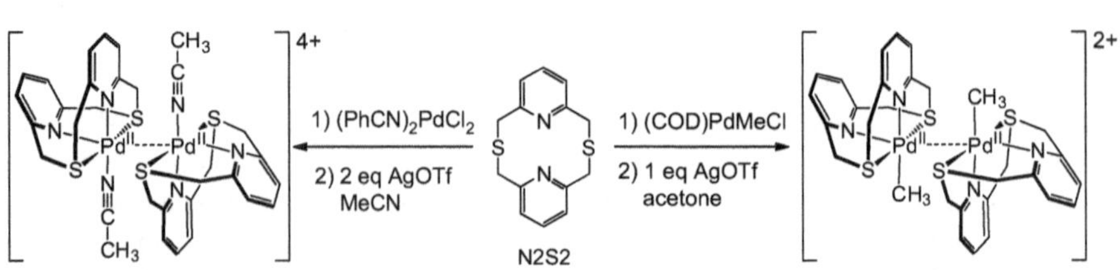

24

Pd...Pd 3.066 Å

25

Pd...Pd 3.113 Å

Scheme 5.24 Dinuclear (N2S2)Pd^{II} complexes with weak d^8–d^8 interactions.

These interactions are not supported by any bridging ligands and were proposed to play a role in the observed oxidatively-induced Me group transfer reactivity leading to the Pd^{IV} species $[(N2S2)Pd^{IV}Me_2]^{2+}$.[73]

5.3.3 Mixed-Valent Pd^{II}–Pd^{III} Complexes

The first isolated mixed-valent Pd^{II}–Pd^{III} complexes were the amidinate-bridged dinuclear complexes **26–28** obtained by chemical or electrochemical oxidation of the corresponding tetrabridged Pd^{II} precursors (Scheme 5.25).[74–77] These paramagnetic complexes were characterized by X-ray crystallography and high-field EPR spectroscopy that confirmed a metal-centered oxidation,[74–77] while DFT calculations suggest a Pd–Pd bond order of 0.5 for such complexes.[77]

Detailed electrochemical studies suggest that the number of ligands bridging the two Pd centers has a pronounced effect on the redox properties of these complexes. Thus, the oxidation of the dibridged complex **29** was shown to be a two-electron process that corresponds to the oxidation of the two non-interacting Pd centers (Scheme 5.26), in line with the long Pd···Pd distance of

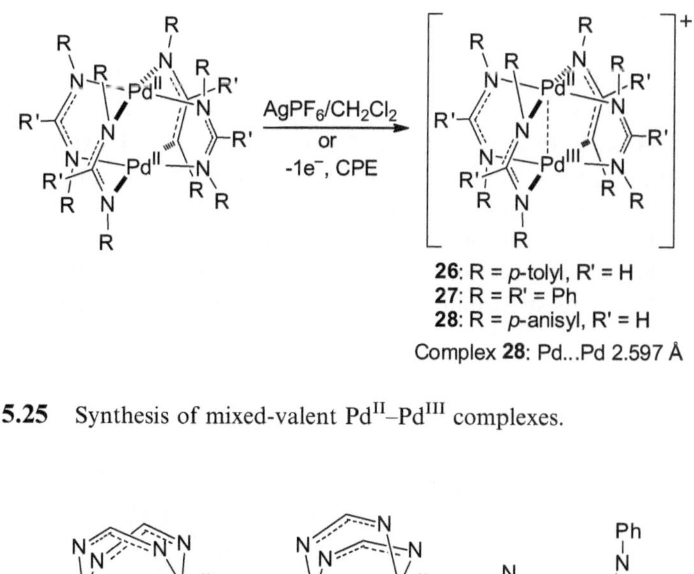

Scheme 5.25 Synthesis of mixed-valent Pd^{II}–Pd^{III} complexes.

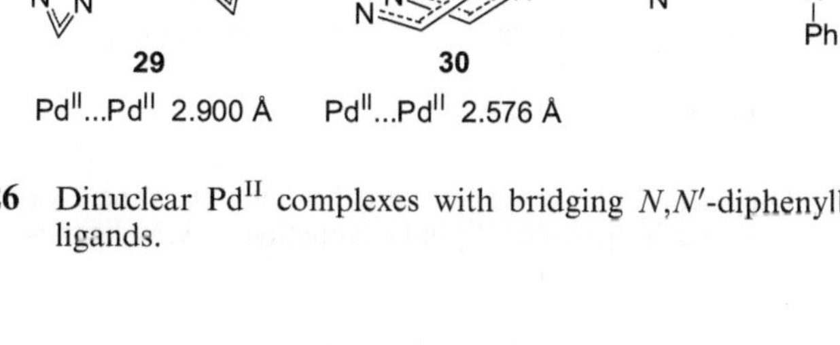

Scheme 5.26 Dinuclear Pd^{II} complexes with bridging *N,N′*-diphenylbenzamidinate ligands.

2.900 Å. By contrast, the tetrabridged complex **30** undergoes a reversible one-electron oxidation corresponding to a $Pd^{II}–Pd^{II}/Pd^{II}–Pd^{III}$ oxidation at a potential that is 0.37 V less positive compared to **29**, and supported by the shorter Pd···Pd distance of 2.576 Å enforced by the four bridging ligands.[76]

5.3.4 Dinuclear $Pd^{III}–Pd^{III}$ Complexes with a $Pd^{III}–Pd^{III}$ Bond

In 1996, Bond, Canty *et al.* reported electrochemical studies of the dinuclear palladacyclic complexes **31** and **32** (Scheme 5.27).[78] These studies suggest a stepwise oxidation to $Pd^{II}–Pd^{III}$ and $Pd^{III}–Pd^{III}$ dinuclear species, respectively, although both oxidation products were unstable and could not be isolated. The CV of complex **32** revealed that the first one-electron oxidation was chemically reversible, indicative of a limited stability of the mixed-valent $Pd^{II}–Pd^{III}$ product, however, electrolysis of this complex results in an overall two-electron oxidation, presumably due to the disproportionation of the mixed-valent intermediate to $Pd^{II}–Pd^{II}$ and $Pd^{III}–Pd^{III}$ species.[78]

Similarly, the CV study of the diacetato-bridged 2-phenylpyridyl complex **23** and its 2-tolylpyridyl analog revealed two one-electron oxidation steps corresponding to sequential formation of $Pd^{II}–Pd^{III}$ and $Pd^{III}–Pd^{III}$ species. The mixed-valent species was proposed to undergo disproportionation to $Pd^{III}–Pd^{III}$ and $Pd^{II}–Pd^{II}$ resulting in a decreased intensity for the second oxidation wave.[72]

In 1998, Cotton reported the first isolated $Pd^{III}–Pd^{III}$ complex **34**, which was obtained by chemical oxidation of the dinuclear $Pd^{II}–Pd^{II}$ precursor **33** with $PhICl_2$ (Scheme 5.28).[79] Complex **34** exhibits a Pd–Pd single bond that is supported by four bridging 1,3,4,6,7,8-hexahydro-2H-pyrimido[1,2-a]pyrimidinate ligands.

In 2011, Ritter *et al.* found that complex **33** catalyzes the aerobic hydroxylation of the α-carbonyl tertiary C–H bonds (Scheme 5.29).[80] Moreover, UV-vis studies suggest formation of a transient $Pd^{III}–Pd^{III}$ intermediate, and

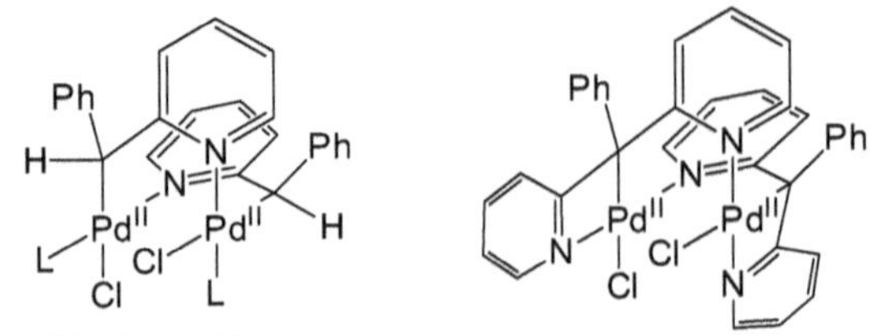

31a: L = 4-Mepy
31b: L = 3,4-Me$_2$py
31c: L = 3,5-Me$_2$py
31d: L = 2-Bnpy
31a: $Pd^{II}...Pd^{II}$ 3.062 Å

32
$Pd^{II}...Pd^{II}$ 2.855 Å

Scheme 5.27 Dinuclear Pd^{II} complexes with bridging pyridine-2-yl(phenyl)methyl-*C*,*N* and bis(pyridine-2-yl)phenylmethyl-*C*,*N*,*N′* ligands.

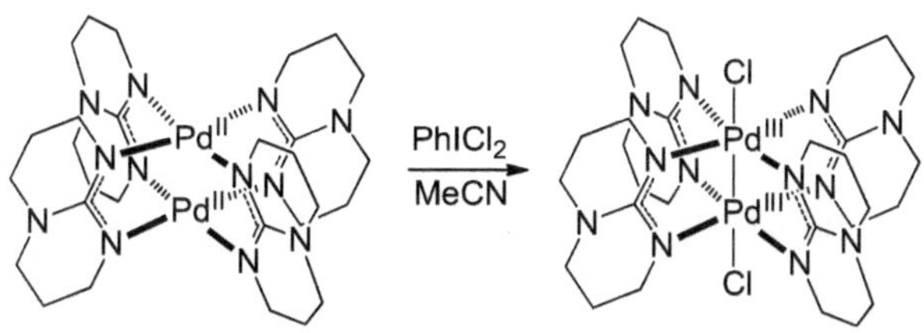

33 Pd...Pd 2.555 Å

34 Pd-Pd 2.391 Å

Scheme 5.28 Synthesis of the first isolated dinuclear Pd^{III}–Pd^{III} complex by Cotton *et al.*

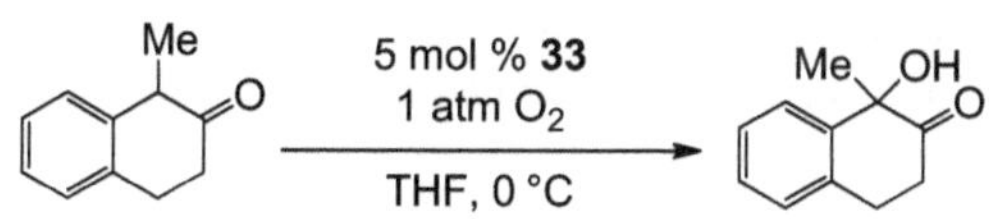

77%

Scheme 5.29 Alpha-hydroxylation of carbonyl compounds catalyzed by **33**.

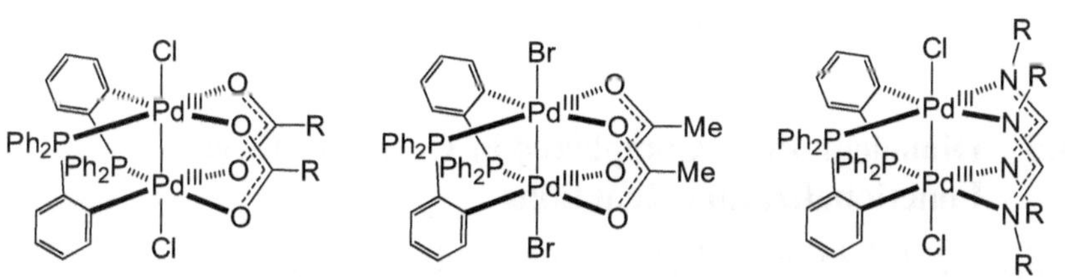

35a: R = Me, Pd-Pd 2.524 Å
35b: R = CF_3,
35c: R = ^tBu

36

37a: R = Ph,
37b: R = p-MeC_6H_4,
37c: R = p-$MeOC_6H_4$

Scheme 5.30 Dinuclear Pd^{III} complexes with bridging orthometalated triphenylphosphine ligands.

O_2 uptake and $^{18}O_2$ labeling experiments imply that both oxygens from O_2 are incorporated into the product. Although the mechanism of this transformation remains to be elucidated, the absence of peroxide or hydroperoxide products and a lack of reactivity for Ph_3CH suggest that a radical metal-mediated autooxidation mechanism is unlikely.

In 2006, Cotton, Lahuerta, Ubeda *et al.* reported the first organometallic Pd^{III}–Pd^{III} complexes **35a–c**, with the Pd^{III} centers being supported by two bridging orthometalated triphenylphosphine ligands and two carboxylate groups (Scheme 5.30).[81] The observed short Pd–Pd distances of 2.524–2.543 Å imply a single bond between the two Pd centers that is supported by DFT calculations. In addition, several analogs (**36–37**) were synthesized using acetate

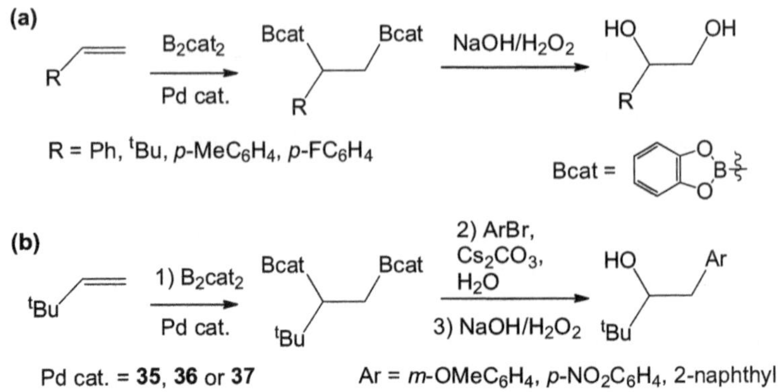

Scheme 5.31 (a) Diboration/oxidation and (b) tandem diboration/arylation/oxidation catalyzed by dinuclear Pd^{III} complexes.

or formamidinate bridging ligands (Scheme 5.30).[82,83] Complexes **35**–**37** were employed as pre-catalysts for the diborylation and tandem borylation/arylation of alkenes (Scheme 5.31).[82,83] The Pd^{III} complexes were shown to be reduced to Pd^{II} by the diboron reagents under the catalytic reaction conditions and thus most likely serve as precursors for the active catalyst.

5.3.5 Dinuclear Pd^{III} Complexes in Catalytic C–H Functionalization Reactions

5.3.5.1 Pd^{III} Intermediates in C–H Acetoxylation

Palladium complexes have been known to catalyze aromatic C–H bond functionalization in the presence of strong oxidants since the 1960s and the involvement of Pd^{IV} intermediates has been proposed.[84–90] General methods for the regioselective ligand-directed functionalization of C_{sp2}–H and C_{sp3}–H bonds to esters, ethers, halides, and amines in the presence of oxidants such as $PhI(OAc)_2$, *N*-chlorosuccinimide (NCS), or *N*-bromosuccinimide (NBS) was reported initially by Sanford *et al.*,[8,11,91] and then by Yu *et al.*,[92,93] Gaunt *et al.*,[94] and others.[10–13,95] For example, the regioselective acetoxylation of 2-phenylpyridine and benzo[h]quinoline was accomplished using $Pd(OAc)_2$ as the catalyst and $PhI(OAc)_2$ as the oxidant (Scheme 5.32a).[8] As no product was formed in the presence of benzoquinone or $Cu(OAc)_2$– typical oxidants used in Pd^{II}/Pd^{0}-mediated reactions – and no reaction occurred in the absence of the oxidant, a Pd^{II}/Pd^{IV} catalytic cycle was proposed that involves C–H activation and formation of a palladacycle, which is oxidized by $PhI(OAc)_2$ to form a Pd^{IV} species followed by reductive C–O elimination (Scheme 5.32b).

In 2009, Ritter *et al.* demonstrated that the two-electron oxidation of the dinuclear palladacycle **23** with $PhI(OAc)_2$ at low temperature yields a structurally characterized dinuclear Pd^{III}–Pd^{III} product **39** that was stable

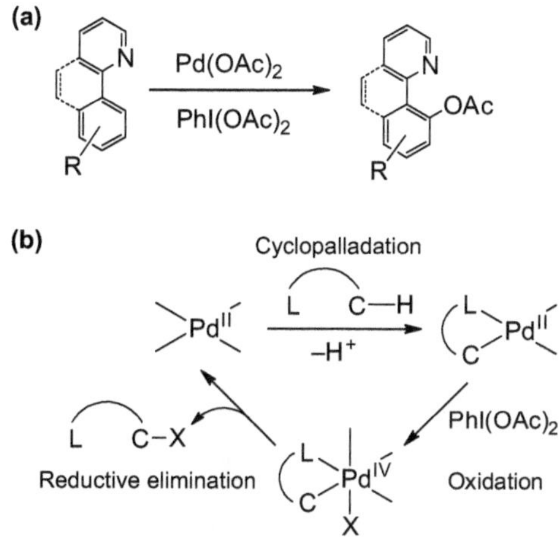

Scheme 5.32 (a) Pd-catalyzed C–H bond acetoxylation and (b) Pd^{II}/Pd^{IV} catalytic cycle proposed by Sanford *et al.*

below −10 °C (Scheme 5.33a).[27] The Pd–Pd distance in **39** is 2.555 Å, 0.307 Å shorter *vs.* the Pd^{II}–Pd^{II} precursor **23**. The complex **39** is diamagnetic, consistent with the presence of a Pd^{III}–Pd^{III} single bond. Warming up the solution of **39** to 40 °C leads to C–O reductive elimination and yields the acetoxylated product (Scheme 5.33a). Based on these results, a catalytic cycle was proposed involving (i) cyclopalladation to give **23**, (ii) bimetallic oxidation of **23** with $PhI(OAc)_2$ to generate **39**, and (iii) bimetallic C–O reductive elimination (Scheme 5.33b). The complex **39** was shown to be a kinetically competent catalytic intermediate, although a potential involvement of a transient Pd^{IV} species could not be excluded.

5.3.5.2 Pd^{III} Intermediates in C–H Halogenation

The oxidation of the palladacycle **40** with $PhICl_2$, leading to formation of the structurally characterized diamagnetic Pd^{III}–Pd^{III} complex **41**, was also reported (Scheme 5.34).[26] Complex **41** is stable at low temperature, yet it reductively eliminates 10-chlorobenzo[h]quinoline at 23 °C in high yield (Scheme 5.34), and the authors proposed a direct C–Cl reductive elimination without prior dissociation to mononuclear Pd^{III} species or disproportionation to Pd^{IV} and Pd^{II} complexes.[26] Detailed experimental and computational studies suggest a synergistic effect of the bimetallic core in both oxidation and C–Cl reductive elimination that lowers the overall energy barrier of these transformations.[29] The complex **41** could also be formed under catalytic conditions, in the presence of excess benzo[h]quinoline.[26]

The mechanism of C–H bond chlorination with *N*-chlorosuccinimde (NCS) as an oxidant was also investigated, and formation of the Pd^{III}–Pd^{III} dinuclear

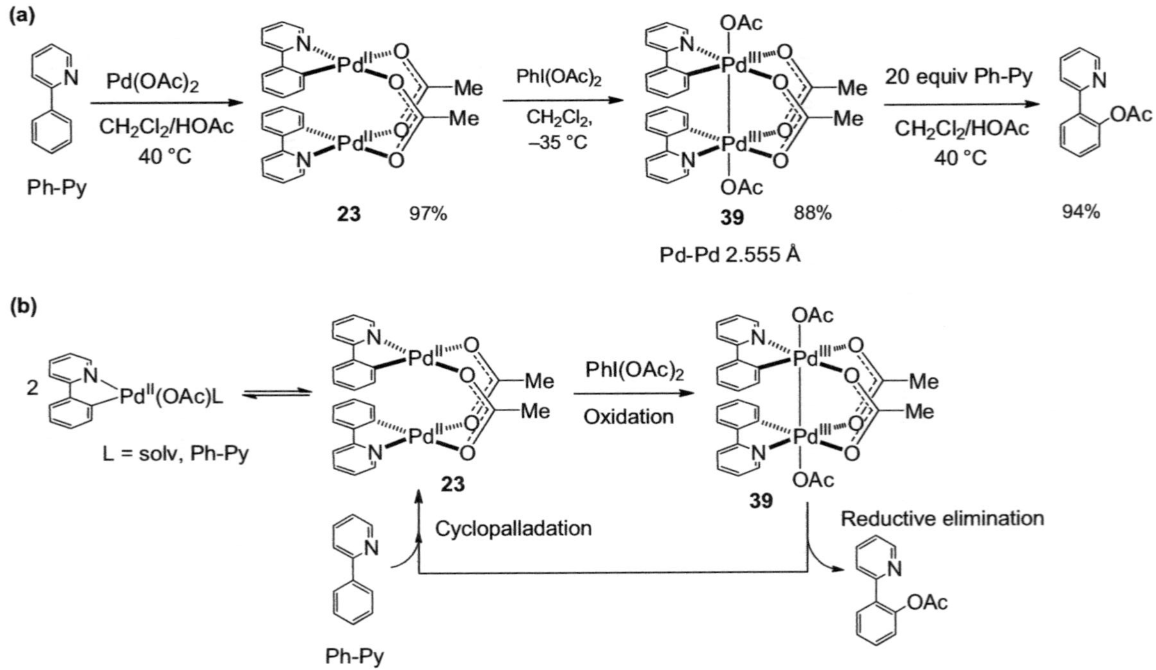

Scheme 5.33 (a) Dinuclear Pd^{II} and Pd^{III} intermediates in C–H bond acetoxylation and (b) bimetallic Pd^{II}/Pd^{III} catalytic cycle proposed by Ritter *et al.*

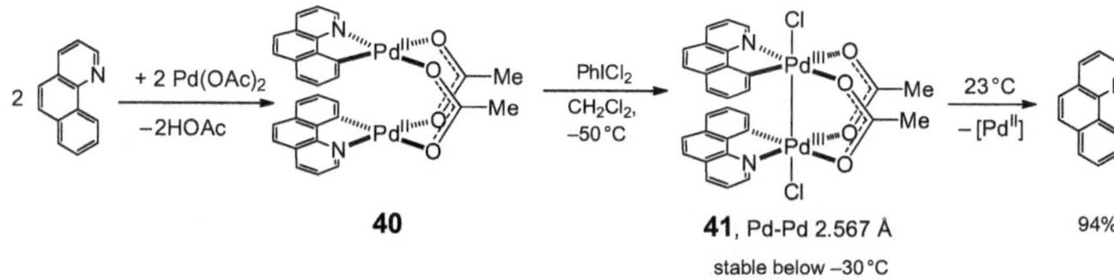

Scheme 5.34 Dinuclear Pd^{II} and Pd^{III} intermediates in C–H bond chlorination.

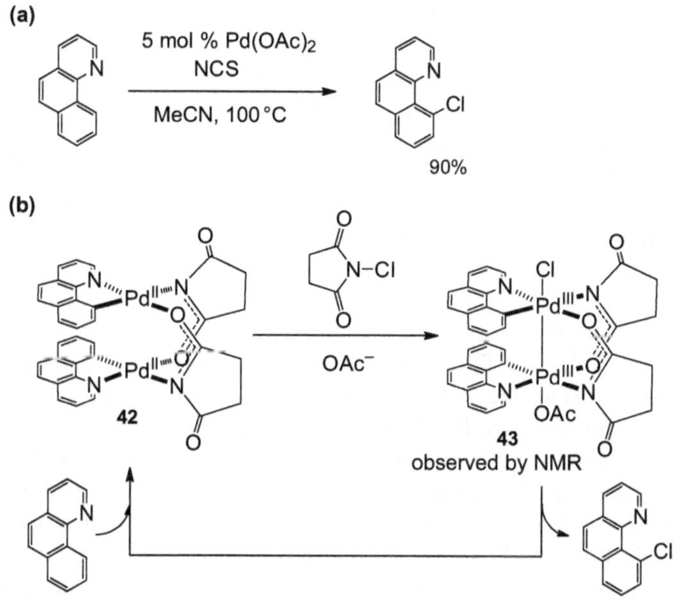

Scheme 5.35 (a) Pd-catalyzed C–H bond chlorination with NCS and (b) proposed bimetallic Pd^{II}/Pd^{III} catalytic cycle.

intermediate **43** was detected by NMR (Scheme 5.35).[96] The resting state of the catalyst is a dinuclear succinimide anion-bridged complex **42**. Mechanistic studies of the catalytic chlorination reveal a turnover-limiting oxidation of **42** by NCS, while the first order of the reaction rate in **42** implies this dinuclear complex is directly involved in an acetate-assisted bimetallic oxidation by NCS (Scheme 5.35b).

Formation of dinuclear and/or mononuclear Pd^{III} intermediates can also be considered in a number of other cases in which Pd^{IV} intermediates were proposed initially, for example, in catalytic chlorination of C–H bonds of arylpyridines or acetanilides with NCS[8,97–99] or $CuCl_2$[100] as an oxidant.

5.3.5.3 Pd^{III} Intermediates in C–H Arylation and Trifluoromethylation

Ortho-phenylation of C–H bonds with diaryliodonium salts catalyzed by $Pd(OAc)_2$ was demonstrated for a number of substrates such as arylpyridines (Scheme 5.36a), quinolines, pyrrolidinones, oxazolidinones, and benzodiazepines.[28,101–103] Mechanistic studies of the ortho-phenylation of 3-methyl-2-phenylpyridine reveal that the mononuclear complex **44** is the catalyst resting state (Scheme 5.36b).[28] The catalytic reaction rate shows a second order dependence on **44**, and the turnover limiting step was proposed to be the oxidation of the dinuclear Pd^{II} species **45** to give the dinuclear Pd^{IV}–Pd^{II} complex **46a**, which can alternatively be formulated as the isoelectronic Pd^{III}–Pd^{III} species **46b**. Subsequent reductive C–C elimination from **46** affords the arylated product and regenerates the Pd^{II} catalyst.

Computational studies were employed to elucidate the electronic structure and reactivity of dinuclear complexes of type **A** and **B** (Scheme 5.37) as a function of ligand environment, as these species are proposed in various Pd-catalyzed C–H functionalization reactions.[104,105] The Pd^{IV}–Pd^{II} bonding character in complex of type **A** was found to be favored when X is a strong σ-donor, while the Pd^{III}–Pd^{III} bonding character was found to be predominant when X is a weak σ-donor ligand.[105]

The arylation of C_{sp2}–H and C_{sp3}–H bonds using aryl iodides in the presence of a stoichiometric amount of Ag salts was also developed (Scheme 5.38).[101,106–109] These reactions were proposed to involve oxidative addition of the aryl iodide to Pd^{II} to give a Pd^{IV}(Ar)(I) product, however, one-electron oxidation to form Pd^{III} intermediates cannot be ruled out.[101,106–109]

Interestingly, the formation of the mononuclear Pd^{IV} complex **47** by treatment of the dinuclear Pd^{II} precursor **40** with 3,3-dimethyl-1-(trifluoromethyl)-1,2-benziodoxole (Togni reagent)[110] was proposed to involve the transient formation of a Pd^{III}–Pd^{III} species **48** followed by Pd^{III}–Pd^{III} bond heterolysis, as suggested by a recent combined experimental and computational study (Scheme 5.39).[111] Thus, such dinuclear Pd^{III} complexes may play a role as intermediates in Pd-catalyzed C–H bond trifluoromethylation reactions using electrophilic trifluoromethylating reagents.[111–113]

Overall, the generation of dinuclear Pd^{III}–Pd^{III} intermediates by two-electron oxidation of palladacycle Pd^{II} precursors and their facile C–O, C–Cl, and C–Br elimination reactivity suggests that Pd^{III}–Pd^{III} species may be involved in a range of Pd-catalyzed C–H oxidative functionalization reactions and should be considered as a viable alternative to Pd^{II}/Pd^{IV} catalytic cycles. The formation of dinuclear Pd^{III}–Pd^{III} species is generally dictated by the ligand environment and the presence of bridging ligands that bring together the two Pd centers. By contrast, the formation of mononuclear Pd^{IV} complexes by two-electron oxidation of the mononuclear Pd^{II} precursors was also observed in many cases.[9] Thus, it can be difficult to assign the observed reactivity to either dinuclear Pd^{III} or mononuclear Pd^{IV} species, as both oxidation states can be generated by an overall two-electron oxidation of two or one mononuclear Pd^{II} precursors, respectively.[9,10,12,104,105,114]

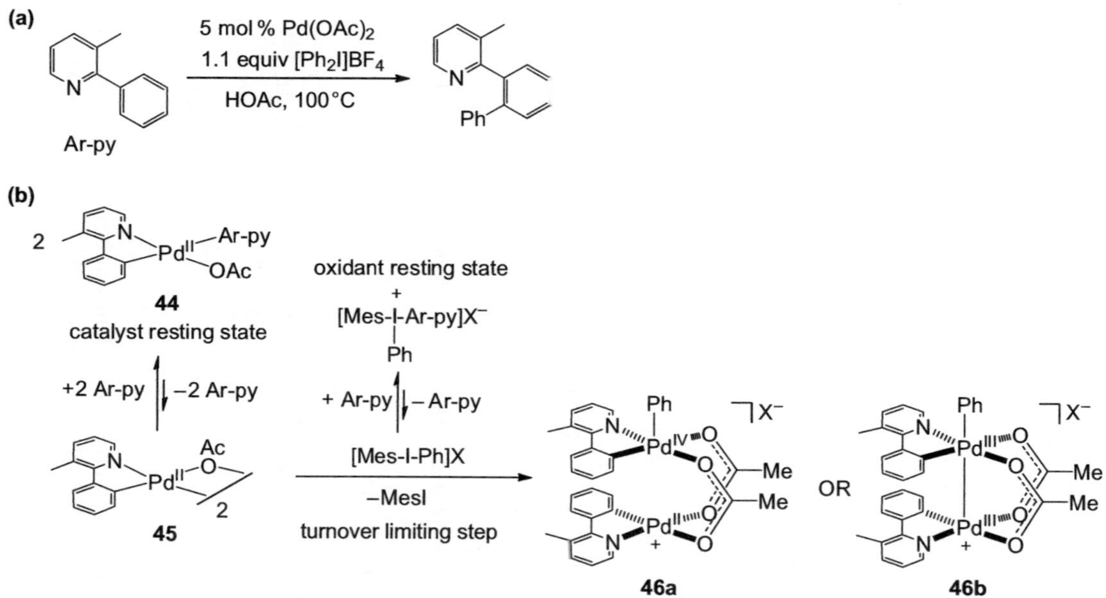

Scheme 5.36 (a) Pd-catalyzed C–H arylation with diaryliodonium salts and (b) proposed mechanism for C–H arylation of 3-methyl-2-phenylpyridine.

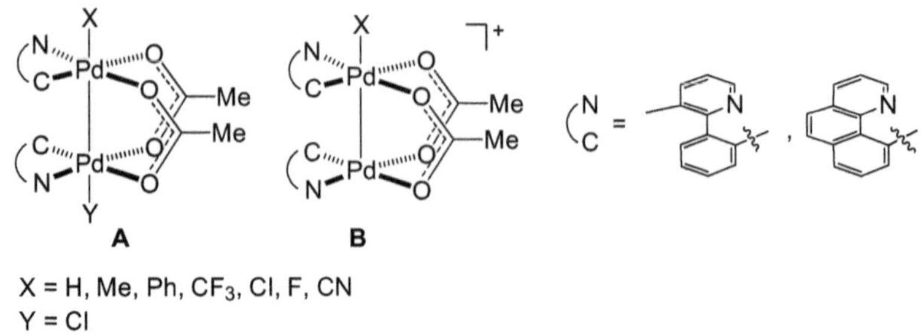

Scheme 5.37 Proposed geometries for dinuclear Pd^{III} diacetato-bridged complexes.

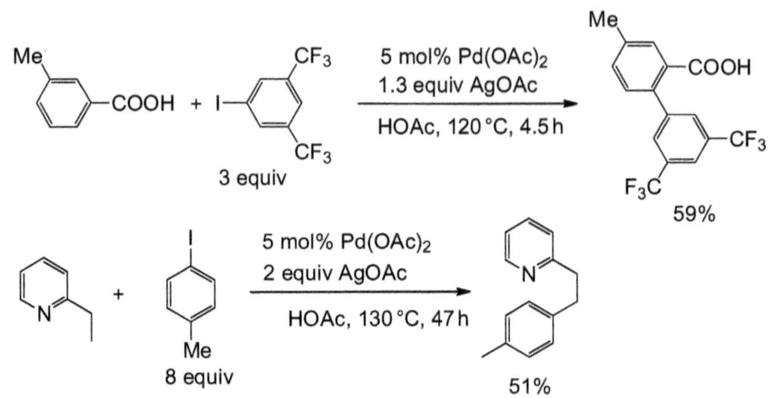

Scheme 5.38 Pd-catalyzed arylation of C–H bonds in presence of Ag^+ salts.

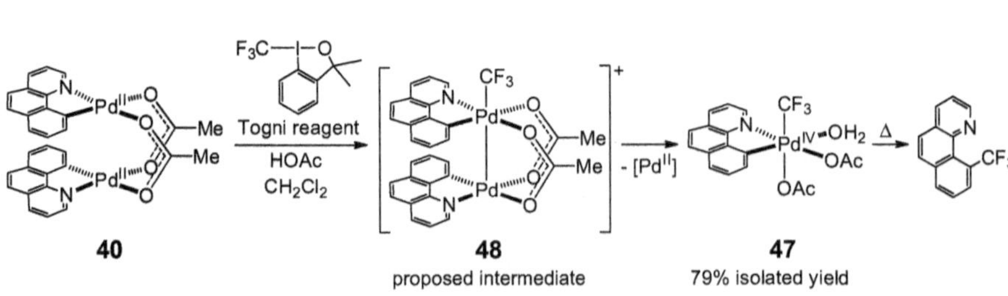

Scheme 5.39 Proposed formation of a dinuclear high-valent Pd intermediate during the oxidation of a Pd^{II} precursor to a mononuclear Pd^{IV} species.

5.3.6 Dinuclear Pd^{III} Complexes Without a Pd–Pd Bond

In 2011, Mirica *et al.* reported a series of dinuclear Pd^{III} complexes **50a–c** that were synthesized by electrochemical oxidation of mononuclear Pd^{II} precursors (Scheme 5.40).[71] These complexes employ the tridentate macrocyclic ligand *N*,*N*′,*N*′-trimethyl-1,4,7-triazacyclononane, Me_3tacn, and the two Pd^{III} centers

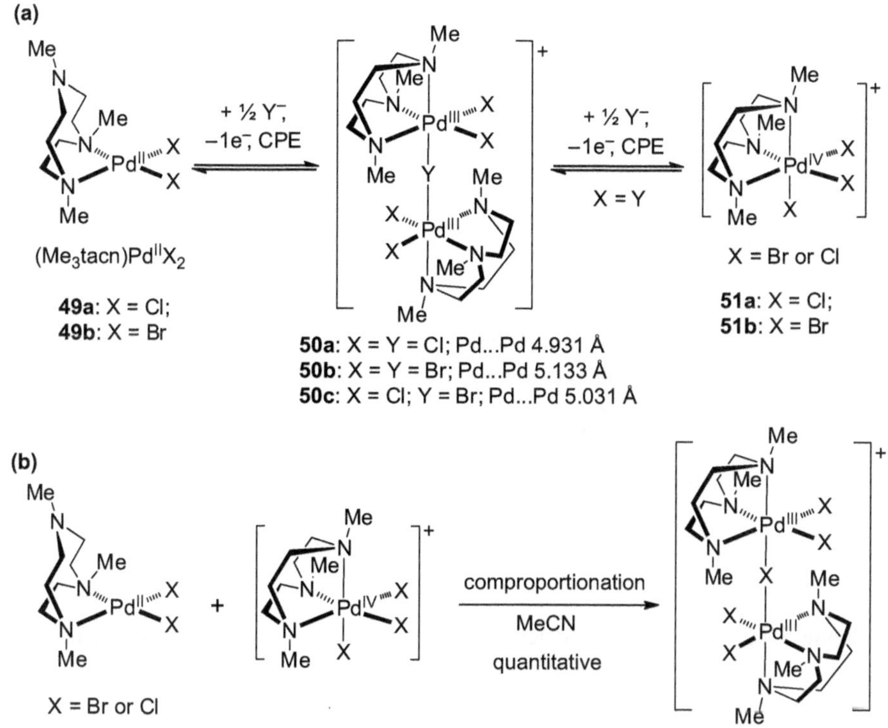

Scheme 5.40 (a) Reversible synthesis of $[(Me_3tacn)_2Pd^{III}{}_2X_4(\mu\text{-}Y)]^+$ and $[(Me_3tacn)Pd^{IV}X_3]^+$ complexes (X, Y = Cl, Br) and (b) comproportionation reactivity.

are bridged by a single halide ligand, representing the first examples of dinuclear Pd^{III} species that are not stabilized by a Pd–Pd bond. The two Pd^{III} centers exhibit a distorted octahedral geometry and are strongly antiferromagnetically coupled through the bridging halide ligand. Further electrochemical oxidation of these complexes in the presence of external halide affords the mononuclear Pd^{IV} complexes **51a–b**, which were structurally characterized (Scheme 5.40a). Interestingly, the interconversion between these $Pd^{II}/Pd^{III}/Pd^{IV}$ species is reversible: reduction of the mononuclear Pd^{IV} complex generates a dinuclear Pd^{III} species and eventually the mononuclear Pd^{II} species. Moreover, the Pd^{IV} complexes react rapidly with 1 equiv. of Pd^{II} complex to quantitatively generate the dinuclear Pd^{III} complex (Scheme 5.40b), suggesting that the latter species is more stable than the corresponding mononuclear Pd^{IV} and Pd^{II} complexes under an analogous ligand environment. The facile interconversion of dinuclear Pd^{III} and mononuclear Pd^{IV} species suggests that for a given ligand environment both types of intermediates can be present, with direct implications into the proposed involvement of analogous intermediates in Pd-catalyzed C–H oxidative functionalization reactions (*vide supra*).[111]

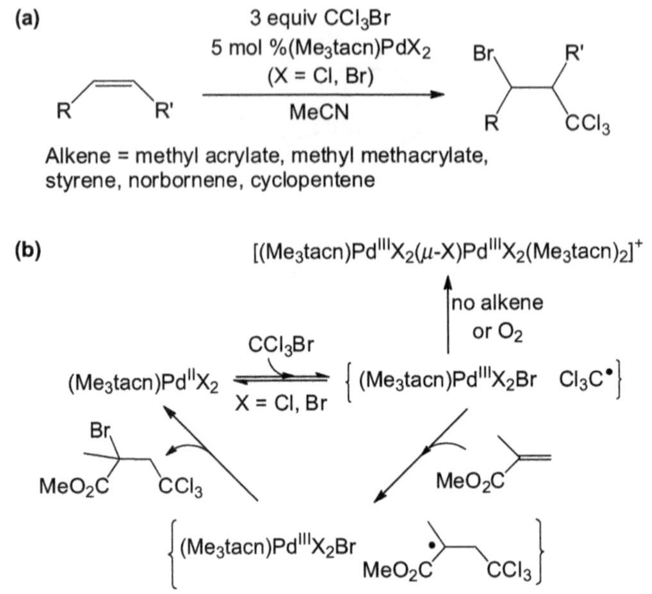

Scheme 5.41 (a) Pd-catalyzed Kharasch radical addition of CCl_3Br to alkenes and (b) proposed mechanism.

Interestingly, the Pd^{II} complexes **49a–b** catalyze the Kharasch radical addition of CCl_3Br to alkenes (Scheme 5.41a).[71] The proposed catalytic cycle involves an electron transfer/halogen transfer from CCl_3Br to Pd^{II} to generate a Pd^{III} species and a CCl_3 radical (Scheme 5.41b). Although no mononuclear Pd^{III} species were detected under catalytic conditions, the formation of dinuclear Pd^{III} species was detected by UV-vis at RT in the absence of the alkene substrate. Moreover, a higher yield of the dinuclear Pd^{III} species was obtained in the presence of O_2, which can act as a radical trap for the CCl_3 radical. As the kinetic analysis revealed a first order dependence of the reaction rate on Pd^{II}, formation of the dinuclear Pd^{III} species could be a means of stabilizing the mononuclear Pd^{III} intermediates involved in the catalytic cycle (Scheme 5.41b).

A recent report showed that the paramagnetic dinuclear Pd^{III} complexes **54** and **55** are active catalysts for asymmetric aza-Claisen rearrangement (Scheme 5.42).[115] When the chloride-bridged pentaphenylferrocene oxazoline (**52**) and imidazoline (**53**) palladacycle precursors were treated with 4 equiv. Ag^+, formation of the dinuclear paramagnetic complexes **54** and **55** was observed (Scheme 5.42a). Characterization of these complexes by XAS, EPR, and Mössbauer spectroscopy suggests that the ligand ferrocene unit remains intact upon oxidation, while two Pd^{II} centers are oxidized to Pd^{III}. The oxidized complexes show an enhanced catalytic activity compared to the Pd^{II} species, presumably[105] due to their higher Lewis acidity.[115]

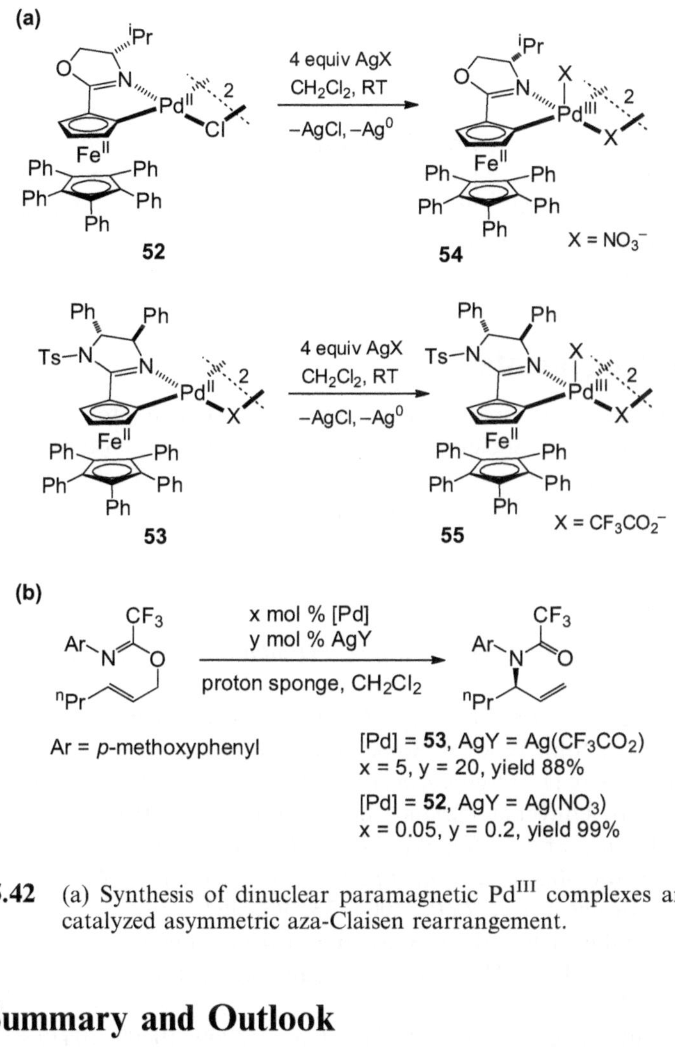

Scheme 5.42 (a) Synthesis of dinuclear paramagnetic Pd^{III} complexes and (b) Pd-catalyzed asymmetric aza-Claisen rearrangement.

5.4 Summary and Outlook

This chapter provides an overview of the organometallic chemistry of isolated Pd^{III} complexes and the proposed involvement of Pd^{III} intermediates in various organometallic transformations. The first mononuclear organometallic Pd^{III} complexes were isolated and characterized in 2010 and shown to exhibit C–C bond formation reactivity. Moreover, mononuclear Pd^{III} intermediates have been observed during the dioxygen-induced C–C bond elimination from Pd^{II} precursors. In addition, the involvement of mononuclear Pd^{III} intermediates has been proposed in a wide range of C–C and C–heteroatom bond formation reactions.

The first isolated dinuclear organometallic Pd^{III} complexes have been reported in 2006, and such dinuclear Pd^{III} species were shown to act as catalytically active intermediates in a range of C–H functionalization reactions, suggesting that Pd^{III} species may be more prevalent than previously anticipated. In addition, dinuclear Pd^{III} complexes that are not stabilized by a Pd–Pd bond and their role in catalytic processes have also been reported. Importantly, the facile interconversion between Pd^{III} and Pd^{IV} complexes has been demonstrated, suggesting that the intermediacy of either Pd^{III} or Pd^{IV} species cannot always be unambiguously confirmed or ruled out. Finally, we envision that in the near future the chemistry of the less-common oxidation states of Pd, such as Pd^{III}, Pd^{III} will lead to novel organometallic reactions that employ odd-electron oxidation states and involve one-electron oxidation or reduction steps.

Acknowledgements

We thank the Department of Chemistry at Washington University, the American Chemical Society Petroleum Research Fund (49914-DNI3), and Department of Energy Catalysis Science Program (DE-FG02-11ER16254) for financial support. L.M.M. is a Sloan Fellow.

References

1. P. M. Henry, *Palladium Catalyzed Oxidation of Hydrocarbons*, D. Reidel, Boston, 1980.
2. E. Negishi, *Handbook of Organopalladium Chemistry for Organic Synthesis*, John Wiley & Sons, Hoboken, NJ, 2002.
3. J. F. Hartwig, *Organotransition Metal Chemistry: From Bonding to Catalysis*, University Science Books, Sausalito, 2010.
4. N. Miyaura and A. Suzuki, *Chem. Rev.*, 1995, **95**, 2457–2483.
5. S. S. Stahl, *Angew. Chem., Int. Ed.*, 2004, **43**, 3400–3420.
6. B. M. Stoltz, *Chem. Lett.*, 2004, **33**, 362–367.
7. K. M. Gligorich and M. S. Sigman, *Chem. Commun.*, 2009, 3854–3867.
8. A. R. Dick, K. L. Hull and M. S. Sanford, *J. Am. Chem. Soc.*, 2004, **126**, 2300–2301.
9. A. Canty, *J. Chem. Soc., Dalton Trans.*, 2009, 10409–10417.
10. K. Muniz, *Angew. Chem. Int. Ed.*, 2009, **48**, 9412–9423.
11. T. W. Lyons and M. S. Sanford, *Chem. Rev.*, 2010, **110**, 1147–1169.
12. P. Sehnal, R. J. K. Taylor and I. J. S. Fairlamb, *Chem. Rev.*, 2010, **110**, 824–889.
13. L.-M. Xu, B.-J. Li, Z. Yang and Z.-J. Shi, *Chem. Soc. Rev.*, 2010, **39**, 712–733.
14. P. K. Byers, A. J. Canty, B. W. Skelton and A. H. White, *J. Chem. Soc., Chem. Commun.*, 1986, 1722–1724.
15. A. J. Canty, *Acc. Chem. Res.*, 1992, **25**, 83–90.
16. P. L. Alsters, P. F. Engel, M. P. Hogerheide, M. Copijn, A. L. Spek and G. van Koten, *Organometallics*, 1993, **12**, 1831–1844.
17. R. van Asselt, E. Rijnberg and C. J. Elsevier, *Organometallics*, 1994, **13**, 706–720.

18. R. van Belzen, H. Hoffmann and C. J. Elsevier, *Angew. Chem., Int. Ed.*, 1997, **36**, 1743–1745.
19. M.-C. Lagunas, R. A. Gossage, A. L. Spek and G. van Koten, *Organometallics*, 1998, **17**, 731–741.
20. Y. Yamamoto, T. Ohno and K. Itoh, *Angew. Chem., Int. Ed.*, 2002, **41**, 3662–3665.
21. Y. Yamamoto, S. Kuwabara, S. Matsuo, T. Ohno, H. Nishiyama and K. Itoh, *Organometallics*, 2004, **23**, 3898–3906.
22. J. Campora, P. Palma, D. del Rio, E. Carmona, C. Graiff and A. Tiripicchio, *Organometallics*, 2003, **22**, 3345–3347.
23. J. Campora, P. Palma, D. del Rio, J. A. Lopez and P. Valerga, *Chem. Commun.*, 2004, 1490–1491.
24. D. C. Powers and T. Ritter, *Top. Organomet.Chem.*, 2011, **35**, 129–156.
25. L. M. Mirica and J. R. Khusnutdinova, *Coord. Chem. Rev.*, 2013, **257**, 299–314.
26. D. C. Powers and T. Ritter, *Nat. Chem.*, 2009, **1**, 302–309.
27. D. C. Powers, M. A. L. Geibel, J. Klein and T. Ritter, *J. Am. Chem. Soc.*, 2009, **131**, 17050–17051.
28. N. R. Deprez and M. S. Sanford, *J. Am. Chem. Soc.*, 2009, **131**, 11234–11241.
29. D. C. Powers, D. Benitez, E. Tkatchouk, W. A. Goddard, III and T. Ritter, *J. Am. Chem. Soc.*, 2010, **132**, 14092–14103.
30. A. Tressaud, S. Khairoun, J. M. Dance, J. Grenncc, G. Demazeau and P. Hagenmuller, *C. R. Acad. Sci.*, 1982, **295**, 183.
31. A. J. Blake, A. J. Holder, T. I. Hyde and M. Schröder, *J. Chem. Soc., Chem. Commun.*, 1987, 987–988.
32. A. J. Blake, L. M. Gordon, A. J. Holder, T. I. Hyde, G. Reid and M. Schröder, *J. Chem. Soc., Chem. Commun.*, 1988, 1452–1454.
33. A. McAuley and T. W. Whitcombe, *Inorg. Chem.*, 1988, **27**, 3090–3099.
34. E. Stephen, A. J. Blake, E. Carter, D. Collison, E. S. Davies, R. Edge, W. Lewis, D. M. Murphy, C. Wilson, R. O. Gould, A. J. Holder, J. McMaster and M. Schröder, *Inorg. Chem.*, 2012, **51**, 1450–1461.
35. J. R. Khusnutdinova, N. P. Rath and L. M. Mirica, *J. Am. Chem. Soc.*, 2010, **132**, 7303–7305.
36. M. S. Remy, T. R. Cundari and M. S. Sanford, *Organometallics*, 2010, **29**, 1522–1525.
37. A. L. Seligson and W. C. Trogler, *J. Am. Chem. Soc.*, 1992, **114**, 7085–7089.
38. M. P. Lanci, M. S. Remy, W. Kaminsky, J. M. Mayer and M. S. Sanford, *J. Am. Chem. Soc.*, 2009, **131**, 15618–15620.
39. J. R. Khusnutdinova, N. P. Rath and L. M. Mirica, *J. Am. Chem. Soc.*, 2012, **134**, 2414–2422.
40. V. V. Rostovtsev, L. M. Henling, J. A. Labinger and J. E. Bercaw, *Inorg. Chem.*, 2002, **41**, 3608–3619.
41. E. M. Prokopchuk, H. A. Jenkins and R. J. Puddephatt, *Organometallics*, 1999, **18**, 2861–2866.

42. E. M. Prokopchuk and R. J. Puddephatt, *Can. J. Chem.*, 2003, **81**, 476–483.
43. F. Tang, Y. Zhang, N. P. Rath and L. M. Mirica, *Organometallics*, 2012, **31**, 6690–6696.
44. J. R. Khusnutdinova, F. Qu, Y. Zhang, N. P. Rath and L. M. Mirica, *Organometallics*, 2012, **31**, 4627–4630.
45. H.-B. Kraatz, M. E. Van der Boom, Y. Ben-David and D. Milstein, *Isr. J. Chem.*, 2001, **41**, 163–171.
46. T. T. Tsou and J. K. Kochi, *J. Am. Chem. Soc.*, 1979, **101**, 7547–7560.
47. J. Hassan, M. Sevignon, C. Gozzi, E. Schulz and M. Lemaire, *Chem. Rev.*, 2002, **102**, 1359–1469.
48. G. Manolikakes and P. Knochel, *Angew. Chem., Int. Ed.*, 2009, **48**, 205–209.
49. A. Jutand and A. Mosleh, *J. Org. Chem.*, 1997, **62**, 261–274.
50. C. Amatore, A. Jutand, F. Khalil and M. F. Nielsen, *J. Am. Chem. Soc.*, 1992, **114**, 7076–7085.
51. C. Amatore and A. Jutand, *Organometallics*, 1988, **7**, 2203–2214.
52. D. Kalyani, K. B. McMurtrey, S. R. Neufeldt and M. S. Sanford, *J. Am. Chem. Soc.*, 2011, **133**, 18566–18569.
53. W.-Y. Yu, W. N. Sit, Z. Zhou and A. S. C. Chan, *Org. Lett.*, 2009, **11**, 3174–3177.
54. W.-Y. Yu, W. N. Sit, K.-M. Lai, Z. Zhou and A. S. C. Chan, *J. Am. Chem. Soc.*, 2008, **130**, 3304–3306.
55. S. J. Reid and M. C. Baird, *Dalton Trans.*, 2003, 3975–3980.
56. Z. Weng, S. Teo and T. S. A. Hor, *Acc. Chem. Res.*, 2007, **40**, 676–684.
57. L. Boisvert, M. C. Denney, H. S. Kloek and K. I. Goldberg, *J. Am. Chem. Soc.*, 2009, **131**, 15802–15814.
58. E. Solomon-Rapaport, A. Masarwa, H. Cohen and D. Meyerstein, *Inorg. Chim. Acta*, 2000, **299**, 41–46.
59. C. Mansano-Weiss, H. Cohen and D. Meyerstein, *J. Inorg. Biochem.*, 2002, **91**, 199–204.
60. C. Mansano-Weiss, A. Masarwa, H. Cohen and D. Meyerstein, *Inorg. Chim. Acta*, 2005, **358**, 2199–2206.
61. G. Ferguson, P. K. Monaghan, M. Parvez and R. J. Puddephatt, *Organometallics*, 1985, **4**, 1669–1674.
62. T. S. Mei, X. S. Wang and J. Q. Yu, *J. Am. Chem. Soc.*, 2009, **131**, 10806–10807.
63. J. J. Neumann, S. Rakshit, T. Droege and F. Glorius, *Angew. Chem., Int. Ed.*, 2009, **48**, 6892–6895.
64. J. Campora, P. Palma, D. del Rio, J. A. Lopez, E. Alvarez and N. G. Connelly, *Organometallics*, 2005, **24**, 3624–3628.
65. S. Takaishi, M. Takamura, T. Kajiwara, H. Miyasaka, M. Yamashita, M. Lwata, H. Matsuzaki, H. Okarnoto, H. Tanaka, S. Kuroda, H. Nishikawa, H. Oshio, K. Kato and M. Takata, *J. Am. Chem. Soc.*, 2008, **130**, 12080–12084.
66. M. Yamashita and S. Takaishi, *Chem. Commun.*, 2010, **46**, 4438–4448.

67. S. Takaishi, H. S. Wu, J. X. Xie, T. Kajiwara, B. K. Breedlove, H. Miyasaka and M. Yamashita, *Inorg. Chem.*, 2010, **49**, 3694–3696.
68. A. N. Vedernikov, M. D. Sayakhov, A. G. Zazybin and B. N. Solomonov, *Russ. J. Org. Chem.*, 1998, **34**, 812–815.
69. A. N. Vedernikov, A. I. Kuramshin, M. D. Sayakhov, A. A. Sobanov and B. N. Solomonov, *Russ. J. Org. Chem.*, 1997, **33**, 29–33.
70. A. S. Dneprovskii, A. A. Ermoshkin, A. N. Kasatochkin and V. P. Boyarskii, *Russ. J. Org. Chem.*, 2003, **39**, 933–946.
71. J. R. Khusnutdinova, N. P. Rath and L. M. Mirica, *Angew. Chem., Int. Ed.*, 2011, **50**, 5532–5536.
72. J. E. Bercaw, A. C. Durrell, H. B. Gray, J. C. Green, N. Hazari, J. A. Labinger and J. R. Winkler, *Inorg. Chem.*, 2010, **49**, 1801–1810.
73. J. Luo, J. R. Khusnutdinova, N. P. Rath and L. M. Mirica, *Chem. Commun.*, 2012, **48**, 1532–1534.
74. F. A. Cotton, M. Matusz and R. Poli, *Inorg. Chem.*, 1987, **26**, 1472–1474.
75. F. A. Cotton, M. Matusz, R. Poli and X. Feng, *J. Am. Chem. Soc.*, 1988, **110**, 1144–1154.
76. C. L. Yao, L. P. He, J. D. Korp and J. L. Bear, *Inorg. Chem.*, 1988, **27**, 4389–4395.
77. J. F. Berry, E. Bill, E. Bothe, F. A. Cotton, N. S. Dalal, S. A. Ibragimov, N. Kaur, C. Y. Liu, C. A. Murillo, S. Nellutla, J. M. North and D. Villagran, *J. Am. Chem. Soc.*, 2007, **129**, 1393–1401.
78. A. M. Bond, A. J. Canty, J. B. Cooper, V. Tedesco, P. R. Traill and D. M. Way, *Inorg. Chim. Acta*, 1996, **251**, 185–192.
79. F. A. Cotton, J. D. Gu, C. A. Murillo and D. J. Timmons, *J. Am. Chem. Soc.*, 1998, **120**, 13280–13281.
80. G. J. Chuang, W. Wang, E. Lee and T. Ritter, *J. Am. Chem. Soc.*, 2011, **133**, 1760–1762.
81. F. A. Cotton, I. O. Koshevoy, P. Lahuerta, C. A. Murillo, M. Sanau, M. A. Ubeda and Q. Zhao, *J. Am. Chem. Soc.*, 2006, **128**, 13674–13675.
82. D. Penno, F. Estevan, E. Fernandez, P. Hirva, P. Lahuerta, M. Sanau and M. A. Ubeda, *Organometallics*, 2011, **30**, 2083–2094.
83. D. Penno, V. Lillo, I. O. Koshevoy, M. Sanau, M. A. Ubeda, P. Lahuerta and E. Fernandez, *Chem.–Eur. J.*, 2008, **14**, 10648–10655.
84. J. M. Davidson and C. Triggs, *Chem. Ind. (London)*, 1966, 457.
85. G. T. Tisue and W. J. Downs, *J. Chem. Soc., Chem. Commun.*, 1969, 410.
86. P. M. Henry, *J. Org. Chem.*, 1971, **36**, 1886–1890.
87. D. R. Fahey, *J. Organomet. Chem.*, 1971, **27**, 283–292.
88. L. M. Stock, K.-t. Tse, L. J. Vorvick and S. A. Walstrum, *J. Org. Chem.*, 1981, **46**, 1757–1759.
89. S. J. Tremont and H. ur Rahman, *J. Am. Chem. Soc.*, 1984, **106**, 5759–5760.
90. T. Yoneyama and R. H. Crabtree, *J. Mol. Catal. A: Chem.*, 1996, **108**, 35–40.
91. J. M. Racowski and M. S. Sanford, *Top.Organomet.Chem.*, 2011, **35**, 61–84.

92. R. Giri, X. Chen and J. Q. Yu, *Angew. Chem., Int. Ed.*, 2005, **44**, 2112–2115.
93. R. Giri, J. Liang, J. G. Lei, J. J. Li, D. H. Wang, X. Chen, I. C. Naggar, C. Y. Guo, B. M. Foxman and J. Q. Yu, *Angew. Chem., Int. Ed.*, 2005, **44**, 7420–7424.
94. E. M. Beck, N. P. Grimster, R. Hatley and M. J. Gaunt, *J. Am. Chem. Soc.*, 2006, **128**, 2528–2529.
95. K. M. Engle, T.-S. Mei, M. Wasa and J.-Q. Yu, *Acc. Chem. Res.*, 2012, **45**, 788–802.
96. D. C. Powers, D. Y. Xiao, M. A. L. Geibel and T. Ritter, *J. Am. Chem. Soc.*, 2010, **132**, 14530–14536.
97. D. Kalyani, A. R. Dick, W. Q. Anani and M. S. Sanford, *Tetrahedron*, 2006, **62**, 11483–11498.
98. D. Kalyani, A. R. Dick, W. Q. Anani and M. S. Sanford, *Org. Lett.*, 2006, **8**, 2523–2526.
99. S. R. Whitfield and M. S. Sanford, *J. Am. Chem. Soc.*, 2007, **129**, 15142–15143.
100. X. Wan, Z. Ma, B. Li, K. Zhang, S. Cao, S. Zhang and Z. Shi, *J. Am. Chem. Soc.*, 2006, **128**, 7416–7417.
101. O. Daugulis and V. G. Zaitsev, *Angew. Chem., Int. Ed.*, 2005, **44**, 4046–4048.
102. J. Spencer, B. Z. Chowdhry, A. I. Mallet, R. P. Rathnam, T. Adatia, A. Bashall and F. Rominger, *Tetrahedron*, 2008, **64**, 6082–6089.
103. D. Kalyani, N. R. Deprez, L. V. Desai and M. S. Sanford, *J. Am. Chem. Soc.*, 2005, **127**, 7330–7331.
104. A. Ariafard, C. J. T. Hyland, A. J. Canty, M. Sharma and B. F. Yates, *Inorg. Chem.*, 2011, **50**, 6449–6457.
105. A. Ariafard, C. J. T. Hyland, A. J. Canty, M. Sharma, N. J. Brookes and B. F. Yates, *Inorg. Chem.*, 2010, **49**, 11249–11253.
106. D. Shabashov and O. Daugulis, *Org. Lett.*, 2005, **7**, 3657–3659.
107. V. G. Zaitsev, D. Shabashov and O. Daugulis, *J. Am. Chem. Soc.*, 2005, **127**, 13154–13155.
108. H. A. Chiong, Q.-N. Pham and O. Daugulis, *J. Am. Chem. Soc.*, 2007, **129**, 9879–9884.
109. F. Yang, Y. Wu, Z. Zhu, J. Zhang and Y. Li, *Tetrahedron*, 2008, **64**, 6782–6787.
110. P. Eisenberger, S. Gischig and A. Togni, *Chem.–Eur. J.*, 2006, **12**, 2579–2586.
111. D. C. Powers, E. Lee, A. Ariafard, M. S. Sanford, B. F. Yates, A. J. Canty and T. Ritter, *J. Am. Chem. Soc.*, 2012, **134**, 12002–12009.
112. X. Wang, L. Truesdale and J.-Q. Yu, *J. Am. Chem. Soc.*, 2010, **132**, 3648–3649.
113. Y. Ye, N. D. Ball, J. W. Kampf and M. S. Sanford, *J. Am. Chem. Soc.*, 2010, **132**, 14682–14687.
114. D. Milstein and J. K. Stille, *J. Am. Chem. Soc.*, 1979, **101**, 4981–4991.
115. S. H. Eitel, M. Bauer, D. Schweinfurth, N. Deibel, B. Sarkar, H. Kelm, H.-J. Krüger, W. Frey and R. Peters, *J. Am. Chem. Soc.*, 2012, **134**, 4683–4693.

CHAPTER 6

Aromatic C–F Activation: Converting Fluoroarenes to Useful Building Blocks

LAUREN KEYES AND JENNIFER A. LOVE*

Department of Chemistry, University of British Columbia, 2036 Main Mall, Vancouver, British Columbia, Canada, V6T 1Z1
*Email: jenlove@chem.ubc.ca

6.1 Introduction

Chemical synthesis is a powerful tool for the construction of new pharmaceuticals and materials with far-reaching applications. One structural unit of particular importance is the carbon–fluorine (C–F) functionality. The carbon–fluorine bond is the strongest single bond in organic chemistry with exceptionally high bond dissociation energies having been reported.[1] In addition, the highly electronegative nature of the fluorine atom has the net effect of strengthening adjacent carbon–carbon (C–C) and carbon–hydrogen (C–H) bonds within a molecule, as well as altering various polar interactions.[2] For example, the incorporation of a fluorine functionality onto a phenolic group is associated with a 0.2–1.2 unit reduction in p*K*a.[2b] Although most often substituted in place of a hydrogen atom, the Van der Waals radius of fluorine (1.47 Å) is more similar to oxygen (1.57 Å) than hydrogen (1.2 Å).[2b] Therefore, through the incorporation of fluorine substituents, the electronic and steric profiles of small organic molecules can be simultaneously altered.

RSC Catalysis Series No. 11
C–H and C–X Bond Functionalization: Transition Metal Mediation
Edited by Xavi Ribas

Published by the Royal Society of Chemistry, www.rsc.org

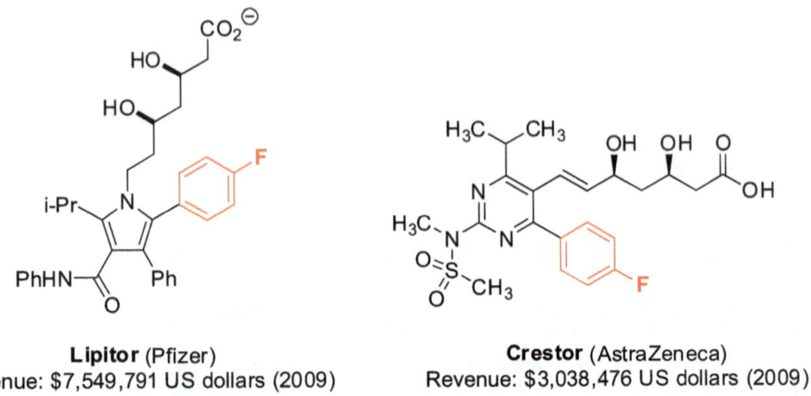

Figure 6.1 Top-selling aryl fluoride-containing pharmaceuticals.

As a direct result of these unique properties, fluorine substituents are emerging as common structural motifs in pharmaceuticals,[3] agrochemicals[3] and materials.[4] Indeed, recent estimates indicate that approximately half of new pharmaceutical agents contain fluoroarene functionalities.[3] Furthermore, as of 2009, two of the five top-selling pharmaceuticals in the US, Pfizer's Lipitor and AstraZeneca's Crestor, contain an aryl fluoride (Figure 6.1).[5]

Despite the numerous applications of fluorine-containing compounds, no known naturally occurring sources of aryl fluorides have been identified to date.[6] As a result, aryl fluorides must be generated through chemical synthesis. Both nucleophilic and electrophilic fluorination strategies have been investigated. However, both have considerable drawbacks. The fluoride ion is highly solvated, lowering its nucleophilicity. The high electronegativity of fluorine has slowed the development of electrophilic fluorinating reagents, although a number of exciting results have recently been forthcoming (*vide infra*). The widespread utility of aryl fluorides necessitates the development of new, versatile technologies for the synthesis of functionalized aryl fluorides. One method involves the use of C–F activation of polyfluoroarenes to access such aryl fluoride building blocks. This chapter will discuss transition metal catalyzed C–F activation as a general approach towards the synthesis of new fluoroarenes.

6.2 Traditional Methods for the Synthesis of Aryl Fluorides

Traditional methods for the synthesis of aryl fluorides include nucleophilic aromatic substitution (the Halex process, Scheme 6.1A)[7] and the pyrolysis of diazonium tetrafluoroborates (the Balz–Schiemann reaction, Scheme 6.1B).[8] Indeed, the majority of simple commercially available aryl fluorides are prepared through such processes. The Halex process is used on an industrial scale for the conversion of simple aryl chlorides to the corresponding aryl

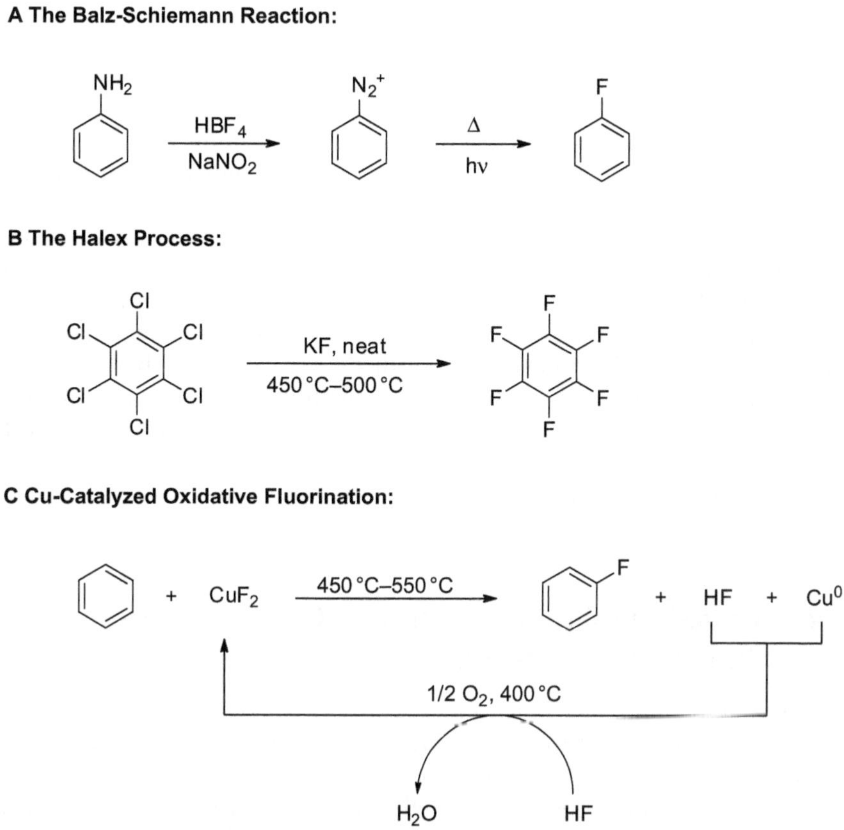

Scheme 6.1 Traditional methods for the preparation of aryl fluorides.

fluorides. An alternative route, developed by DuPont, involves the copper catalyzed oxidative fluorination of simple arenes (Scheme 6.1C).[9] Importantly, stoichiometric by-products as well as the copper catalyst are readily recycled. However, all of these methods require the use of high temperatures and harsh conditions which restrict the functional group tolerance of these reactions. Furthermore, these processes are not applicable for late-stage fluorination of complex molecules, for which there is an increasing need. Clearly, there exists a need for the development of robust synthetic methods for the construction of functionalized aryl fluorides.

6.3 Modern Methods for the Synthesis of Aryl Fluorides

Given the relatively few methodologies for the construction of aryl fluorides and their associated limitations, much effort has gone towards the development of new methodologies for the construction of fluorinated molecules. This has led to the development of different strategies for the construction of polyfluoroarenes. One general strategy has been the selective incorporation of fluorine into a previously non-fluorinated position on an aromatic ring using either electrophilic[10] or nucleophilic[11] fluorinating agents in conjunction with transition metals. However, thus far these methodologies are limited to the construction of monofluoroarenes. Catalytic fluorination technologies have been the subject of several recent reviews[12] and will not be discussed in this chapter.

Alternatively, methodologies involving the catalytic functionalization of aromatic C–F bonds have also emerged. Indeed, the turn of the millennium has seen the development of technologies for both the hydrodefluorination and/or cross-coupling of aryl fluorides. Both strategies can provide functionalized aryl fluorides that have the potential for further use as synthetic building blocks. A similar approach involves the direct activation of C–H bonds of fluoroarenes. However, this approach is complicated by the challenge of selective activation of C–H bonds in the presence of C–F bonds. As examples of this reactivity remain limited,[13] these methodologies will not be discussed in detail.

6.4 Transition Metal Catalyzed C–F Activation – Stoichiometric Studies

Selective activation and functionalization of carbon–element bonds is of fundamental importance to organic and organometallic chemistry. To this end, transition metal complexes have been extensively utilized to activate strong carbon–element bonds. Not surprisingly, transition metal mediated stoichiometric C–F activation has also been widely explored.[14] Indeed, metal fluoride complexes, resulting from aromatic C–F activation, have been reported for the early[15–17] and late d-block transition metals[19–29] as well as the lanthanide and actinides.[18] More recently examples of C–F activation involving main group complexes have also emerged.[30] Stoichiometric transition metal C–F activation has been reviewed by Kiplinger *et al.*,[14a] Burdeniue *et al.*,[14b] Torrens[14c] and Perutz and Braun.[14d] Herein specific examples are presented as they relate to the development of methodologies for the catalytic functionalization of aryl fluorides. For clarity, throughout this chapter electron-deficient transition metal complexes (d^0f^n or d^1f^n) include group 3, 4 and 5 transition metals as well as the lanthanides and actinides. Electron-rich transition metal complexes (d^n, $n > 6$) consist of transition metal complexes of groups 6–12.

6.4.1 Stoichiometric C–F Activation – Electron-Deficient Metal Complexes

One of the earliest examples of transition metal mediated activation of an aromatic C–F bond was reported by Stone and coworkers in 1964 (eqn (6.1)).[15] During the course of their studies of early transition metal pentafluorophenyl complexes it was observed that prolonged pyrolysis of $Cp_2Ti(C_6F_5)_2$ (**1**) resulted in formation of small quantities of $Cp_2Ti(C_6F_5)_2F$ (**2**). It was speculated that complex **2** was formed through intramolecular F migration.

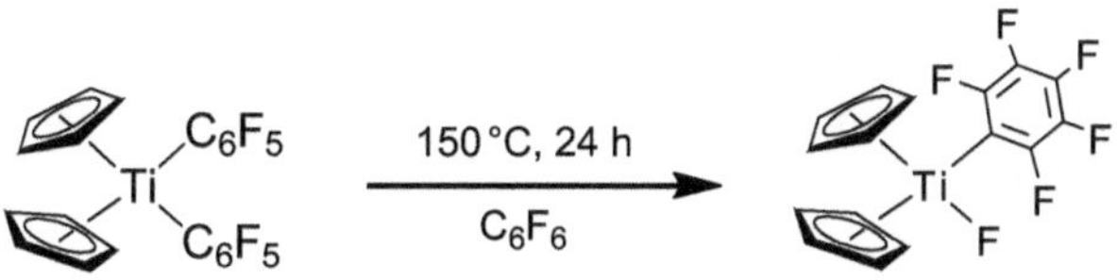

(**1**) (**2**)

8.5% yield

(6.1)

The reactivity of group 3 hydride complexes with hexafluorobenzene derivatives has also been extensively explored. Jones and co-workers have made several important contributions in this area.[16] In particular, in 2001 the first examples of aromatic C–F activation with $Zr^{(IV)}$ dihydride complexes were reported.[16] Prolonged heating of a solution mixture of $Cp^*_2ZrH_2$ (**3**) with monofluorobenzene resulted in a mixture of organic and organometallic products – one of which was $Cp^*_2Zr(C_6H_5)F$ complex **6** (eqn (6.2)).

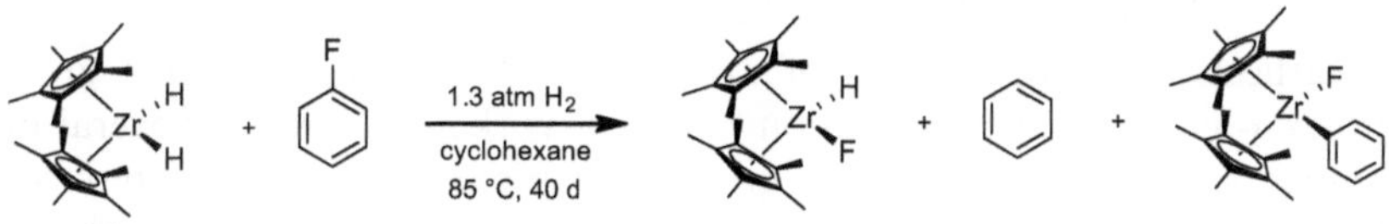

(**3**) (**4**) (**6**)

ratio: 1 : 1 : 0.75

(6.2)

On the basis of the product distribution as well radical trapping experiments it was concluded that $Cp^*_2Zr(C_6H_5)F$ (**6**) was formed as a result of β-F elimination from $Cp^*_2Zr(C_6H_5)H$ complex **5** to form a benzyne intermediate, which undergoes rapid insertion into the aromatic C–F bond to generate $Cp^*_2Zr(C_6H_5)F$ complex **6** (eqn (6.3)). Formation of such a complex is likely driven by the formation of a particularly strong Zr–F bond. Consistent with this, formation of complex **6** was found to be irreversible.

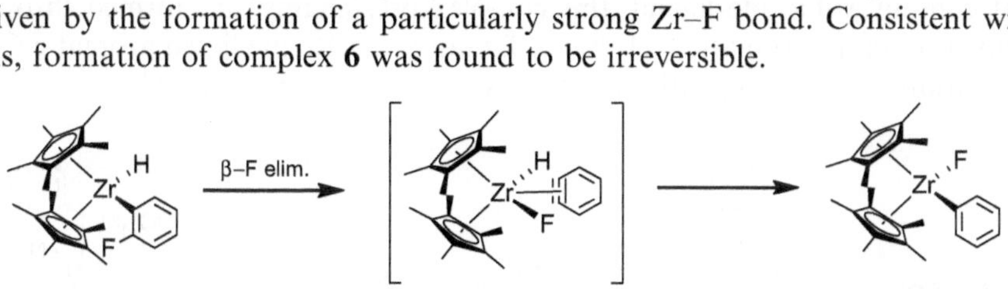

(**5**) (**6**)

(6.3)

Group 5 complexes have also been utilized in the activation of C–F bonds but to a much lesser extent.[17] Furthermore, the few reported examples involve the activation of either aliphatic or alkynyl C–F bonds.[17] To date, there are no reported examples of aromatic C–F activation involving group 5 transition metal complexes. In contrast, the lanthanide and actinindes have demonstrated robust reactivity towards aromatic C–F bonds.[18] Early work in this area was pioneered by Deacon[18a–c] and Bergman[18d] who investigated the chemistry of Yb and U complexes, respectively. However, as the chemistry of the f-block has been limited to single electron reduction processes, this work will not be discussed in detail.

6.4.2 Stoichiometric C–F Activation – Electron-Rich Metal Complexes

Late transition metal complexes, particularly of group 9 and 10, have been extensively applied towards the activation of both aliphatic and aromatic C–F bonds as these complexes are expected to have broad functional group tolerance. Such systems offer considerable potential for further development into catalytic methodologies with broad substrate scope (*vide infra*).

6.4.2.1 Group 6 and 7 Complexes

The earliest reported example of aromatic C–F activation with a group 6 complex was reported by Richmond and co-workers in 1987.[19] Several other examples of C–F activation by group 6 complexes have been reported.[19] However, as group 6 transition metal complexes have not been employed to date in the catalytic functionalization of aryl fluorides, this work will not be discussed within the context of this chapter.

On the other hand, group 7 Mn^0 complexes have been employed in stoichiometric activation of aryl fluorides as well methodologies for the catalytic functionalization of these substrates (*vide infra*). In 1975, Bruce and co-workers reported the first examples of stoichiometric aromatic C–F activation involving group 7 complexes.[20] Reaction of Mn^0 carbonyl compounds, such as **7**, with pentafluoroazobenzene resulted in Mn^{II} complex **8** – albeit in particularly low yields despite the use of harsh conditions (eqn (6.4)). Although complex **8** does not contain a fluoride ligand, this complex presumably was formed through activation of an aromatic C–F bond. Examples of Re-mediated aromatic C–F activation remain rare, with only two systems reported to date.[21]

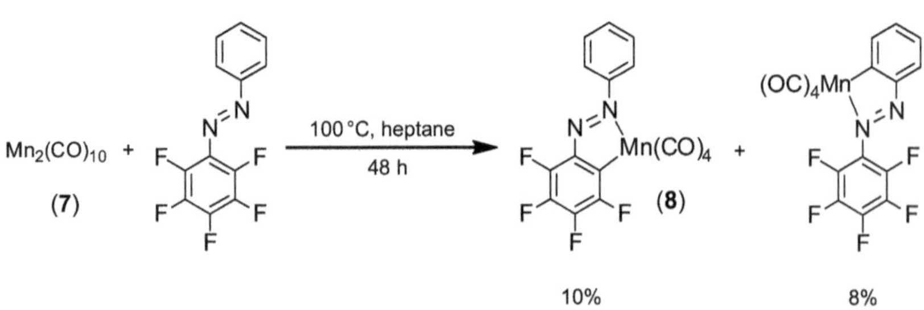

(6.4)

6.4.2.2 Group 8 Complexes

Given the tendency for Fe-complexes to undergo a variety of disproportionation reactions, the majority of the investigations of group 8 mediated C–F activation of aryl fluorides has centered on the use of electron-rich Ru[22] or Os[23] complexes. The first example of aromatic C–F activation using a group 8 complex was reported by Bruce and co-workers in 1976 and involved the reaction of nucleophilic Ru complex **9** with the same azafluorobenzene ligands (eqn (6.5)) as in their previously reported work with Mn complexes (*vide supra*).[22a] Once again, the intermediacy of a M–F complex was invoked to rationalize formation of metal complex **10**. The observed preference for C–F activation over weaker aromatic C–H bonds was rationalized on the basis of the large steric differences between the two aryl rings.

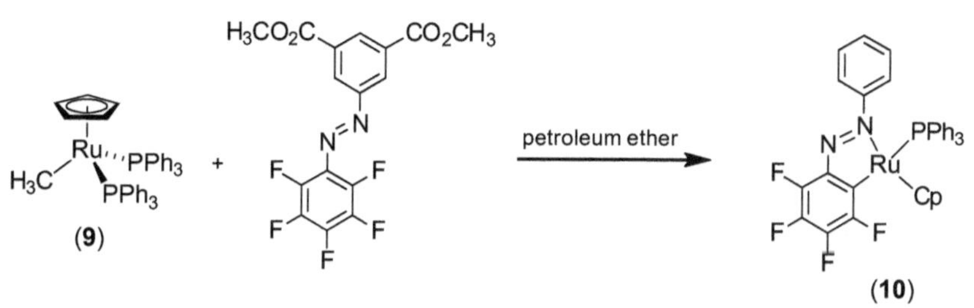

(6.5)

Interestingly despite the growing interest in the development of Fe-based catalysis, only a single example of discrete, isolable Fe–F complexes has been reported. In 2011, Li and co-workers reported an interesting study of Fe^0-mediated C–F activation of aromatic C–F bonds.[24] Using a similar strategy as was utilized by Bruce and co-workers in their studies of Ru-mediated C–F activation,[22a] the Li group found that imine-based substrates (**11**) facilitated intramolecular C–F activation with the imine moiety acting as directing group, resulting in the formation of cyclometalated bis-chelate iron(II) complex **14** (eqn (6.6)).

The formation of bis-chelate complexes resulting from double C–F activation is particularly noteworthy as it highlights the redox-active nature of Fe metal centers. A potential mechanism for the formation of these complexes was proposed to involve initial C–F activation to generate intermediate Fe^{II} complex **12**. A subsequent redox reaction involving loss of difluorotrimethylphosphorane (F_2PMe_3) and reaction with a second equivalent of imine **11** resulted in the formation of cyclometalated product **14** (Scheme 6.2). Consistent with their proposed mechanism, the formation of F_2PMe_3 was verified using ^{19}F and ^{31}P NMR spectroscopy.

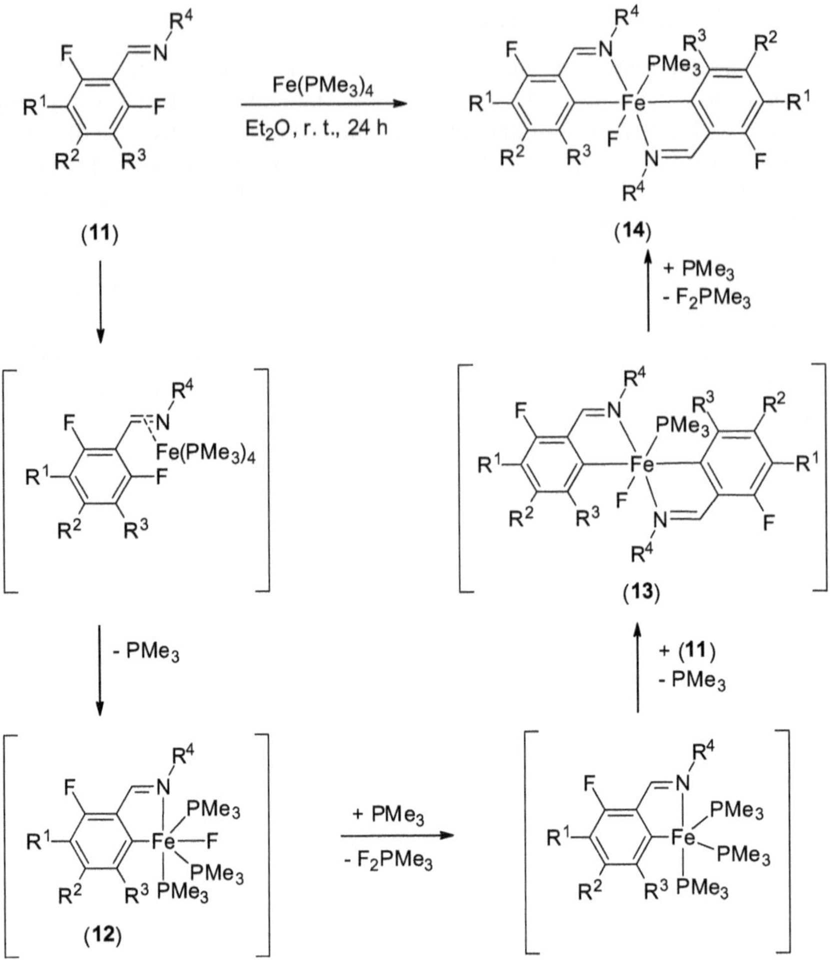

Scheme 6.2 Mechanism for the formation of cyclometalated iron(II) complex (**14**).

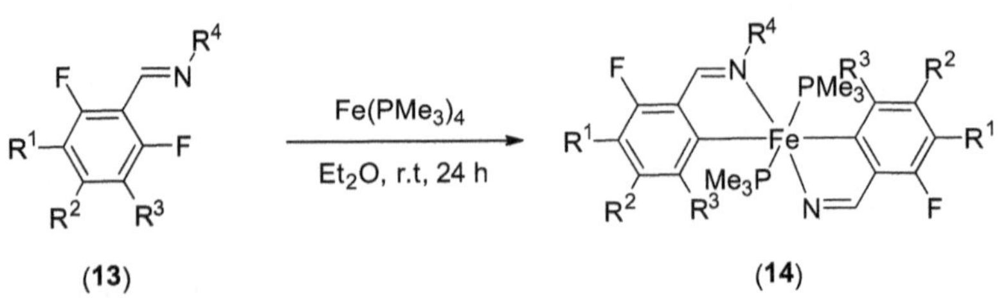

(6.6)

R = Ph Ph Me
R^1 = H F F
R^2 = H F HNMe
R^3 = H F F

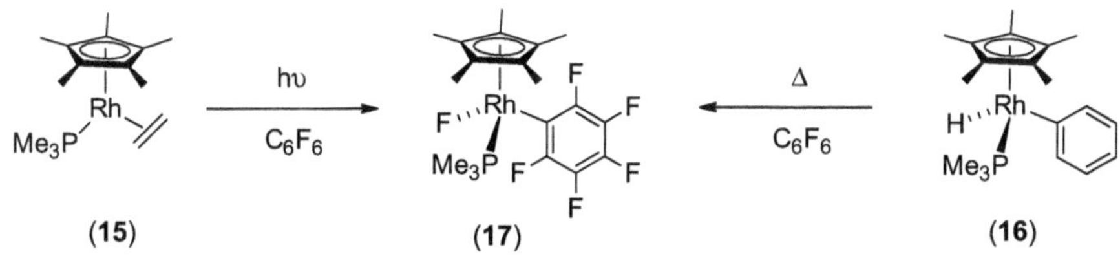

Scheme 6.3 C–F activation of hexafluorobenzene with Cp*Rh(PMe_3)(C_2H_4) (**15**) and Cp*Rh(PMe_3)(H)(C_6H_5) (**16**).

6.4.2.3 Group 9 Complexes

The application of group 9 complexes towards aromatic C–F activation was pioneered by Milstein, Jones and Perutz. Indeed, as a result of their detailed investigations of the stoichiometric reactivity of various metal complexes, several important catalytic methodologies have been developed (*vide infra*).[25] Studies of the reactivity of group 9 complexes towards aromatic C–F bonds began with a report by Jones and Perutz on the intermolecular C–F activation of hexafluorobenzene with cyclopentadienylrhodium(I) complexes (Scheme 6.3).[25h] Interestingly, during the course of their investigations, they found that the nature of the ancillary ligands on the metal center had a direct effect on the reactivity of the metal complex. Specifically, Cp*Rh(PMe_3)(C_2H_4) complex **15** required photochemical activation whereas related Cp*Rh(PMe_3)(C_6H_5) complex **16** reacted under thermal conditions to give the same Rh^{III} complex **17**. The mechanism of formation and reaction kinetics associated with these two related, yet distinct, transformations were studied in detail.[25h]

Since this initial study, Braun and Perutz have expanded the C–F chemistry of Rh^{I} complexes to include fluorinated *N*-heterocycles.[25i] In 2004, Braun and co-workers reported the reaction of RhH(PEt_3)$_2$ (**18**) with pentafluoropyridine to generate activated Rh^{III} complex **19** – [Rh(4-C_5NF_4)(PEt_3)$_3$]$^{2+}$ (eqn (6.7)).[25h] The regioselectivity associated with the reaction is consistent with C–F activation proceeding *via* either nucleophilic substitution or electron transfer pathways as opposed to a concerted oxidation addition mechanism which would favour substitution at the 2-position. It is noteworthy that this regioselectivity is in direct contrast to what is observed with Ni or Pd systems (*vide infra*).

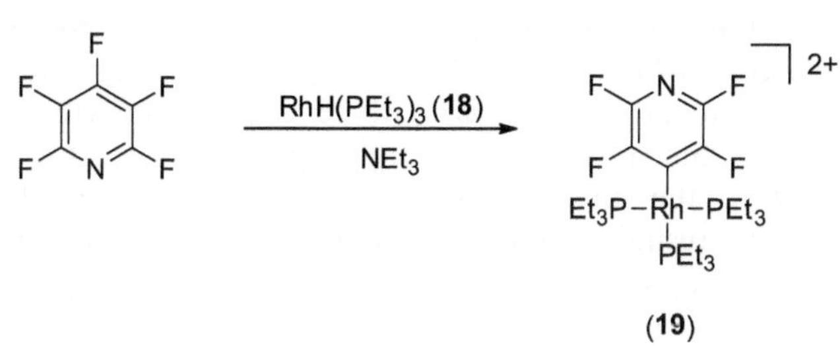

(6.7)

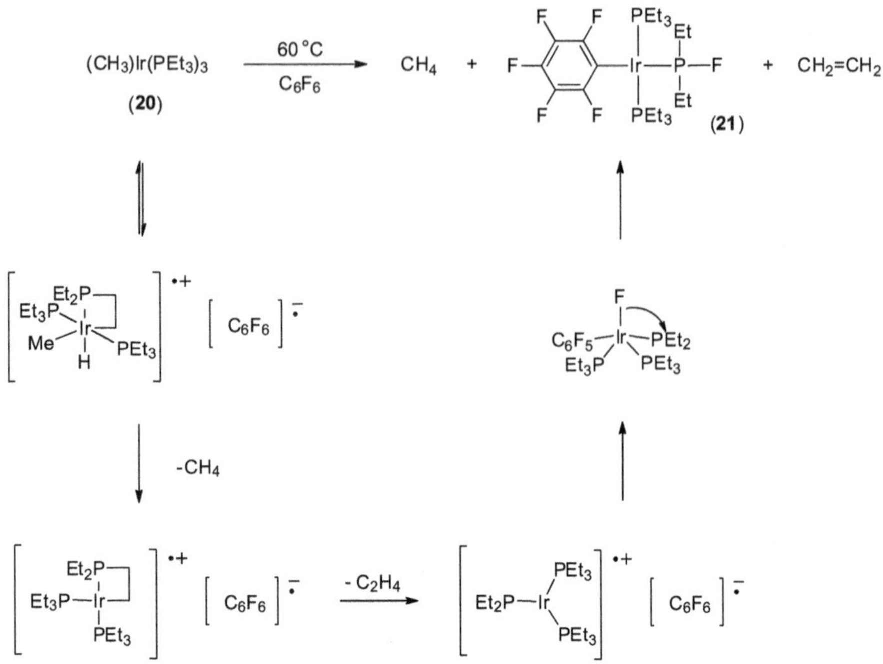

Scheme 6.4 Ir-catalyzed activation of hexafluorobenzene.

In 1991, Milstein and co-workers disclosed the first examples of Ir-mediated C–F activation of aryl fluorides.[25j] In this report it was found that thermolysis of $(CH_3)Ir(PEt_3)_3$ (**20**) in hexafluorobenzene resulted in the formation of $Ir(PEt_3)_2(PEt_2F)(C_6F_5)$ (**21**) (Scheme 6.4). The formation of Ir complex **21** is intriguing as the formation of a P–F bond is seemingly preferred over strong C–F and P–C bonds. Further investigations lead the authors to propose that this product is formed through a radical mechanism (Scheme 6.4).

Co complexes have also demonstrated utility in the activation of aromatic C–F bonds. However, the chemistry of these complexes has been considerably less explored than the other group 9 metal complexes. In 2006, Li and co-workers reported some of the first examples of discrete Co–F complexes (**23**) which are formed through intramolecular C–F activation of a fluorinated azine ligand using electron-rich $CoMe(PMe_3)_4$ (eqn (6.8)).[25c] Similarly to other cyclometalated complexes (*vide supra*) these complexes were found to be highly stable. In a later report, the same group reported the use of the same Co-complex **22** to the intermolecular C–F activation of perfluorotoluene (eqn (6.9)).[25n] The resulting benzyne complex **24** was trapped, isolated and characterized by X-ray crystallography.

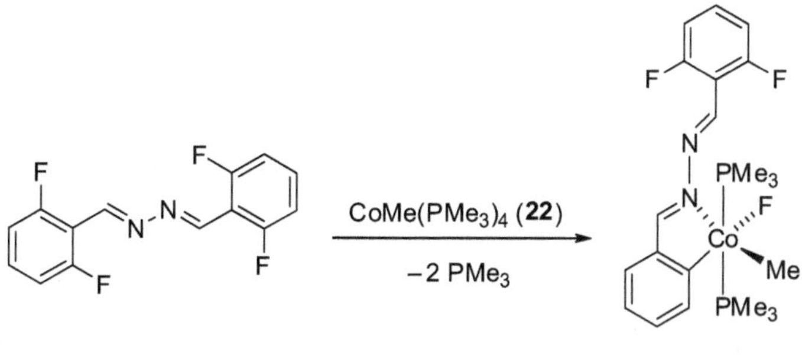

(23)

(6.8)

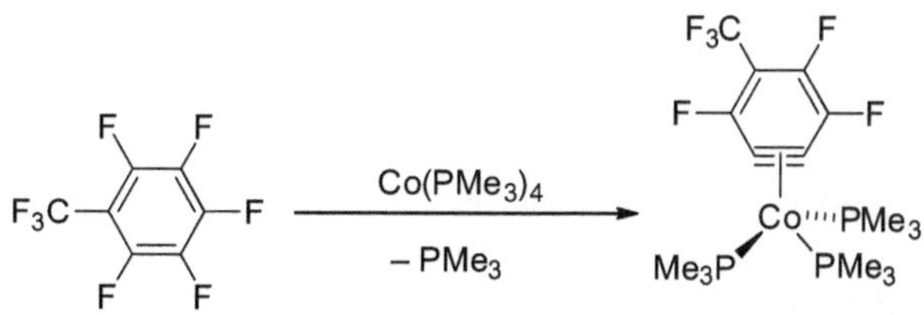

(24)

(6.9)

Much more recently, Holland and co-workers reported a unique example of low-coordinate Co–F complexes **25** and **26** formed through intermolecular C–F activation reactions with monofluorobenzene (eqn (6.10)).[25o] Detailed kinetic and mechanistic investigations have revealed that these complexes are formed through the concerted oxidative addition of C–F bonds.[25o]

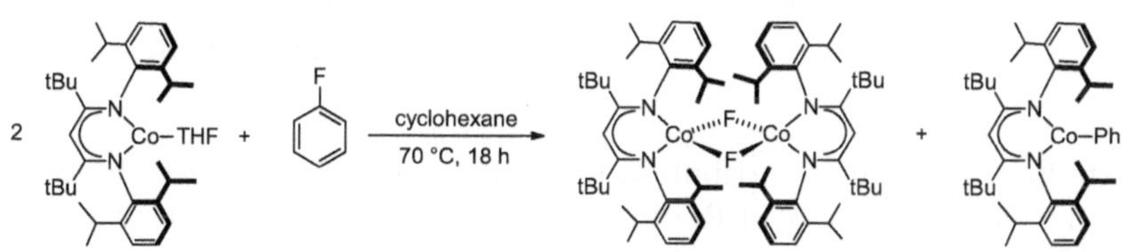

(25) (26)

(6.10)

6.4.2.4 Group 10 Complexes

The C–F activation chemistry of group 10 transition metal complexes has been extensively studied. In particular, Ni complexes have demonstrated robust reactivity towards the activation of aromatic C–F bonds.[26] In 1977, Fahey and Mahan reported the first examples of Ni(0) mediated C–F activation of hexafluorobenzene.[26a] However, it should be noted that the reaction of $Ni(PEt_3)_2(COD)$ (**27**) and hexafluorobenzene to generate $(PEt_3)_2Ni(F)(C_6F_5)$ (**28**) required long reaction time (several days) and resulted in a very low yield (7%) of Ni^{II}-F complex **28** (eqn (6.11)).

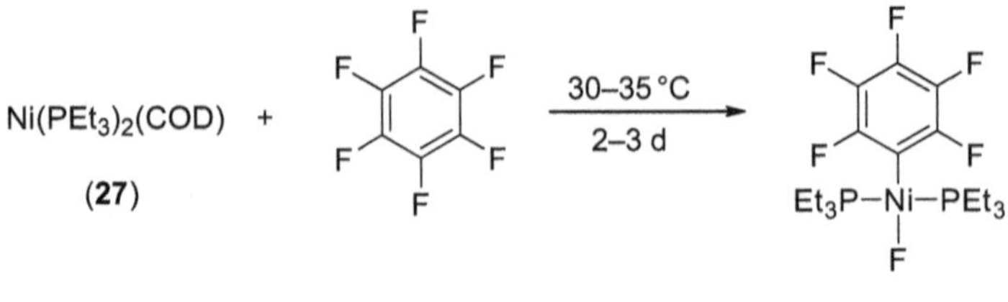

(27) (6.11)

Building on their original investigations of group 9 mediated aromatic C–F activation, Perutz and Braun jointly reported the application of a Ni^0 complex – $Ni(COD)_2$ – to the stoichiometric C–F activation of aromatic *N*-heterocycles (eqn (6.12)).[26b–e] It is worth noting that the reaction of $Ni(COD)_2$ with pentafluoropyridine proceeded with exclusive selectivity for activation at the 2-position of the heterocycle. This result is in direct contrast to the selectivity that was observed with Rh^I complexes (eqn (6.7)) and has been suggested to result from a concerted oxidative addition process in which the regioselectivity is dictated by pre-coordination of the Ni-complex to the nitrogen atom of the pyridine ring.[26b–e] Recent computational studies have indicated that the phosphine ligands are likely non-innocent and are involved in C–F activation.[26k]

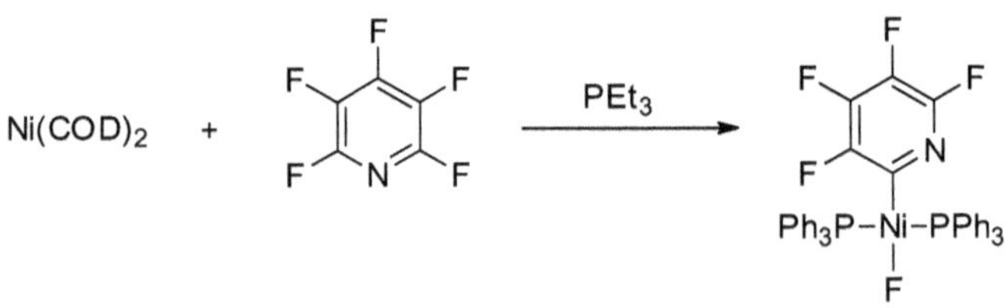

(6.12)

Another important contribution to the field of Ni-mediated C–F activation includes the recent work of Radius and co-workers who, in 2004, reported the rapid, room temperature C–F activation of hexafluorobenzene with Ni^0 carbene complex **29** (eqn (6.13)).[26h]

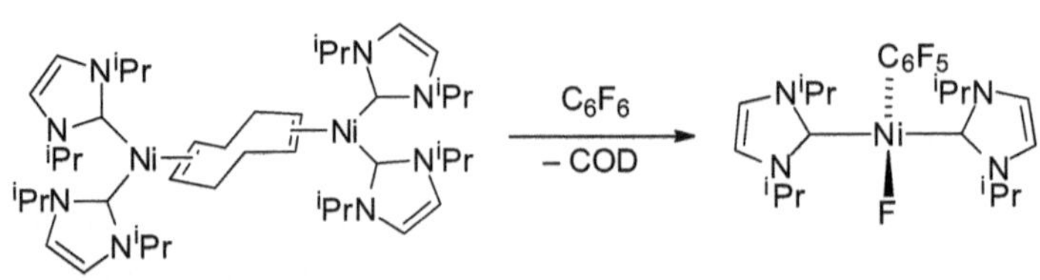

(29) (6.13)

As a logical extension of their work with $Ni^0(0)$ complexes, Braun and Perutz also investigated the chemistry of Pd^0 and Pt^0 complexes.[27] It is interesting to note that unlike the Ni^0 system, both Pd^0 and Pt^0 complexes **30** and **31** were found to react preferentially at the 4-position of pentafluoropyridine (eqn (6.14) and (6.15)). This is the same selectivity that was exhibited by Rh^I complexes and, as such, is believed to result from a nucleophilic substitution process.

These studies clearly show that transition metal reactivity varies not only across the d-block but also within metal complexes with the same d-electron configuration.

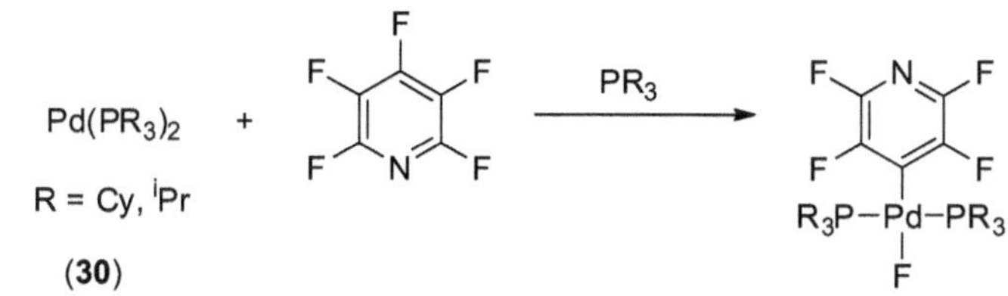
(6.14)

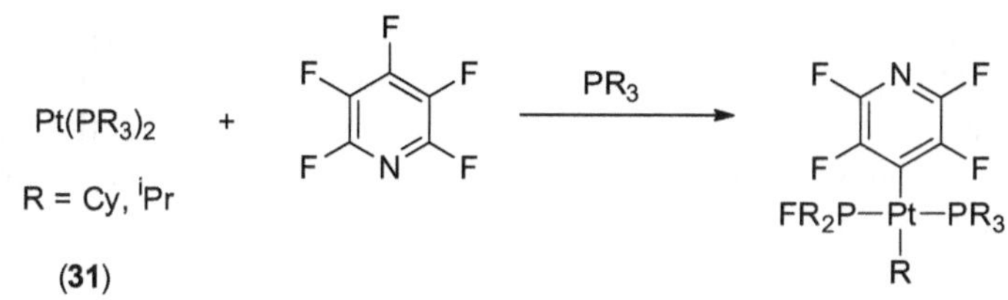
(6.15)

Pt^0 and Pt^{II} complexes have also been used to activate aryl fluorides.[28,29] The first examples of Pt^{II} mediated C–F activation of aryl fluorides were reported by Puddephatt and co-workers.[29e] Taking advantage of a perfluorinated aryl imine motif (**32**), intramolecular C–F activation with bis-platinum(II) dimer **33** proceeded in high yields to generate the corresponding cyclometalated Pt^{IV} complexes (**34**) (eqn (6.16)). Interestingly, reductive elimination from **34** to generate ethane was not observed.

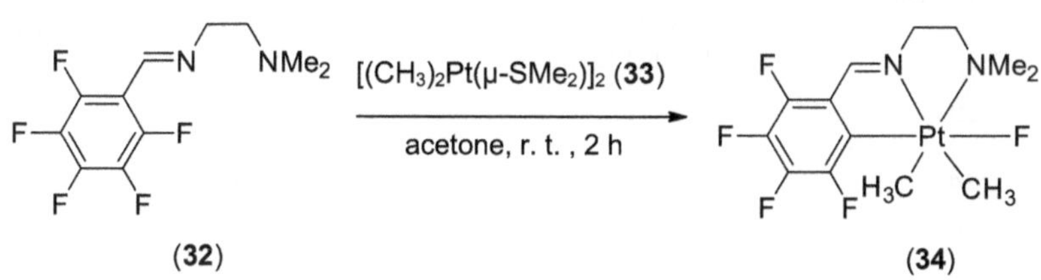
(6.16)

Building on this original report, Crespo and Martinez have reported on the C–F and C–H activation chemistry of the same bis-platinum dimer **33**.[29g] Notably, it was found that C–F activation occurred exclusively in the *ortho*-position relative to the imine functionality. Furthermore, it was found that C–F activation is favoured for more electron-deficient fluoroaryl rings (eqn (6.17) and (6.18)). Even more significant is that aromatic C–F activation occurred in the presence of weaker carbon–halogen bonds (eqn (6.18)). Pt^0 complexes have also been studied by Crespo and Martinez but to a lesser extent.[29h]

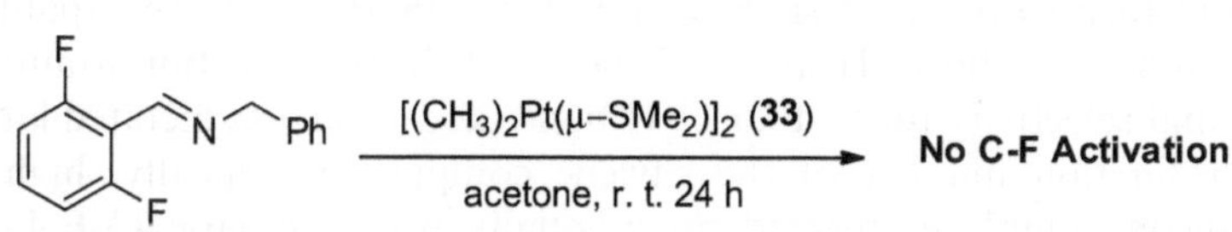
(6.17)

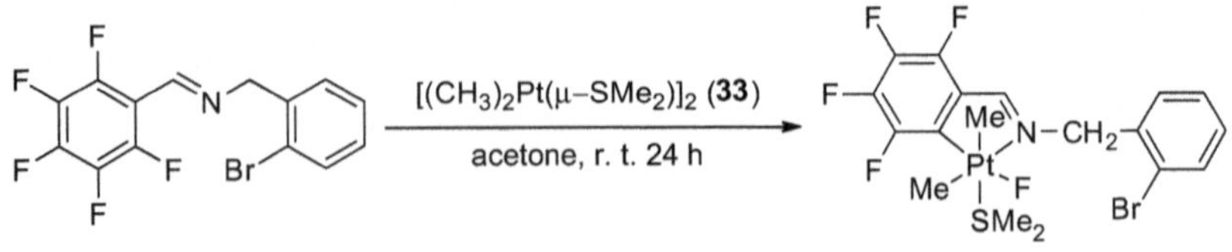

(6.18)

6.4.2.5 Group 11 Complexes

In 2011, the first example of stoichiometric defluorination of aryl fluorides using a group 11 complex was reported by Ribas and co-workers.[30] During the course of their investigations of Cu-catalyzed halide exchange reactions it was discovered that the same Cu^{I} complex, $Cu(CH_3CN)_4OTf$, that promoted catalytic C–X (X = Cl, Br, I) reductive elimination was also capable of stoichiometric C–F activation and subsequent halide exchange. Presumably the unique macrocyclic ligand (**34**) plays a critical role in stabilizing the resulting Cu^{III} complex **35** (eqn (6.19)). Interestingly, the choice of solvent was found to be critical to the success of this reaction. Furthermore, to date this represents the only example of an aromatic defluorination reaction mediated by a group 11 metal complex.

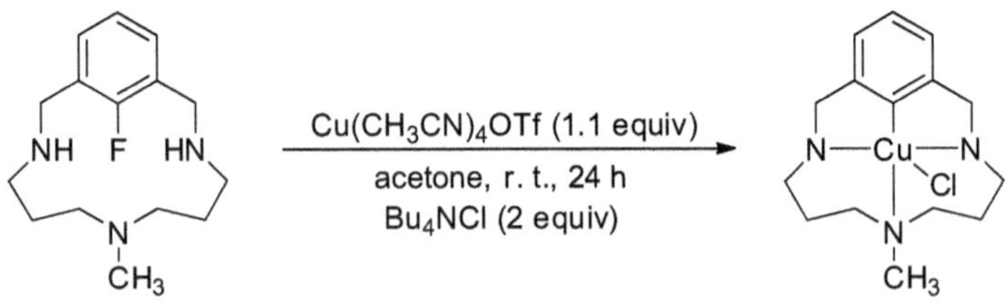

(**34**) (**35**)

(6.19)

6.4.2.6 Stoichiometric C–F Activation – Main Group Complexes

In contrast to transition metal catalyzed C–F activation, examples of either stoichiometric aliphatic[31] or aromatic C–F activation[32] with main group complexes remain relatively rare. Indeed, it wasn't until recently that reports of stoichiometric aromatic C–F activation with main group complexes emerged. In 2010, Roesky and co-workers reported a fundamental study evaluating the reactivity of low-valent silylenes with aromatic compounds containing C–H and/or C–F bonds.[32a] During the course of their investigations it was found that two and three-coordinate silylene complexes **36** and **37** were capable of activating both aromatic C–H and C–F bonds (Scheme 6.5). Importantly, it was found that selectivity for C–F or C–H bond activation was dictated largely by the coordination number of the silylene complex. Specifically, high co-ordinate silicon complexes reacted preferentially with aromatic C–F bonds

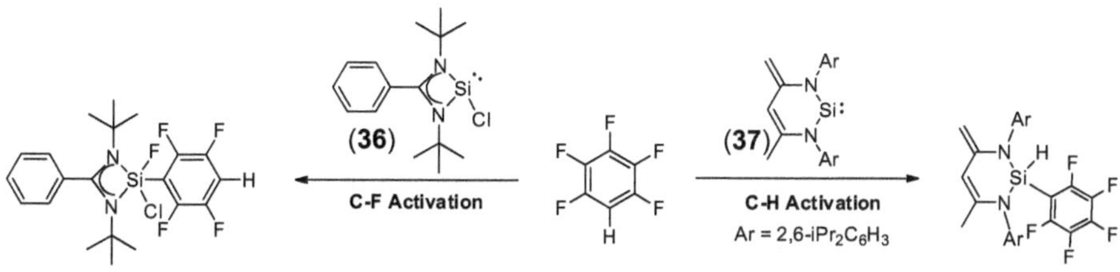

Scheme 6.5 C–F versus C–H activation with three- and two-coordinate silicon complexes, respectively.

whereas the opposite selectivity was observed with low coordinate complexes. The stoichiometric reactivity of group 14 complexes with aromatic C–F and C–H bonds has been recently reviewed.[32c]

6.5 Transition Metal Catalyzed Functionalization of Aryl Fluorides

With the ever increasing need for rapid access to a wide variety of aryl fluoride building blocks, transition metal catalyzed functionalization of aryl fluorides has become a promising approach towards the synthesis of functionalized fluoroaromatic compounds. Indeed, transition metal mediated stoichiometric C–F activation has proven to be an effective method for the formation of M–F bonds. Furthermore, by careful selection of both the transition metal and ligand framework, the reactivity of the resulting metal complex, and M–F bond, can be tuned for the specific synthetic methodology. On the basis of the remarkable progress in stoichiometric C–F activation, the number of reports on catalytic transformations of aryl fluorides continues to increase. Herein, catalytic reactions based on transition metal promoted C–F bond activation are presented.

6.6 Transition Metal Catalyzed Hydrodefluorination of Aryl Fluorides

Dehalogenation of organohalogen compounds is a powerful method for the detoxification of organohalogen pollutants, particularly polychlorobiphenyls (PCBs) and chlorofluorocarbons (CFCs). In addition, the products of these transformations can undergo a number of further transformations, such as hydrogenation, to be converted into other valuable agrochemical feedstocks.

6.6.1 Catalytic Hydrodefluorination – Early Work

The earliest examples of catalytic hydrodefluorination of monofluoroarenes utilized a combination of hydrogen gas and a heterogeneous metal catalyst.[33] The first systems were developed in the 1920s and made use of Pt and Ni alloys.

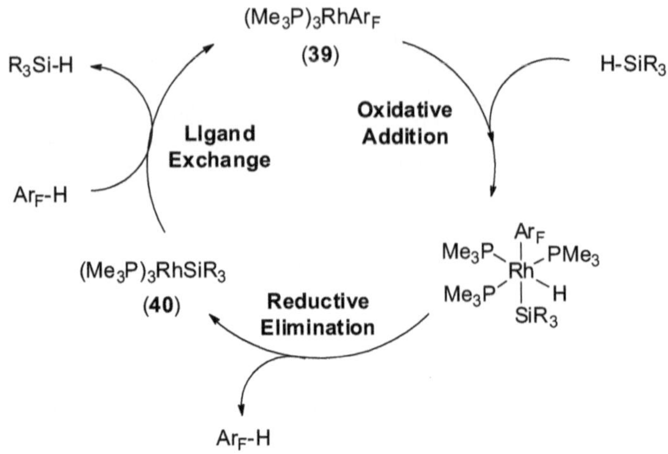

Scheme 6.6 Mechanism of Rh(I)-catalyzed hydrodefluorination.

However, the requirement for high temperatures, ranging from 180–200 °C, and high gas pressures limited the synthetic utility of these methods. In 1994, Milstein and co-workers reported the first homogeneous system for the catalytic dehydrofluorination of aryl fluorides.[34a] Rh-phosphine complex **38** reacts readily with the C–F bonds of heavily fluorinated aromatic molecules such as hexafluorobenzene and pentafluorobenzene (eqn (6.20)). When used in conjunction with a silane reagent, EtO_3Si–H, catalytic C–F activation and hydrodefluorination proceed in high yields and with high regioselectivity. The catalytic cycle is believed to proceed through initial C–F activation, to generate Rh(I)–F complex **39**, followed by Si–H oxidative addition and C–H reductive elimination to generate the defluorinated products and Rh-silyl complex **40** which is capable of further catalysis (Scheme 6.6). Since this landmark discovery, several other systems for the catalytic hydrodefluorination of aryl fluorides have emerged.

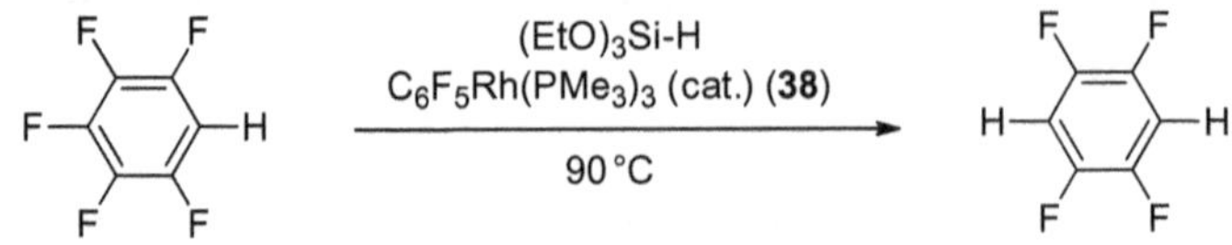

(6.20)

6.6.2 Catalytic Hydrodefluorination

Following Milstein's initial report, several other systems for the catalytic hydrodefluorination of aryl fluorides using late-transition metal complexes have been reported. In 1999, Grushin reported a method for the catalytic hydrodefluorination of highly unactivated monofluoroarenes using a similar Rh(I)-hydride complex (**41**) (eqn (6.21)).[34n] This is in contrast to the Milstein system which required heavily fluorinated aromatic substrates. The mechanism of this process has been suggested to be similar to the mechanism proposed by Milstein and co-workers.

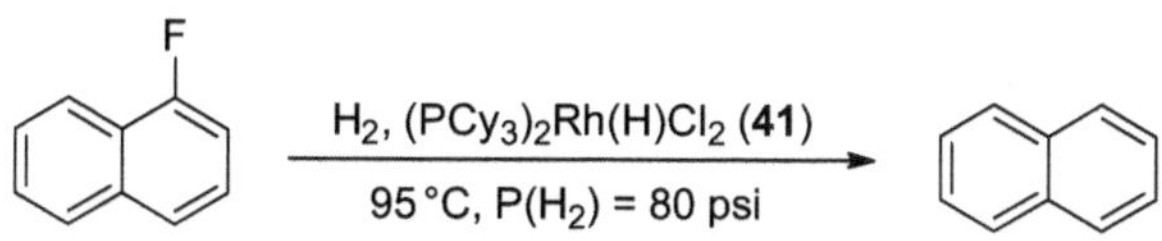

(6.21)

More recent work in this area includes exciting work by Holland and co-workers who reported on the reactivity of an iron-hydride complex.[34h] Importantly, this was the first example of the application of Fe-complexes to the catalytic functionalization of aryl fluorides. Furthermore, this remains one of the few reports of Fe-catalyzed C–F activation. Holland and co-workers developed a low-valent Fe–F complex (**42**) that reacts with a silane reagent, Et_3SiH, to generate an active Fe-hydride complex. The resulting Fe-hydride complex catalyzes the hydrodefluorination of hexafluorobenzene, pentafluoropyridine and perfluorotoluene (Scheme 6.7). Notably, the reaction with perfluorotoluene proceeded with complete selectivity for activation of aromatic C–F bonds. Also noteworthy is that catalytic hydrodefluorination of pentafluoropyridine also proceeded with high regioselectivity for reactivity in the 4-position of the pyridine ring – consistent with the stoichiometric reactivity observed by Braun and Perutz (*vide supra*).

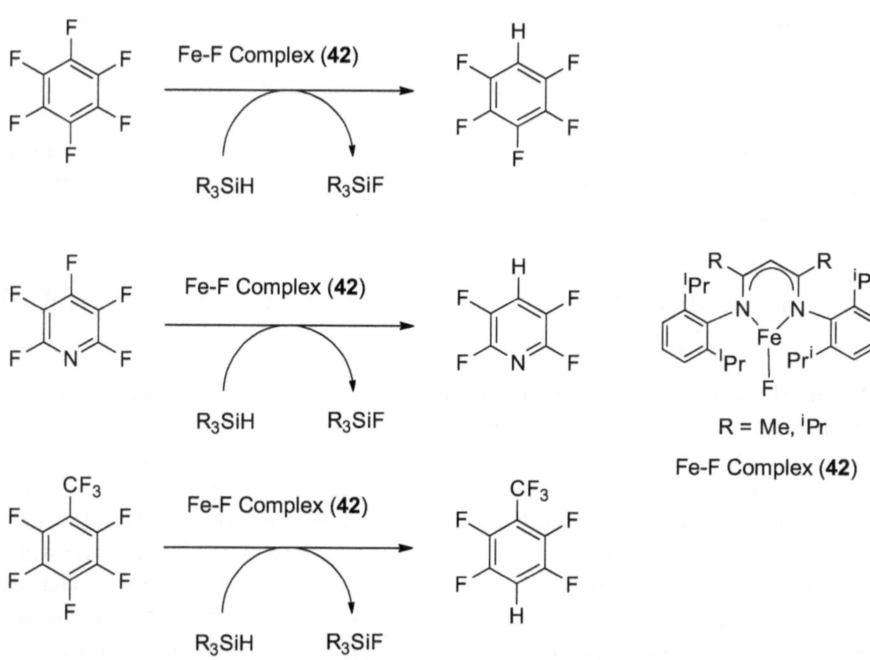

Scheme 6.7 Fe-catalyzed hydrodefluorination of perfluoroarenes.

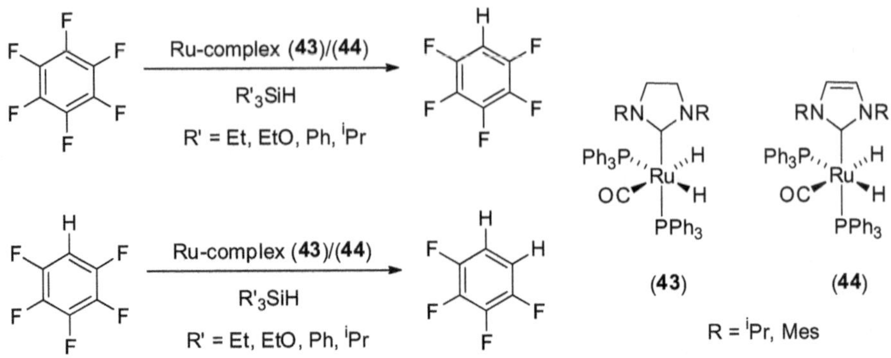

Scheme 6.8 Ru-catalyzed hydrodefluorination of aryl fluorides.

An example of Ru-catalyzed hydrodefluorination of aryl fluorides comes from the work of Whittlesey and co-workers who reported the application of Ru *N*-heterocyclic carbene complexes **43** and **44** to the hydrodefluorination of heavily fluorinated fluoroarenes (Scheme 6.8).[34m] It is interesting to note that hydrodefluorination of pentafluorobenzene proceeded with exclusive selectivity to give 1,2,3,4-tetrafluorobenzene as the sole organic product. To account for the unusual *ortho*-regioselectivity the authors postulated the involvement of a tetrafluorobenzyne intermediate. Detailed mechanistic and computational studies have added further support for the formation of such a complex.[34m]

Several important contributions involving the application of main group complexes to the field of catalytic hydrodefluorination have come from Muller[35a] and Ozerov.[35b,c,d] However, as these systems have only been applied, thus far, to the selective catalytic activation and defluorination of aliphatic Csp^3–F bonds,[35] this work will not be discussed within this chapter.

6.7 Transition Metal Catalyzed C–C Cross-Coupling of Aryl Fluorides

Despite the various reports of stoichiometric C–F activation, examples of C–C bond forming reactions are still relatively rare. Indeed, transition metal based C–F activation and subsequent cross-coupling reactions have received considerably less attention compared to other carbon–halogen bonds, presumably due to the strength of the carbon–fluorine bond. Over the past decade, both early and late-transition metal complexes have been successfully applied towards the catalytic functionalization of aryl fluorides. Herein, pertinent examples are presented. Additional reviews on methods for the functionalization of aryl fluorides have been reported by Amii and Uneyama[36a] and Sun and Love.[36b]

6.7.1 Cross-Coupling of Aryl Fluorides – Early Work

In 1973, Kumada and co-workers reported the first examples of a catalytic C–C cross-coupling reaction involving aryl fluorides.[37] It was found that Ni(II) phosphine complex **45** catalyzed the reaction of a Grignard reagent, isopropylmagnesium chloride, with monofluorobenzene generating the corresponding cross-coupled products (eqn (6.22)). It is worth noting that significant amounts of isomerization were observed. Several years later, in 1999, the first catalytic C–C bond-forming reaction with a polyfluoroarene was reported by Jones and co-workers (eqn (6.23)).[38]

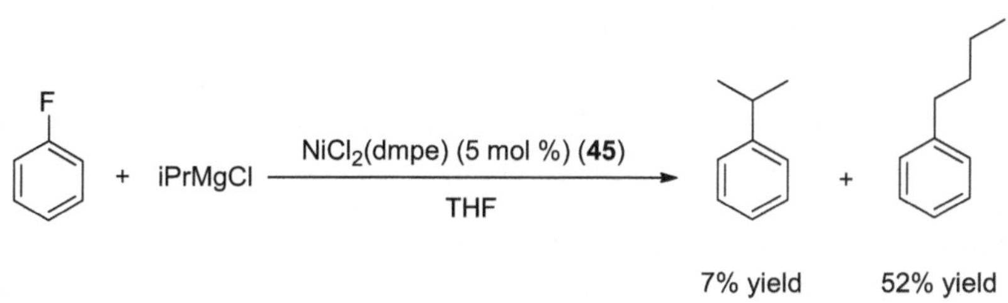

(6.22)

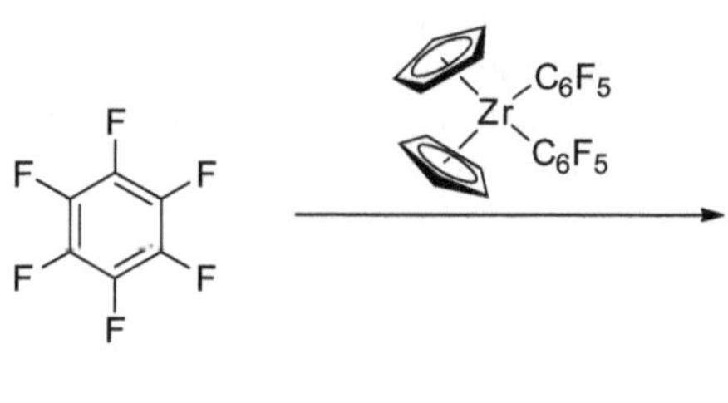

(6.23)

6.7.2 Cross-Coupling of Aryl Fluorides – Electron-Deficient Metal Complexes

The majority of early investigations in the field of transition metal mediated C–F activation involved the use of highly electron-deficient early transition metal complexes (*vide supra*). However, there is only a single example of a catalytic C–C cross-coupling reaction involving an early transition metal complex. In 2006, Takahashi and co-workers reported the successful cross-coupling of monofluoroarenes with alkyl Grignard reagents catalyzed by Ti(IV) or Ta(IV) complexes (eqn (6.24)).[39] Consistent with the original report by Kumada and co-workers, the major product of this reaction is the result of isomerization of the alkyl Grignard reagent.

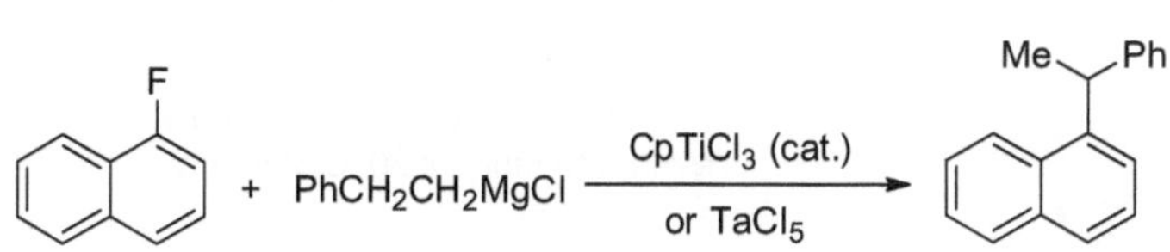

(6.24)

6.7.3 Cross-Coupling of Aryl Fluorides – Electron-Rich Metal Complexes

The majority of the examples of catalytic cross-coupling of aryl fluorides have utilized late-transition metal complexes. In particular, systems based on group 9 and group 10 complexes have received the most attention presumably due to the extensive investigations and well understood cross-coupling reactions that have been developed for the other aryl halides.

6.7.3.1 Group 7 and 8 Complexes

In 1999, Cahoez and co-workers reported the manganese-catalayzed cross-coupling between aryl halides and alkyl Grignard reagents.[40] While the reaction of monofluoroarenes bearing an imine motif did proceed, in the absence of catalyst, to give the cross-coupled products, the reaction was found to be significantly enhanced by the addition of a $MnCl_2$ catalyst (Scheme 6.9).

Despite the growing interest in the development of catalysis using Fe- and Ru-based catalysts, to date there have been no reported examples of catalytic cross-coupling reactions of aryl fluorides using group 8 complexes. The C–F activation chemistry of group 8 complexes has been largely focused on the development of hydrodefluorination technologies (*vide supra*).

6.7.3.2 Group 9 Complexes

The development of methodologies for the catalytic cross-coupling of aryl fluorides with group 9 complexes has largely focused on the use of Rh[41] and Co[42] complexes. In 2006, Knochel and co-workers reported the first and only example of Co-catalyzed C–C cross-coupling of polyfluoroarenes.[42] The reaction of a variety of polyfluoroaryl ketones with aryl copper reagents, catalyzed by $Co(acac)_2$, proceed to give the corresponding cross-coupled products in high yields (eqn (6.25)). The use of certain additives such as Bu_4NI and 4-fluorostyrene were found to significantly increase the overall yields, presumably by stabilizing catalytic intermediates.

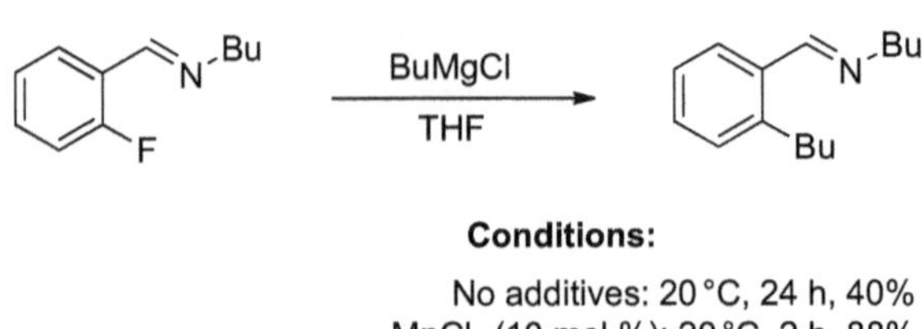

Scheme 6.9 Mn-catalyzed C–C cross-coupling.

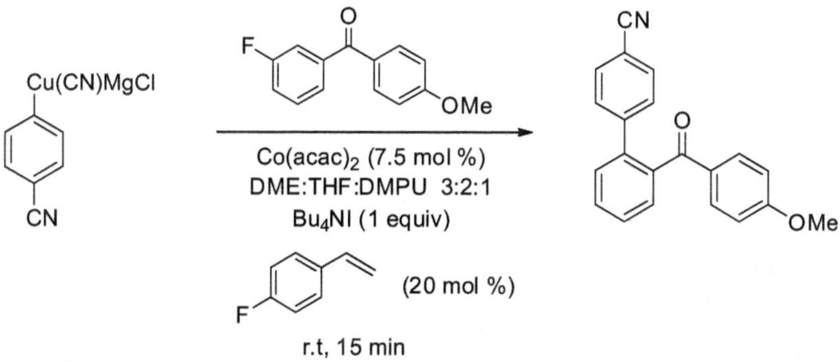

(6.25)

The first example of a Rh-catalyzed C–F cross-coupling reaction was reported by Murai and Chatani who reported a method for the C–Si functionalization of polyfluoroarenes.[41b] In particular, the reaction of fluorinated aromatic ketones and oxazolines with hexamethyldisilane in the presence of a catalytic amount of cationic Rh complex **46** – $[Rh(COD)_2]BF_4$ – gave the *ortho*-functionalized products in high yields and with complete regioselectivity (eqn (6.26)). The ketone and oxazoline functionalities serve as directing groups which coordinate to the Rh-complex and direct C–F activation and subsequent C–Si bond formation to the *ortho*-position of the fluoroaryl ring.

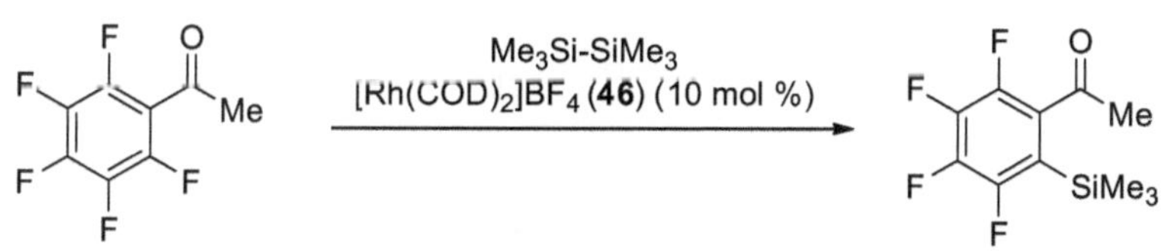

(6.26)

Following this study, Rh complexes have also been successfully applied to the thiolation of perfluoroarenes. The first examples were reported in 2008 by Yamaguchi and co-workers.[43a] They developed the polyarylthiolation of perfluoroarenes with diaryl disulfides catalyzed by Rh-hydride complex **47** and 1,2-bis(diphenylphosphino)benzene (**48**) (Scheme 6.10). Indeed, monofluoroarenes were found to be unreactive under these conditions, clearly indicating the need for a highly activated fluoroaryl ring. Interestingly, the extent of

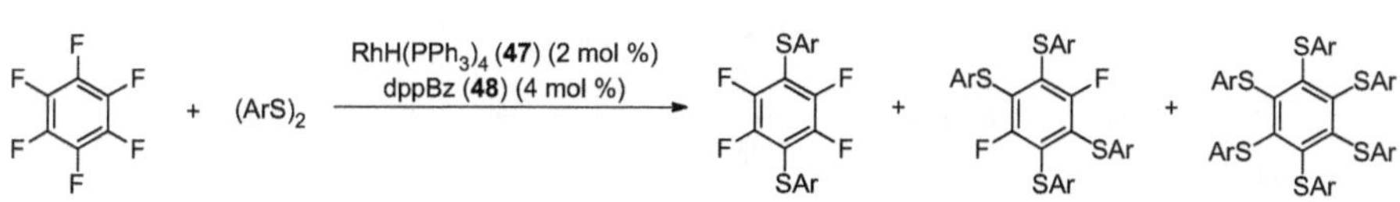

C_6F_6	$(ArS)_2$	bis(SAr)	tris(SAr)	tetrakis(SAr)
2.0 equiv	1.0 equiv	95%	5%	0%
0.5 equiv	1.0 equiv	0%	95%	5%
0.33 equiv	1.0 equiv	3%	3%	92%

Scheme 6.10 Rh-catalyzed polyarylthiolation of hexafluorobenzene.

substitution as well as the substitution pattern could be controlled by the ratio of substrate to disulfide. The preference for para-substitution remains unknown.

6.7.3.3 Group 10 Complexes

The majority of developments in the catalytic cross-coupling of aryl fluorides have involved the use of group 10 transition metal catalysts. In particular, several examples of Ni-catalyzed cross-coupling reactions of aryl fluorides have been reported.[43] In 2001, the Hermann group reported one of the earliest examples of a Ni-catalyzed Kumada cross-coupling reaction of monofluoroarenes.[43i] Key to their particular strategy was the use of electron-rich Ni(0) *N*-heterocyclic carbene complex **49** to facilitate the reaction between monofluoroarenes and Grignard reagents and afford a variety of biaryl products (eqn (6.27)).

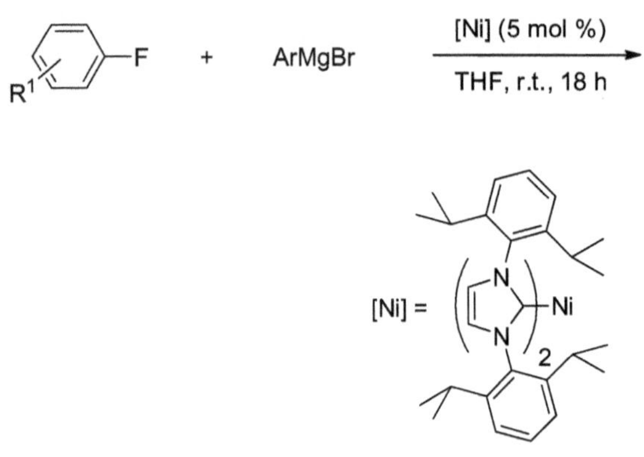

(49)

(6.27)

Following this report, Ackermann and co-workers developed an analogous reaction using $Ni(acac)_2$ supported with a unique phosphine oxide ligand (**50**) as the catalyst to facilitate the Csp^2–Csp^2 cross-coupling of a variety of monofluoroarenes with alkyl and aryl Grignard reagents (eqn (6.28)).[43j] Of note is that this method could be extended to the cross-coupling of monofluorinated heterocycles, such as 3-fluoropyridine, to afford a series of synthetically diverse biaryl products.

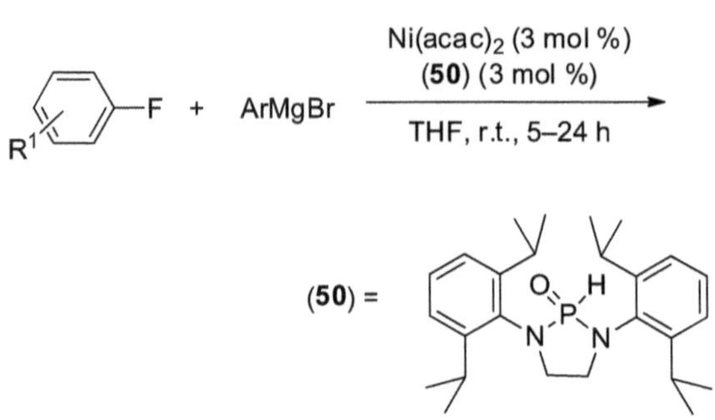

(6.28)

Building on their extensive study of stoichiometric C–F activation with Ni(0) complexes, Braun and Perutz reported the first examples of Ni-catalyzed C–C cross-coupling of polyfluoroarenes.[43a] Using a catalytic amount of the Ni(II)–F complex (**50**) that had been the focus of their previous studies, they reported the successful coupling of pentafluoropyridine and vinyl tin reagent **51** (eqn (6.29)). The regioselectivity of the cross-coupling reaction is the same as the regioselectivity associated with Ni-catalyzed C–F activation of *N*-heterocycles (*vide supra*).

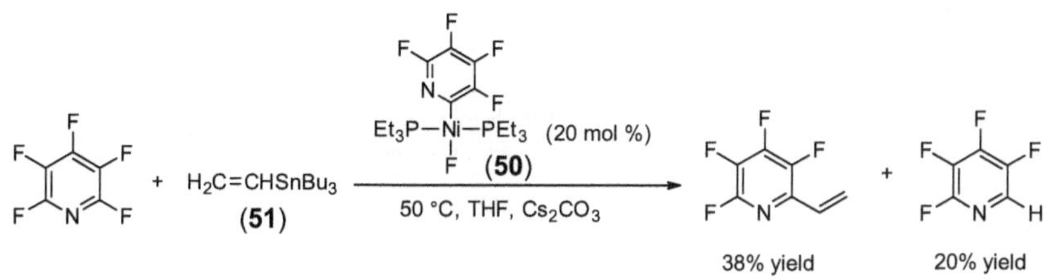

(6.29)

In a subsequent study, Braun reported the successful cross-coupling between polyfluorinated *N*-heterocycles and aryl boronic acids using Ni complex **52**.[43b] Of particular interest is that the reaction of 5-chloro-2,4,6-trifluoropyrimidine proceeded with complete selectivity for functionalization of an aryl C–F bond (eqn (6.30)). This work represents the first example of selective functionalization of an aryl C–F bond in the presence of weaker carbon halogen bonds.

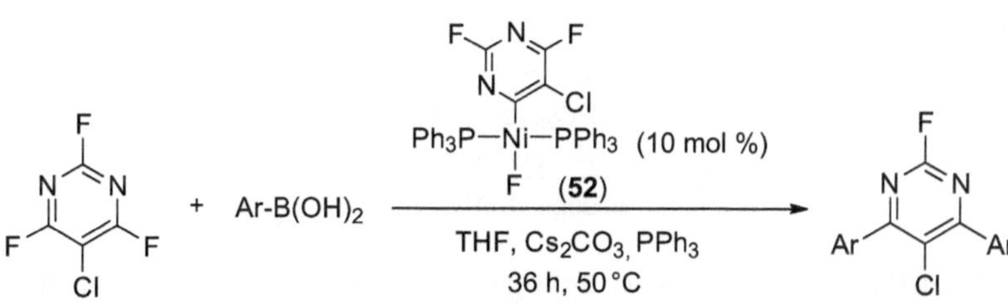

(6.30)

The first examples of Ni(0) catalyzed Suzuki cross-coupling of perfluoroarenes came from Radius and co-workers in 2006.[43d] As an extension of their prior investigations of Ni(0) catalyzed stoichiometric C–F activation of hexafluorobenzene, they reported that the same Ni(0)-carbene complex **29** could be employed in the catalytic functionalization of the same substrate (eqn (6.31)).

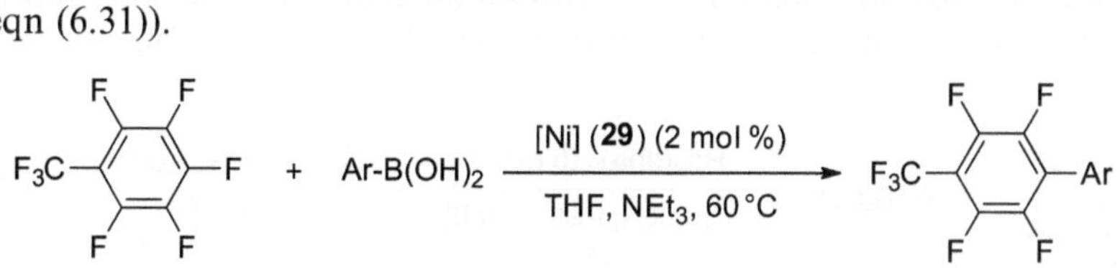

(6.31)

As part of a research program focused on the development of new catalytic methodologies for the synthesis of functionalization aryl fluorides, we recently reported methodologies for the Ni-catalyzed Suzuki cross-coupling of polyfluoroaryl imines (eqn (6.32)).[43f] The imine functionality directs C–F activation and subsequent functionalization to the *ortho*-position of the fluoroaryl ring. Most importantly, this work represents the first examples of catalytic Csp^2–Csp^2 cross-coupling of polyfluoroarenes to generate functionalized products that still contain aromatic C–F bonds. That a variety of substitution patterns on the fluoroaryl ring can be tolerated further highlights the versatility of this approach. In particular, both methodologies appear to have exclusive selectivity for preferential C–F activation in the presence of other weaker, C–H bonds thus greatly increasing the scope of this transformation. This same selectivity is not observed in other related catalytic systems (*vide infra*).

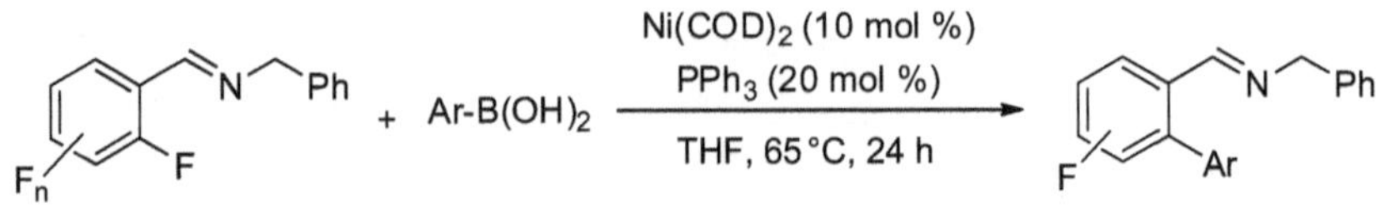

(6.32)

In 2011, Chatani and co-workers reported a related system for the Suzuki cross-coupling reaction to include monofluoroarenes.[43g] Notably, using a simple Ni(0) catalyst system along with CsF the corresponding cross-coupled products were obtained in high yields for a range of aryl fluorides and aryl boronic acids (eqn (6.33)). Furthermore, the reaction could tolerate several directing group functionalities, such as pyridines, pyrmidines and oxazolines. The addition of a catalytic amount of ZrF_4 was found to increase the yields of the reaction.

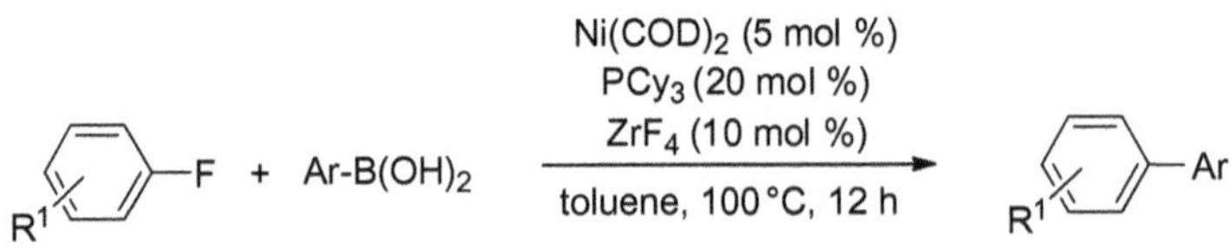

(6.33)

Reports of Pd-catalyzed cross-coupling of aryl fluorides have been much less forthcoming. Indeed, only a few examples have been reported to date.[44] In 1999, Widdowson and co-workers reported the first examples of Pd-catalyzed cross-coupling of monofluoroarenes. $Pd_2(dba)_3$ proved to be a highly effective catalyst for the functionalization fluorobenzene (eqn (6.34)).[44c] Tricarbonylchromium(0) was a necessary additive in order to obtain high yields. This is presumably due to coordination of the chromium complex to the aryl fluoride which serves to activate the fluoroaryl ring towards C–F activation.

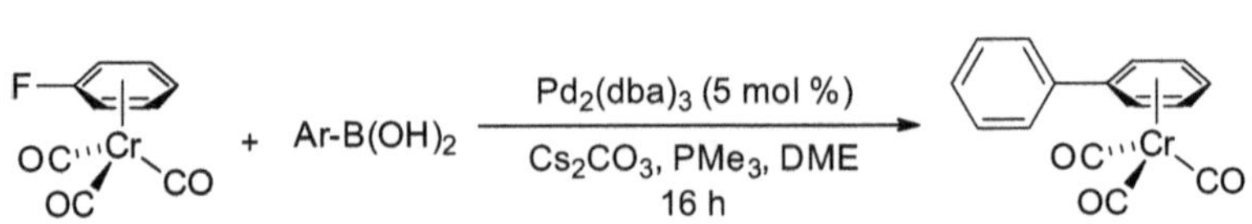

(6.34)

A particularly interesting example of Pd-catalyzed cross-coupling was reported by the Yu group in 2003.[44d] They reported the Pd(0) catalyzed coupling of electron-deficient monofluoroarenes to generate a variety of functionalized products. Importantly, this methodology has been successfully applied to the Stille and Suzuki cross-coupling as well as the amination of monofluoroarenes (Scheme 6.11).

One of the first examples of Pd-catalyzed cross-coupling of polyfluoroarenes with aryl Grignard reagents was reported in 2008 by Manabe and co-workers (Scheme 6.12).[44b] Furthermore, this is one of the only methodologies in which hydroxy-groups were successfully employed as directing group functionalities to give partially functionalized aryl fluorides.

In 2007, we reported the first examples of Pt-catalyzed methylation and methoxylation of polyfluoroaryl imines (Scheme 6.13).[45] This methodology is based on the report by Crespo and Martinez that a Pt-dimer,

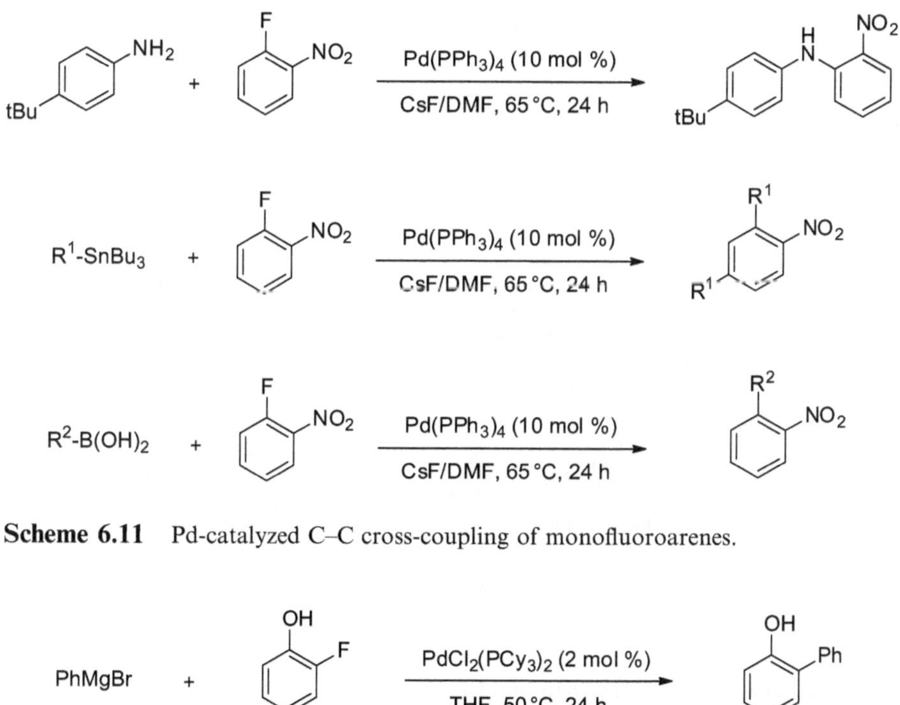

Scheme 6.11 Pd-catalyzed C–C cross-coupling of monofluoroarenes.

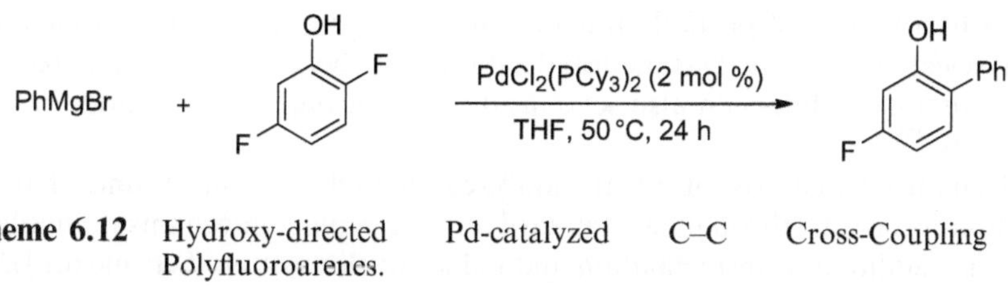

Scheme 6.12 Hydroxy-directed Pd-catalyzed C–C Cross-Coupling of Polyfluoroarenes.

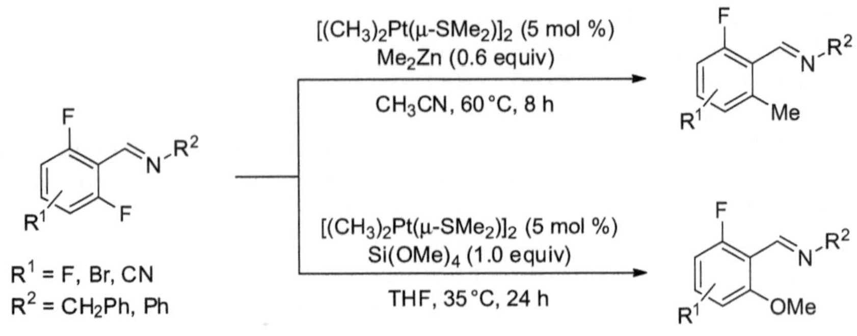

Scheme 6.13 Pt(II) catalyzed cross-coupling of polyfluoroaryl imines.

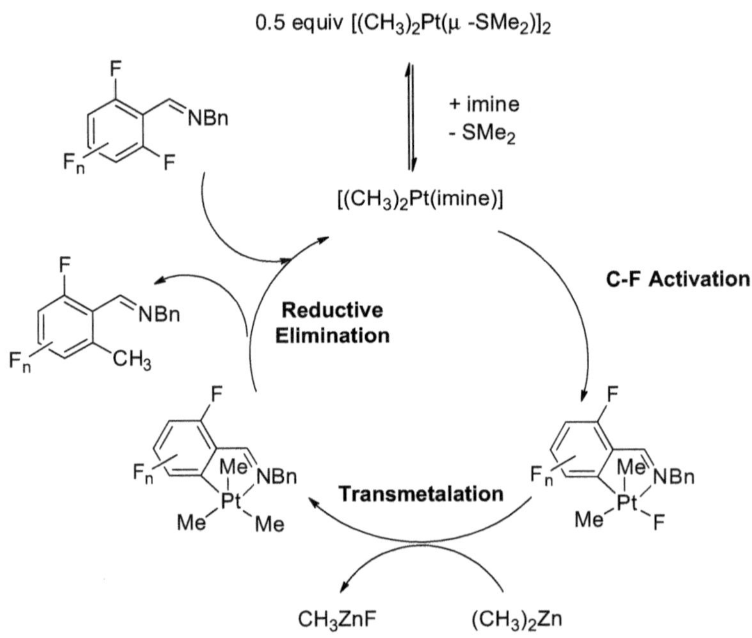

Scheme 6.14 Proposed mechanism of Pt(II) catalyzed methylation.

$[(CH_3)_2Pt(\mu\text{-}SMe_2)]_2$, effects the stoichiometric C–F activation of a series of polyfluoroaryl imines (eqn (6.17) and (6.18)). These methodologies provide access to a variety of partially functionalized aryl fluorides in high yields with high selectivity for *ortho*-functionalization. Furthermore, we have recently found that oxazolines can also serve as directing group functionalities in these reactions.[45f]

Mechanistic analysis of Pt(II) catalyzed methylation has found that the reaction proceeds through a standard cross-coupling mechanism involving oxidative addition, transmetalation and reductive elimination (Scheme 6.14).[45e,g]

It should be noted that a directing group is required to promote intramolecular C–F activation. In addition, all substrates must contain a 2,6-disubstitution pattern as the Pt-catalyst exhibits preferential activation of aromatic C–H bonds. This is in direct contrast to the selectivity displayed by Ni-catalysts (*vide supra*). Consistent with the stoichiometric investigations conducted by Crespo and Martinez, a minimum of three electron-withdrawing functionalities in addition to the imine substrate are required for catalytic reactivity. Furthermore, additional investigations have revealed the C–F activation is rate determining for the majority of substrates.[45g]

6.7.3.4 *Group 11 Complexes*

Reports of methods for the functionalization of aryl fluorides catalyzed by group 11 complexes have only recently emerged. In 2012, Shi and co-workers reported the first examples of Cu-catalyzed direct thiolation of perfluoroarenes.[46] Using CuBr as a catalyst, a variety of diaryl disulfides or aryl thiols underwent cross-coupling with pentafluorobenzene (Scheme 6.15). The addition of 1,10-phenanthroline and DDQ (2,3-dichloro-5,6-dicyano-1,4-benzoquinone) significantly increased the yields of the reactions. As was observed with analogous Rh-systems (*vide supra*), both mono and di-substituted products could be obtained depending on the equivalents and electronic nature of the thiol/disulfide.

6.7.3.5 *Main Group Complexes*

In contrast to transition metal catalyzed C–F cross-coupling reactions with group 10 complexes, catalytic C–F activation reactions based on main group complexes are particularly rare. The first reported example was that of Boruah and co-workers who, in 2005, developed a novel Sonogashira cross-coupling reaction between monofluoroarenes and phenylacetylene catalyzed by $InCl_3$ salts (eqn (6.35)).[47] Not only is this one of the few examples of Csp–Csp^2 cross-coupling involving aryl fluorides, but the use of non-toxic In salts renders this reaction heavy metal and phosphine free – significant advantages from an industrial standpoint.

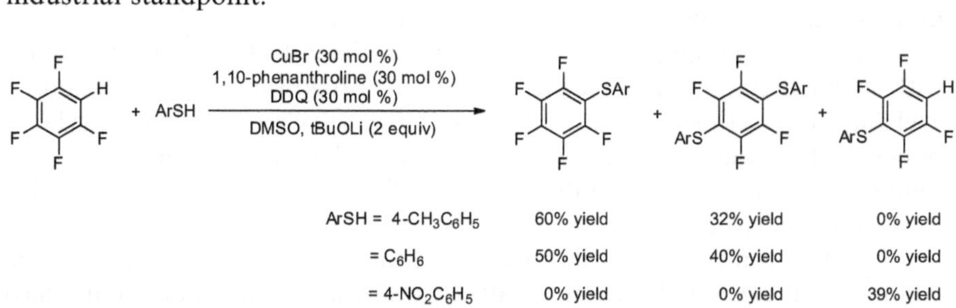

Scheme 6.15 Cu-catalyzed direct polyarylthiolation.

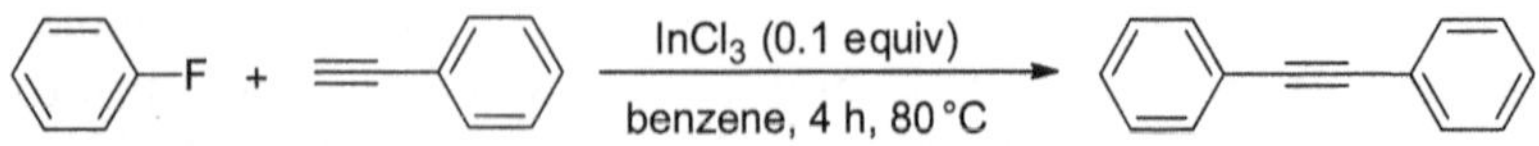 (6.35)

Very recently, a particularly elegant approach towards the functionalization of aryl fluorides was reported by Siegel and co-workers.[48] Inspired by the successful application of silane reagents to the hydrodefluorination of aliphatic and aromatic C–F bonds, the Siegel group applied this reactivity to the development of a novel intramolecular Freidel–Crafts coupling of polycyclic monofluoroarenes (eqn (6.36)). The reaction is catalyzed by silyl cations paired with weakly coordinating carborane counterions and enthalpically driven by the formation of strong Si–F bonds.

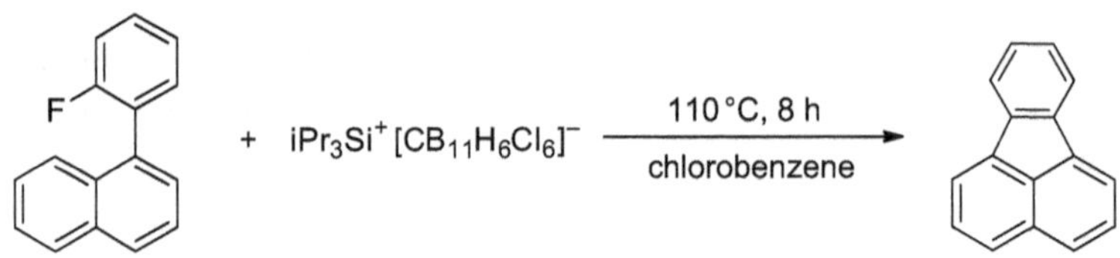 (6.36)

6.8 Conclusions and Outlook

In this chapter, recent advances in the transition metal catalyzed functionalization of aryl fluorides have been described. The increasing incorporation of aryl fluoride moieties into pharmaceuticals, agrochemicals and materials has necessitated the development of a variety of synthetic strategies for the facile construction of a variety of aryl fluoride building blocks. To this end, selective transition metal catalyzed C–F activation of polyfluorinated arenes has emerged as a promising approach towards the preparation of functionalized aryl fluorides. Indeed, despite the many challenges associated with the activation of exceptionally strong C–F bonds, many elegant catalytic methodologies using early and late transition metal catalysis have been developed. Nevertheless, many significant challenges still remain. Many of the current catalytic systems rely on the use of activated substrates to facilitate C–F activation; systems allowing for the direct C–F activation of non-activated fluoroaromatics would be of great interest. Furthermore, the majority of catalytic cross-coupling methodologies are restricted to C–C bond formation with simple arene coupling partners. Given the broad range of applications of aryl fluorides, methodologies allowing for the selective replacement of aryl C–F bonds with a range of carbon or heteroatom-based functionalities would significantly enhance the overall synthetic utility of these transformations. The focus of future synthetic efforts will likely include the development of methods for the trifluoromethylation and trifluoromethoxylation of aryl fluorides. The addition of these complementary synthetic transformations will no doubt have a remarkable effect of the scope of organofluorine compounds in practical synthetic applications.

References

1. F. A. Cotton and G. Wilkinson, *Advanced Inorganic Chemistry*, Wiley Interscience, 1980, p. 375.
2. (a) K. Muller, C. Faeh and F. Diederich, *Science*, 2007, **317**, 1881–1886; (b) K. B. Park, N. R. Kitteringham and P. M. O'Neill, *Annu. Rev. Pharmacol. Toxicol.*, 2001, **41**, 443–470.
3. H.-J. Bohm, D. Banner, S. Bendels, M. Kansy, B. Kuhn, K. Muller, U. Obst-Sanch and M. Stahl, *ChemBioChem*, 2004, **5**, 637–643.
4. For recent examples see: (a) S. Hong, J. Lee, M. Kim, Y. Park, C. Park, M.-H. Kim, S.-S. Jew and H. Park, *J. Am. Chem. Soc.*, 2011, **133**, 4924–4929; (b) G. C. Welsh and G. Bazan, *J. Am. Chem. Soc.*, 2011, **133**, 4632–4644; (c) S. C. Price, A. C. Stuart, L. Yang, H. Zhou and W. You, *J. Am. Chem. Soc.*, 2011, **133**, 4625–4631; (d) J. D. Nguyen, J. W. Tucker, M. D. Konieczynska and R. J. Stephenson, *J. Am. Chem. Soc.*, 2011, **133**, 4160–4163; (e) M. M. Ali, S. D. Aguirre, H. Lazim and Y. Li, *Angew. Chem., Int. Ed.*, 2011, **50**, 3751–3754; (f) H. Zhou, L. Yang, A. C. Stuart, S. C. Price, S. Liu and W. You, *Angew. Chem., Int. Ed.*, 2011, **133**, 3051–3054; (g) Q. Zhao, M. Yu, L. Shi, S. Liu, C. Li, M. Shi, Z. Zhou, C. Huang and F. Li, *Organometallics*, 2010, **29**, 1085–1091; (h) K. Rangan, M. Fianchini, S. Singh and R. H. V. Dias, *Inorg. Chim. Acta*, 2009, **362**, 4347–4352; (i) B. Zheng, M. O. Miranda, A. G. DiPasquale, J. J. A. Golen, A. C. Rheingold and L. H. Doerrer, *Inorg. Chem.*, 2009, **48**, 4274–4276; (j) E. J. Schelter, P. Yang, B. L. Scott, J. D. Thompson, R. L. Martin, J. P. Hay, D. E. Morris and J. C. Kiplinger, *Inorg. Chem.*, 2007, **46**, 7477–7488; (k) R. Hughes, J. M. Smith, C. D. Incarvito, K.-C. Lam, B. Rhatigan and A. C. Rheingold, *Organometallics*, 2002, **21**, 2136–2144. For reviews see: (l) J.-F. Carpentier, *Dalton Trans.*, 2010, **39**, 37–38; (m) M. Kuhnel and D. Lentz, *Dalton Trans.*, 2010, **39**, 9745–9759.
5. M. Bartholow, *Pharmacy Times*, May 2010, 35–36.
6. F.-H. Vaillancourt, E. Yeh, D. A. Vosburg, S. Garneau-Tsodikova and C. T. Walsh, *Chem. Rev.*, 2006, **106**, 3364–3378.
7. (a) D. J. Adams and J. H. Clark, *Chem. Soc. Rev.*, 1999, **28**, 225–231; (b) H. Sun and S. G. DiMagno, *Angew. Chem., Int. Ed.*, 2006, **45**, 2720–2725.
8. (a) G. Balz and G. Schiemann, *Ber. Dtsch. Chem. Ges.*, 1927, **60**, 1186–1190; (b) W. J. Feast, W. K. R. Musgrave and R. G. Weston, *J. Chem. Soc., Section C: Organic*, 1971, **5**, 1547–1549.
9. M. A. Subramanian and L. E. Manzer, *Science*, 2002, **297**, 1665–1667.
10. (a) T. Furuya, A. E. Strom and T. Ritter, *J. Am. Chem. Soc.*, 2009, **131**, 1662–1663; (b) P. Tang, T. Furuya and T. Ritter, *J. Am. Chem. Soc.*, 2010, **132**, 12510–12154; (c) T. Furuya, J. E. M. N. Klein and T. Ritter, *Synthesis*, 2010, 1804–1821; (d) T. Furuya and T. Ritter, *Org. Lett.*, 2009, **11**, 2860–2863; (e) K. L. Hull, W. Q. Anani and M. S. Sanford, *J. Am. Chem. Soc.*, 2006, **128**, 7134–7135; (f) N. D. Ball and M. S. Sanford, *J. Am. Chem. Soc.*, 2009, **131**, 3796–3797; (g) A. W. Kaspi, I. Goldberg and A. Vigalok,

J. Am. Chem. Soc., 2010, **132**, 10626–10627; (h) A. Vigalok, *Chem.–Eur. J.*, 2008, **14**, 5102; (i) X. Wang, T.-S. Mei and I.-Q. Yu, *J. Am. Chem. Soc.*, 2009, **131**, 7520.

11. D. A. Watson, M. Sui, G. Teverovskiy, Y. Zhang, J. G. Fortanet, T. Kinzel and S. L. Buchwald, *Science*, 2009, **325**, 1661–1664.
12. (a) T. Furuya, A. S. Kamlet and T. Ritter, *Nature*, 2011, **473**, 470–378; (b) A. Vigalok, *Organometallics*, 2011, **30**, 4802–4810.
13. (a) M. Lafrance, C. N. Rowley, K. Woo and K. Fagnou, *J. Am. Chem. Soc.*, 2006, **128**, 8754–8756; (b) M. Lafrance, D. Shore and K. Fagnou, *Org. Lett.*, 2006, **8**, 5097–5100; (c) S. I. Gorelsky, D. Lapointe and K. Fagnou, *J. Am. Chem. Soc.*, 2008, **130**, 10848–10849; (d) H.-Q. Do and O. Daugulis, *J. Am. Chem. Soc.*, 2008, **130**, 1128–1129; (e) H.-Q. Do, R. M. Khan and O. Daugulis, *J. Am. Chem. Soc.*, 2008, **130**, 15185–15192.
14. (a) J. L. Kiplinger, T. G. Richmond and C. E. Osterbergy, *Chem. Rev.*, 1994, **94**, 373–431; (b) J. Burdeniuc, B. Jedlicka and R. H. Crabtree, *Chem. Ber.*, 1997, **30**, 145–154; (c) H. Torrens, *Coord. Chem. Rev.*, 2005, **249**, 1957–1985; (d) R. N. Perutz, T. Braun, Transition Metal-mediated C–F bond Activation in *Comprehensive Organometallic Chemistry*, ed. D. M. P. Mingos and R. H. Crabtree, Elsevier, 2007, vol. 1, pp. 725–758.
15. (a) P. M. Treichel, M. A. Chaudhari and F. G. A. Stone, *J. Organomet. Chem.*, 1963, **1**, 98–100; (b) M. A. Chaudhari, P. M. Treichel and F. G. A. Stone, *J. Organomet. Chem.*, 1964, **2**, 206–212.
16. B. M. Kraft and W. D. Jones, *J. Organomet. Chem.*, 2002, **658**, 132–140.
17. (a) J. Sala-Pala, J. Amaudrut, J. E. Guerchais, R. Mercier and M. Gerutti, *J. Fluorine Chem.*, 1979, **14**, 269–271; (b) K. Fuchibe and T. Akiyama, *J. Am. Chem. Soc.*, 2006, **128**, 1434–1435; (c) S. A. Strazisar and P. T. Wolozanski, *J. Am. Chem. Soc.*, 2001, **123**, 4728–4740.
18. (a) G. B. Deacon, P. I. Mackinnon and T. D. Tuong, *Aust. J. Chem.*, 1983, **36**, 43–53; (b) M. L. Cole, G. B. Deacon, C. M. Forsyth, P. C. Junk, K. Konstas and J. Wang, *Chem.–Eur. J.*, 2007, **13**, 8092–8110; (c) G. B. Deacon, C. M. Forsyth, P. C. Junk and J. Wang, *Chem.–Eur. J.*, 2009, **15**, 3082–3092; (d) M. Weydert, R. A. Andersen and R. G. Bergman, *J. Am. Chem. Soc.*, 1993, **115**, 8837–8838; (e) E. L. Werkema, E. Messines, L. Perrin, L. Maron, O. Eisenstein and R. A. Andersen, *J. Am. Chem. Soc.*, 2005, **127**, 7781–7795; (f) J. T. Lyon and L. Andrews, *Eur. J. Inorg. Chem.*, 2008, **49**, 1047–1058.
19. (a) T. G. Richmond, C. E. Osterberg and A. M. Arif, *J. Am. Chem. Soc.*, 1987, **109**, 8091–8092; (b) W. Liu, K. Welch, C. O. Trindle, M. Sabat, W. H. Myers and W. D. Harman, *Organometallics*, 2007, **26**, 2589–2597; (c) C. E. Osterberg and T. G. Richmond, *ACS Symp. Ser.*, 1994, **555**, 392–404.
20. M. J. Bruce, B. L. Goodall, G. L. Sheppard and F. G. A. Stone, *Dalton Trans.*, 1977, 591–595.
21. (a) A. H. Klahn, M. H. Moore and R. N. Perutz, *Chem. Commun.*, 1992, 1699–1701; (b) W. Mohr, G. A. Stark, H. Jiao and J. A. Gladysz, *Eur. J. Inorg. Chem.*, 2001, **4**, 925–933.

22. (a) M. L. Bruce, R. C. F. Gardner and F. G. A. Stone, *Dalton Trans.*, 1976, 81–89; (b) M. S. Kirkham, M. F. Mahon and M. K. Whittlesey, *Chem. Commun.*, 2001, 813–814.
23. (a) P. Barrio, R. Castoslenas, M. A. Esteruelas, A. Lledos, F. Maseras, E. Onate and J. Tomas, *Organometallics*, 2001, **20**, 442–452; (b) M. Arroyo, S. Bernes, M. Ceron, V. Cartina, C. Mendoza and H. Toerrns, *Inorg. Chem.*, 2007, **46**, 4857–4867; (c) A. J. Arce, M. R. Chierotti, Y. D. Sanctis, R. Gobetto, T. González, R. Machado and M. Márquez, *Inorg. Chim. Acta*, 2009, **362**, 291–294.
24. X. Xu, H. Sun, Y. Shi, J. Jia and X. Li, *Dalton Trans.*, 2011, **40**, 7866–7872.
25. (a) R. L. Hunt and G. Wilkinson, *Inorg. Chem.*, 1965, **4**, 1270–1272; (b) B. K. Bennet, R. G. Harrison and T. G. Richmond, *J. Am. Chem. Soc.*, 1994, **116**, 11165–11166; (c) X. Li, H. Sun, F. Yu, U. Florke and H.-F. Klein, *Organometallics*, 2006, **25**, 4695–4697; (d) S. Camadanli, R. Beck, U. Floerke and H.-F. Klein, *Dalton Trans.*, 2008, 5701–5704; (e) W. D. Jones, M. G. Partridge and R. N. Perutz, *J. Chem. Soc., Chem. Commun.*, 1991, 264–266; (f) S. T. Belt, M. Helliwell, W. D. Jones, M. G. Partridge and R. N. Perutz, *J. Am. Chem. Soc.*, 1993, **115**, 1429–1440; (g) B. L. Edelbach and W. D. Jones, *J. Am. Chem. Soc.*, 1997, **119**, 7734–7742; (h) C. Noveski, T. Braun, B. Neumann, A. Stammler and H.-G. Stammler, *Dalton Trans.*, 2004, 4106–4119; (i) R. J. Lindup, T. B. Marder, R. N. Perutz and A. C. Whitwood, *Chem. Commun.*, 2007, 3664–3666; (j) O. Blum, F. Frolow and D. J. Milstein, *J. Chem. Soc., Chem. Commun.*, 1991, 258–259; (k) R. P. Hughes, R. B. Laritchev, L. N. Zakharov and A. L. Rheingold, *J. Am. Chem. Soc.*, 2004, **126**, 2308–2309; (l) G. Doyle, *J. Organomet. Chem.*, 1982, **224**, 355–362; (m) S. Erhardt and S. A. Macgregor, *J. Am. Chem. Soc.*, 2008, **130**, 15490–15498; (n) T. Zheng, H. Sun, Y. Chen, X. Li, S. Durr, U. Radius and K. Harms, *Organometallics*, 2009, **28**, 5771–5776; (o) T. M. Dugan, J. M. Goldberg, W. W. Brennessel and P. L. Holland, *Organometallics*, 2012, **31**, 1349–1360.
26. (a) D. R. Fahey and J. E. Mahan, *J. Am. Chem. Soc.*, 1977, **99**, 2501–2508; (b) T. Braun and R. N. Perutz, *Chem. Commun.*, 2002, 2749–2757; (c) L. Cronin, C. L. Higgitt, R. Karch and R. N. Perutz, *Organometallics*, 1997, **16**, 4920–4928; (d) S. J. Archibald, T. Braun, J. A. Gaunt, J. E. Hobson and R. N. Perutz, *J. Chem. Soc., Dalton Trans.*, 2000, 2013–2018; (e) T. Braun, S. P. Foxon, R. N. Perutz and P. H. Walton, *Angew. Chem., Int. Ed.*, 1999, **38**, 3326–3329; (f) S. Burling, P. I. P. Elloitt, N. A. Jasim, R. J. Lindup, J. McKenna, R. N. Perutz, S. J. Archibald and A. C. Whitwood, *Dalton Trans.*, 2005, 3686–3695; (g) T. Schaub and U. Radius, *Chem.–Eur. J.*, 2005, **11**, 5024–5030; (h) T. Schaub, P. Fischer, A. Steffen, T. Braun, U. Radius and A. Mix, *J. Am. Chem. Soc.*, 2008, **130**, 9304–9317; (i) T. Yamamoto and M. Abla, *J. Organomet. Chem.*, 1997, **535**, 209–211; (j) S. A. Johnson, C. W. Huff, F. Mustafa and M. Saliba, *J. Am. Chem. Soc.*, 2008, **130**, 17278–17280; (k) E. Clot, O. Eisenstein,

N. Jasim, S. A. Macgregor, J. E. McGrady and R. N. Perutz, *Acc. Chem. Res.*, 2011, **44**, 333–348.

27. (a) T. Braun, A. Steffen, V. Schorlemer, B. Neumann and H.-G. Stammler, *Dalton Trans.*, 2005, 3331–3336; (b) T. Braun, V. Schorlemer, B. Neumann and H.-G. Stammler, *J. Fluorine Chem.*, 2006, **127**, 367–372; (c) N. A. Jasim, R. N. Perutz, A. C. Whitwood, T. Braun, J. Izundu, B. Neumann, S. Rothfeld and H.-G. Stammler, *Organometallics*, 2004, **23**, 6140–6149.
28. (a) P. Hofmann and G. Unfried, *Chem. Ber.*, 1992, **125**, 659–661; (b) M. Crespo, J. Granell, M. Font-Bardía and X. Solans, *J. Organomet. Chem.*, 2004, **689**, 3088–3092; (c) L. Schwartsburd, R. Cohen, L. Konstantinovshi and D. Milstein, *Angew. Chem., Int. Ed.*, 2008, **47**, 3603–3606.
29. (a) S. Park, M. Pontier-Johnson and D. M. Roundhill, *Inorg. Chem.*, 1990, **29**, 2689–2697; (b) S. Park, M. Pontier-Johnson and D. M. Roundhill, *J. Am. Chem. Soc.*, 1989, **111**, 3101–3103; (c) N. A. Jasim and R. N. Perutz, *J. Am. Chem. Soc.*, 2000, **122**, 8685–8693; (d) L. Villanueva, M. Arroyo, S. Bernès and H. Torrens, *Chem. Commun.*, 2004, 1942–1943; (e) C. M. Anderson, R. J. Puddephatt, G. Ferguson and A. J. Lough, *J. Chem. Soc., Chem. Commun.*, 1989, 1297–1298; (f) C. M. Anderson, M. Crespo, G. Ferguson, A. J. Lough and R. J. Puddephatt, *Organometallics*, 1992, **11**, 1177–1181; (g) M. Crespo, M. Martinez and J. Sales, *Organometallics*, 1993, **12**, 4297–4164; (h) M. Crespo, M. Martinez and J. Sales, *J. Chem. Soc., Chem. Commun.*, 1992, 822–823.
30. A. Casitas, M. Canta, M. Sola, M. Costas and X. Ribas, *J. Am. Chem. Soc.*, 2011, **133**, 19386–19392.
31. (a) C. B. Caputo and D. W. Stephan, *Organometallics*, 2012, **31**, 27–30; (b) D. Ferraris, C. Cox, R. Anand and T. Lectka, *J. Am. Chem. Soc.*, 1997, **119**, 4319–4320; (c) J. Terao, S. A. Begum, Y. Shinohara, M. Tomita, Y. Naiton and N. Kambe, *Chem. Commun.*, 2007, 855–857.
32. (a) A. Jana, P. P. Samuel, G. Tavcar, H. W. Roesky and C. Schulzke, *J. Am. Chem. Soc.*, 2010, **132**, 10164–10170; (b) A. Jana, H. W. Roesky, C. Schulze and P. P. Samuel, *Organometallics*, 2010, **29**, 4837–4841; (c) S. K. Mandal and H. W. Roesky, *Acc. Chem. Res.*, 2012, **45**, 298–307.
33. (a) F. Swarts, *Bull. Acad. Roy. Belg.*, 1920, 399; (b) F. Swarts, *Bull. Sci. Acad. Roy. Belg.*, 1936, **22**, 122; (c) M. W. Renoll, *J. Am. Chem. Soc.*, 1946, **68**, 1159; (d) F. L. M. Pattison and B. C. Saunders, *J. Chem. Soc.*, 1949, 2745; (e) L. D. Freedman, G. O. Doak and E. L. Petit, *J. Am. Chem. Soc.*, 1955, **77**, 4262; (f) J. Blum, A. Rosenfeld, F. Gelman, H. Schumann and D. Avnir, *J. Mol. Catal. A: Chem.*, 1999, **146**, 117.
34. (a) M. Aizenberg and D. Milstein, *Science*, 1994, **265**, 359–361; (b) M. Aizenberg and D. Milstein, *J. Am. Chem. Soc.*, 1995, **117**, 8674–8675; (c) B. L. Edelbach and W. D. Jones, *J. Am. Chem. Soc.*, 1997, **119**, 7734–7742; (d) J. L. Kiplinger and T. G. Richmond, *Chem. Commun.*, 1996, 1115–1116; (e) J. L. Kiplinger and T. G. Richmond, *J. Am. Chem. Soc.*, 1996, **118**, 1805–1806; (f) H. Yang, H. Gao and R. J. Angelici,

Organometallics, 1999, **18**, 2285–2287; (g) N. Yu and V. F. Starichenko, *Mendeleev Commun.*, 2000, **10**, 60–61; (h) J. Vela, J. M. Smith, Y. Yu, N. A. Ketterer, C. J. Flaschenriem, R. J. Lachicotte and P. L. Holland, *J. Am. Chem. Soc.*, 2005, **127**, 7857–7870; (i) T. Braun, D. Noveski, M. Ahijado and F. Wehmeier, *Dalton Trans.*, 2007, 3820–3825; (j) K. Fuchibe, Y. Ohshima, K. Mitomi and T. Akiyama, *J. Fluorine Chem.*, 2007, **128**, 1158–1167; (k) U. Jäger-Fiedler, M. Klahn, P. Arndt, W. Baumann, A. Spannenberg, V. V. Burlakov and U. Rosenthal, *J. Mol. Catal., A Chem.*, 2007, **261**, 184–189; (l) G. Meier and T. Braun, *Angew Chem., Int. Ed.*, 2009, **48**, 1546–1548; (m) S. P. Reade, M. F. Mahon and M. K. Whittlesey, *J. Am. Chem. Soc.*, 2009, **131**, 1847–1861; (n) R. J. Young and V. V. Grushin, *Organometallics*, 1999, **18**, 294–296.

35. For examples of catalytic activation of Csp^3–F bonds see: (a) R. Panish, M. Bolte and T. Muller, *J. Am. Chem. Soc.*, 2006, **128**, 9679–9682; (b) C. Douvris and O. V. Ozerov, *Science*, 2008, **321**, 1188–1190; (c) V. J. Scott, R. Celenligil-Cetin and O. V. Ozerov, *J. Am. Chem. Soc.*, 2005, **127**, 2852–2853; (d) W. Gu, M. R. Haneline, C. Douvris and O. V. Ozerov, *J. Am. Chem. Soc.*, 2009, **131**, 11203–11212.
36. (a) H. Amii and K. Uneyama, *Chem. Rev.*, 2009, **109**, 2119–2183; (b) A. D. Sun and J. A. Love, *Dalton. Trans.*, 2010, **39**, 10362–10374.
37. Y. Kiso, K. Tamao and M. Kumada, *J. Organomet. Chem.*, 1973, **50**, C12–C14.
38. B. L. Edelbach, B. M. Kraft and W. D. Jones, *J. Am. Chem. Soc.*, 1999, **121**, 10327–10331.
39. H. Guo, F. Kong, K.-i. Kanno, J. He, K. Nakajima and T. Takahashi, *Organometallics*, 2006, **25**, 2045–2048.
40. G. Cahoez, F. Lepifre and P. Ramiandrasoa, *Synthesis*, 1999, 2138.
41. (a) M. Arisawa, T. Suzuki, T. Ishikawa and M. Yamaguchi, *J. Am. Chem. Soc.*, 2008, **130**, 12214–12215; (b) Y. Ishii, N. Chatani, S. Yorimitsu and S. Murai, *Chem. Lett.*, 1998, 157–158.
42. T. J. Korn, M. A. Schade, S. Wirth and P. Knochel, *Org. Lett.*, 2006, **8**, 725–728.
43. (a) T. Braun, R. N. Perutz and M. I. Sladek, *Chem. Commun.*, 2001, 2254–2255; (b) N. Yoshikai, H. Mashima and E. Nakamura, *J. Am. Chem. Soc.*, 2005, **127**, 17978–17979; (c) A. Steffen, M. I. Sladek, T. Braun, B. Neumann and H.-G. Stammler, *Organometallics*, 2005, **24**, 4057–4064; (d) T. Schaub, M. Backes and U. Radius, *J. Am. Chem. Soc.*, 2006, **128**, 15964–15965; (e) T. Saeki, Y. Takashima and K. Tamao, *Synlett*, 2005, **7**, 1771–1774; (f) A. D. Sun and J. A. Love, *Org. Lett.*, 2011, **13**, 2750–2753; (g) M. Tobisu, T. Xu, T. Shimasaki and N. Chatani, *J. Am. Chem. Soc.*, 2011, **133**, 14952–14955; (h) V. P. W. Böhm, C. W. K. Gstöttmayr, T. Weskamp and W. A. Herrmann, *Angew. Chem., Int. Ed.*, 2001, **40**(3387–3389); (i) L. Ackermann, R. Born, J. H. Spatz and D. Meyer, *Angew. Chem., Int. Ed.*, 2005, **44**, 7216–7219.
44. (a) T. Braun, J. Izundu, A. Steffen, B. Neumann and H.-G. Stammler, *Dalton Trans.*, 2006, 5118–5123; (b) K. Manabe and S. Ishikawa, *Synthesis*,

2008, 2645–2649; (c) D. A. Widdowson and R. Wilhelm, *Chem. Commun.*, 1999, 2211–2212; (d) Y. M. Kim and S. Yu, *J. Am. Chem. Soc.*, 2003, **125**, 1696–1697.

45. (a) T. Wang, B. Alfonso and J. A. Love, *Org. Lett.*, 2007, **9**, 5629–5631; (b) H. Buckley, A. D. Sun and J. A. Love, *Organometallics*, 2009, **28**, 6622–6624; (c) A. D. Sun and J. A. Love, *J. Fluorine Chem.*, 2010, **131**, 1237–1240; (d) H. Buckley, T. Wang, O. Tran and J. A. Love, *Organometallics*, 2009, **28**, 2356–2359; (e) T. Wang and J. A. Love, *Organometallics*, 2008, **27**, 3290–3296; (f) L. Keyes, A. D. Sun and J. A. Love, *Eur. J. Org. Chem.*, 2011, **20**, 3985–3994; (g) T. Wang, L. Keyes and J. A. Love, *Organometallics*, 2012, **31**, 1397–1407.
46. C. Yu, C. Zhang and X. Shi, *Eur. J. Org. Chem.*, 2012, **10**, 1953–1959.
47. H. N. Borah, D. Prajapati and R. C. Boruah, *Synlett*, 2005, 2823.
48. O. Allemann, S. Duttwyler, P. Romanato, K. K. Baldridge and J. S. Siegel, *Science*, 2011, **332**, 574–579.

CHAPTER 7

Strategies Towards Challenging Fluorination Reactions

CHARLOTTE HOLLINGWORTH AND VÉRONIQUE GOUVERNEUR*

University of Oxford, Chemistry Research Laboratory, 12 Mansfield Road, Oxford, OX1 3TA, UK
*Email: veronique.gouverneur@chem.ox.ac.uk

7.1 Introduction

Fluorine incorporation into an organic molecule can have a profound effect on its physical, chemical and biological properties.[1] Consequently, fluorinated compounds are ubiquitous in agrochemical, medicinal and material science.[2–4] For clinical applications, the exceptional sensitivity of ^{19}F NMR to molecular and microenvironmental changes has prompted the development of ^{19}F-based reporter molecules to interrogate physiological phenomena *in vivo*.[5] Complementing ^{19}F NMR, biomarkers labeled with the radioactive ^{18}F isotope are also commonly used for Positron Emission Tomography (PET), a molecular imaging technology allowing for the early detection of diseases such as Parkinson's and cancer.[6] The scarcity of naturally occurring organofluorine compounds has led to a reliance on synthetic chemistry for the production of molecules containing one or more fluorine substituents. Over recent years, the field has progressed enormously with the successful development of transition metal mediated fluorination and trifluoromethylation of a large variety of activated and poorly activated substrates. This chapter will focus on the use of transition metals in C_{sp^2}–F and C_{sp^3}–F bond formation.

RSC Catalysis Series No. 11
C–H and C–X Bond Functionalization: Transition Metal Mediation
Edited by Xavi Ribas

Published by the Royal Society of Chemistry, www.rsc.org

7.2 Metal Mediated Fluorination

7.2.1 Aryl Fluorides

The fluorination of (hetero)aryl derivatives is a critically important transformation for the pharmaceutical industry, encouraging substantial research in this area. Many reactions that do not involve a transition metal are known and have been extensively reviewed.[7–9] However, substrates displaying weak electrophilicity typically require the involvement of a transition metal complex for activation. Catalytic cross-coupling reactions allowing for aryl–S, aryl–N and aryl–O bond formation served as a solid foundation for the development of new transformations for the preparation of fluorinated aromatics.[10] The most challenging elementary step of such catalytic cross-coupling reaction is the reductive elimination event. Careful mechanistic investigations by Hartwig and co-workers revealed that the reductive elimination is easier for the more nucleophilic heteroatom with the following order of reactivity established: C–P > C–S > C–N > C–O.[11] In 2001, the same group examined the feasibility of a reductive elimination from a Pd^{II} complex for aryl halide bond formation.[12] Reductive elimination was disfavoured for aryl iodide but the formation of aryl chlorides and aryl bromides was possible in the presence of an excess of the strongly electron donating phosphine $P(tBu)_3$. These results suggest that reductive elimination is plausible for the preparation of aryl fluorides although this was not substantiated by experimental evidences. The high electronegativity of fluorine however implies that reductive elimination from a palladium complex leading to aryl–F bond formation is most likely a challenging process.

Grushin and co-workers investigated the formation of fluoro palladium complexes and the feasibility of reductive elimination of C–F and C–CF_3 from a Pd^{II} species. In 1997, the group disclosed the synthesis of the first isolated organopalladium tertiary phosphine fluoride, $[(Ph_3P)_2Pd(Ph)F]$ (**1**) prepared from $[(Ph_3P)_2Pd(Ph)I]$ and AgF (Scheme 7.1).[13] This complex was found to be stable in air with no dimerisation observed in solution. Initial efforts towards reductive elimination disappointingly gave only F–P and C–P type products.[14] The fluoro palladium complex [(Xantphos)Pd(Ph)F] (**2**) (structure in Scheme 7.2) featuring the bulky bidentate phosphine ligand XantPhos was also resistant to C–F reductive elimination.

Interestingly, trifluoromethylated complexes prepared upon treatment of the corresponding fluoride complex [(L)Pd(Ph)F] with the Ruppert's reagent (CF_3SiMe_3) were more reactive. The complexes $[(dppe)Pd(Ph)CF_3]$ and $[(dppp)Pd(Ph)CF_3]$ were stable under rigorously dry conditions up to temperatures of 130–135 °C but $PhCF_3$ was detected (10–30%) after 6–64 h at 145 °C in xylenes.[15] Importantly, $PhCF_3$ was formed quantitatively from $[(Xantphos)Pd(Ph)CF_3]$ (**3**) at 50–80 °C in benzene.[16]

In 2007, Yandulov undertook a computational study of the activation enthalpies associated with aryl–F bond formation (Scheme 7.3).[17] The data confirm that C–P reductive elimination and $Pd^{II}F$ dimer formation from complex **4** are more facile processes than C–F bond formation.

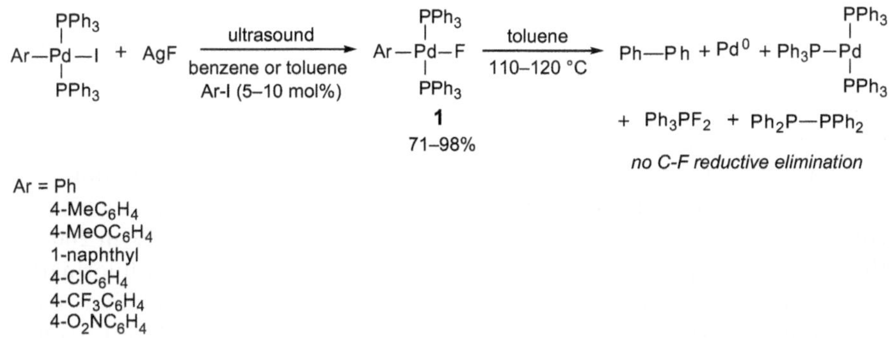

Scheme 7.1 Synthesis of $ArPd^{II}F$.

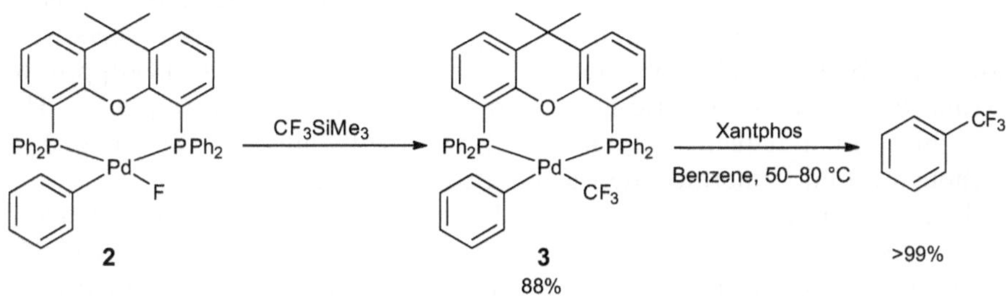

Scheme 7.2 Reductive elimination of a C–CF_3 bond from $ArPd^{II}(Xantphos)(CF_3)$.

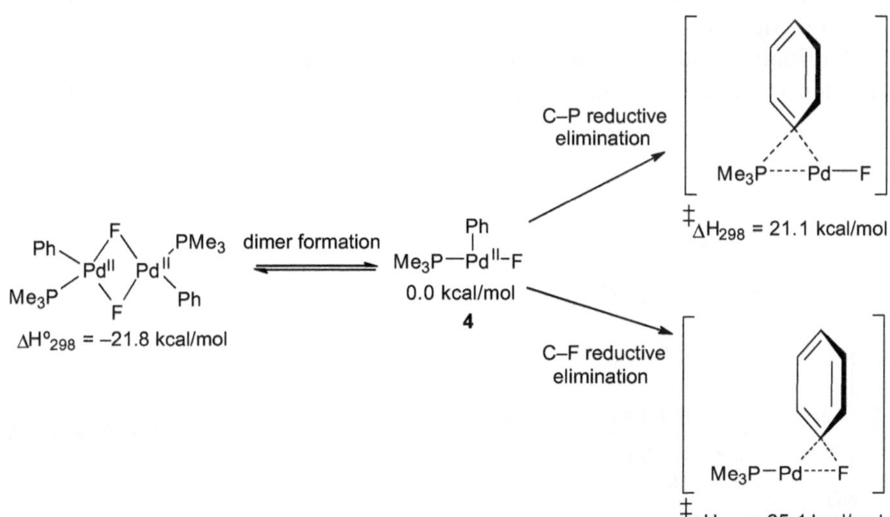

Scheme 7.3 Feasibility of C–F reductive elimination from a Pd^{II} complex.

The use of electron withdrawing groups on the aryl substrates decrease the energy required to reach the C–F reductive elimination transition state. Notably, the use of a bulky phosphine ligand developed in the Buchwald group, *t*-BuXPhos, along with the presence of a nitro group positioned *para* on the arene, gave the desired aryl fluoride in ~10% yield (Scheme 7.4).

A major conceptual breakthrough emerged with the recognition that reductive elimination from a higher oxidation state Pd^{IV} complex may be more efficacious to generate aryl fluoride; this was verified experimentally in 2006 by the Sanford group who reported the $Pd(OAc)_2$ catalysed oxidative fluorination of 2-phenylpyridine in a microwave reactor with *N*-fluoropyridinium tetrafluoroborate (**N-F1**), a reagent acting both as an oxidant and fluorine source (Scheme 7.5).[18] The proposed mechanism involved a cyclopalladation event directed by the proximal nitrogen followed by oxidative fluorination at the metal with **N-F1**. Subsequent reductive elimination from the resulting Pd^{IV} fluoride intermediate gave the aryl fluorides. While investigating the synthesis of diphosphine difluoropalladium^{II} complexes, Vigalok and co-workers were able to obtain 4-difluorobenzene on exposure of Pd^{II}(*p*-FPhI)(dppp) to *N*-fluoro-2,4,6-trimethylpyridinium tetrafluoroborate in ~10% yield.[19] The possibility of a reductive elimination from a Pd^{IV} fluoride was suggested, although no Pd^{IV} intermediate was isolated for formal verification.

In 2008, Yu and co-workers applied the same principles using triflamide-protected benzylamines to direct *ortho*-fluorination in the presence of 10 mol% $Pd(OTf)_2{\cdot}2H_2O$ and an excess of *N*-fluoro-2,4,6-trimethylpyridinium triflate, **N-F2**.[20] The reaction gave higher yield of product when carried out in the presence of 0.5 equivalents of NMP (NMP = *N*-methyl-2-pyrrolidone). The use of $Pd(OAc)_2$ led to side reactions involving acetate as a competing nucleophile. The reaction is compatible with a wide range of functionalities and tolerates electron rich and electron poor substitutions on the aryl (Scheme 7.6). Difluorination did occur under the reaction conditions, so the scope of this reaction is limited to *ortho*-substituted substrates. *Meta*-substituted aromatics

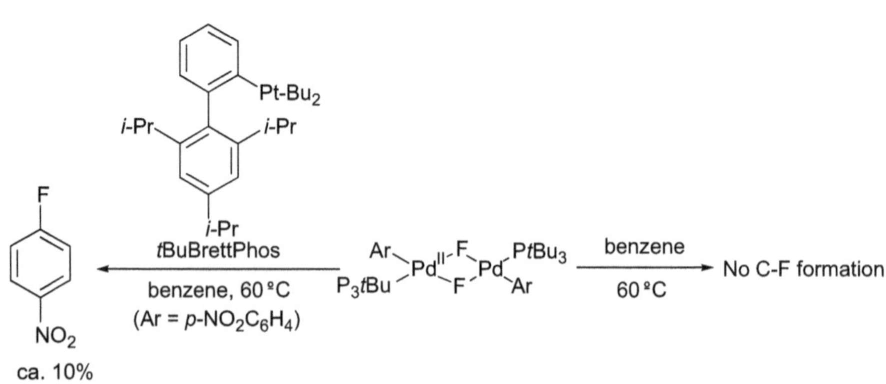

Scheme 7.4 Reductive elimination of an aryl–F bond from a Pd^{II} centre.

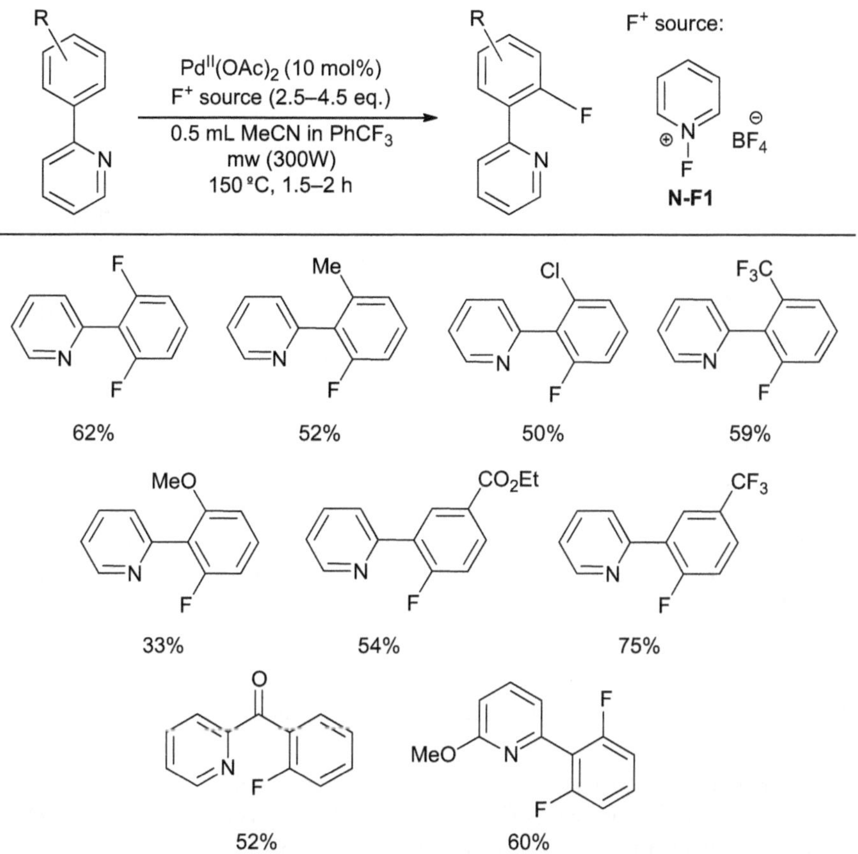

Scheme 7.5 Pyridine-directed Pd-catalysed electrophilic fluorination of arenes.

underwent monofluorination as the major pathway, however difluorination was detectable leading to a mixture of products that were difficult to separate.

The problem of undesired difluorination was overcome using an alternative *ortho*-directing group.[21] The reaction of 4-(*tert*-butyl)-*N*-(2,3,5,6-tetrafluoro-4-(trifluoromethyl)phenyl)-benzamide **5** with 10 mol% $[Pd(OTf)_2(MeCN)_4]$ with **N-F2** in the presence of NMP in acetonitrile gave the desired fluoroaromatic **6** in 82% yield along with only 5% of the difluorinated product **7** (Scheme 7.7). Substrates with electron-donating alkyl and alkoxy substituents on the aryl required shorter reaction times to secure good yields, while electron-donating groups required extended reaction times. Solvent exchange to (trifluoromethyl)benzene and an increased excess of **N-F2** led to difluorinated arenes as the major products. Fluorinated benzoic acids that are useful precursors for many pharmaceuticals were released in excellent yields upon hydrolysis under basic conditions.

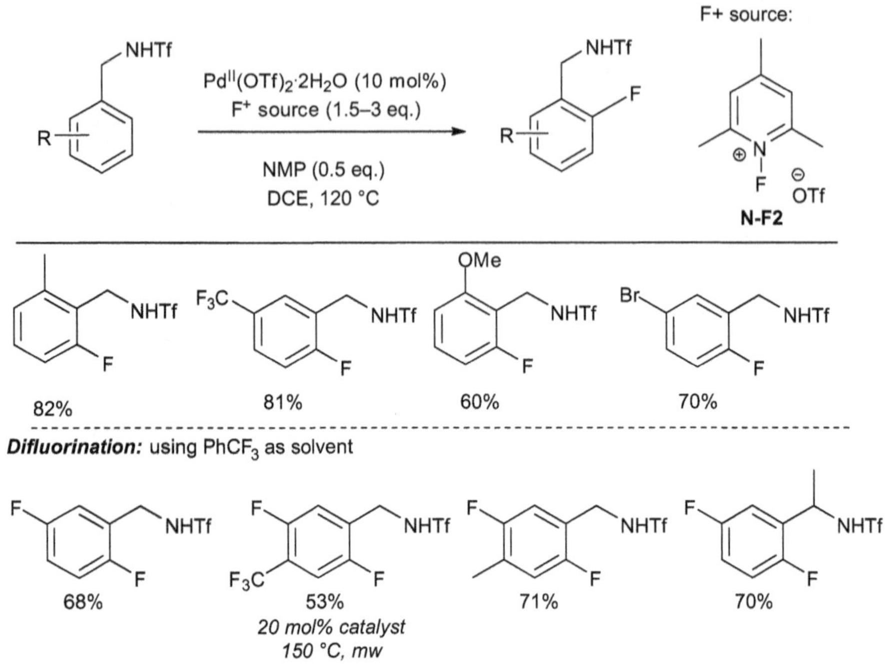

Scheme 7.6 Pd-catalysed *ortho*-fluorination in presence of NMP.

In 2009, Sanford provided experimental evidence for C–F reductive elimination from discrete Pd^{IV} complexes (Scheme 7.8).[22] The {(*t*-Bu-bpy)Pd^{II}(*p*-XC$_6$H$_4$)(F)} complexes **9** (*t*-Bu-bpy = 4,4′-di-*t*-butyl-2,2′-bipyridine) featuring a carefully chosen rigid bidentate nitrogen ligand known to stabilize Pd^{IV} centres were synthesized from the corresponding Pd^{II} iodides upon sonification with AgF. Exposure of these complexes to XeF$_2$ in nitrobenzene at 90 °C gave the corresponding aryl fluorides in 57%, 60% and 45% yield for X = F, CF$_3$ and OMe, respectively. When this reaction was carried out at 70°C for 2.5 minutes, a new fluorinated species was isolated and unambiguously characterised as the Pd^{IV} species **10** by ^{19}F-NMR spectroscopy analysis at −70 °C and X-ray crystallography. The thermolysis of this complex only gave trace amounts of aryl fluoride with the major product identified as the corresponding biaryl product; however, when an excess of XeF$_2$ was used, the aryl fluorides were formed in good to excellent yields. The formation of C–C bonds in this transformation indicates that the rate-limiting step is the C–F coupling, a process competing with aryl exchange between metal centres.

Ritter and co-workers accomplished the fluorination of arylboronic acids in 2008.[23] Aryl fluorides were obtained in high yields upon transmetalation of a Pd^{II} acetate species **11** bearing a pyridyl-sulfonamide ligand with aryl boronic acids and subsequent oxidation with the electrophilic fluorinating reagent Selectfluor (Scheme 7.9). The high valent Pd^{IV} complex **12** was suspected as the reactive intermediate for C–F bond formation, but no such species could be detected by ^{19}F-NMR spectroscopy.

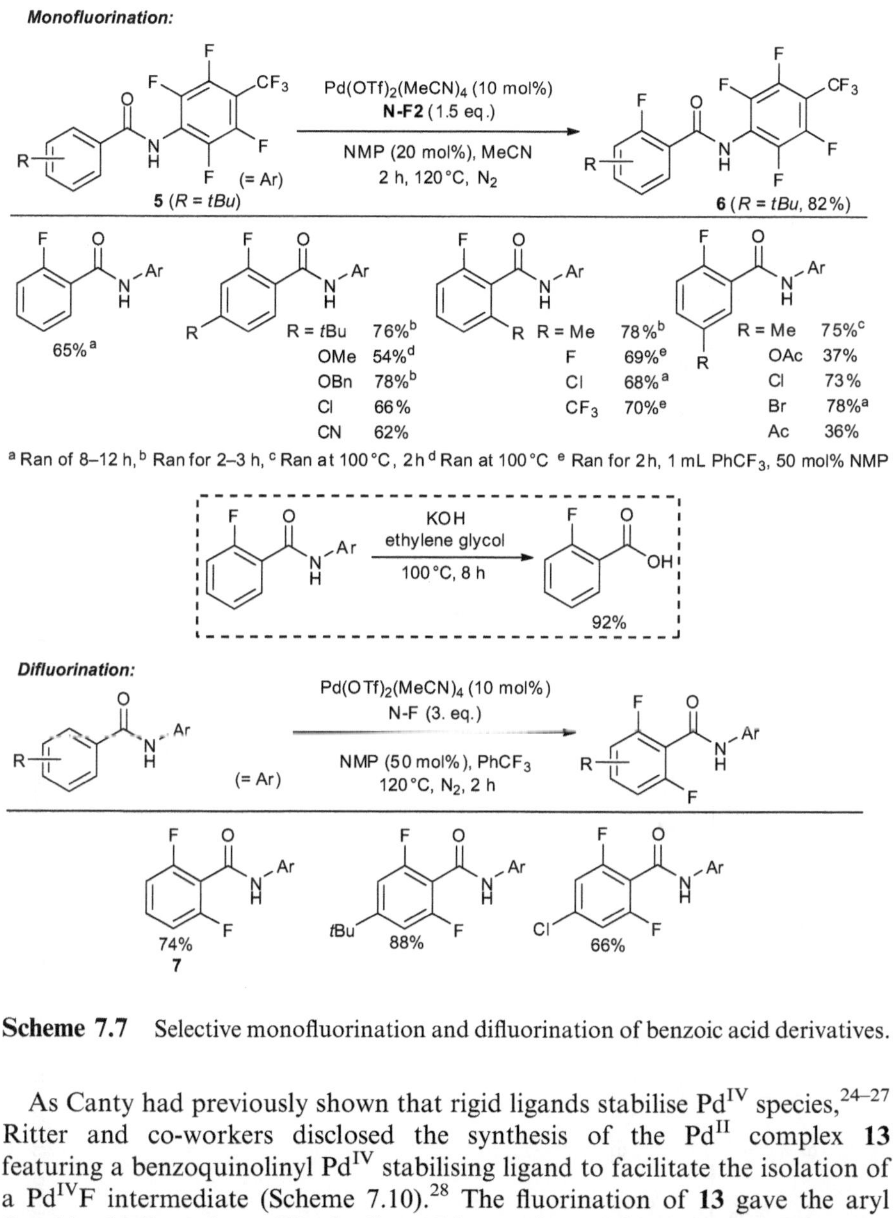

Scheme 7.7 Selective monofluorination and difluorination of benzoic acid derivatives.

As Canty had previously shown that rigid ligands stabilise Pd^{IV} species,[24–27] Ritter and co-workers disclosed the synthesis of the Pd^{II} complex **13** featuring a benzoquinolinyl Pd^{IV} stabilising ligand to facilitate the isolation of a $Pd^{IV}F$ intermediate (Scheme 7.10).[28] The fluorination of **13** gave the aryl fluoride, 10-fluorobenzo[*h*]quinoline **15** in excellent yield (94%). The presence of the Pd^{IV} intermediate **14** was detected by ^{19}F-NMR (δ –278 ppm for $Pd^{IV}F$) at room temperature in acetonitrile, however this could not be isolated for further analysis. Unlike the Sanford system, the thermolysis of this complex in solution did yield aryl fluoride in excellent yields. The reaction of this fluorobenzo[*h*]quinolinyl palladium^{IV} pyridine-sulfonamido tetrafluoroborate **14** with (TMAF·$3H_2O$) gave the Pd^{IV} difluoride **16**, fully characterized by X-ray crystallography. The thermolysis of this complex in DMSO at 150 °C gave aryl fluoride **15** in 97% within 10 minutes. This reaction has not been extended to a

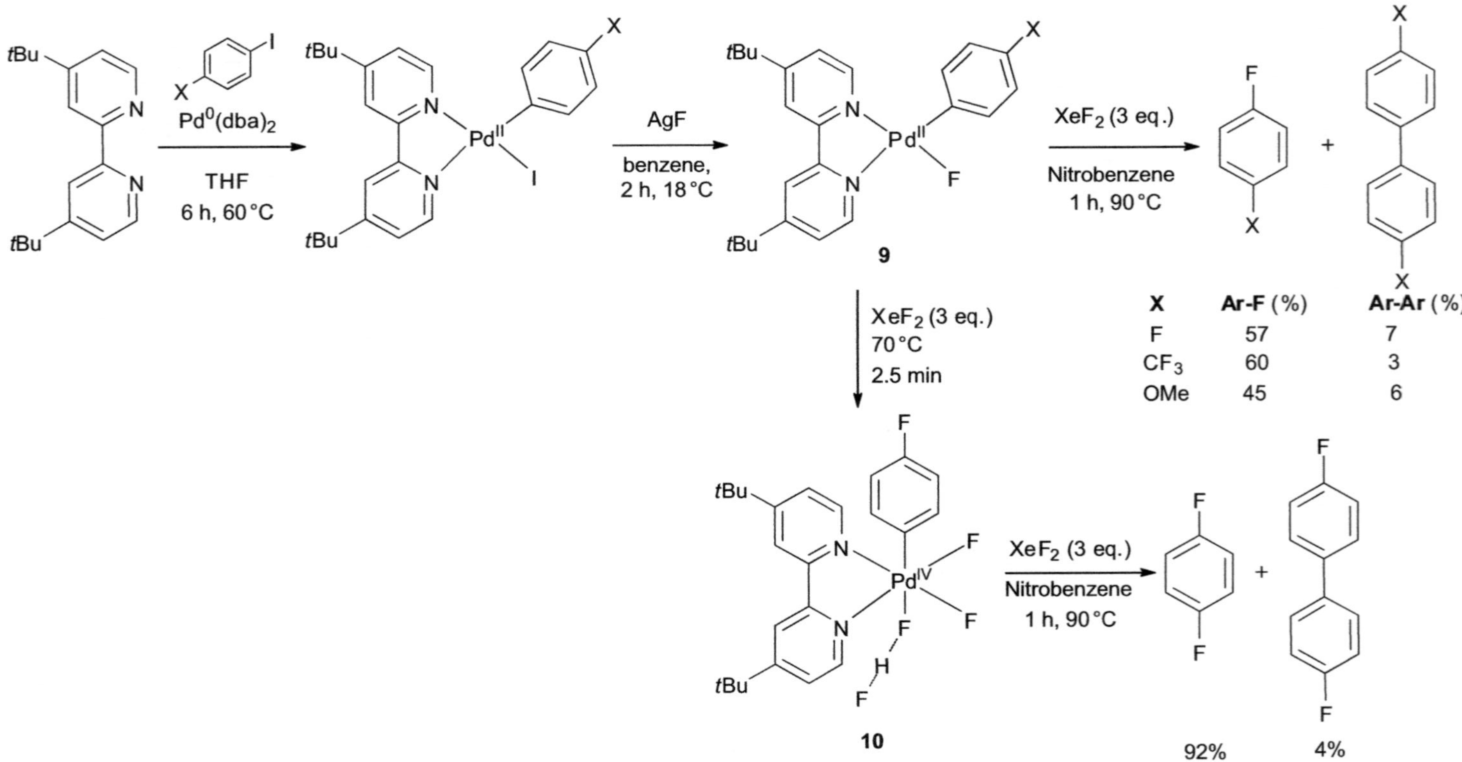

Scheme 7.8 Mechanistic studies into a Pd^{II}/Pd^{IV} mediated aromatic fluorination.

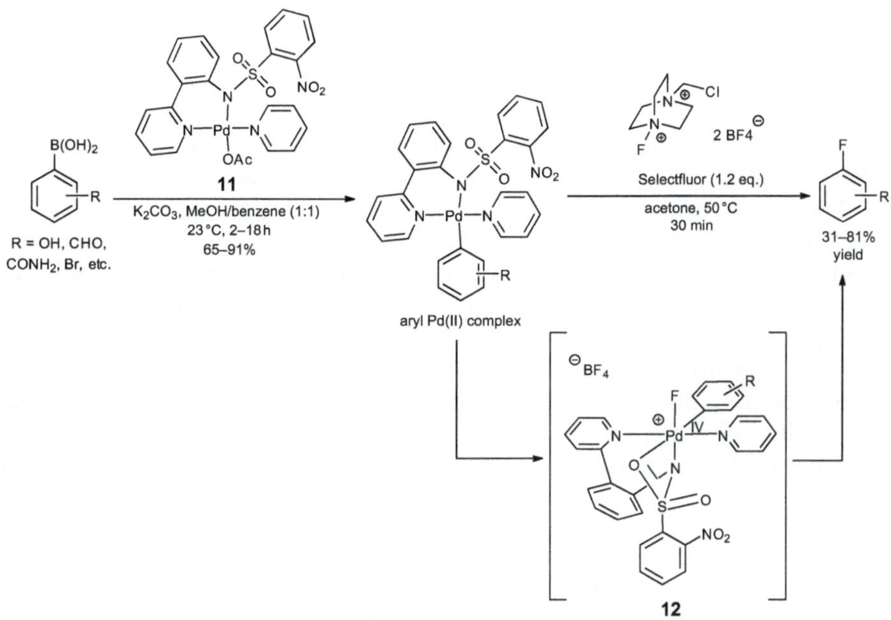

Scheme 7.9 Pd mediated electrophilic fluorination of arylboronic acids using Selectfluor.

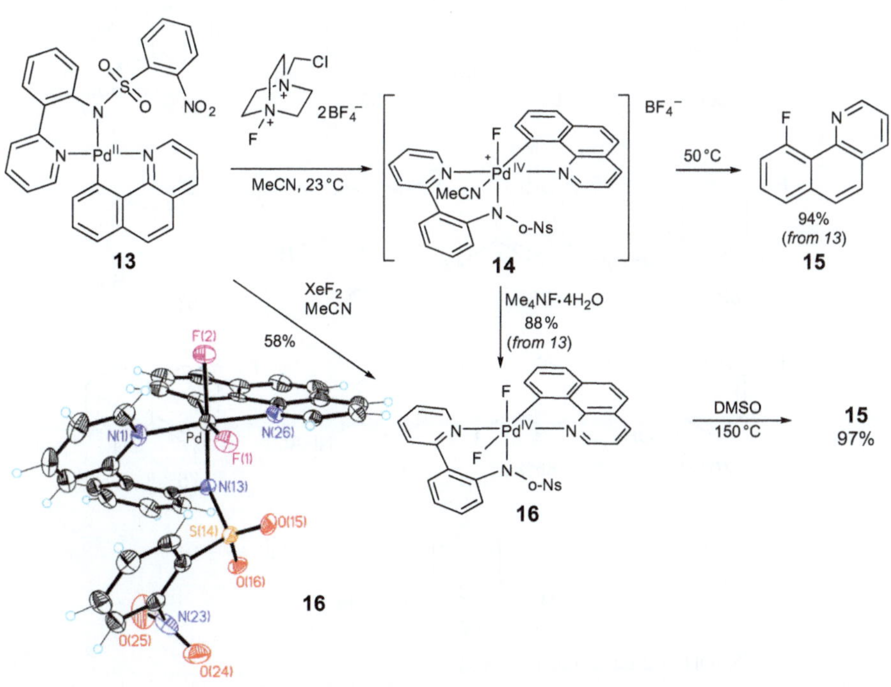

Scheme 7.10 Identification and isolation of the $Pd^{IV}F$ intermediate **16**.

catalytic system but the substrate scope is vast. The fluorination tolerates electron-rich and electron-poor aryls and heterocycles in addition to protic functionalities typically not suitable for late stage nucleophilic fluorination. Further mechanistic studies revealed that the pyridyl-sulfonamide ligand is crucial in the positioning of the aryl and fluoride ligands both in the Pd^{II} and Pd^{IV} complexes to facilitate C–F bond reductive elimination.[29]

Arylboronic acids can also be converted to aryl fluorides with stoichiometric $Ag^{I}OTf$ and base (NaOH) in methanol/acetone at room temperature; the transmetalation (aryl–B to aryl–Ag) was conducted in methanol followed by solvent exchange to acetone and addition of Selectfluor at room temperature.[30] The authors postulate a bimetallic Ag^{II} complex bearing a fluoro ligand as the reactive intermediate. The reaction shows a high tolerance to various functionalities and is also applicable to arylboronic esters, although fluorination yields are significantly lower for these precursors. A drawback to this reaction is the use of superstoichiometric transition metal (2 equivalents of AgOTf) (Scheme 7.11).

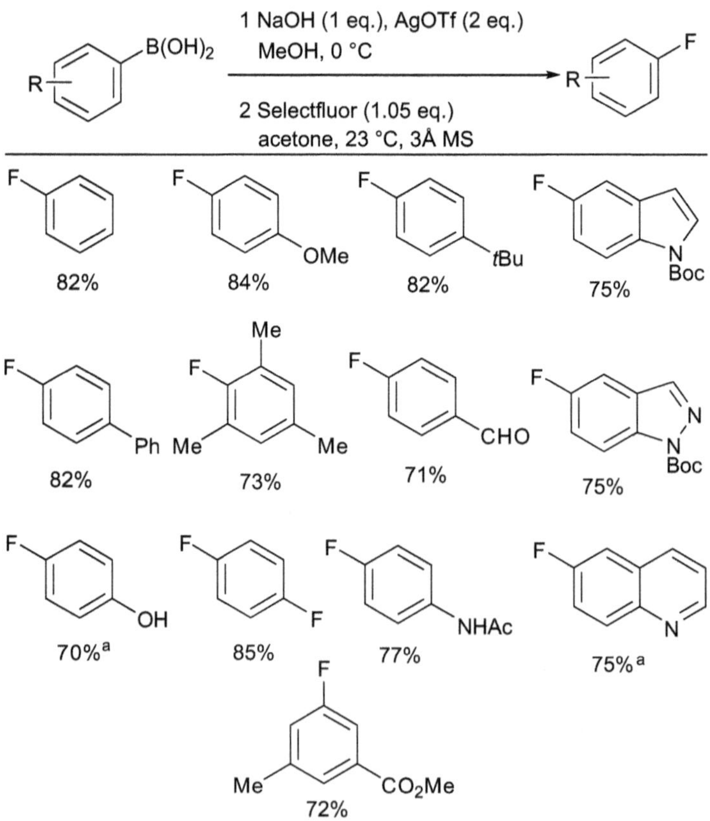

aNaOH (1.2 eq.), AgOTf (3 eq.)

Scheme 7.11 Ag-mediated fluorination of arylboronic acids.

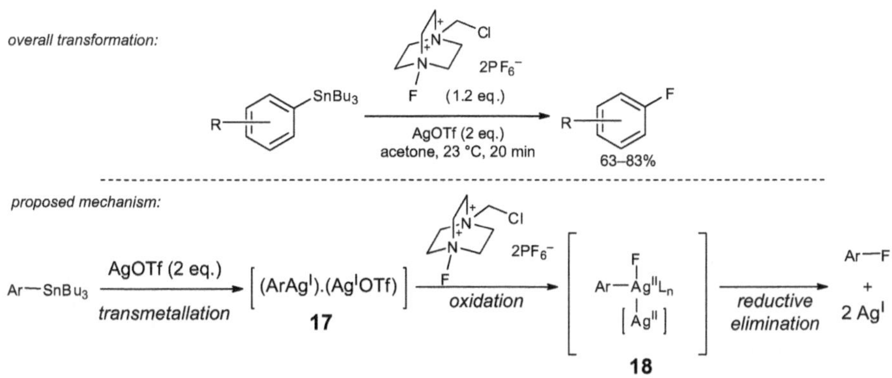

Scheme 7.12 Putative mechanism for the Ag mediated aryl fluorination *via* a high valent bimetallic species.

The Ag-mediated fluorination is applicable to arylstannanes without the need for a base to aid transmetalation (Scheme 7.12).[31] For this class of substrates, the counter-ion of the N–F reagent was found to be important; PF_6^- (in place of BF_4^-) increased significantly the product yield, an experimental observation rationalised by transmetalation of the arylstannanes with BF_4^-. The substrate scope of this reaction is broad, with a range of bioactive compounds made accessible by applying this methodology. However, the use of toxic organotin reagents along with byproducts resulting from competitive protodestannylation limits the value of this transformation. Mechanistic studies on these Ag-mediated fluorinations identified (ArAg)(AgOTf) **17** by ^{19}F and ^{1}H NMR spectroscopy. The authors propose a transmetalation leading to **17** followed by oxidation with **N-F3** to form the bimetallic Ag^{II} species **18** that undergoes subsequent reductive elimination with concomitant release of the aryl fluoride product. The presence of two Ag atoms is necessary for reductive elimination as this is a two electron process and Ag has one-electron redox chemistry.

A catalytic variant of this reaction was developed using 5 mol% of Ag_2O.[32] The problematic protodestannylation suggested to occur with *in situ* formed triflic acid was overcome by the addition of a base ($NaHCO_3$). The addition of NaOTf was found to increase the rate of fluorination most likely due to the formation of a more soluble Ag^I complex from Ag_2O which is poorly soluble. The fluorination was tolerant to a range of functional groups, remained regioselective and gave good to excellent yields; however basic functional groups, sulfides and carboxylic acids were not tolerated. Nonetheless the methodology does have a vast substrate scope seemingly more tolerant than the palladium systems. However, a disadvantage still remains with the use of arylstannanes which must be synthesised and are highly toxic. This catalytic variant was applied to the late stage fluorination of highly functionalized targets inclusive of a series of important pharmaceuticals and biologically active targets such as ezetimibe and strychnine (Scheme 7.13). A proposed catalytic cycle is shown in Scheme 7.14.

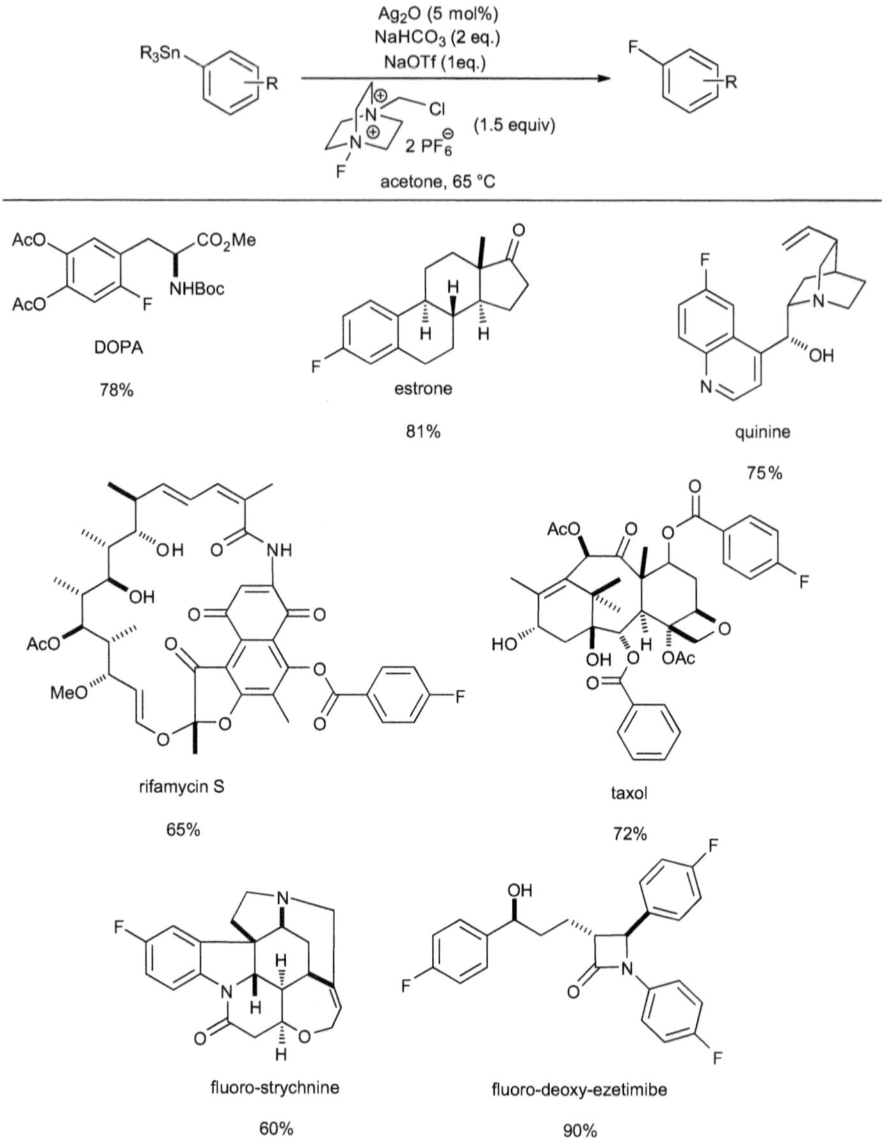

Scheme 7.13 Ag-catalyzed fluorination of complex biologically active molecules.

As limitations were identified both with the use of aryl stannanes, which are highly toxic, and the use of boronic acids, which require a change of solvent between transmetalation and fluorination, arylsilanes were investigated as alternative susbtrates. Aryl trihydroxysilanes and aryl trialkoxysilanes were found to be optimal for fluorination with Ag_2O and Selectfluor in acetone at 90 °C.[33] The addition of BaO reduced protodesilylation. Other electrophilic fluorinating reagents or solvents other than acetone did not lead to

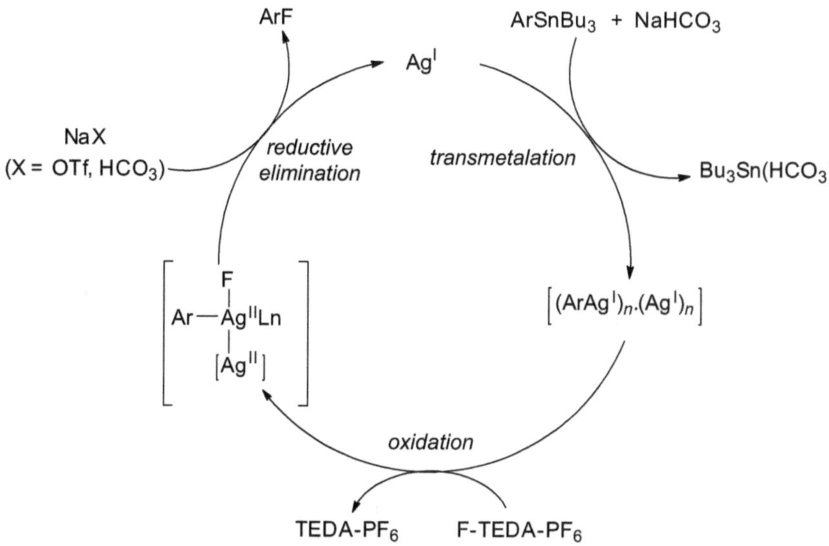

Scheme 7.14 Proposed catalytic cycle for the Ag-catalyzed fluorination of arylstannanes.

fluorination. The reaction tolerates electron-rich/poor aryl precursors, various heterocyclic motifs and 2,6-substituted aryls. In the case of electron-rich ring systems, unwanted double fluorination was suppressed by addition of 2,6-lutidine. Mechanistically, a high valent bimetallic silver complex is proposed as a putative intermediate but this hypothesis was not substantiated experimentally.

Buchwald and co-workers disclosed the first example of reductive elimination from a {$ArPd^{II}F$} complex leading to aryl fluorides in 2009.[34] This breakthrough contribution built on the observation that dimer formation may be a limiting step in the C–F bond formation and that bulky phosphine ligands are promising candidates to address this challenge. In previous work, Buchwald reported the C–N reductive elimination from a monomeric complex {LPd(Ar)(X)} (X = Cl or Br) to give aryl amines.[35] The nature of the ligand, monophosphine BrettPhos (2-(dicyclohexylphosphino)3,6-dimethoxy-2′,4′,6′-triisopropyl-1,1′-biphenyl) was critical to effect this transformation successfully. The corresponding monomeric {**L1**Pd^{II}(Ar)(F)} (**L1** = BrettPhos) complex **19** was prepared *via* halogen exchange with AgF and its structural identity unambiguously confirmed by X-ray crystallography as a monomeric tricoordinated species. Upon thermolysis in toluene, the corresponding aryl fluorides were formed in moderate yields (Scheme 7.15). At this point, the three essential events necessary for the development of a catalytic variant were established: the oxidative addition of Pd^0 into aryl bromides, the halogen exchange and the sought after aryl–F reductive elimination from a Pd^{II} centre.

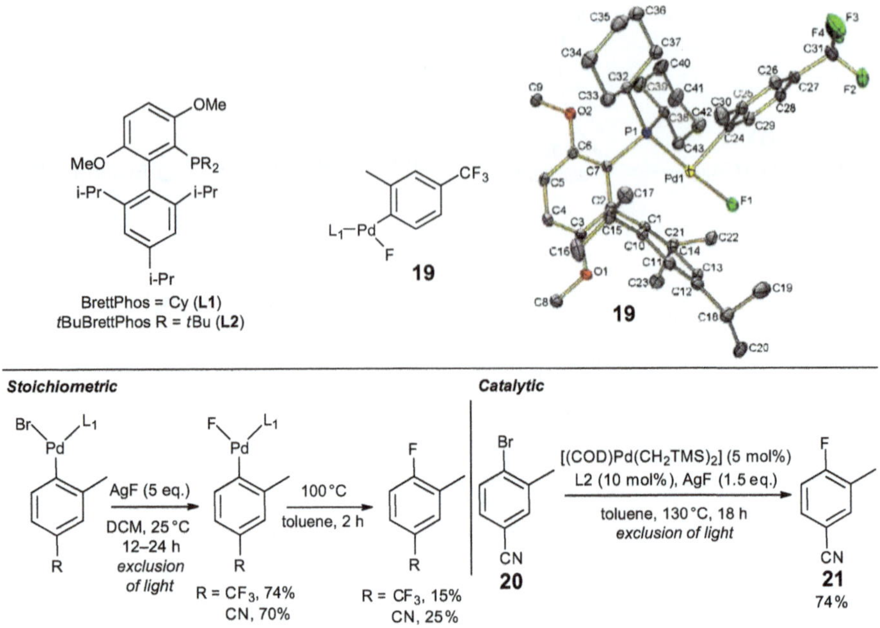

Scheme 7.15 Stoichiometric and catalytic aromatic fluorination *via* reductive elimination of a C–F bond from a Pd^{II} centre.

Initial optimisation showed that the use of 5 mol% [(COD)Pd(CH_2TMS)$_2$], 10 mol% BrettPhos and 1.5 equivalents of AgF in toluene for 18 h at 130 °C afforded aryl fluoride **21** from aryl bromide **20** in 74% yield. Under these conditions, the reaction had severe substrate limitations, as only electron-poor aromatics with an *ortho* substituent gave fluorinated products. The fluorination of 1-naphthyl triflate was initially inefficient, however on varying the fluoride source to CsF and changing the precatalyst to [(cinnamyl)PdCl]$_2$ a moderate yield of 30% could be achieved. A ligand screening identified the use of *t*BuBrettPhos **L2** as optimal, affording 1-fluoronaphthalene in 79% yield under anhydrous conditions. Control reactions ruled out the possibility of nucleophilic aromatic substitution (S_NAr) or aryne-type mechanisms. In comparison to the reaction with aryl bromide, the fluorination of aryl triflates shows a large substrate scope (Scheme 7.16). The reaction tolerates highly substituted aryls, electron deficient systems and a range of heteroaryl motifs including indoles, quinolines and flavones. More forcing conditions (130 °C) were necessary for electron rich arenes and the reaction does not seem to accommodate protic groups. For electron rich aryl bromides, mechanistic studies suggested the *in situ* formation of an arylated phosphine ligand.[36] Therefore, the yields increase upon addition of excess aryl bromides. In certain cases, a mixture of regioisomers was observed, however the formation of undesired regiosiomers was suppressed by the use of apolar solvents such as cyclohexane. Further studies revealed that the use of RockPhos in place of *t*BuBrettPhos gives comparable yields of

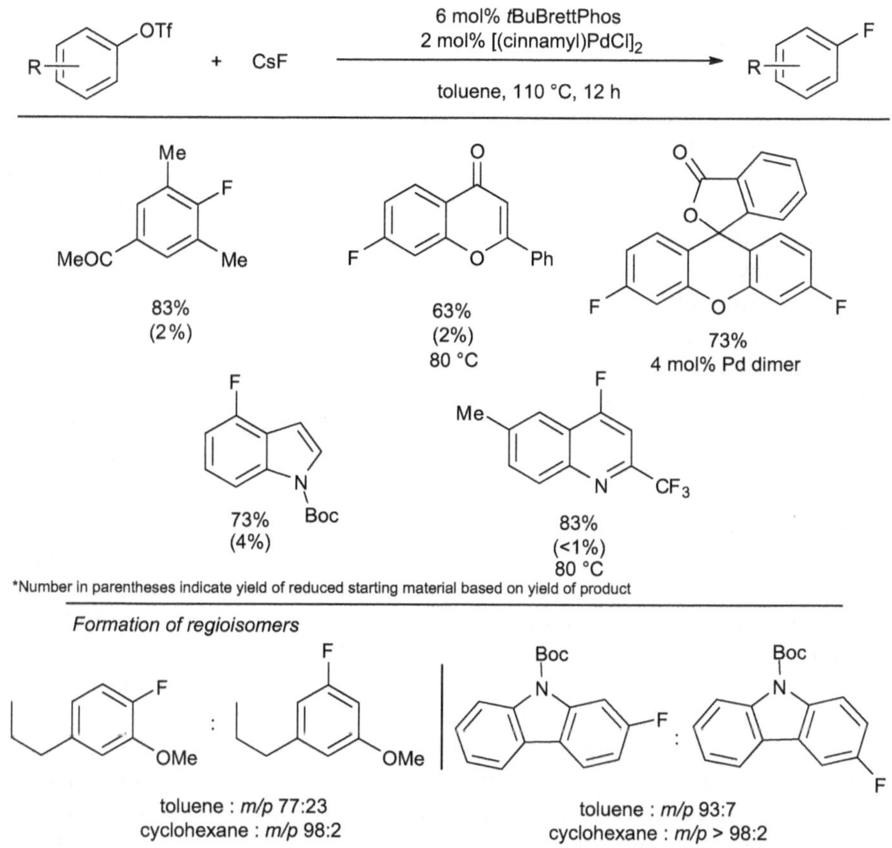

Scheme 7.16 Pd^{II} catalysed aromatic fluorination of aryl triflates.

fluorination; advantageously, the complexes synthesised with this ligand are easier to isolate and analyse for use in mechanistic investigations.

As fluoroaromatics are commonly used in Positron Emission Tomography (PET), the prospect of applying this methodology to access [^{18}F]aryl fluorides is very attractive. To comply with the various constraints of [^{18}F]radiochemistry for PET, the reaction must be complete in less than 1 hour (^{18}F $t_{1/2}=109.7$ min)[37] with the fluoride as the limiting reagent by a very large margin. Attempts in cold mode (^{19}F) to shorten the reaction time were successful (~30 min) but only with a large excess of the fluoride source and in the presence of a solubilising agent. These conditions are not suitable for the [^{18}F]radiolabelling of an aryl fluoride, but the finding that large excesses of CsF increases reaction rates encouraged the authors to adapt this methodology for use in a microflow reactor.[38] Excellent yields were obtained when the reactions were carried out at 120 °C with a 20 minute residence time. Continuous mode production was also demonstrated with 3 mmol of 1-naphthyl triflate converted to aryl fluoride over 8 h with no depreciation in yield.

In 2011, Ritter and co-workers disclosed a protocol for the [^{18}F]radiolabelling of [^{18}F]aryl fluorides using a novel [^{18}F]PdF complex prepared from [^{18}F]fluoride but acting as an electrophilic [^{18}F]fluorine source.[39] Careful consideration in the design of complex **23** was necessary to ensure a sufficient level of fluorophilicity (fluoride concentration 10^{-4} M for radiochemistry) and stability. A high valent Pd^{IV} centre was chosen due to its ability to act as an oxidant. Benzo(*h*)quinolyl and tetrapyrazole borate ligands were used to stabilise the Pd^{IV} centre and avoid surplus reductive eliminations. The complex would ideally act as an electrophile in a S$_N$2 reaction where Pd is the leaving group requiring selective attack of the nucleophile at the fluorine and not at the ligands; calculation of LUMO energy levels identified only the fluorine as susceptible to attack in complex **23**. The reaction of the Pd^{IV}[^{19}F] complex with Pd^{II}aryl precursors prepared from the corresponding boronic acids, provided a route to electron neutral and electron rich fluoroaromatics not accessible by direct aromatic nucleophilic substitution. The radiochemical variant of this reaction is significant as [^{18}F]fluoride is easier to handle and more readily available than [^{18}F]F_2. Its use also allows for the production of [^{18}F]molecules with higher specific activity, an important parameter at least for some applications. Several Pd^{II}aryl precursors were fluorinated using this methodology inclusive of a selection of substrates bearing functionalities commonly found in PET tracers. The reaction of complex **22** with [^{18}F]KF/18-crown-6 in acetonitrile gave **23** in 10 minutes which was directly reacted with the Pd^{II}aryl precursors giving decay-corrected radiochemical yields (RCY) up to 33%, the overall reaction time being no longer than 60 minutes (Scheme 7.17).

The groups of Subramanian, Grushin and Dolbier reported that Cu^{II}F$_2$ is a competent reagent for the nucleophilic fluorination of aromatics. Subramanian reported the production of fluorobenzene in up to 30% yield reacting CuF_2 with benzene at 450–550 °C.[40] No products resulting from perfluorination were observed. In this protocol, CuF_2 can be regenerated by treatment with HF/O$_2$ at 400 °C. In 2008, Dolbier used a copper aluminum fluoride species ($CuAl_2F_8$) for the fluorination of aryl chlorides at 500 °C.[41] In 2007, Grushin reported the reaction of aryl iodides and aryl bromides with CuF_2 in the presence of *N,N,N′,N′*-tetramethylethylenediamine at 150–180 °C; this reaction gave low yields of aromatic fluorides.[42] Although the mechanism has not been fully investigated, the involvement of aryne intermediates was ruled out since the fluorination of 2-iodonaphthalene gave 2-fluoronaphthalene as the only regiosiomeric product.

Ribas *et al.* reported the Cu^{I}/Cu^{III} catalysed aromatic fluorination of aryl halides with AgF (Scheme 7.18).[43] The proposed reaction mechanism consists of three archetypical elemental steps, an oxidative addition leading to the Cu^{III} species **25**, a halide exchange affording **26** and a final reductive elimination releasing the product **27** and the Cu^{I} catalyst. The Cu^{III}–F complex **26** could not be detected experimentally unlike with other halides (Cl, Br and I), and as such the halide exchange was proposed as the rate-limiting step followed by a rapid reductive elimination. This is in contrast to the Buchwald Pd system where the reductive elimination is the limiting step. The use of a *N*-permethylated

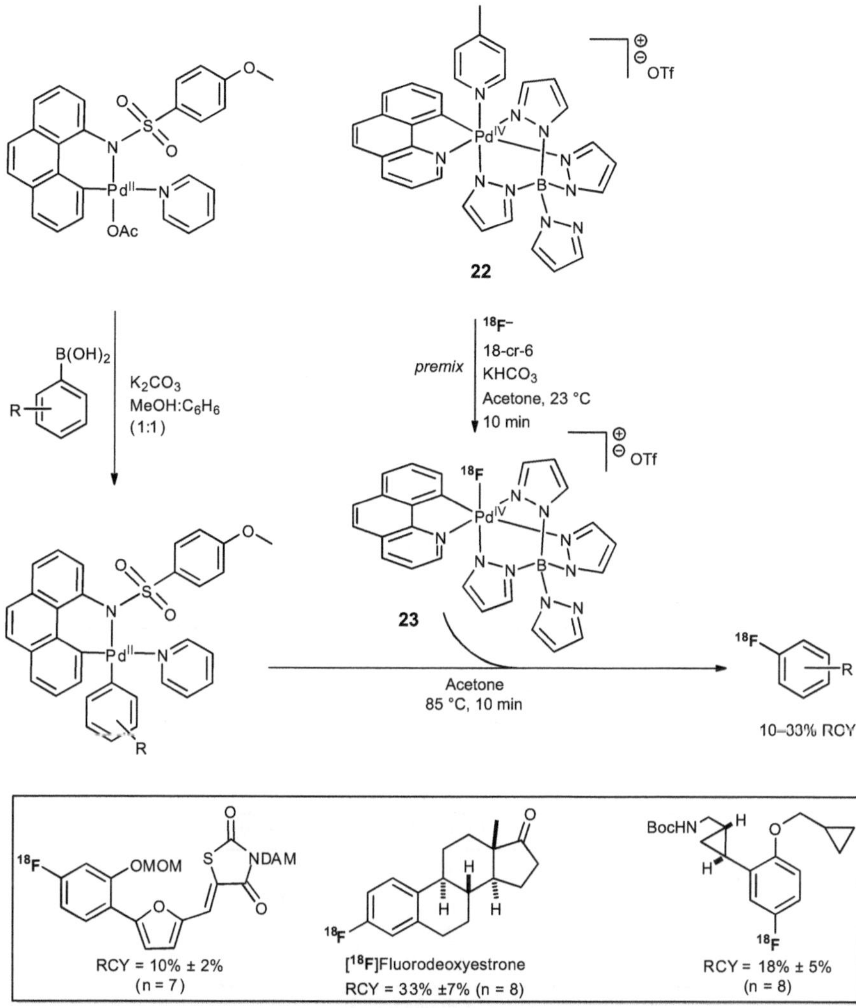

Scheme 7.17 Late-stage fluorination with a $Pd^{IV}[^{18}F]$ species, an electrophilic fluorinating reagent synthesized from $[^{18}F]F^{-}$.

macrocycle led to a higher performance catalytic system possibly due to the ability of fluoride to deprotonate non-methylated secondary amines. For this reaction, it was important to use acetonitrile and acetone as a mixture of solvent, as the use of acetonitrile over enhances the stability of Cu^{I} species.

A gold catalysed cyclization–fluorination leading to fluorinated pyrazoles from alkynyl phenylhydrazones was reported in 2011.[44] Optimisation on 1-phenyl-2-(4-phenylbut-3-yn-2-ylidene)hydrazine led to a high yielding (81%) tandem aminofluorination when the reaction was performed in the presence of $Ph_3PAuNTf_2$, Selectfluor and a base ($NaHCO_3$). The choice of solvent was crucial, with acetonitrile found to be the optimum; when water is added

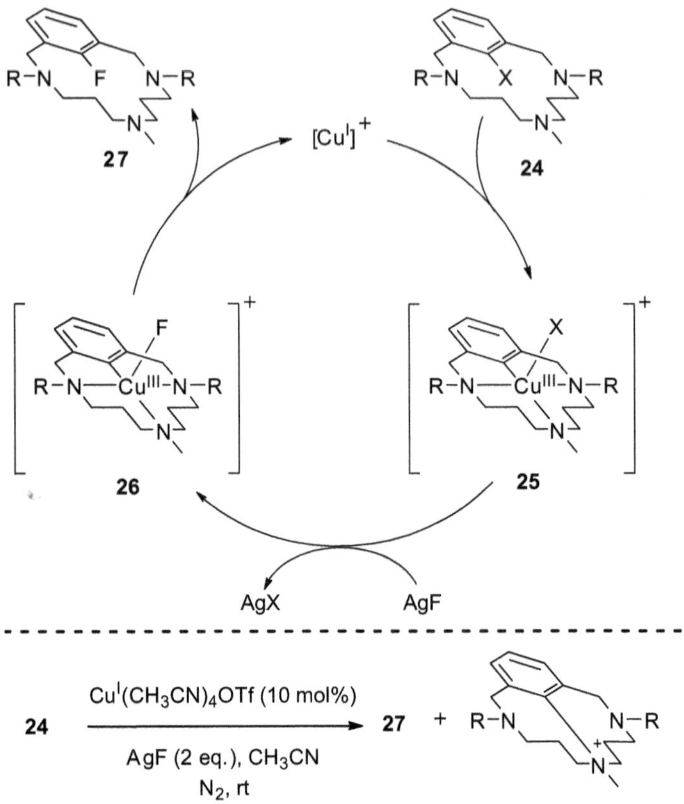

R	X	Time (h)	Yield of Ar-F
H	Cl	6	76%
	Br	4	71%
CH_3	Cl	12	98%[a]
	Br	24	97%[a]

[a]Solvent system is CH_3CN:Acetone (1:3)

Scheme 7.18 Cu^I/Cu^{III} catalysed aromatic fluorination of aryl chlorides.

(MeCN:H_2O, 20:1), only nonfluorinated pyrazole was obtained presumably *via* cyclization–protodeauration. Substitution at the alkyne unit was well tolerated with electron-rich aryls giving moderate to good yields, electron-poor aryls giving slightly reduced yields and a range of alkyl and heterocyclic substituents well tolerated (Scheme 7.19). The substrate scope was further extended by substitution at the hydrazone motif. Several mechanistic pathways are possible for this transformation involving both Au^I and Au^{III} intermediates; no conclusive evidence has been found to rule out either mechanism. The observed formation and consumption of the nonfluorinated pyrazole during the reaction suggested this as a possible reactive intermediate. Mechanistic investigations showed the

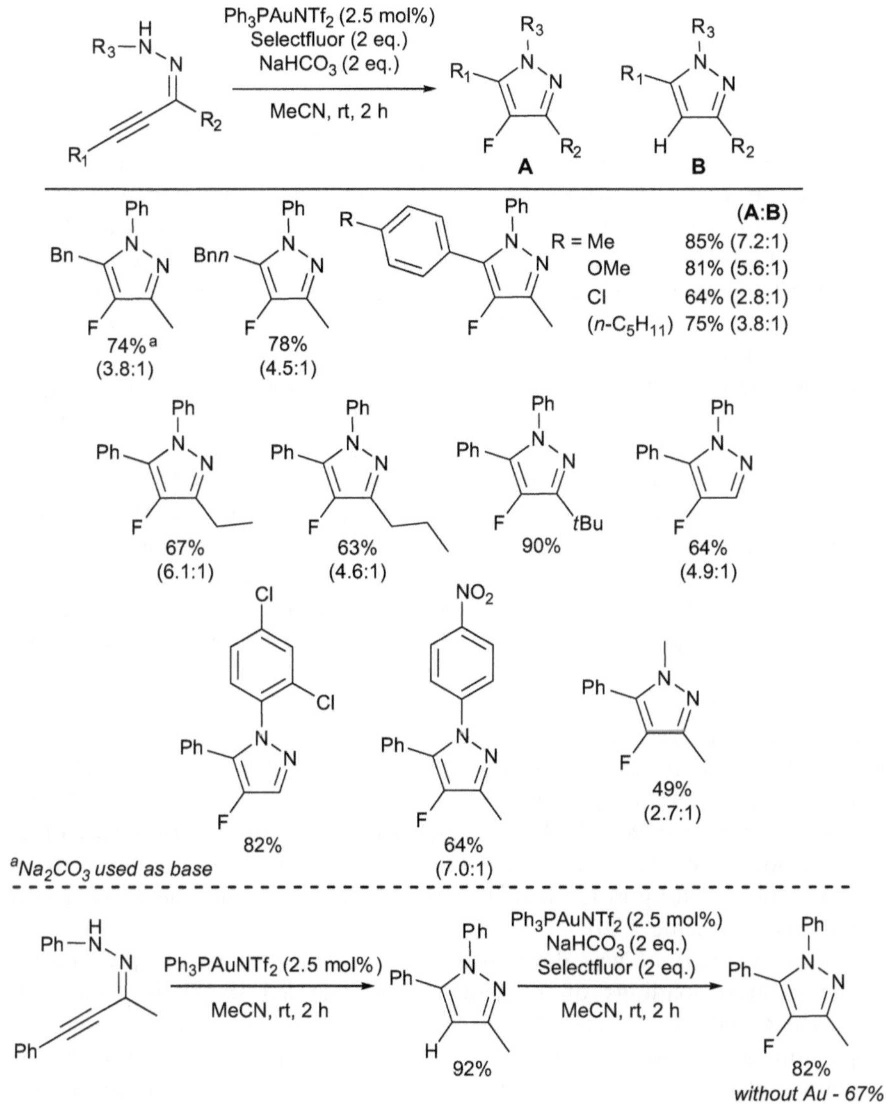

Scheme 7.19 Au-catalysed tandem aminofluorination of alkynes to yield fluorinated pyrazoles.

pyrazole resulting from the Au-catalysed cyclisation of 1-phenyl-2-(4-phenylbut-3-yn-2-ylidene)hydrazine can undergo subsequent fluorination with Selectfluor in the presence and absence of $Ph_3PAuNTf_2$ to give the fluorinated pyrazole in 82% and 67%, respectively. The improved yield in the presence of the Au^I catalyst can be rationalized evoking an aminoauration of 1-phenyl-2-(4-phenylbut-3-yn-2-ylidene)hydrazine leading to a vinylAu^I intermediate that could undergo either fluorodeauration or undesired protodeauration.

7.2.2 Alkenyl Fluorides

Alkenyl fluorides are particularly useful as probes for structural and mechanistic studies. The synthesis of fluoroalkenes *via* late stage fluorination is challenging, as control over regiochemistry and *E*/*Z* geometry may be required in addition to C–F bond formation. While methodologies exist that can fluorinate an activated double bond without the need for a transition metal, these are not commonly used for downstream applications as the alkenyl fluorides are formed as a mixture with the corresponding *gem*-difluoroalkanes resulting from over-fluorination. This complication has encouraged the development of transition metal catalysed methods for the preparation of this important class of fluorinated targets.

In 2005, Sadighi and co-workers were able to isolate the stable $Au^{I}F$ complex **28**.[45] The *N*-heterocyclic carbene ligand SIPr was found to be optimal for stabilising the Au^{I} fluoride motif. The reaction of (SIPr)AuCl (SIPr = 1,3-bis(2,6-diisopropylphenyl)imidazolin-2-ylidene) with NaO*t*Bu in benzene at room temperature and subsequent exposure of the resulting (SIPr)AuO*t*Bu complex to 1 equivalent of triethylamine trishydrofluoride gave **28** in 48% yield. While this complex was stable in the solid state, the addition of CH_2Cl_2 resulted in partial decomposition to unidentified by-products. Exposure of complex **28** to excess 3-hexyne in CH_2Cl_2 gave the β-(fluorovinyl)gold complex **29** in a reversible reaction where **28** can be completely recovered upon the removal of solvent and excess alkyne.[46] For mechanistic investigation, the cationic complex $\{(SIPr)Au[\eta^2\text{-(3-hexyne)}]\}^+ BF_4^-$ **30** was prepared upon treatment of (SIPr)AuCl with $AgBF_4$ followed by addition of 3-hexyne. When **30** was reacted with the soluble fluoride $[(Me_2N)_3P]_2N^+F^-$ the equilibrium between **28** and **29** was re-established. However, with the use of the mildly acidic $Et_3N{\cdot}3HF$ as a source of fluoride, hydrofluorination of the alkene was possible resulting in the formation of (*Z*)-3-fluoro-3-hexene in 64% yield (^{19}F-NMR) (Scheme 7.20).

A catalytic variant was developed using complex **30** as the pre-catalyst. Using catalyst loadings of 1–5 mol%, 3-hexyne led to **31** in 53% yield (^{19}F-NMR) with $Et_3N{\cdot}3HF$. The nature of the ligand was shown to be very important, as less sterically hindered NHCs were unable to stabilise the low oxidation state gold complex; this was diagnosed by precipitation of gold metal. 4,5-Dichloro-1,3-bis(2,6-diisopropylphenyl)imidazolin-2-ylidene was also found to be a competent ligand for fluorination. Other counter-ions in the pre-catalyst were tolerated, with O*t*Bu as ligand giving comparable conversion. The addition of acidic additives such as $KHSO_4$ and $PhNMe_2{\cdot}HOTf$ led to the desired fluoroalkenes in increased yields. Aryl, alkyl and heteroaryl substituted alkynes were tolerated with good levels of regioselectivity seen on unsymmetrical alkynes (Scheme 7.21). Arenes bearing electron-withdrawing groups showed complete control over regioselectivity in favour of β-fluorination; the same sense of regiocontrol was seen with the thienyl substituent. The preference for the β-fluorostyrene products is consistent with a mechanism featuring nucleophilic attack of fluoride on complex **30** forming **29**, which

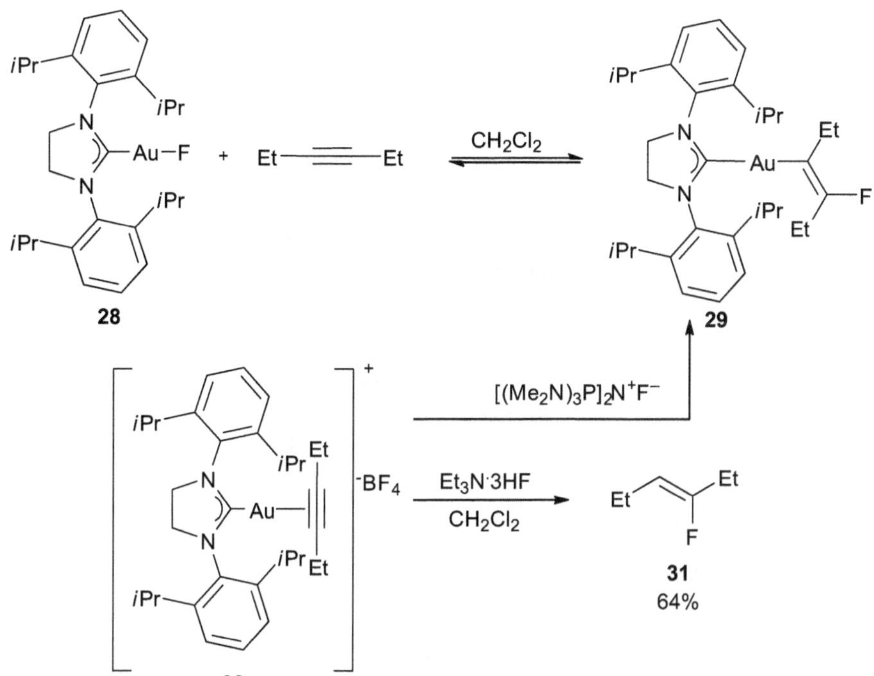

Scheme 7.20 Isolation of a Au^{I}fluoride complex and *trans* hydrofluorination of alkynes.

undergoes a subsequent protodeauration to give the fluoroalkene in an overall net *trans* HF addition across the alkyne.

The regioselectivity of this reaction could be reversed by the use of an ester or nitrogen-containing directing group on the alkyne (Scheme 7.22). The carbamate 2,2,2-trichloroethoxycarbonyl (Troc) group gave superior regioselectivity for 1,2-dialkyl- and 1,2-alkylaryl alkynes.[47] The cationic Au^I precatalyst is proposed to coordinate to both the alkyne and directing group to allow for the formation of the energetically more favourable cyclic intermediate **33** shown in Scheme 7.22. Protonation of the gold in complex **33** followed by a putative reductive protodeauration could explain the preferential formation of the major *Z*-vinyl fluoride regioisomer.

In 2008, Gouverneur *et al.* reported the Au catalysed cyclization-fluorination of β-hydroxy-α,α-difluoroynones conducted in the presence of the electrophilic fluorine source Selectfluor; this reaction afforded trifluorinated dihydropyranones **35a** (Scheme 7.23).[48] Other sources of electrophilic fluorine were not successful, however the iodination and bromination were possible using *N*-iodo- or bromosuccinimide, respectively. The reaction tolerates the presence of both alkyl and aryl substituents, but in all cases the formation of **35b** resulting from unwanted protodeauration could not be suppressed.

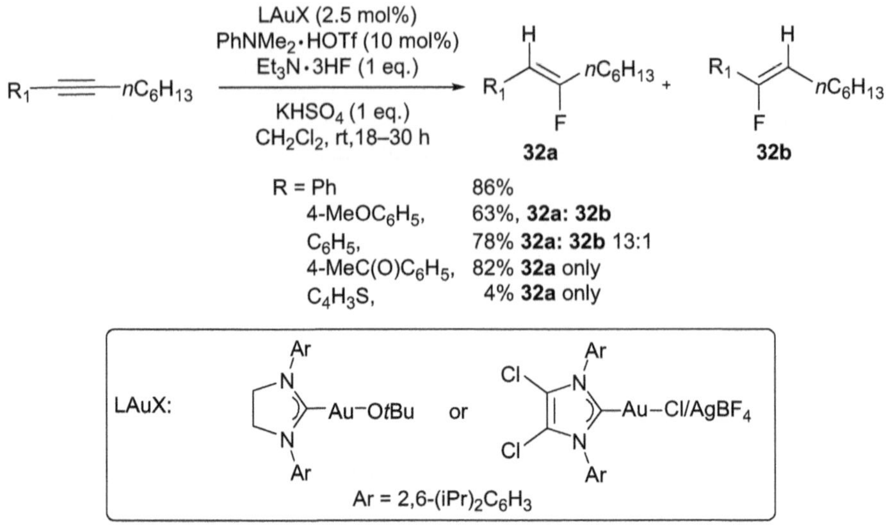

Scheme 7.21 Au^{I} catalysed hydrofluorination.

Mechanistic investigations revealed that a transient vinylgold species **36** is the likely intermediate for this reaction (Scheme 7.24). Control reactions verified that the electrophilic fluorination of the difluorinated dihydropyranone **35b** with Selectfluor in the absence or presence of Au^{I} catalyst is not taking place, an observation suggesting that the presence of the *gem*-difluoro group is deactivating **35b** towards electrophilic fluorination. The formation of **35a** is therefore best explained by electrophilic fluorodeauration of **36**. The fluorination of propargyl acetates to produce α-fluoroenones has also been demonstrated in the presence of Au^{I} catalysts and Selectfluor.[49,50]

The tandem fluorination–cyclization of various esters and amides bearing pending alkenyl and alkynyl chains was investigated with the electrophilic fluorination reagent NFSI (*N*-fluorobenzenesulfonimide) under Pd catalysis ($Pd(TFA)_2$) (TFA = trifluoroacetate)) (Scheme 7.25).[51] Treatment of enynes with a 5 mol% load of Pd catalyst used with the bidentate ligand bathocuproine (7.5 mol%) and NFSI in *N,N*-dimethylacetamide at 50 °C allowed for the formation of cyclized fluorinated products as *E* and *Z* geometrical isomers. The addition of 4-nitrophenol improved the yield of the reaction as well as the *E*:*Z* ratio in favour of the *E* isomer. Under these optimized reaction conditions, the fluorination of enynes bearing aryl-terminated alkynes with electron-withdrawing and donating substituents on the aryl group gave the product in moderate to good yields, respectively. Substitution at the nitrogen atom for amides was tolerated; interestingly, the *Z* isomer was formed as the major product for *N*-benzylated substrates. Internal alkenes gave moderate level of fluorination (37%); for alkyl-terminated alkynes and disubstituted alkenes, only a trace amount of fluorinated products was observed. Deuterium-labelling studies carried out to identify the source of the proton incorporated into the product gave further insight into the

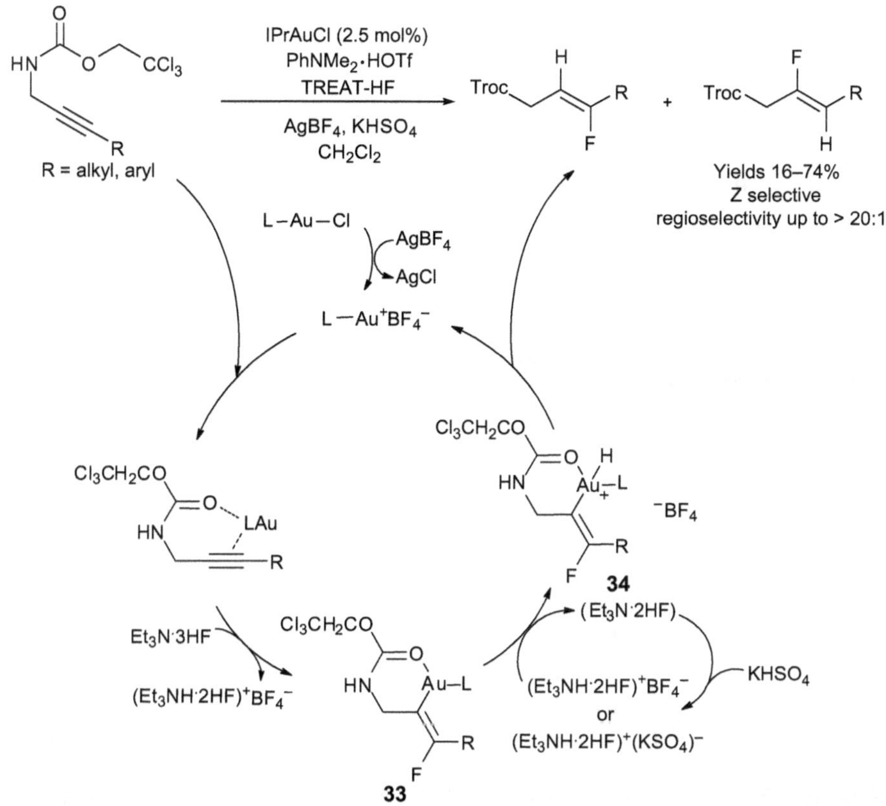

Scheme 7.22 Use of Troc directing groups in the Au^{I} catalysed hydrofluorination of alkynes.

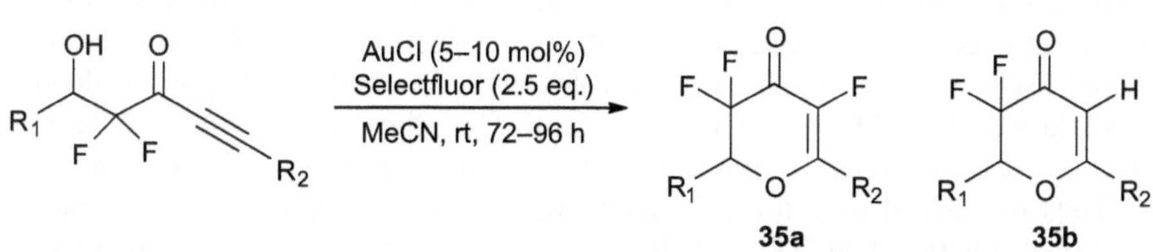

R_1 =	R_2 =	Yield of 35a:35b
Ph	Ph	20:33
p-MeOPh	Ph	26:33
Cy	Ph	20:15
Cy	C_3H_7	27:37
CH_2OBn	CH_3	24:43

Scheme 7.23 Au^{I} catalysed alkoxyfluorination of β-hydroxyynones.

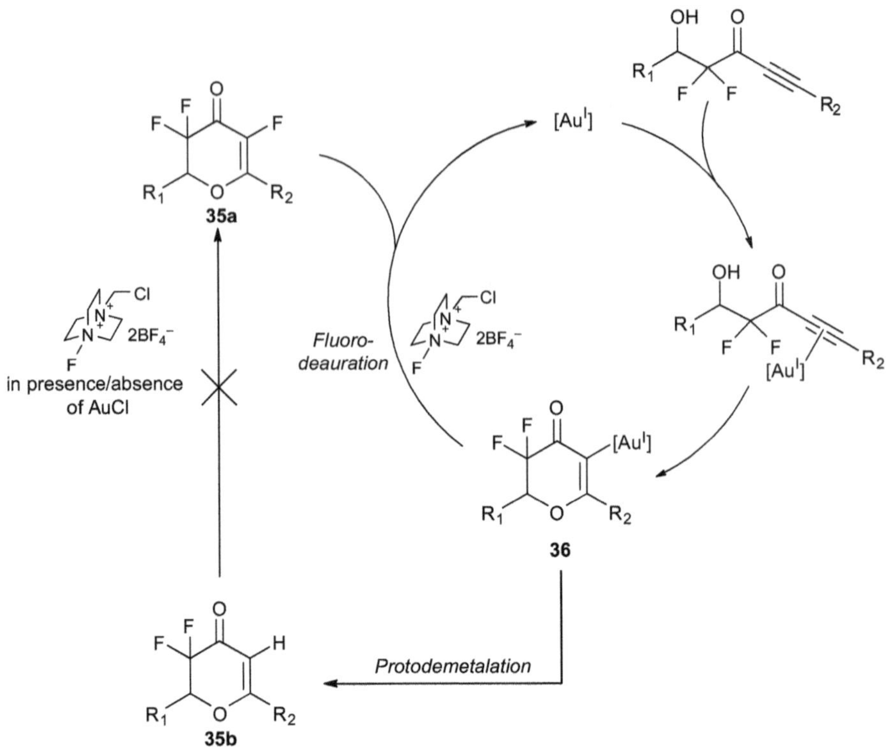

Scheme 7.24 Proposed mechanism for the formation of ß-hydroxy-α,α-difluoroynones.

mechanism. No incorporation of deuterium in the product was seen when *i*PrOD is used as the additive. However the use of *i*PrOH-d_8 showed incorporation of deuterium, even in the presence of H_2O. The authors proposed a fluoropalladation (**38**) of the alkyne with **37** and subsequent alkene insertion of the resulting vinyl–Pd **39** species to give **40**. The reduction of **40** affords the fluorinated cyclic product *via* hydride transfer with concomitant isopropyl alcohol oxidation, followed by reductive elimination. The origin of the *E* and *Z* isomers is accounted for evoking an unselective *cis* and *trans* fluoropalladation at **38**.

The synthesis of vinyl fluorides from vinylstannane with XeF_2 in the presence of stoichiometric Ag(OTf) in dichloromethane was first demonstrated by Tius and Kawakami in 1992.[52–54] While the fluorination took place slowly without Ag, the metal mediated reaction was typically completed within 10 minutes. The reaction was found to be regio- and stereoselective affording vinyl fluorides bearing a wide variety of functionalities in moderate to good yields. Yields of fluorination could be further improved by adding 2,6-di-*tert*-butyl-4-methylpyridine, a soluble proton scavenger. The loading of Ag could be reduced to 50 mol% with no decrease in yields (Scheme 7.26).

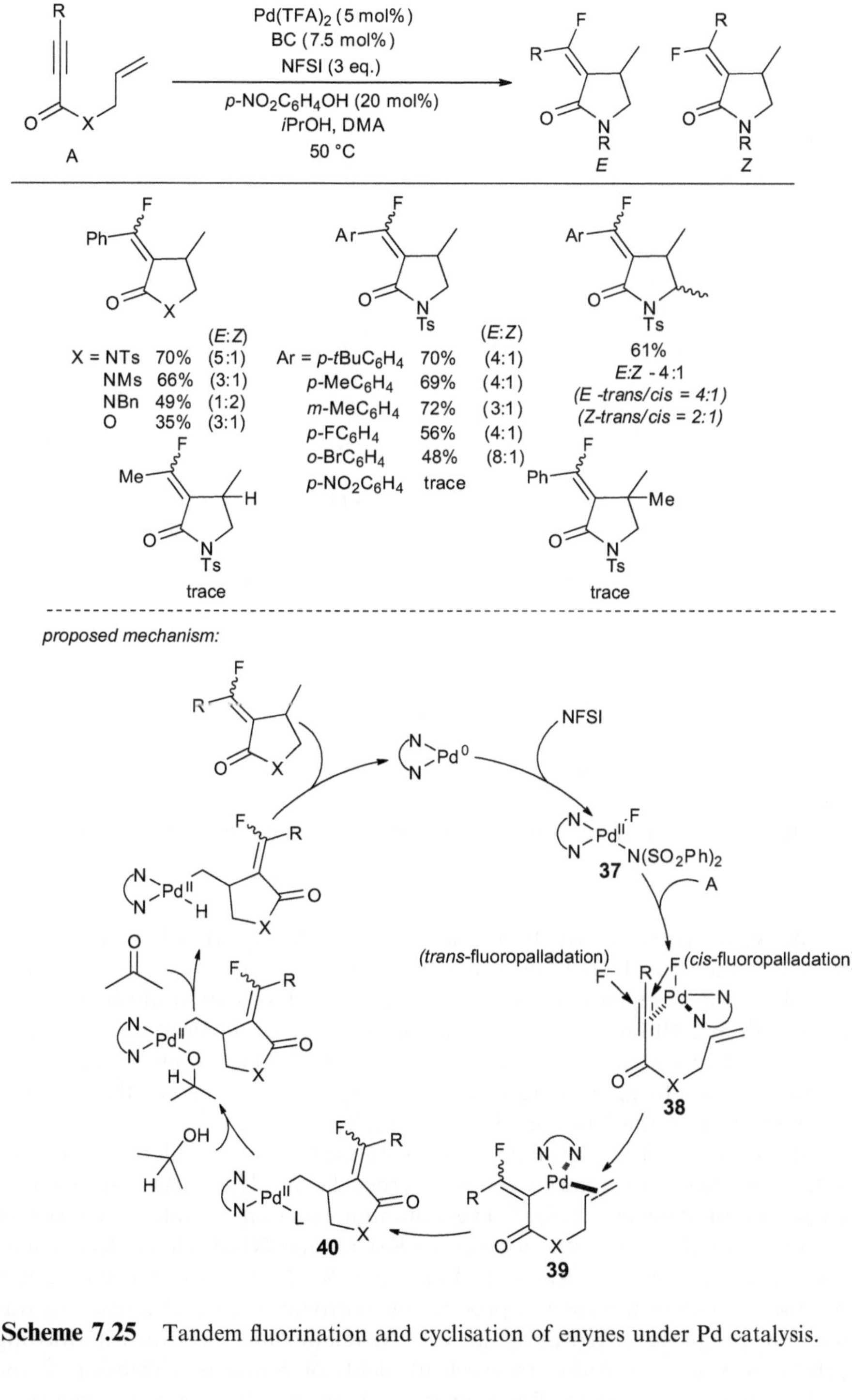

Scheme 7.25 Tandem fluorination and cyclisation of enynes under Pd catalysis.

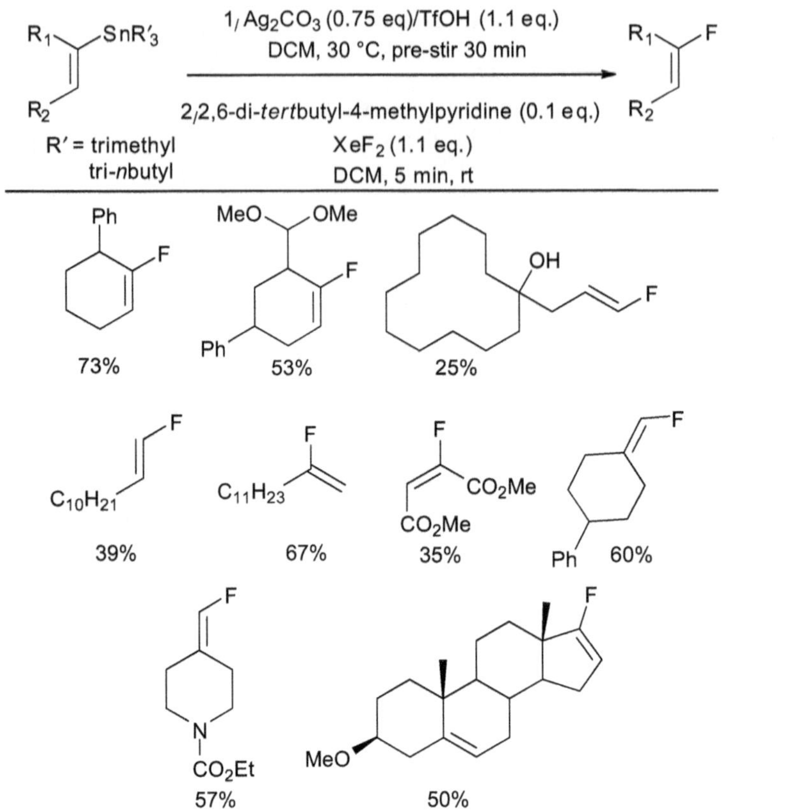

Scheme 7.26 Formation of vinylfluorides from vinylstannanes with XeF_2 mediated by Ag^I.

The Ag-mediated fluorination of arylboronic acids previously discussed has been extended to alkenyl substrates (Scheme 7.27). The reaction maintains complete *E*/*Z* stereochemical integrity unlike the fluorination of alkylboronic acids with Selectfluor, where complete racemisation of the fluorinated product is observed in the absence of a transition metal.[55] This would suggest that transmetalation from boron to silver is stereospecific, as would be the reductive elimination from the fluorinated silver complex.

Liu and co-workers developed a Ag-catalysed intramolecular aminofluorination of allenes to access various heterocyclic products featuring a fluoro alkene motif (Scheme 7.28).[56] The optimum reaction conditions consist of 20 mol% $AgNO_3$ used with an excess of K_2CO_3 and NFSI. Other electrophilic fluorinating reagents (Selectfluor, XeF_2 and **N-F2**) did not give fluorinated products, yielding instead the product of hydroamination. The reaction was tolerant to allene substitution at C3. Aromatic and electron-withdrawing substituents gave moderate to excellent yields of products. Extension of the carbon chain gave access to fluorinated tetrahydropyridines in moderate yields.

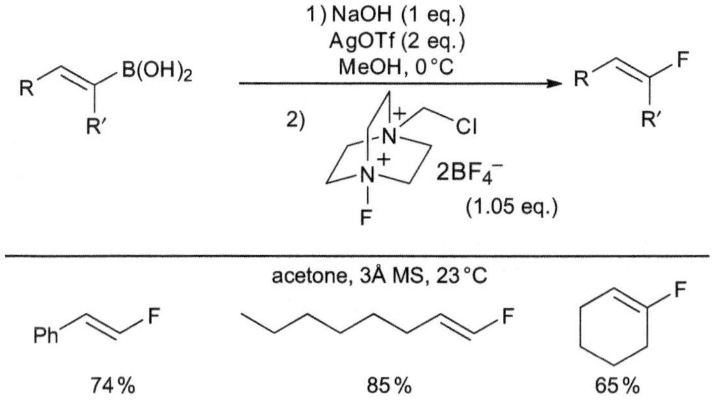

Scheme 7.27 Ag-mediated electrophilic fluorination of alkenylboronic acids.

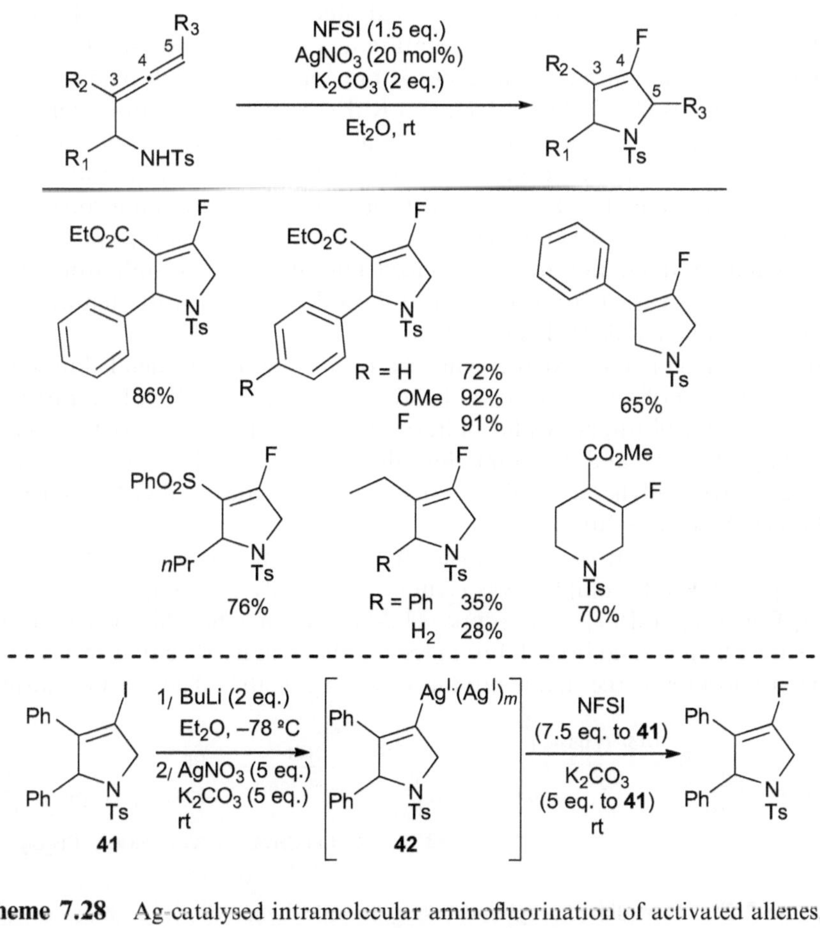

Scheme 7.28 Ag-catalysed intramolecular aminofluorination of activated allenes.

Alkyl groups at C3 were less tolerated as the products were formed in significantly reduced yields (~30%). The nature of the protecting group on the amine is critical as only 4-toluenesulfonyl led to successful fluorination. The products were further reacted with either KOtBu or DDQ/EtONa to access fluorinated pyrrole derivatives without loss of fluorine. The formation of by-products resulting from hydroamination is consistent with a catalytic cycle beginning with an initial Ag-catalysed amination of the allene followed by oxidation with NFSI and a final C–F reductive elimination event releasing the product. This putative mechanism is supported by the observation that the independently synthesized cyclic vinyl–Ag complex **42** reacts with NFSI to give the desired fluorinated product in 32%.

7.2.3 Alkyl and Benzyl Fluorides

Methods relying on transition metals for the synthesis of alkyl and benzyl fluorides have appeared since 1998. Grushin *et al.* investigated the use of $[PdF(Ph)(PPh_3)_2]$ for the release of 'naked' fluoride.[57] Halide metathesis with CH_2Cl_2 was possible at room temperature to yield a mixture of CH_2ClF and CH_2F_2. This reaction has been reported previously with high oxidation state transition metals (ReF_6, OsF_6, IrF_6 and RuF_5) but harsh reactions conditions and/or extended reaction times were required.[58] The same halide metathesis was reported by Bergman using $Cp^*(PMe_3)Ir(Ph)F$ **43**, an iridium(III) complex formed by adding TASF (tris(dimethylamino)sulfonium difluorotrimethylsilicate) to $Cp^*(PMe_3)Ir(Ph)(OTf)$.[59] In all cases, the reaction gave >90% of fluorinated product; this includes benzyl fluoride and chlorofluoromethane. Reaction times varied from minutes to days depending on the nature of the starting organic halide (Scheme 7.29).

In 2000, Togni reported the synthesis of an alternative metal fluoride for the fluorination of a range of alkyl halides.[60] The complex $\{[RuF(dppp)_2]PF_6\}$ **44** (dppp = 1,3-bis(diphenylphosphino)propane), synthesised from the corresponding chloride and thallium fluoride was successfully reacted stoichiometrically with allylic, as well as tertiary and secondary alkyl bromides and chlorides (Scheme 7.30).

A catalytic variant of this transformation relying on *in situ* formation of a competent Ru–F complex was examined; a catalyst loading of 1–10 mol% of $\{[RuCl(dppe)_2]PF_6\}$ **45** (dppe = 1,2-bis(diphenylphosphino)ethane) was reacted with alkyl (and benzyl) bromides and thallium fluoride to give alkyl (and benzyl) fluorides as the major products (Scheme 7.30).[61] The reaction displays

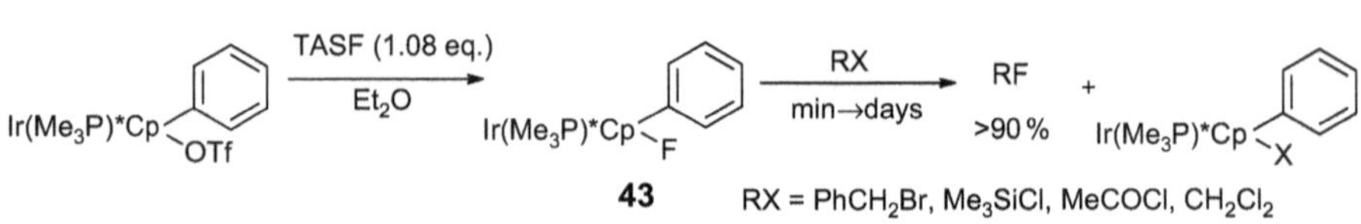

Scheme 7.29 Ir mediated alkyl and benzyl fluoride formation.

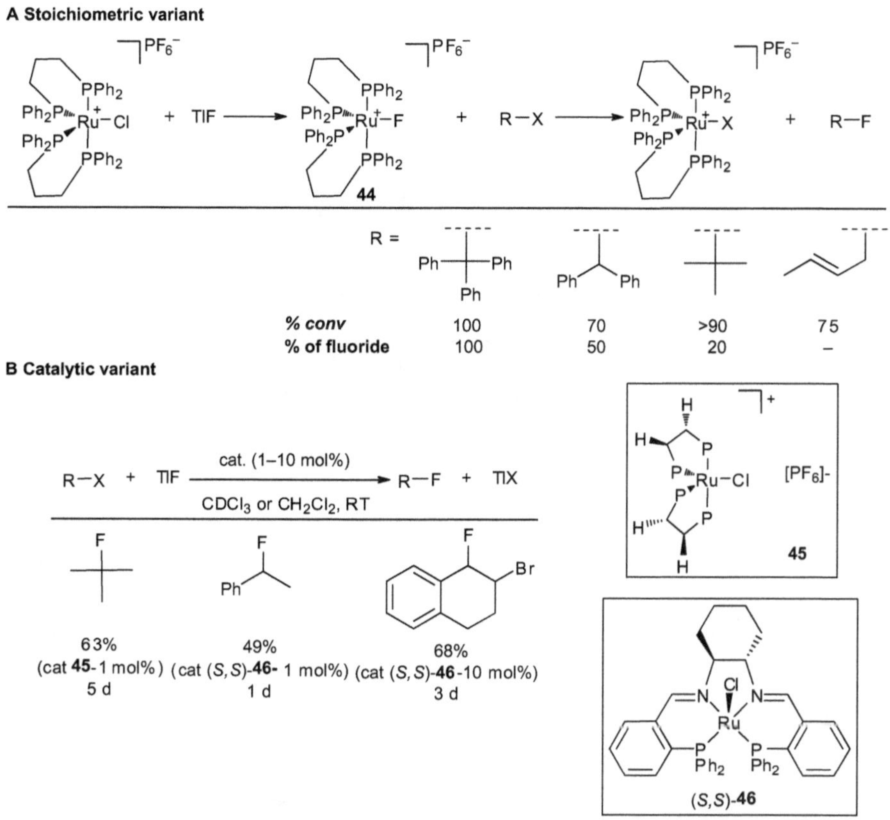

Scheme 7.30 Alkyl fluorination. A: Ru–F mediated fluorination of alkyl halides. B: Ru catalysed fluorination of alkyl halides using TlF.

narrow substrate scope with primary alkyl halides and α-halogenated ketones were found to be unreactive, an observation suggesting that the transition state involves charge separation for halogen transfer. The tolerance for functional groups is low since molecules with protic groups did not undergo fluorination (fluoride sequestration). The use of chiral ligand PNNP (1*S*,2*S*-*N*,*N′*-bis[2-diphenylphosphino)benzylidene]diamino-cyclohexane) [catalyst = (*S*,*S*)-**46**] led to successful fluorination but the enantiomeric excess for the products was low (ee < 20% for 1-phenyl-1-fluoroethane at 1% conversion).

While no definitive mechanism has been established, two plausible routes were tentatively proposed (Scheme 7.31). The first involved the σ-bond metathesis promoted by the polarisation of the C-halogen *via* coordination to Ru (**Path I**). Alternatively, TlF could coordinate the RuF complex and the resulting adduct **47** acts both as a fluoride source and halide scavenger (**Path II**). Notably, the Ru complex **47** was observed by ^{19}F-NMR spectroscopy [74.5 ppm (**PF**$_6$), −296 ppm (**RuF**$_2$)] and ^{205}Tl NMR [1055 ppm (F$_2$**Tl**)] and its structure unambiguously characterized by X-ray crystallography.

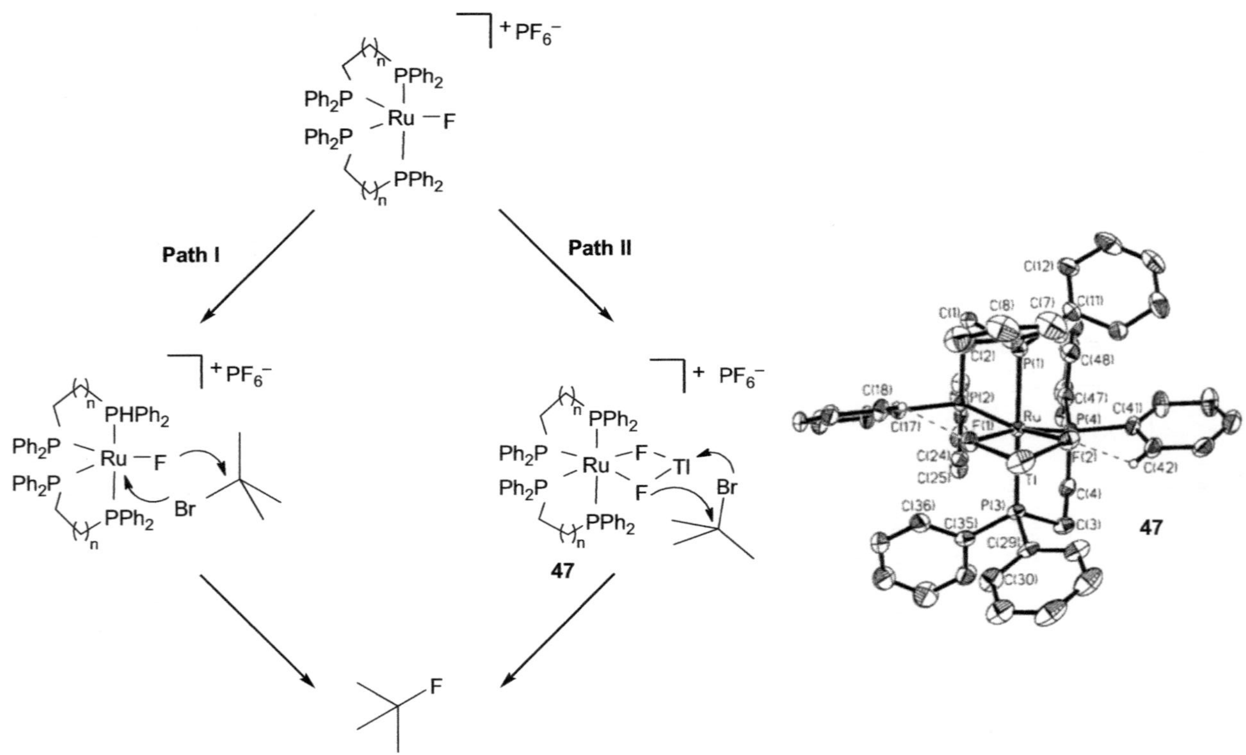

Scheme 7.31 Postulated mechanism of Ru-catalysed fluorination of alkyl bromides with TlF.

Alkyl fluorides can be prepared under transition metal catalysis from alkenes. Liu and co-workers reported the Pd-catalyzed intramolecular oxidative fluoroamination of unactivated alkenes bearing a pending *N*-tosyl amino group (Scheme 7.32).[62] These substrates were inert upon treatment with the oxidizing electrophilic fluorinating reagents (*e.g.* Selectfluor or NFSI) in the presence of 10 mol% of $Pd(OAc)_2$. In contrast, the use of AgF (5 equivalents), the hypervalent iodine reagent $PhI(OPiv)_2$ (2 equivalents) and the catalyst $Pd(OAc)_2$ (10 mol%) gave a wide range of piperidine derivatives in moderate to good yields with excellent regiocontrol under mild conditions (acetonitrile, room temperature). Some structural restrictions do however apply in terms of substrate scope; for example, the use of a 1,2-disubstituted *Z*-alkene gave the product of elimination and no trace of fluorinated product.

Mechanistic studies with deuterium-labeled substrates led the authors to propose a catalytic cycle with the following elementary steps: (i) *trans*-aminopalladation of the alkene with $Pd(OAc)_2$, (ii) oxidation of the resulting intermediate by $PhI(OPiv)_2$/AgF to a Pd^{IV} complex and (iii) reductive elimination to generate the C–F bond. This final step is competing with an S_N2 attack by fluoride that gives the minor *cis*-fluorinated product (Scheme 7.33).

Liu developed an intermolecular variant of this reaction using NFSI as the fluorinating reagent instead of AgF/$PhI(OPiv)_2$; in this reaction, NFSI also acts as the amine source (Scheme 7.34).[63] The aminofluorination of styrene was possible with $Pd(OAc)_2$ (5 mol%) and the ligand bathocuproine (BC) (5 mol%) with NFSI in dioxane at 50 °C in a 65% yield. Product formation was found

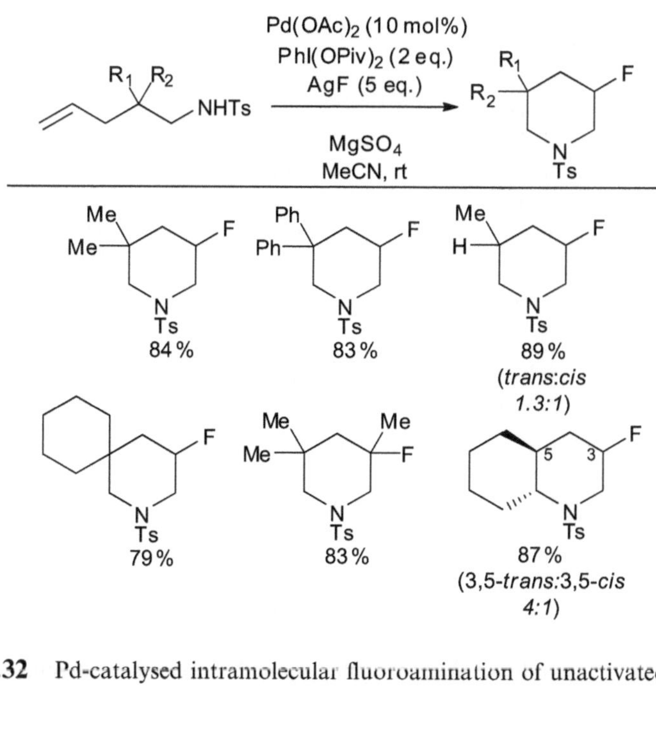

Scheme 7.32 Pd-catalysed intramolecular fluoroamination of unactivated alkenes.

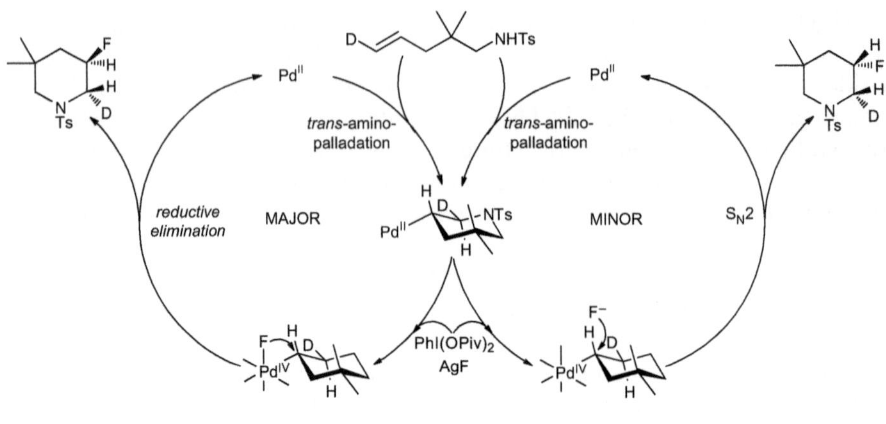

Scheme 7.33 Suggested mechanism of the intramolecular aminofluorination of alkenes.

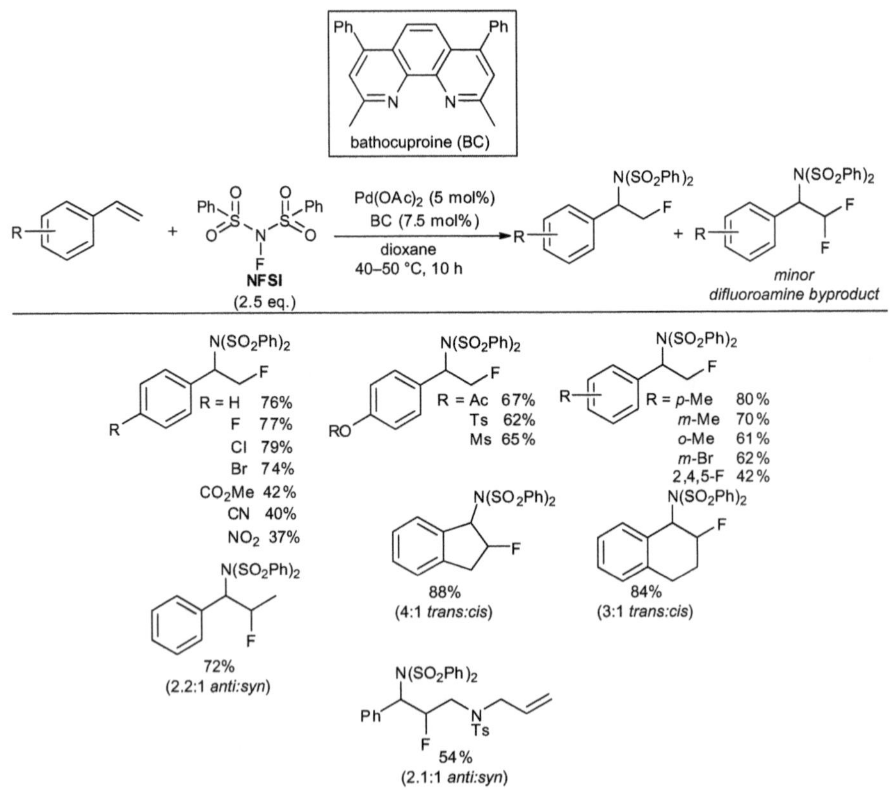

Scheme 7.34 Pd-catalyzed intermolecular aminofluorination of styrenes.

to be highly dependent on the presence of the bidentate-nitrogen ligand, bathocuproine (BC); other bidentate nitrogen ligands gave a reduced yield. An increased loading of BC (7.5 mol%) gave the desired product in 80% yield. A minor difluoroamine by-product was seen with the reaction of styrene. For styrene derivatives, substitutions at the *p*, *m* and *o* positions on the aryl group, halide and protected alcohols were tolerated. The yields were reduced in the case of electron-withdrawing groups. The reaction displays good regio- and chemoselectivity, but only moderate levels of diastereoselectivity were achievable.

A tentative mechanism was proposed based on the successful independent formation of the $Pd^{II}F$ complex **48** (from Pd^0, BC and NFSI) characterized by ^{19}F-NMR [broad singlet −381 ppm (PdF)], ^1H-NMR (3.61 ppm and 3.06 ppm) and ESI-MS spectrometry. Initial oxidation with NFSI of a Pd^0 species bearing the BC ligand would generate complex **48** *in situ*, which is proposed to undergo fluoropalladation with styrene to give **50**. This can then undergo nucleophilic attack by the amine to release the aminofluorinated product, **Path 1**. Alternatively the product of fluoropalladation **50** could be oxidised with NFSI to give the Pd^{IV} complex **49** with subsequent C-N reductive elimination, **Path 2** (Scheme 7.35).

An intramolecular variant of this reaction using substrates with a pending tosyl-protected amine was subsequently reported using the same $Pd(OAc)_2$/BC

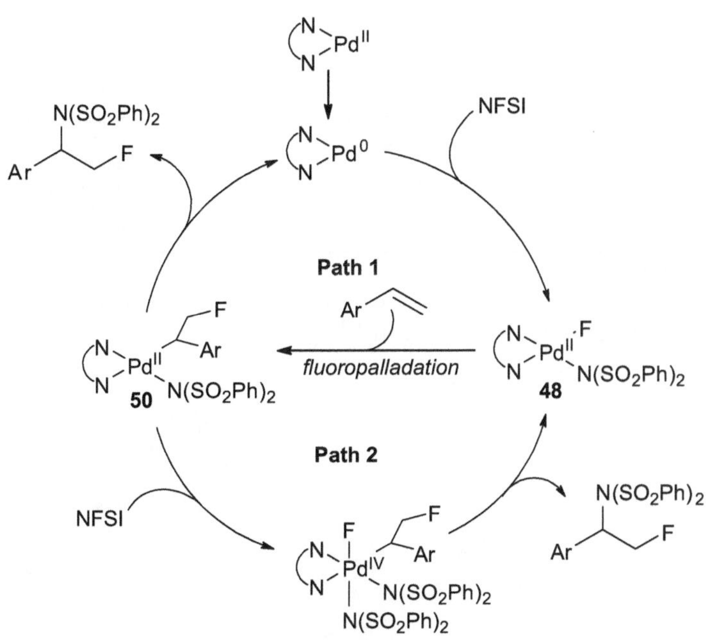

Scheme 7.35 Proposed mechanistic pathways to give aminofluorinated products.

catalytic system; the reaction requires an excess of NFSI (3 equivalents).[64] The *Z*-alkene (*3Z*)-*N*-(tosyl)-4-phenylbut-3-en-1-amine gave superior yield than the corresponding *E* isomer (59% and 43%, respectively), so further studies used *Z* precursors. Other bidentate ligands and fluorinating reagents were not suitable. The reaction is limited in scope as it tolerates only aryl substituents on the *Z*-alkene, with alkyl groups giving no fluorinated cyclic product. The diastereoselectivity of these intramolecular reactions is low, possibly due to poor control over selectivity in the proposed *cis* and *trans*-fluoropalladation event (Scheme 7.36). A chiral pincer ligand gave product 3-fluoro-2-phenyl-1-tosylpyrrolidine with enantiomeric excess of 21% and 44% for the *cis* and *trans*-product, respectively (Scheme 7.37).

Gagné disclosed cases of stereoretentive fluorination of Pt^{II}-intermediates with a range of electrophilic fluorinating reagents.[65] The isolated alkyl-Pt complex **52** ligated with bis(2-diphenylphosphinoethyl)phenylphosphine (PPP) was reacted with various commercially available N–F reagents and XeF_2 (Table 7.1). All reagents, with the exception of trimethylfluoropyridinum tetrafluoroborate, gave the fluorinated product in good to excellent yield; the reaction times and temperatures were however variable, with XeF_2 shown to be the most reactive. In its presence, the desired fluoride was obtained in 87% yield after less than 3 minutes reaction time at 0 °C. The product of β-hydride elimination was consistently seen as a by-product.

The sterics of the alkyl group of the Pt^{II} complexes have a significant impact on the efficiency of fluorination. Bulky alkyl groups gave the best results with

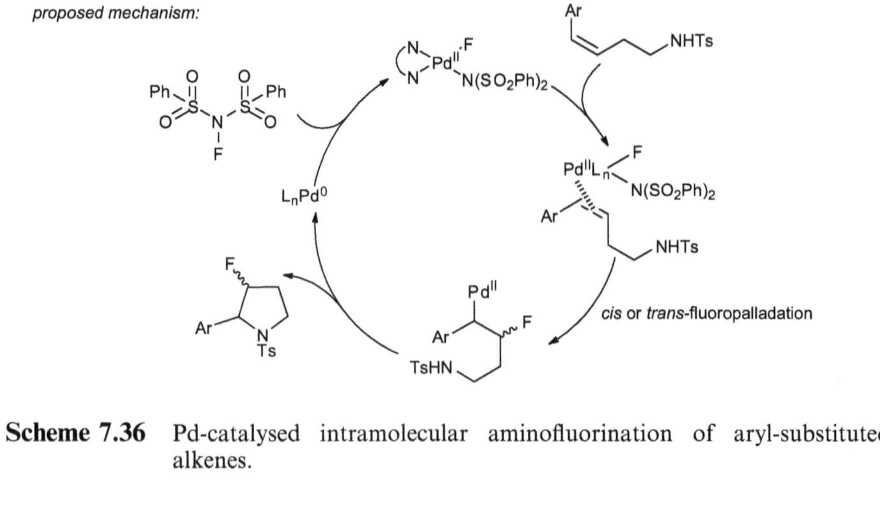

Scheme 7.36 Pd-catalysed intramolecular aminofluorination of aryl-substituted alkenes.

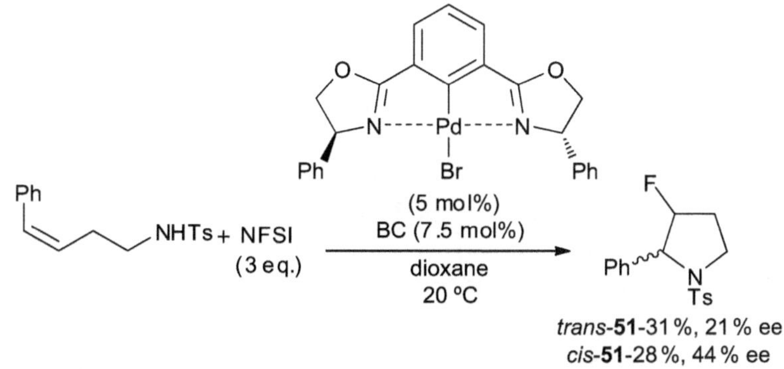

Scheme 7.37 Aminofluorination with chiral Pd pincer ligand.

Table 7.1 Fluorination of isolated Pt complex **52** with electrophilic fluorine.

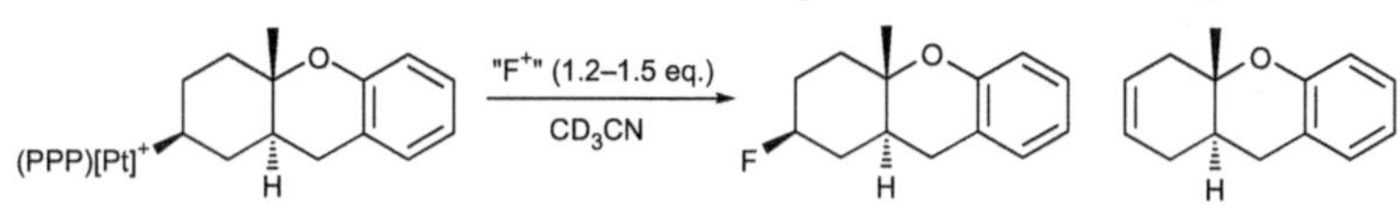

52

F^+	*Temp (°C)*	*Time (h)*	*Yield of fluorinated product (%)*
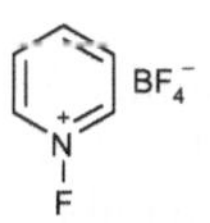	70	28	93
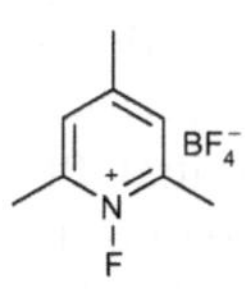	70	72	0
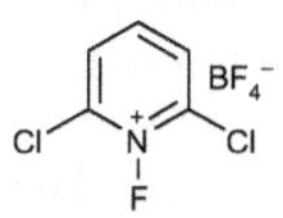	25	35	91
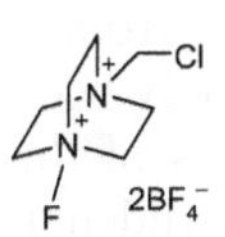	25	6	88
XeF_2	0	< 3 min	87 (85 isolated)
NFSI	70	17	81

yields diminished as the sterics of the group were reduced, with products of β-hydride elimination and C–N bond formation (from the counterion of the N–F reagent, *i.e.* $N(SO_2Ph)_2$) increasingly formed. Benzyl fluoride was formed in >95% yield using this methodology (Scheme 7.38).

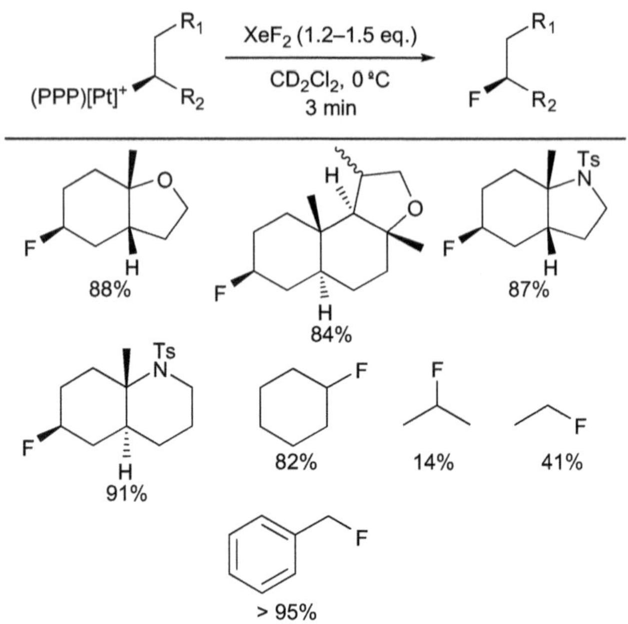

Scheme 7.38 Alkyl and benzyl fluoride formation from isolated Pt complexes.

Careful NMR studies for the fluorination of the benzyl(PPP)Pt^{II} complex **53** revealed the formation of an intermediate with spectroscopic data characteristic of a $Pt^{IV}F$ species, **54**. Such species were prepared and unambiguously characterized by Vigalok in 2010.[66] These findings along with the retentive stereochemical pattern suggest that the reaction proceeds *via* a putative dicationic $Pt^{IV}F$ **56** formed upon attack of F^+ on the initial Pt^{II} complex **55** (Scheme 7.39). The reaction of the structurally analogous Pd complex **57** with NFSI gave predominantly the alkyl fluoride (retention) in 76% yield along with alkene resulting from competitive β-hydride elimination (Scheme 7.40).

Alkyl fluorides are also accessible under a Au^{I}/Au^{III} mediated process. The Toste group reported in 2012 the first C_{sp3}–F reductive elimination from a *cis*-F_2Au^{III}(alkyl)(IPr) complex **59**.[67] Until then, $Au^{III}F$ species had only been speculated as intermediates for C–F bond formations.[49,50] Complexes of the type (IPr)Au(R) **58** (IPr = 1,3-bis(2,6-diisopropylphenyl)imidazol-2-ylidene), where R is an alkyl group, were synthesised and reacted with XeF_2 to form the corresponding *cis*-difluoride Au^{III} species. As steric crowding has been shown to facilitate reductive elimination, complexes bearing bulky alkyl groups were prepared. In the case of the more sterically hindered alkyls, fluorinated products were identified as the major product (Table 7.2). When bulkier cyclic alkyls were used, β-hydride elimination was observed but was completely suppressed in the case of (–)-menthyl. No reductive elimination was seen with *cis*-$F_2Au(CH_3)$(IPr). β-Hydride elimination outcompetes C–F reductive elimination for linear

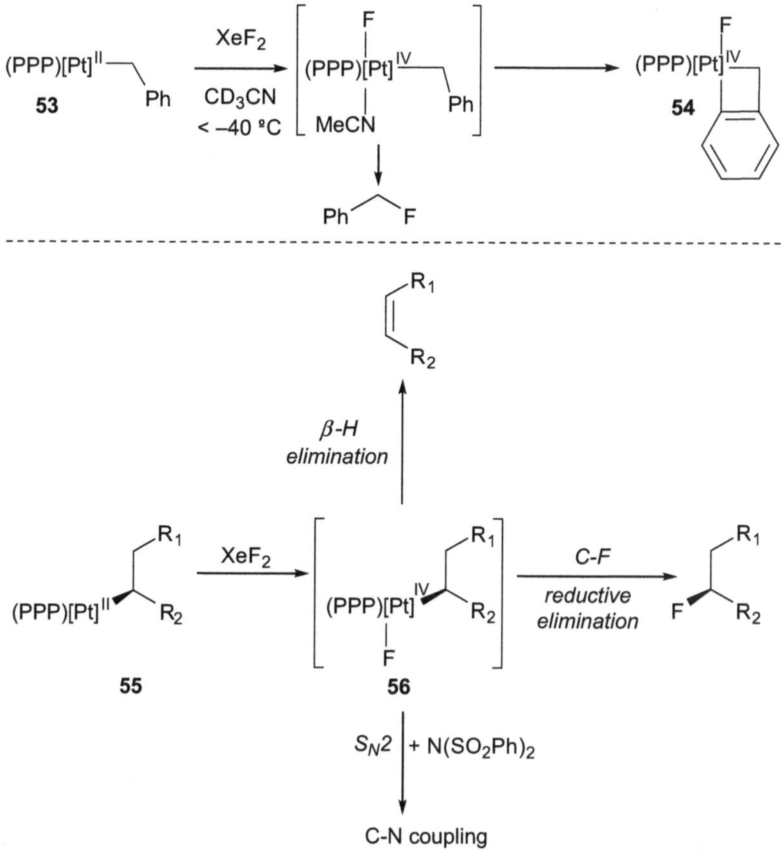

Scheme 7.39 Mechanistic studies into the formation of a Pt–F complex.

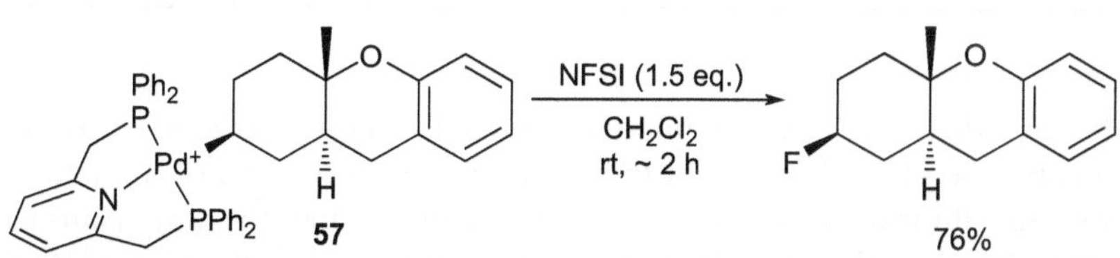

Scheme 7.40 Pd-mediated alkyl fluorination with NFSI.

elongated alkyl substituents, affording alkenes as major products. An interesting observation was the production of rearrangement products and of an allyl fluoride from a cyclopropyl-substituted precursor. Benzyl fluoride was formed in 85% yield using this methodology (Table 7.2 final entry).

Mechanistic studies carried out on complex **60** suggest initial dissociation of fluoride to give the cationic gold species **61**. In the final step, complex **61** can

Table 7.2 Reductive Elimination of C–F from Au^{III} fluorides.

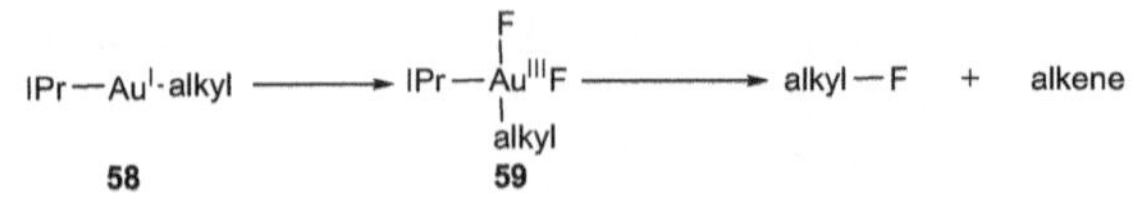

alkyl	*Products*

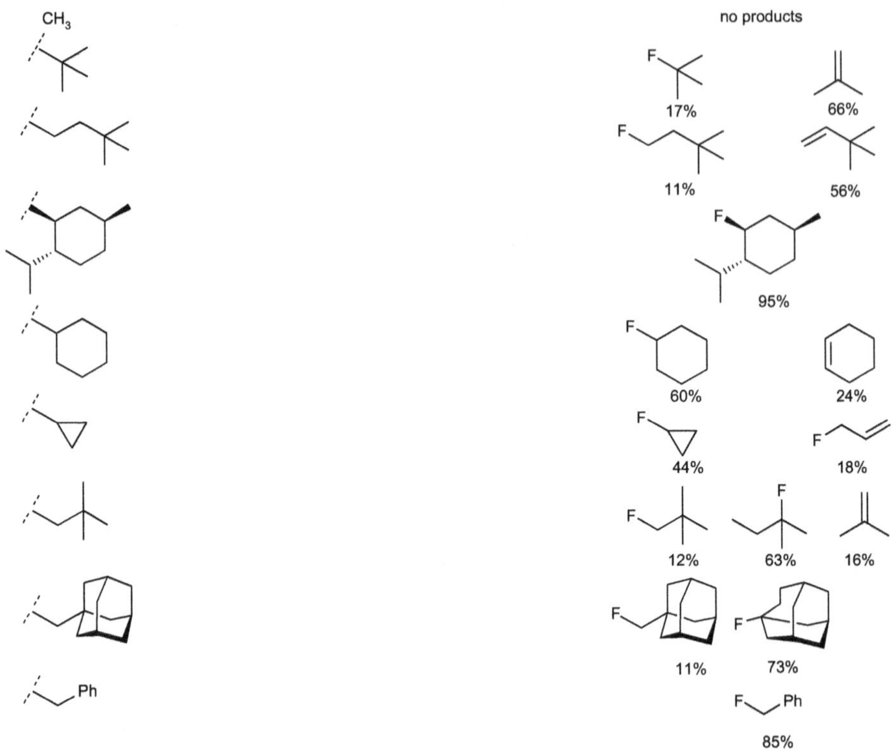

undergo either reductive elimination to yield alkyl fluorides or β-hydride elimination to give alkenes with the released $[(IPr)Au]^+$ species trapped by the dissociated fluoride (Scheme 7.41). Data showed that the disappearance of complex **60** obeyed first-order kinetics and an inverse first-order relationship with fluoride concentration. The authors propose polarisation of the Au-alkyl bond to account for the formation of rearrangement products. A highly ordered transition state or the solvent reorganisation necessary to solvate the fluoride is proposed to explain the large negative $\Delta S^{\ddagger}$, uncommon for a dissociative process.

The isolation of an alkyl$Pd^{IV}F$ species, namely $\{[CH_2CMe_2\text{-}o\text{-}C_6H_4]\text{-}Pd^{IV}F[OH][bpy]\}$ (bpy = bipyridine) **64** was disclosed by Sanford in 2012.[68] The complex was formed by the reaction of **62** with *N*-fluoro-2,4,6-trimethylpyridinium triflate in CH_2Cl_2. The exchange of the triflate ligand

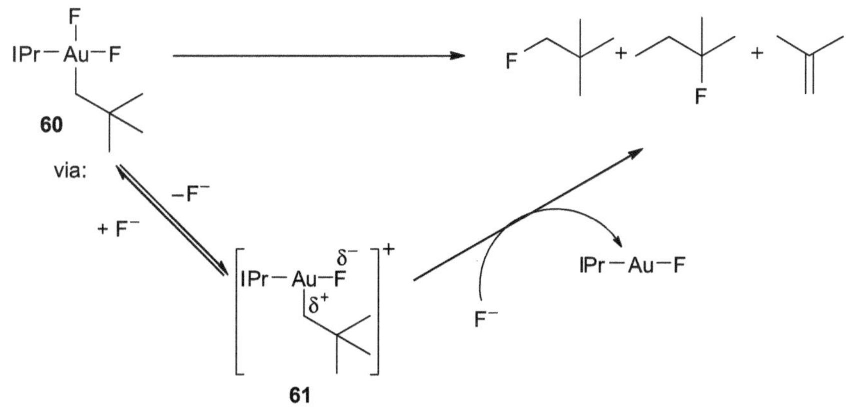

Scheme 7.41 Proposed mechanism for the Au^{I}/Au^{III} mediated fluorination.

with H_2O was possible and allowed the characterisation of the resulting cationic species **64** by X-ray crystallography. The exchange of triflate for pyridine gave complex **65**. Further reaction with MeN_4F gave the difluoride complex **68**, however only NMR analysis of this species was possible due to its instability. C_{sp3}–F formation was possible at 80 °C in CH_2Cl_2 within 30 minutes to give **66** in 93% yield (Scheme 7.42).

The addition of pyridine to the reaction slowed the formation of **66** drastically and further studies found the reaction to have inverse first-order dependence on the concentration of pyridine. Though the exact mechanism of this transformation is not yet fully understood, based on this data, the dissociation of the pyridine ligand from the complex is proposed to occur prior to the rate-limiting step. The C–F bond formation could then occur by a reductive elimination or by dissociation of the fluoride ligand from the Pd^{IV} complex **67** and direct S_N2-type attack of fluoride on the alkyl of the dicationic Pd complex. The authors suggest that the reductive elimination is more likely due to the unfavourable formation of a dication Pd^{IV} species and the poor nucleophilicity of fluoride.

Sanford also reported the formation of benzylic fluorides using the $Pd(OAc)_2$ catalysed oxidative fluorination methodology (*vide supra*, aryl fluorides). Three examples of benzylic fluorination using *N*-fluoro-2,4,6-trimethylpyridinium tetrafluoroborate (oxidant and fluorine source) were reported giving the products with moderate yields (Scheme 7.43).

Vigalok and co-workers reported a benzylic fluorination involving platinum complexes (Scheme 7.44). Complexes **69** have previously been shown to give the products of C–C reductive elimination when exposed to XeF_2 or *N*-fluoro-2,4,6-trimethylpyridinium tetrafluoroborate and after subsequent cyclometalation and HF elimination led to product **73**.[69,70] However, Vigalok reported an interesting development in 2010 demonstrating the formation of a benzylic C–F bond.[66] When the aryl group was changed to a mesityl group and exposed to *N*-fluoro-2,4,6-trimethylpyridinium tetrafluoroborate, C–C bond formation

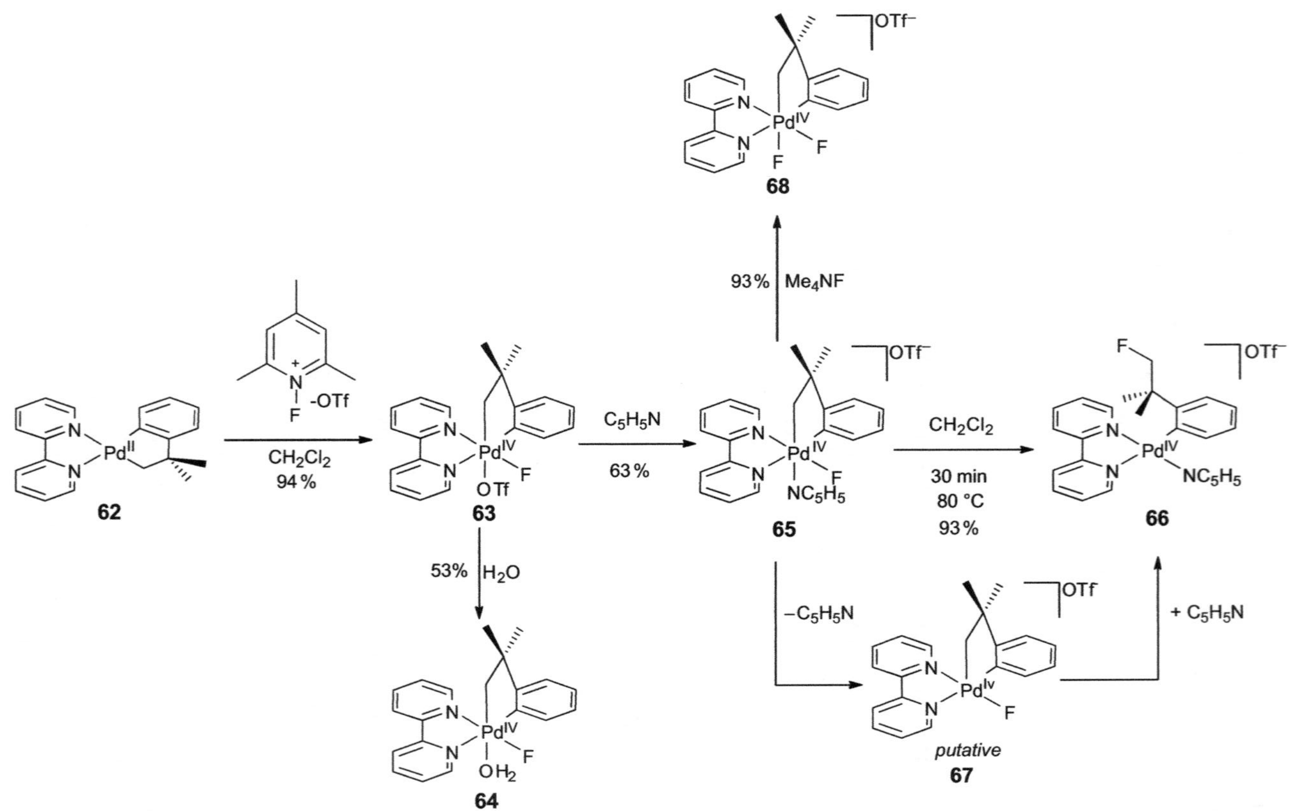

Scheme 7.42 Alkyl C–F bond formation from a Pd^{IV} centre.

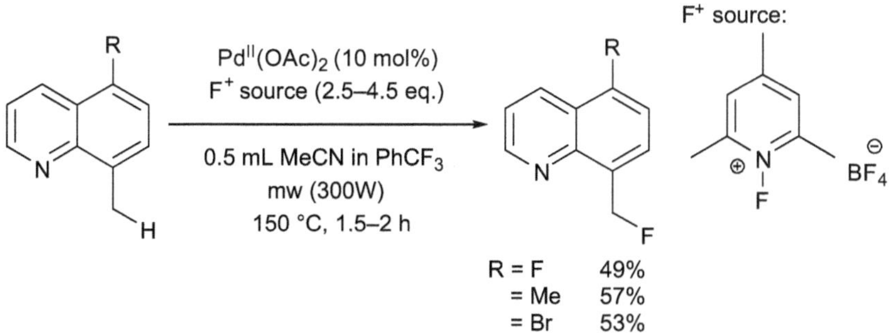

Scheme 7.43 Electrophilic fluorination at the benzylic position catalysed by $Pd(OAc)_2$.

was observed; however, in the presence of XeF_2, the formation of **71** with a benzylic C–F at the *ortho* CH_3 of the mesityl group was observed. The structure of this new fluorinated product was confirmed by NMR analysis; subsequent ligand exchange of the pyridine with PPh_3 allowed for single crystal X-ray analysis of **72**. The authors propose a Pt^{IV} difluoride as the reactive intermediate based on the isolation of a Pt^{IV} difluoride **70** from the reaction of the *p*-fluorophenyl complex with XeF_2. This intermediate could then allow for the fluoride assisted metalation of one of the *ortho* methyl groups of the mesityl, assisted by the large steric crowding at Pt. Reductive elimination of the C–F bond would lead to the final benzylic fluoride product.

The Sanford group developed a transition metal mediated route to benzylic fluorides using nucleophilic fluoride.[71] This reaction, which proceeds *via* a Pd^{II}/Pd^{IV} mechanism, requires an oxidant in addition to the fluoride source. Liu *et al.* had previously demonstrated the use of $PhI(OPiv)_2$ as a competent oxidant in conjunction with AgF.[62] This combination of reagents was reacted with 8-methylquinoline **74** in dichloromethane for 24 hours affording the benzylic fluoride **75** in 58% yield. A byproduct identified as the C–H oxygenation product **76** was isolated in 25% yield. The authors propose that a $Pd^{IV}F$ species **78** formed *via* two plausible routes is the key intermediate for this transformation. The first involves the oxidation of the Pd^{II} complex **77** by $ArI(OPiv)_2$ followed by a ligand substitution of pivalate for fluoride. Alternatively, $ArI(OPiv)_2$ would react with AgF to form $ArIF_2$ which can then oxidatively fluorinate the Pd^{II} complex **77** (Scheme 7.45).

7.2.4 Allylic Fluorides

The fluorination of allylic alcohols with the nucleophilic fluorinating reagent DAST (diethylaminosulfur trifluoride) is the most commonly used uncatalyzed reaction to access allylic fluorides.[72] The fluorination of activated alcohols and halides by tetraalkylammonium fluorides and metal fluorides has also been reported. However, with all these methodologies, limitations are still apparent in substrate scope and selectivity.[73] While many selectivity issues have been

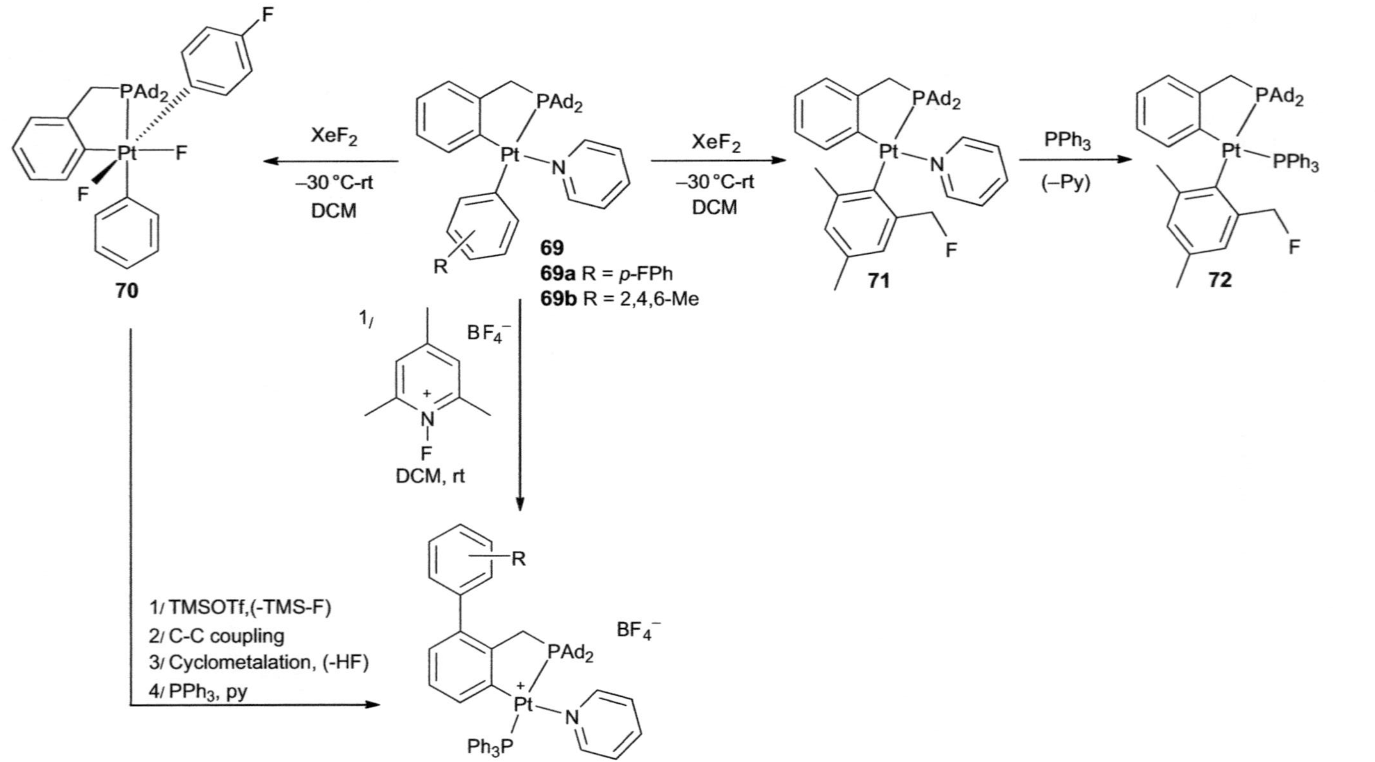

Scheme 7.44 Formation of benzylic C–F bond on Pt complexes with XeF_2.

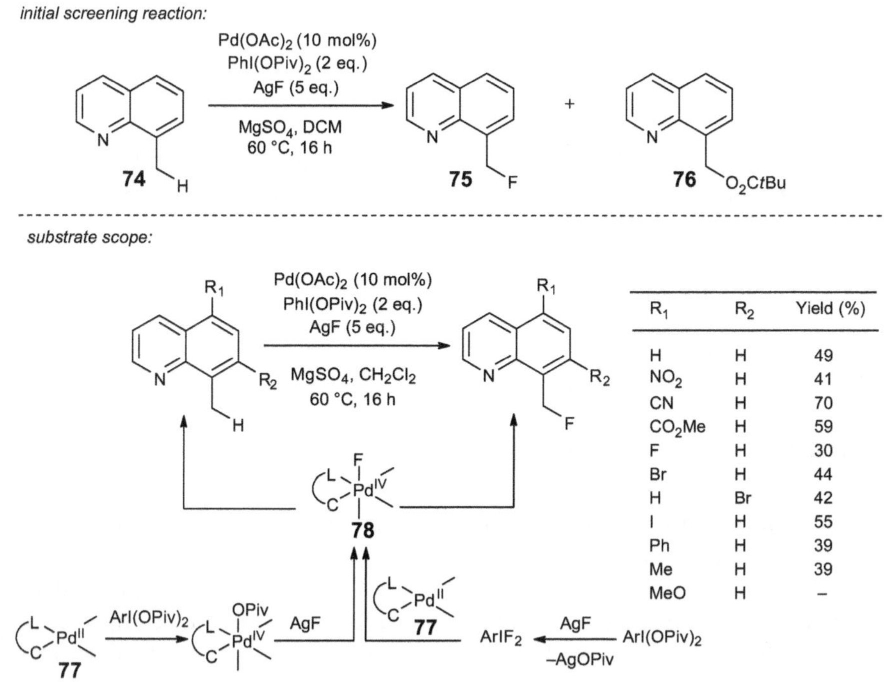

R_1	R_2	Yield (%)
H	H	49
NO_2	H	41
CN	H	70
CO_2Me	H	59
F	H	30
Br	H	44
H	Br	42
I	H	55
Ph	H	39
Me	H	39
MeO	H	–

Scheme 7.45 Pd-catalysed benzylic fluorination with AgF and ArI(OPiv)$_2$.

addressed by the electrophilic fluorination of allylsilanes using NFSI or Selectfluor,[74] the synthesis of allylic fluorides utilizing transition metals has received a lot of attention in recent years. Togni reported the first transition metal mediated allylic fluorination in 1999.[75] The allylic fluoride 1,1′-(*E*)-3-fluoroprop-1-ene-1,3-diyl)dibenzene was accessible in quantitative yield within 1 minute by reacting 1,1′-(*E*)-3-bromoprop-1-ene-1,3-diyl)dibenzene with the ruthenium fluoride complex, [RuF(dppp)$_2$]PF$_6$ (dppp = 1,3-bis(diphenylphosphino)propane).[76] The demonstration that nucleophilic fluoride could be used to access allylic fluorides under mild conditions encouraged further research. The Tsuji–Trost reaction has been extensively utilized in the synthesis of allylic carbon–carbon bonds and more recently carbon–heteroatom bond formation.[77] Furthermore, the value of the Tsuji–Trost reaction has vastly increased with the development of asymmetric variants.[78] In the case of hard nucleophiles, such as fluoride, complications with this reaction can arise from the affinity of the nucleophile to the metal centre hindering catalysis or from the oxidative addition of the metal into the allylic fluoride products themselves. Indeed, the oxidative addition of various metals into the carbon–fluorine bond is well documented.[79] For many years, fluoride has been more commonly used as an additive in Tsuji–Trost allylic alkylation to enhance selectivity, not as a competent nucleophile.[80,81] Togni reported the first investigation into the possibility of a Pd-catalyzed allylic nucleophilic

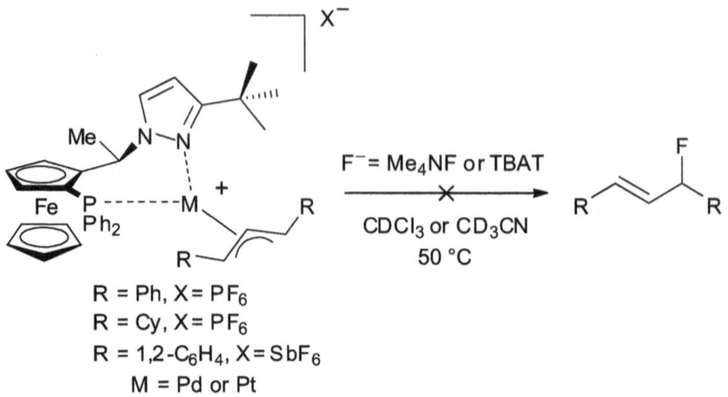

Scheme 7.46 Investigations into the feasibility of Pd-catalysed allylic fluorination.

fluorination in 2006.[82] The exposure of isolated cationic Pd^{II} and Pt^{II} allyl complexes to a variety of fluoride sources at room temperature and at 50 °C gave no allylic fluoride (Scheme 7.46). In the case of substrates able to undergo elimination reactions, diene formation prevailed. With the Pt^{II} complexes, an anion exchange of PF_6^- with F^- was observed. The exposure of an independently synthesised allylic fluoride (RuF fluorination of the corresponding bromide) to $Pd(dba)_2$ found the oxidative addition of Pd^0 into the C–F to be facile.

Building on these results, the Gouverneur and Brown groups examined the activation of allylic C–F bonds under Pd^0 catalysis to establish the leaving group propensity of this functionality in traditional Pd-catalysed allylic alkylations with malonate.[83] Using a cyclic bifunctional model system for investigation, methyl carbonate was identified as a superior leaving group to fluoride but fluoride was substituted in preference to acetate. These findings suggest that allylic fluorination is more likely to succeed if one uses allylic carbonate instead of allylic acetate. Indeed, the treatment of a representative allylic carbonate with a nucleophilic fluoride source under Pd catalysis led to the formation of the desired allylic fluoride.[84] Yields could be greatly improved by the use of a *p*-nitrobenzoate as a leaving group in place of methyl carbonate, with $Pd(dba)_2$ (5 mol%) and PPh_3 (15 mol%) found to be the optimal catalytic system. $TBAF(tBuOH)_4$, a reagent of low hygroscopicity and good nucleophilicity, was identified as the only competent fluoride source.[85,86] Various 2-substituted propenyl fluorides bearing substitution on the aromatic ring could be prepared in high yields (Scheme 7.47). The addition of electron withdrawing groups at the *para* position resulted in lower reactivity. Electron-rich substrates were more reactive but the yields remained moderate because of the propensity of these products to undergo decomposition upon work-up. Various linear fluorides were accessible from the corresponding linear carbonates allowing for the isolation of cinnamyl fluoride in quantitative yield, a result demonstrating the mildness and selectivity pattern of this reaction.[87] As found by Togni, the

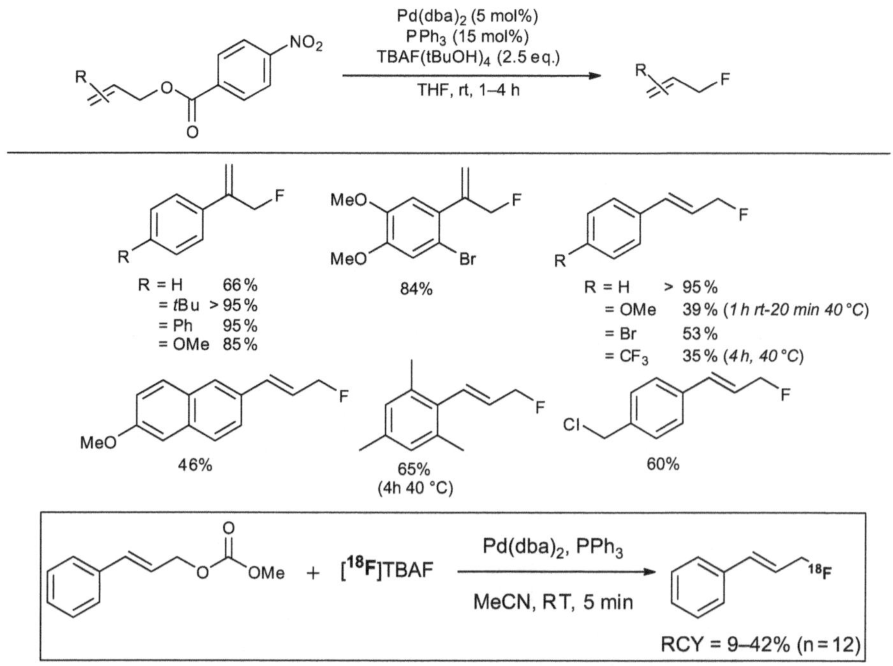

Scheme 7.47 Pd-catalyzed fluorination of allylic *p*-nitrobenzoates and allylic carbonate.

use of substrates capable of elimination reactions predominantly gave the diene and no fluorination of ethyl-1,3-diphenylallyl *p*-nitrobenzoate was seen. This Pd-catalysed allylic fluorination was further extended to radiochemistry using [^{18}F]TBAF and cinnamyl methyl carbonate. The [^{18}F]-labelled cinnamyl fluoride was obtained within 5 minutes at room temperature with radiochemical yields of 9–42% ($n = 12$). This reaction is the first transition metal mediated synthesis of an ^{18}F-radiolabelled compound.

In an alternative approach, Doyle *et al* reported the asymmetric fluorination of cyclic allylic chlorides under Pd^0 catalysis using AgF and the commercially available Trost ligand,[88] (1*R*,2*R*)-(+)-1,2-diaminocyclohexane-*N*,*N′*-bis(2-diphenylphosphinobenzoyl) **L3**.[89] The products were formed in moderate to high yields and variable enantiomeric excesses (Scheme 7.48). More often, no side product resulting from elimination was observed. The reaction is tolerant of a range of functionalities, even those vulnerable to fluoride attack but the size of the ring was shown to be important with reduced yields and *ee*'s for five and seven membered cyclic precursors. Notably, conventional leaving groups in Pd^0-catalysed allylic alkylations such as methyl carbonate and acetate remained inert under these conditions, allowing therefore for orthogonal manipulations. The lack of reactivity of other leaving groups suggests that the formation of AgCl contributes to the driving force of the reaction. The reaction proceeds with overall retention of configuration normally seen with soft

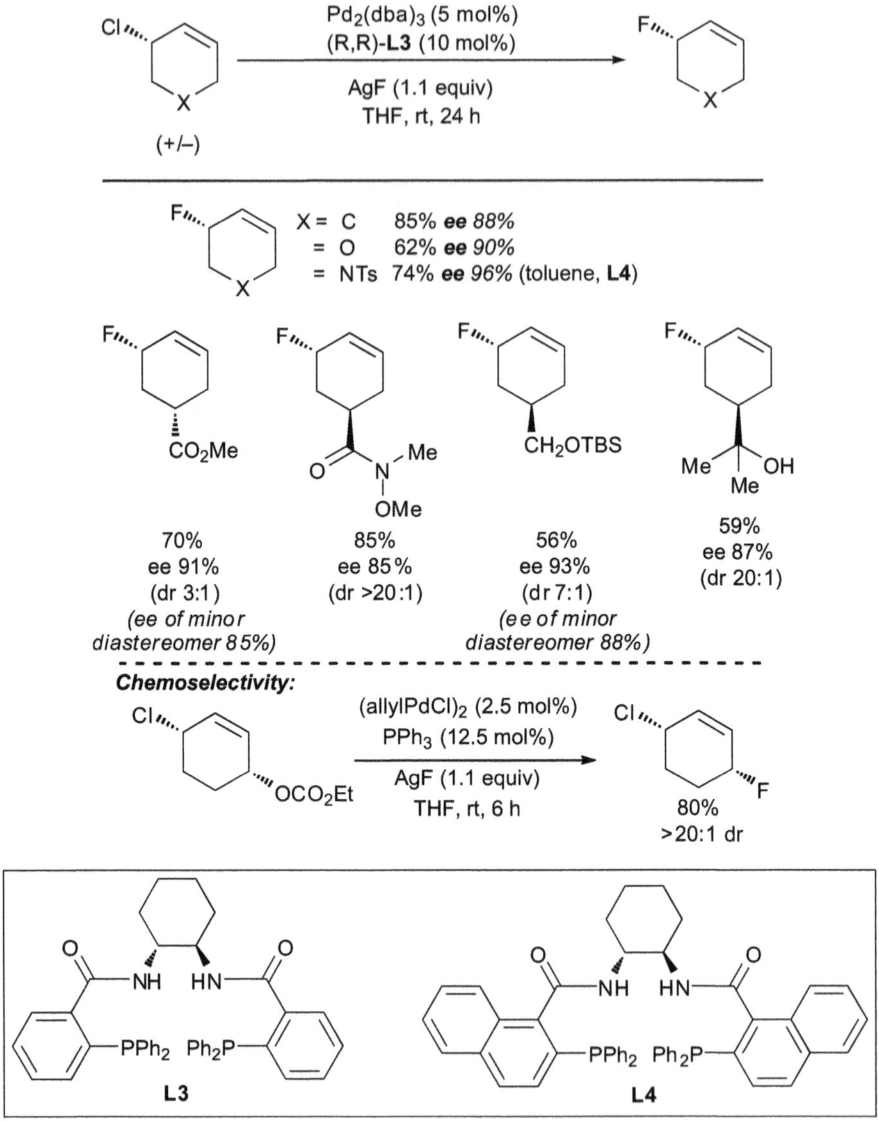

Scheme 7.48 Enantioselective allylic fluorination under Pd catalysis.

nucleophiles. The authors propose an initial inversion in the oxidative addition of Pd^0 into the allylic chloride followed by a second inversion upon fluoride addition on the π-allyl intermediate, chloro[(1*R*,2*R*)-(+)-1,2-diaminocyclohexane-*N*,*N*′-bis(2-diphenylphosphinobenzoyl][R-cyclohexenyl]Pd, *via* an S_N2-type outer-sphere attack.

Extension of this work to acyclic linear chlorides was subsequently reported using the naphtyl substituted Trost ligand (1*R*,2*R*)-(+)-1,2-diamino-cyclohexane-*N*,*N*′-bis(2-diphenylphosphino-1-naphthoyl), **L4**. Branched allylic

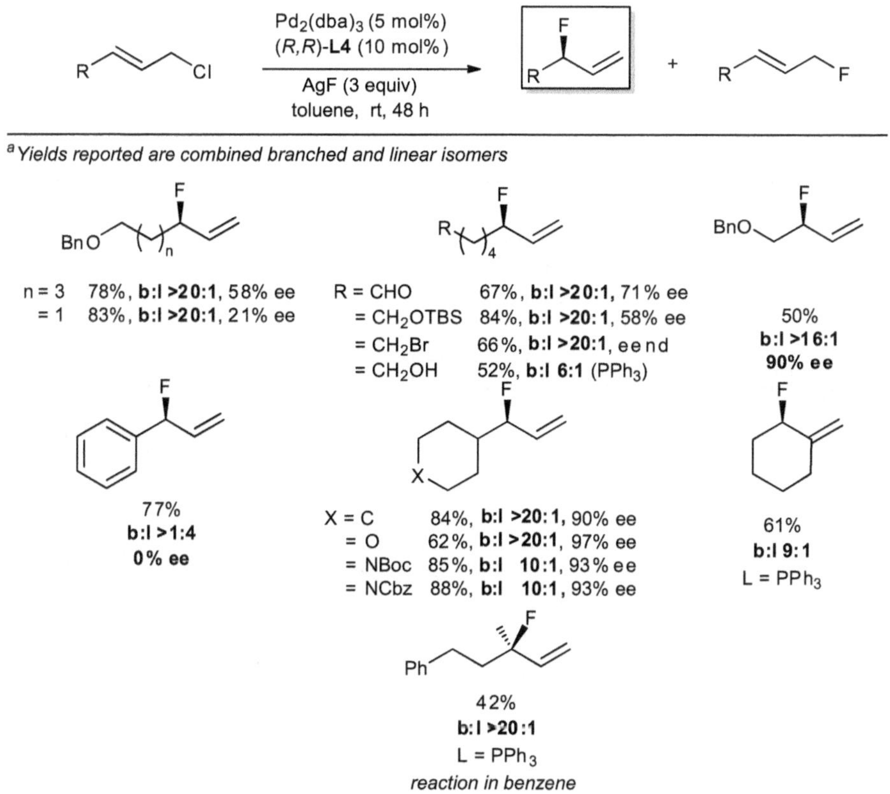

Scheme 7.49 Pd-catalysed fluorination of linear allylic chlorides.

fluorides were formed predominantly from linear precursors with good to high branched/linear selectivity and poor to high enantiomeric excesses (Scheme 7.49).[90] It was necessary to use toluene (non-polar solvent) to minimize uncatalyzed S_N2-type reactions that could compromise the enantiomeric excess. The reaction was tolerant of various functionalities other than a free alcohol. Reduced yields were seen due to by-product formation resulting from competitive intramolecular attack of the alcohol on the Pd-allyl intermediate formed. Access to tertiary fluorides was shown to be limited to low yields due to competitive elimination.

Nguyen developed an alternative route to branched allylic fluorides in 2011. Under iridium catalysis, branched allylic trichloroacetimidates were transformed with satisfactory yields into the corresponding secondary and tertiary fluorides with the use of triethylamine trihydrofluoride (TREAT·3HF).[91] Tertiary fluorides were accessible by this methodology but no isolation was possible due to polymerization or decomposition of the products (Scheme 7.50). Enantiopure starting material led to nearly complete racemization upon fluorination, so this method is not suitable for the preparation of enantioenriched branched allylic fluorides. Along with good functional group

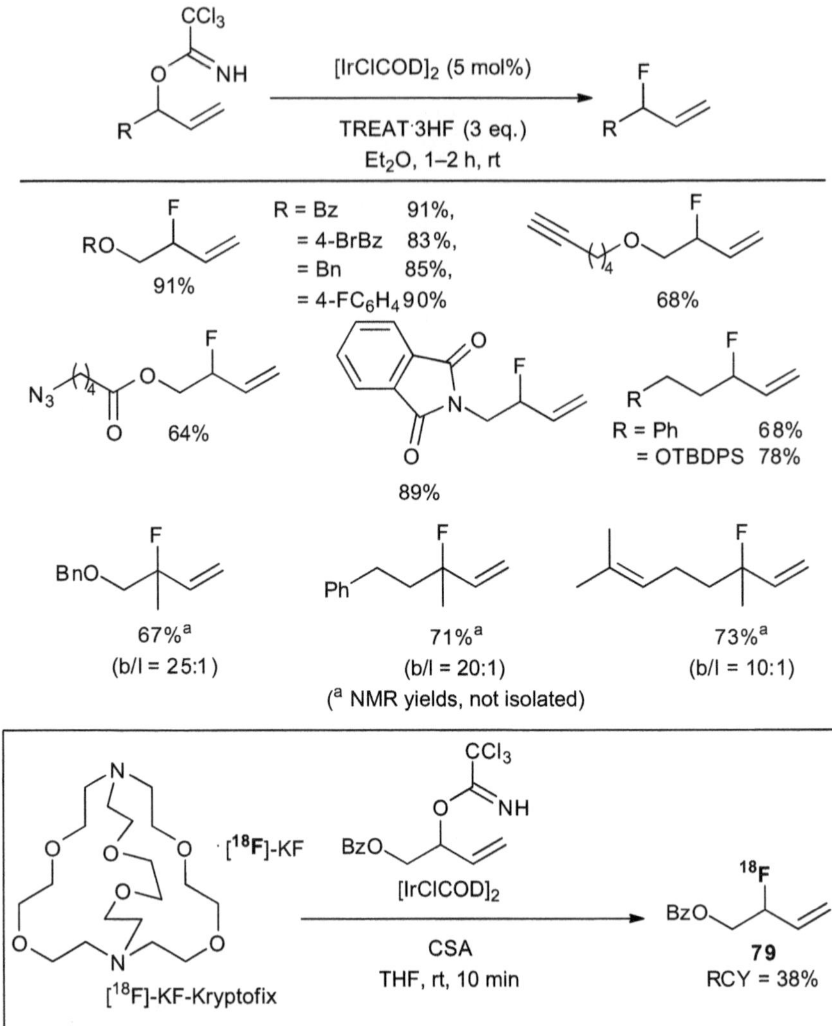

Scheme 7.50 Ir-catalysed allylic fluorination of allyl trichloroacetimidates.

tolerance, the reaction was shown to be highly regioselective and very rapid with reactions complete within 1–2 hours. This short reaction time allowed for the development of the radiochemical variant of this reaction. A representative branched [^{18}F]allylic fluoride **79** was obtained in 38% RCY upon fluorination with [^{18}F]KF·Kryptofix.

7.2.5 Fluorohydrins

The Haufe group initiated studies on the enantioselective ring opening of epoxides with HF-reagents in the presence of a stoichiometric amount of the

Jacobsen's enantiopure (salen)chromium chloride complex **80**. In 2000, Haufe presented the first enantioselective nucleophilic fluorination of *meso*- and racemic epoxide precursors (enantioselective desymmetrization and kinetic resolution, respectively).[92] The reaction was conducted in DMF with KHF_2 in the presence of 18-crown-6 (1,4,7,10,13,16-hexaoxacyclooctadecane), a set of conditions minimizing the formation of unwanted chlorohydrins as side products. For the racemic unsymmetrical epoxide cyclohexene oxide **81**, the reaction is regioselective leading to (*R*,*R*)-(–)-2-fluorocyclohexanol **82** with good to excellent enantioselectivity (Scheme 7.51). Racemic epoxide styrene oxide gave **84** as the major product of fluorination, where fluoride attack occurs at the less substituted carbon and the other possible regioisomer **85** is only seen as a minor product. The *ee* was severely diminished upon reduction of the amount of complex **80** to 10 mol%. Subsequent studies revealed that silver fluoride could be used as an alternative fluoride source leading to improved

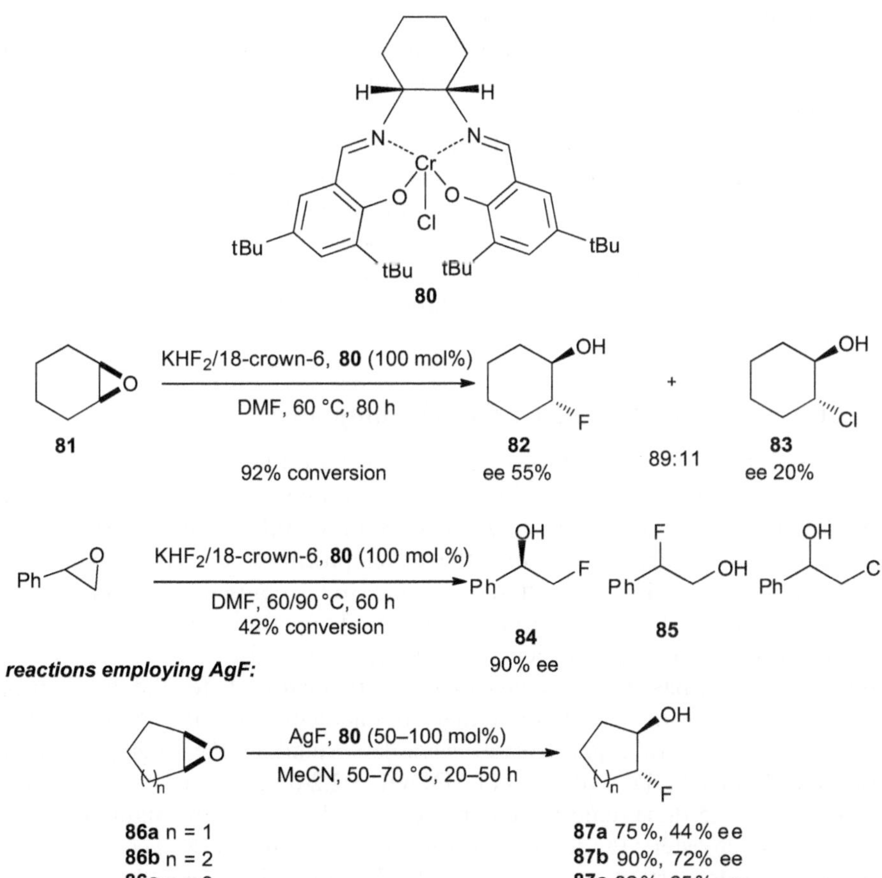

Scheme 7.51 Fluorination of epoxides using KHF_2 and the Jacobsen's (*S*,*S*)-(+)-(salen)CrCl complex.

enantioselectivity (**87a–c**). The synthesis of a fluorinated analogue of the prostaglandin biosynthesis inhibitor lasiodiplodin was synthesized using this methodology.[93,94]

The Doyle group revisited the opening of *meso* epoxides with an HF.amine complex formed *in situ* upon addition of 1,1,1,3,3,3-hexafluoroisopropanol (HFIP) to benzoyl fluoride in the presence of an amine.[95] Using both the chiral (salen)Co complex **88**, structurally related to the Jacobsen's chromium catalyst and the chiral amine (–)-tetramisole as catalysts, the fluorination did occur as programmed but required extended reaction times (up to 120 h). When an achiral or chiral amine is used as a catalyst in the absence of the (salen)Co complex, no fluorination took place. Control reactions revealed that the use of a mismatched Lewis acid/amine system led to lower enantiomeric excesses. This observation supports the idea of a cooperative cocatalyst effect merging the two chiral environments. This dual catalyst system was applied to a substrate library with cyclic epoxides containing alkene, ester and protected amine functionalities and gave good conversions to a large range of fluorohydrins with excellent enantioselectivity. The kinetic resolution of racemic terminal epoxides proceeded with excellent regiocontrol and *ee* up to 99% ($k_{rel} > 300$) (Scheme 7.52).

Based on kinetic studies, Doyle and co-workers proposed a (salen)Co(III) fluoride as the active nucleophilic fluorine, a complex that could form a resting state dimer. Axial ligation of the amine to the (salen)Co is proposed to facilitate dimer dissociation and account for the observed cooperativity effect. Additionally, the fluorinations carried out with the dimeric catalyst **89** were vastly improved with reaction times dramatically reduced and substrate scope increased (Scheme 7.53).[96]

7.2.6 α-Fluorination of Carbonyl Compounds

Stereoselective $C_{sp}{}^3$–F bond formation is very desirable for the construction of biologically and pharmaceutically relevant compounds. Research has focused on the use of chiral fluorinating reagents or chiral starting materials but more recently redirected towards the development of catalytic asymmetric fluorinations.[97–101] For such studies, ß-ketoesters represent a useful class of substrates due to their ability to react at room temperature with N–F reagents *via* the formation of a transient enol. Pioneering work in the Togni group revealed that Ti-based Lewis acids are competent catalysts for the fluorination of ß-ketoesters with Selectfluor. Other commercially available N–F reagents (NFSI, 1-fluoropyridinium triflate) were less suitable. Pleasingly, the non-racemic chiral TADDOL-derived (TADDOL = 2,2-dimethyl-α,α,α′,α′-tetraaryl-1,3-dioxolan-4,5-dimethanol) Ti complexes provided a route to enantioenriched α-fluoro-ß-ketoesters; the steric parameters of the ligand on the Ti catalyst were found to be important with 1-naphthyl-substituted TADDOL ligand giving superior enantiomeric excesses (catalyst **90B**). The nature of the ester also influenced enantioselectivity with optimal results achieved with mesityl esters (R_2 = mesityl). The reaction is proposed to proceed through a reactive chiral

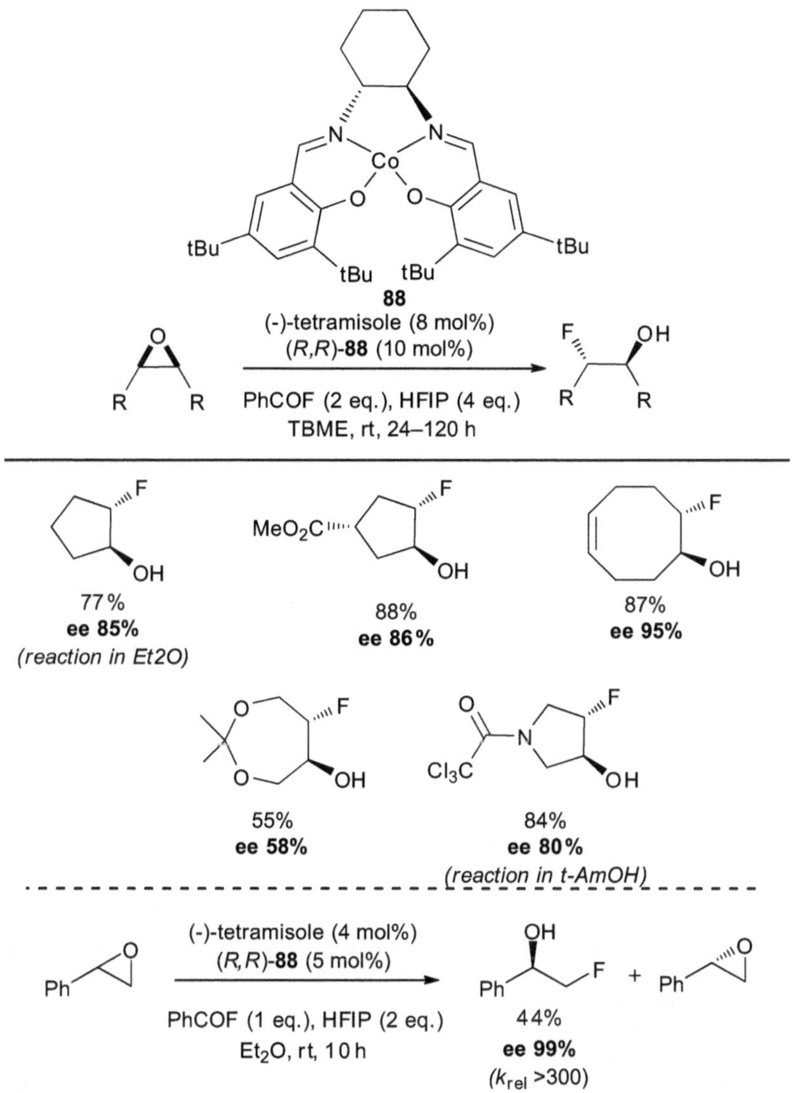

Scheme 7.52 Asymmetric fluorination of *meso* epoxides *via* cooperative co-catalysis.

enolate intermediate formed between the ß-ketoester and cat **90B** that reacts with Selectfluor to give the desired fluoride (Scheme 7.54).[102]

Isolation of the enolate intermediates gave Ti(TADDOLato) complexes containing two carbonylenolato units predominantly present in two diastereoisomeric forms.[103] In the major isomer the naphthyl substituents of the (*S*,*S*)-TADDOL ligand are found to be responsible for shielding the *Si* face of the coordinated carbonylenolato ligand so that the electrophile (F^+) attacks preferentially at the opposite *Re* face (Scheme 7.55).

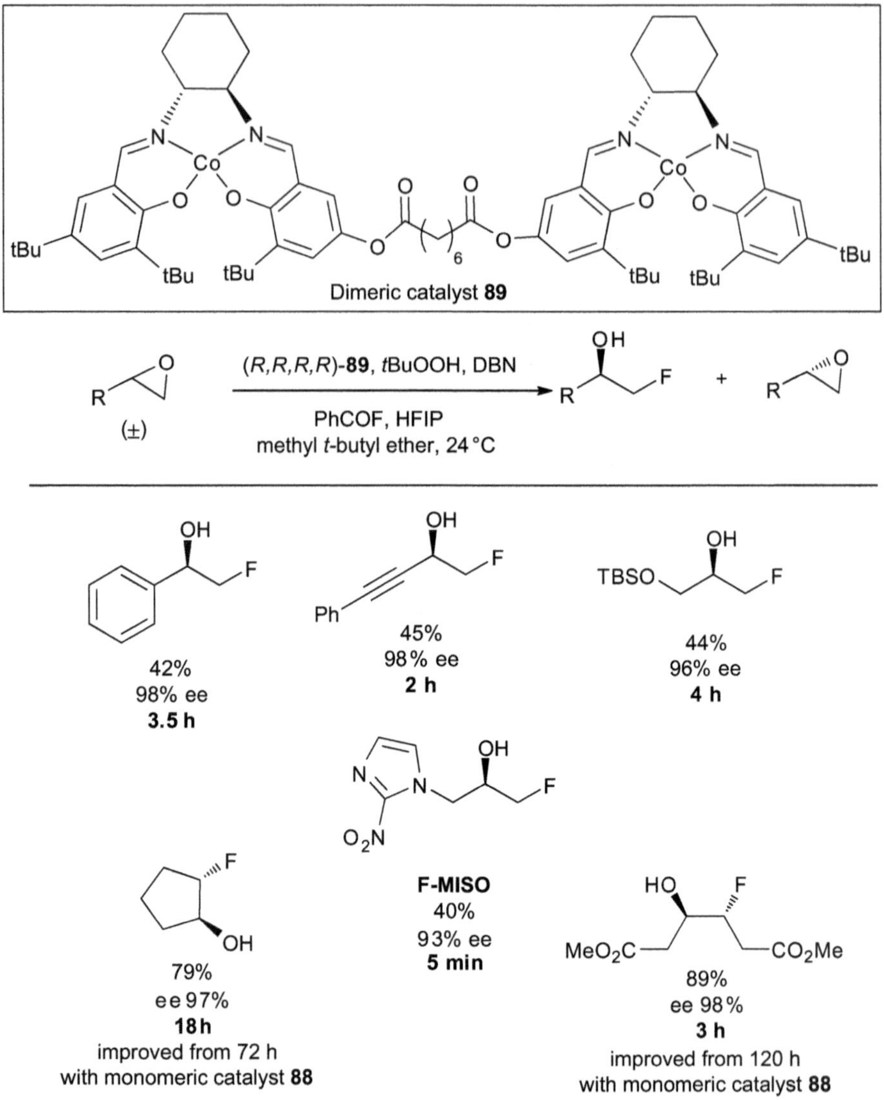

Scheme 7.53 Improved fluorination of meso and racemic epoxides with a dimeric catalyst.

The reaction was extended to ß-ketoamides, ß-ketothioester, ß-ketophosphonates and α-nitroesters. The geminal heterohalogenation of ß-ketoesters was performed using *N*-chlorosuccinimide (NCS) and Selectfluor with the sense of stereoselectivity defined by the order of addition of reagents. A chlorination–fluorination sequence led to the formation of the *S* enantiomer; reciprocally, the reverse fluorination–chlorination gave the *R* enantiomer.[104] Sequential fluorination–amination is also possible leading to α-fluoro-α-amino acids precursors with ee up to 94% (Scheme 7.56).[105]

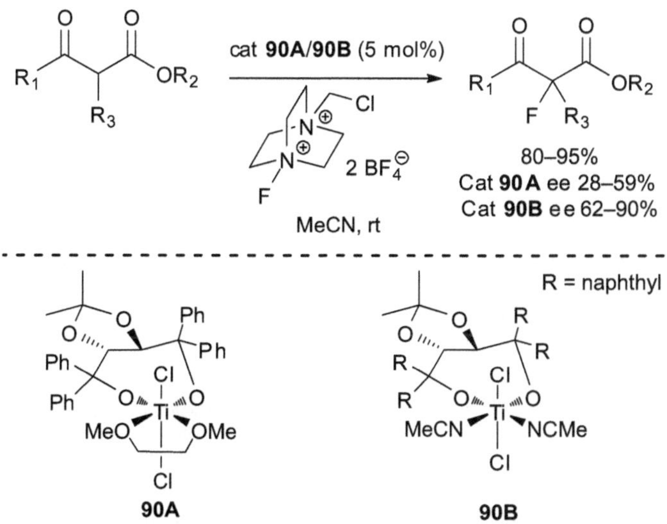

Scheme 7.54 Ti-catalyzed enantioselective fluorination of ß-ketoesters.

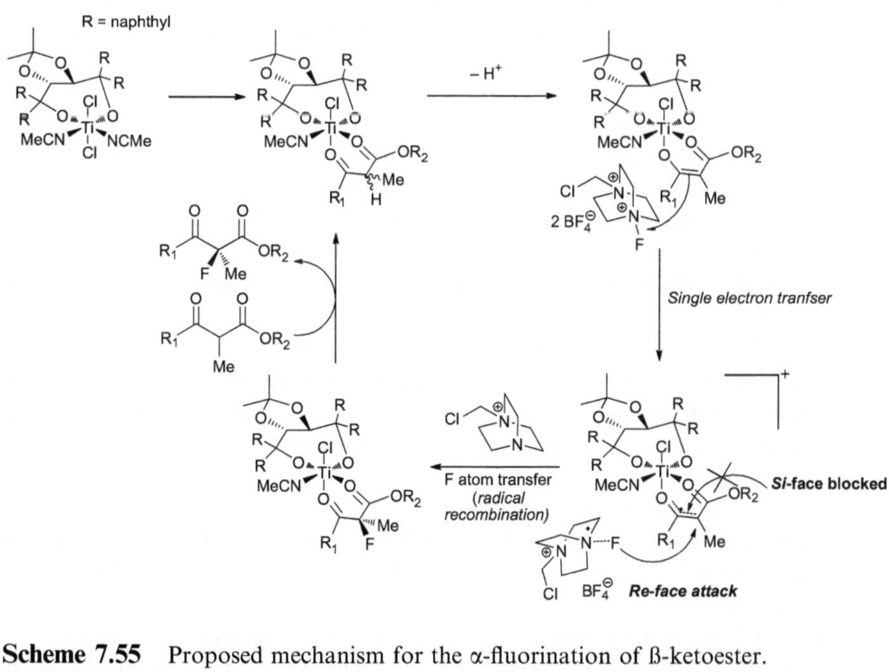

Scheme 7.55 Proposed mechanism for the α-fluorination of ß-ketoester.

Togni also disclosed the use of the *in situ* formed $[Ru(Et_2O)_2(PNNP)]$[106] (PNNP = (1*S*,2*S*)-*N*,*N*′-bis(*o*-(diphenylphosphino)benzylidene)cyclohexane-1,2-diamine) complex **91** to catalyse the fluorination of acyclic and cyclic ß-ketoesters with NFSI.[107,108] The use of a fluorinating reagent less reactive

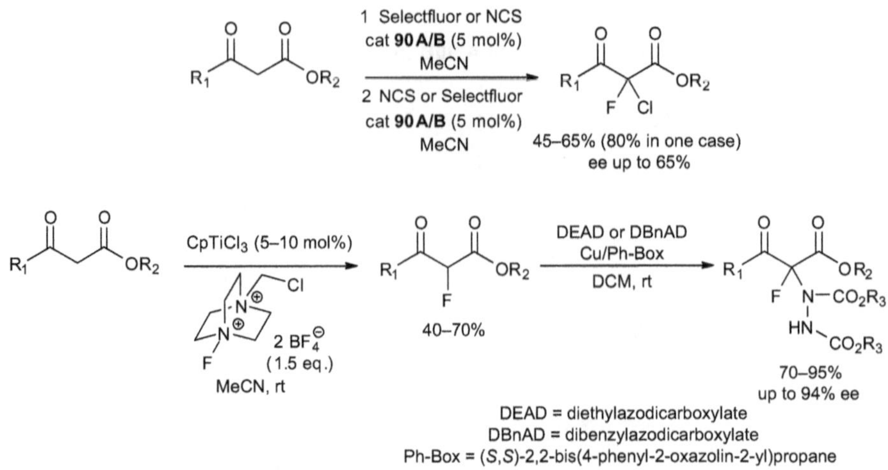

Scheme 7.56 Ti-catalysed α,α-difunctionalisation of ß-ketoester.

than Selectfluor allowed for the substrate scope to be extended compared to the Ti(TADDOLato) catalytic system. The Ru catalyst was found to be more tolerant of bulky groups at the α-position but at the expense of extended reaction times. Oxygenated solvents were found to reduce control over enantioselectivity likely due to coordination to the metal (Scheme 7.57).

Sodeoka and co-workers interrogated the reactivity of Pd enolates as these softer intermediates could be advantageous over traditional metal enolates. Early investigations found [Pd(*R*)-BINAP(OH_2)$_2$] (BINAP = (*R*)-(+)-(1,1′-binaphthalene-2,2′-diyl)bis(diphenylphosphine) (**92A-II**) in conjunction with NFSI to be an optimal system for fluorination.[109] The replacement of BINAP with DM-BINAP (3,5-dimethyl) or DTBM-SEGPHOS (DTBM-SEGPHOS = (*R*)-(−)-5,5′-bis[di(3,5-di-*tert*-butyl-4-methoxyphenyl)phosphino]-4,4′-bi-1,3-benzodioxole) improved enantioselectivity. The reaction was equally successful with the Pd μ-hydroxo complex **93** ligated with (*R*)-DTBM-SEGPHOS (catalyst **93B-I**). Polar solvent drastically reduced the reaction time and improved yield with ethanol found to be the optimum. The reaction was tolerant of both cyclic and acyclic ß-ketoesters, with *ee* maintained even at catalyst loadings as low as 1 mol%, and could be scaled-up without loss of efficiency or erosion of *ee* (Scheme 7.58). The reaction was further improved if conducted in ionic liquids; this variation allowed for the catalyst to be recycled up to ten times.[110]

The catalytic fluorination was extended to ß-ketophosphonates.[111,112] Catalyst **92** ligated to (*R*)-DM-BINAP (**92A-I**) gave the best results when the reaction was performed in ethanol. DTBM-SEGPHOS was a competent ligand for enantiocontrol but the products were isolated in lower yields possibly due to overwhelming steric repulsion upon fluorination. The reaction was successful for a range of cyclic substrates including tetralone and indanone derivatives. Acyclic substrates underwent fluorination with similar enantioselectivity but

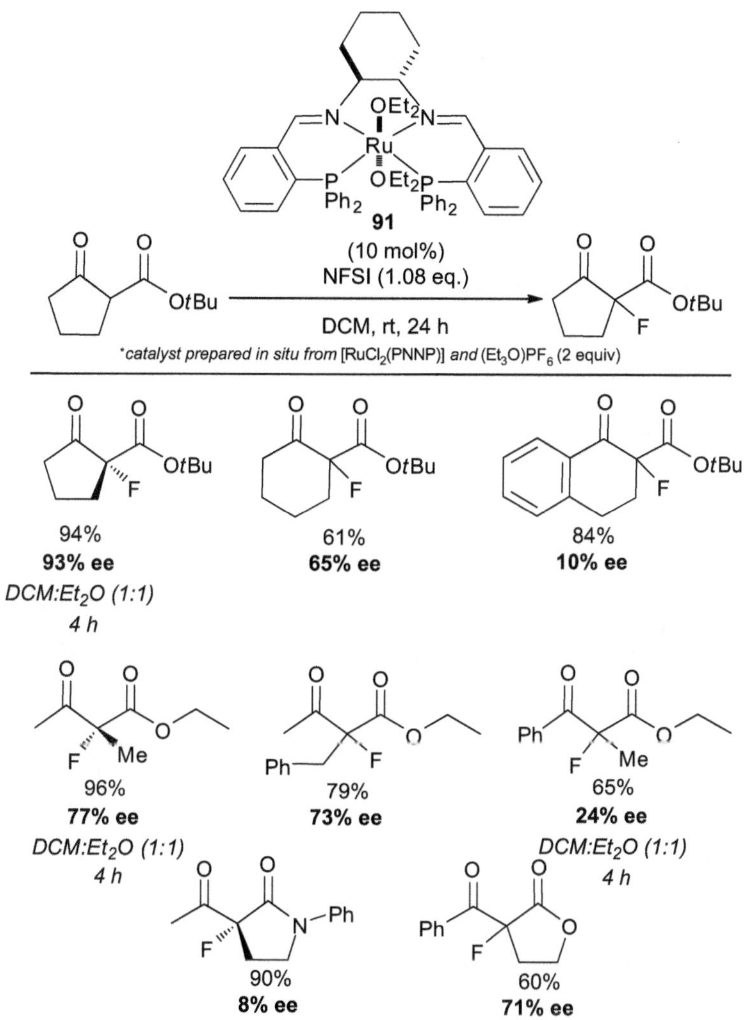

Scheme 7.57 Ru-catalysed fluorination of 1,3-dicarbonyl compounds.

these substrates required extended reaction time. N-Boc protected oxindoles were susceptible to deprotection with catalyst **92**; a problem circumvented with the use of the less acidic catalyst complex **93**. The reaction tolerates substitutions on the oxindole aryl sub-motif and various alkyl or aryl R_1 substituents (Scheme 7.59).[113]

Lactones and lactams were also accessed using this methodology; for the less acidic lactams, 2,6-lutidine is required as a co-catalyst to facilitate enolisation (**94**) (Scheme 7.60).[114] This system was subsequently applied to the enantioselective monofluorination of α-keto esters; the resulting products were transformed into ß-fluoro-α-hydroxy esters and ß-fluoro-α-amino esters.[115]

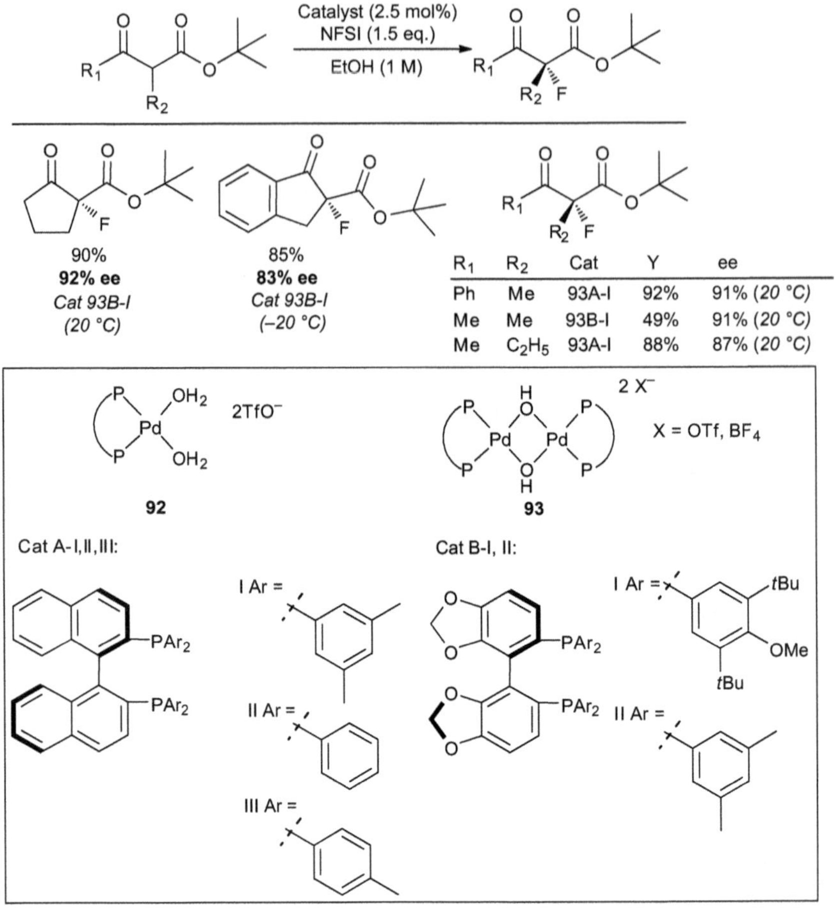

R1	R2	Cat	Y	ee
Ph	Me	93A-I	92%	91% (20 °C)
Me	Me	93B-I	49%	91% (20 °C)
Me	C_2H_5	93A-I	88%	87% (20 °C)

Scheme 7.58 Pd catalysed α-fluorination of ß-ketoesters.

Kim and co-workers reported the fluorination of ß-keto phosphonates with a slightly modified catalyst bearing an acetonitrile ligand in place of one hydroxy ligand and tetrafluoroborate as the counter ion; this system allowed for the efficient fluorination of cyclic and acyclic ß-ketoesters with NFSI.[116–118]

In 2004, the Shibata group reported an interesting example of metal dependent enantiodivergence for the fluorination of 1-indanone-2-carboxylate with NFSI (Scheme 7.61).[119] These reactions were performed in the presence of 10 mol% of (*S*,*S*)-2,2-bis(4-phenyl-2-oxazolin-2-yl)propane (Ph-BOX). With $Cu(OTf)_2$, the *S* product was formed, whereas the *R* product is obtained with the use of $Ni(ClO_4)_2{\cdot}6H_2O$. This reverse sense of enantiocontrol is proposed to arise from a change of the geometry of the metal enolate with the Cu and Ni complex adopting a distorted square planar and square pyramidal molecular geometry, respectively.

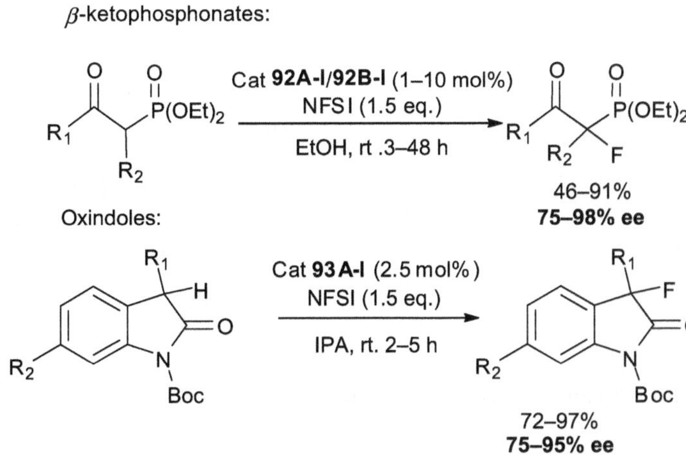

Scheme 7.59 Pd-catalysed α-fluorination of ß-ketophosphonates and oxindoles.

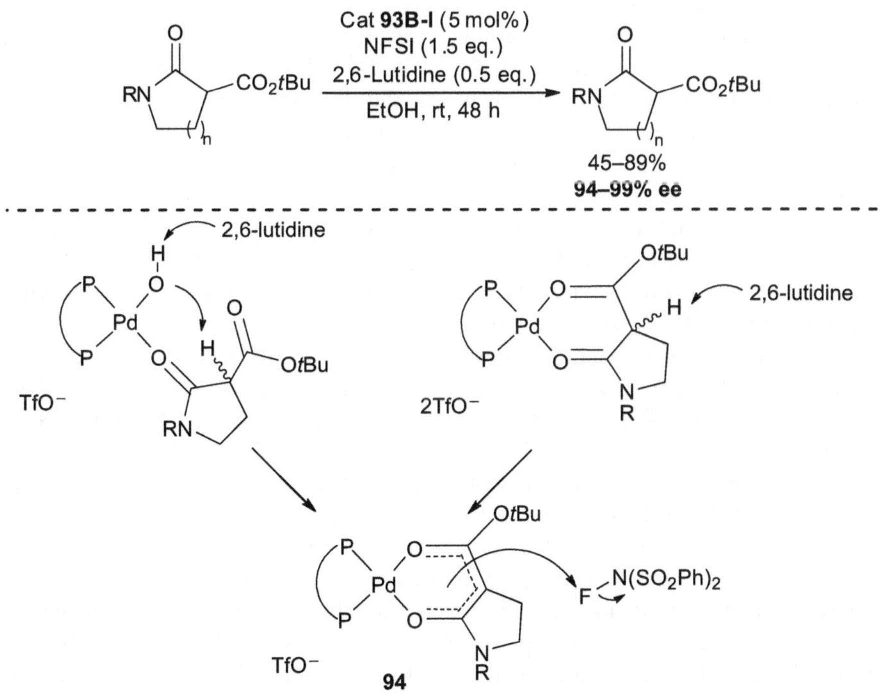

Scheme 7.60 Fluorination of lactams with the Pd catalyst **93B-I**.

The reaction was extended to substrates capable of two-point binding such as oxindoles.[120] Excellent control over enantioselectivity was possible with the DBFOX-Ph ligand (2 mol%) and $Ni(ClO_4)_2 \cdot 6H_2O$ even at low catalyst loadings (2 mol%) (Scheme 7.62). The sense of enantiocontrol is consistent with

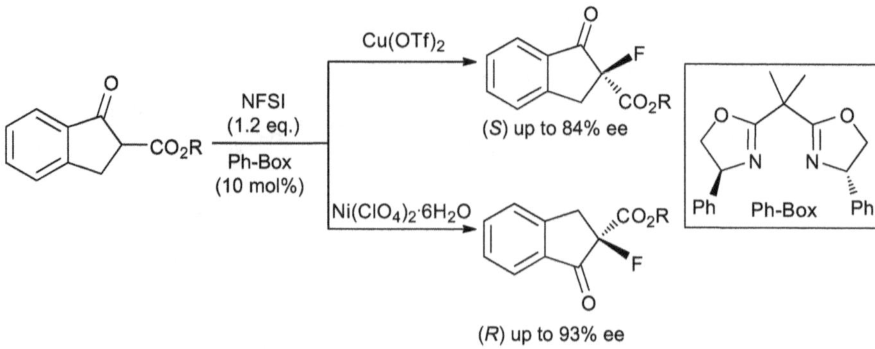

Scheme 7.61 Enantiodivergent fluorination of ß-ketoesters catalysed by Ni^{II} or Cu^{II} salts.

an attack of NFSI from the *Si* face of the Ni-bound enolate intermediate. Shibata also reported the first asymmetric amplification in halogenation reactions of ß-ketoesters with this system.[121]

Cahard reported cases of $Cu(OTf)_2$ catalysed fluorination of ß-ketoesters using the Ph-BOX ligand and NFSI.[122,123] Moderate to good levels of enantioselectivity were observed when the fluorination was performed in toluene or diethyl ether. The addition of strong base to neutralise the highly acidic by-product $[(PhSO_2)_2NH]$ was found to be detrimental to the enantioselectivity. In contrast, the use of 1,1,1,3,3,3-hexafluoro-2-propanol was found to improve enantiocontrol.[124] The alcohol is proposed to aid in the release of the fluorinated product from the catalyst to assist with the turnover of the reaction. The fluorination of acyclic and cyclic substrates is possible under these optimised conditions (Scheme 7.63).

The majority of carbonyl α-fluorinations are performed with NFSI as the electrophilic fluorinating reagent. In 2006, Inanaga reported the first catalysed fluorination of ß-ketoesters with 1-fluoropyridinium triflate **(N-F1)** (Scheme 7.64).[125] With the Lewis acid $Sc[(R)F_8\text{-}BNP]_3$, the reaction was efficient in terms of product yield but suffered from low *ee*'s. For this Sc-based catalyst, NFPy-OTf was found to be a superior electrophilic fluorinating reagent compared to NFSI and Selectfluor. Optimisation of solvent and temperature resulted in improved enantioselectivity when the reactions were carried out at 0 °C in toluene. Another important divergence with respect to previous contributions is the possibility to use cyclic and acyclic methyl and ethyl ß-ketoesters (instead of the bulky *tert*-butyl ß-ketoesters) with no compromise on yield and enantiomeric excesses.

Shibata reported the fluorination of malonates in 2008.[126] The use of $Ni(ClO_4)_2 \cdot 6H_2O$ (10 mol%) with NFSI in the presence of 11 mol% of (*R*,*R*)-DBFOX-Ph (DBFOX = 4,6-dibenzofurandiyl-2,2′-bisoxazoline) gave excellent yields but slightly reduced *ee* in comparison with the structurally related ß-ketoesters. Further optimisation identified a superior catalytic system; $Zn(OAc)_2$/(*R*,*R*)-DBFOX-Ph in dichloromethane at reflux in the presence of

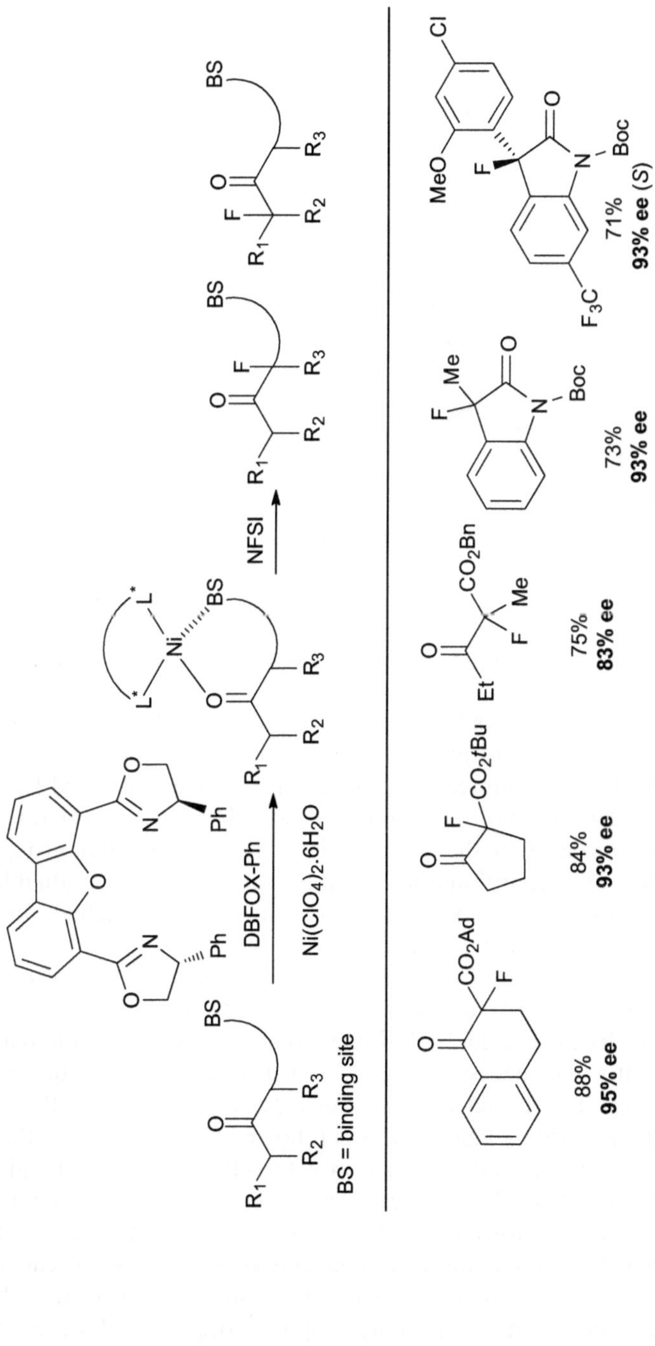

Scheme 7.62 Fluorination of two-point binding substrates catalyzed by Ni^{II} and dbfox-Ph ligand.

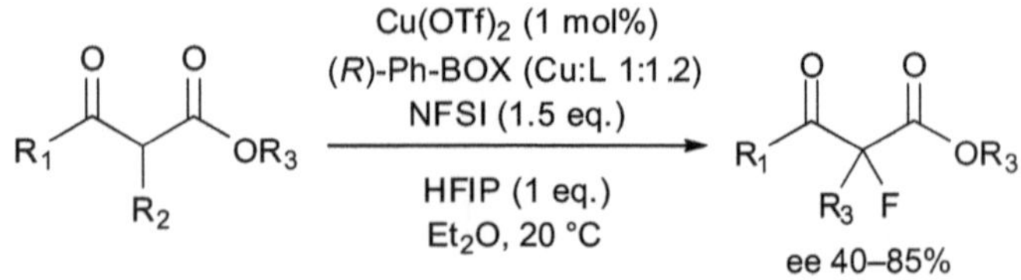

Scheme 7.63 Cu catalysed fluorination of ß-ketoesters.

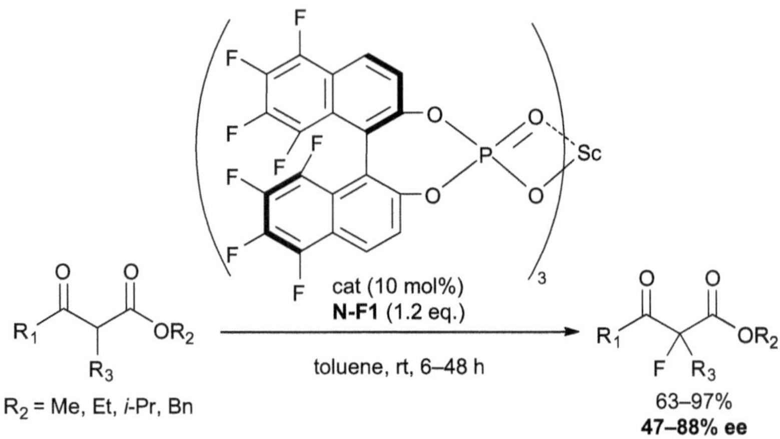

Scheme 7.64 Sc[(*R*)F_8-BNP]$_3$ catalysed fluorination of ß-ketoesters with **N-F1**.

molecular sieves (4 Å) allows for excellent enantiocontrol of the reaction while maintaining almost quantitative yields. The sense of enantiocontrol with ligand (*R*,*R*)-DBFOX-Ph is accounted for evoking the approach of NFSI from the *Si* face of the chiral Zn-bound ligated intermediate. The reaction tolerates a wide range of substitutions, with alkyl, aryl, sulfide, alkoxy and amino groups giving excellent yields and enantiomeric excesses. A selection of valuable pharmaceutical precursors was prepared from the enantioenriched fluorinated malonate products (Scheme 7.65).

The Ni-catalyzed fluorination of α-aryl acetic acid derivatives was reported in 2007.[127] The optimised conditions identified the need for the presence of $NiCl_2$, BINAP, R_3SiOTf and 2,6-lutidine along with NFSI for the reaction to proceed efficiently. The necessity to carry out this reaction with the co-catalyst R_3SiOTf suggests that the active Ni species is generated *in situ* by the exchange of chloride (**95**) for triflate ligands; this is followed by formation of the desired enolate with 2,6-lutidine and fluorination with NFSI, whose electrophilicity is likely enhanced by R_3SiOTf. The reaction was conducted at low temperatures in toluene to avoid the formation of the silyl enol ether that would diminish enantioselectivity. This reaction is limited to α-aryl derivatives (Scheme 7.66).

In 2008, Leckta *et al.* applied a dual activation protocol to acyl chlorides, using both a metal and a chiral nucleophilic organocatalyst, to generate

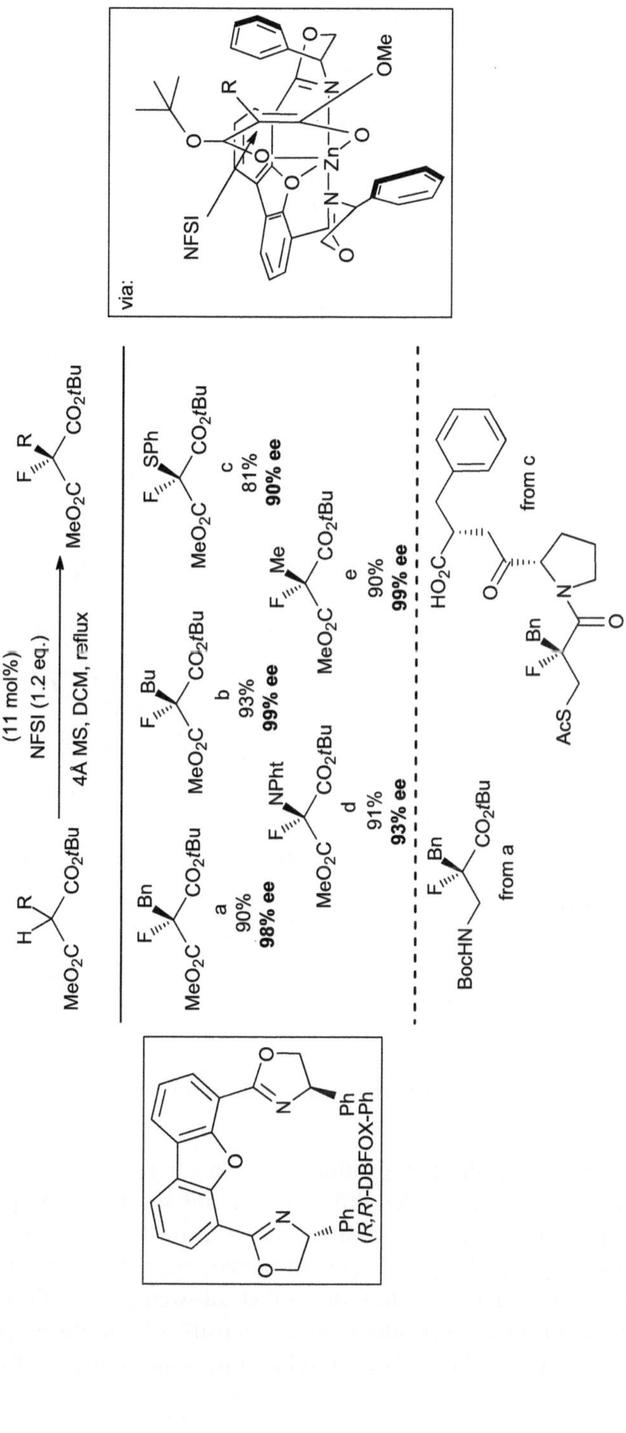

Scheme 7.65 Zn-catalyzed fluorination of malonates.

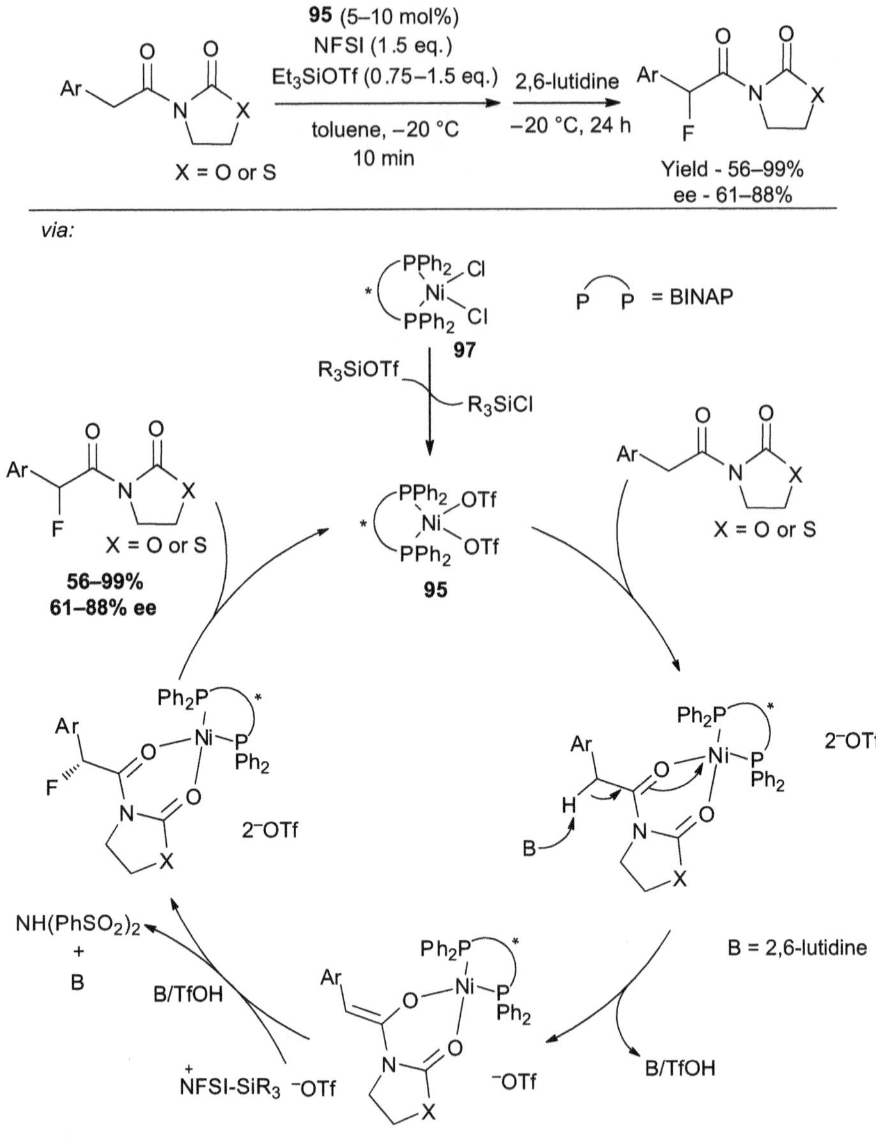

Scheme 7.66 Fluorination of α-aryl acetic acid derivatives.

enantioenriched α-fluorinated products.[128] This reaction, conducted in the presence of a competent nucleophile, stands out as it affords products that would be difficult to access by direct enolisation/fluorination. The use of *trans*($(PPh_3)_2PdCl_2$ (or (dppp)$NiCl_2$) with benzoylquinidine (BQd), the Hunig's base (*N*,*N*-diisopropylethylamine) and NFSI allowed for the fluorination of a wide range of products dependent on the nature of the nucleophile used to quench the reaction. Aryl, heterocyclic and aminoalkyl substitution are

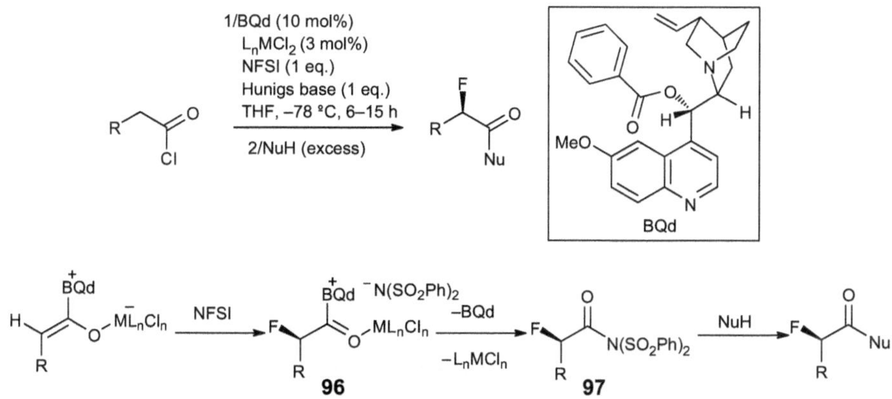

Scheme 7.67 Asymmetric fluorination of acid chlorides.

tolerated on the starting acid chloride and the formation of esters, thioesters, acids and amides is possible with the appropriate nucleophile. The BQd and metal are proposed to dually activate the acyl chloride into a chiral enolate that undergoes fluorination to give the acyl ammonium salt **96**. The reaction of **96** with the dibenzenesulfonimide anion produces the active amide intermediate **97** that undergoes transacylation with the external nucleophile to give the fluorinated product (Scheme 7.67). Further studies led to the discovery of the use of a Lewis acid co-catalyst $LiClO_4$ that allowed improved yields of aliphatic products through activation of NFSI.[129]

Mezzetti identified the major product of the reaction of *cis*-stilbene oxide with $AgHF_2$ in the presence of complex $\{[Ru(OEt_2)_2(PNNP)][PF_6]_2\}$ (5 mol%) as fluorodiphenylacetaldehyde; this product is likely formed *via* a [1,2]-phenyl shift followed by fluorination of the resulting aldehyde.[130] In 2011, Martín-Matute reported a tandem Ir-catalyzed isomerization/fluorination of allylic alcohols leading to α-fluoroketones.[131] The reaction of various allylic alcohols with Selectfluor in THF in the presence of 1 mol% of $(IrCp^*Cl_2)_2$ (Cp^* = 1,2,3,4,5-pentamethylcyclopentadiene) provided two regioisomeric products. A mixture of THF:H_2O (5:1) was found to suppress isomerisation and give moderate to high yields of a single regioisomer. The side-product of this reaction is the unwanted non-fluorinated ketone typically isolated in low yield. Terminal and 1,2-disubstituted alkenes were tolerated under the reaction conditions along with *gem*-disubstituted allylic alcohols leading to α-fluorinated ketones with the fluorine substituent on a quaternary carbon. The authors propose a mechanism that involves oxidation of the allylic alcohol into an unsaturated ketone ligated to an Ir–H species. Deuterium labeling experiments suggest the next step to be hydride addition affording an enolate that may bind to the Ir in an η^1-mode through the oxygen atom, an η^3-mode as an oxallyl or directly to the methylenic carbon. Subsequent reaction with the Selectfluor would generate the α-fluoroketone products (Scheme 7.68).

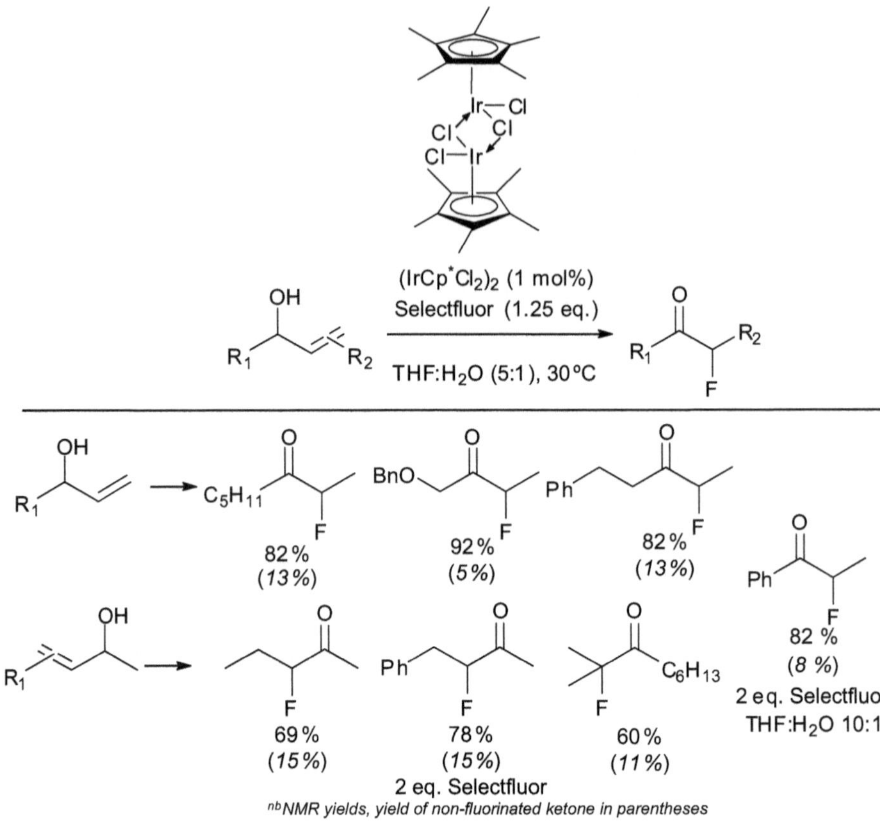

Scheme 7.68 Ir-catalysed formation of α-fluoroketones from allylic alcohols.

7.3 Conclusions

The developments in transition metal mediated/catalysed reactions to date have been vast. The large substrate scope, selectivity and ease of use of this chemistry has allowed for the construction of large complex molecules in far fewer steps than previously needed. This chapter displays the next step in its development, now allowing for the formation of the synthetically difficult yet ubiquitous C–F bonds. The employment of both electrophilic and nucleophilic fluorine sources has allowed access to a plethora of synthetically useful fluorinated compounds. Of particular importance are the identification of carbon–fluorine reductive eliminations from both Pd^{II} and Pd^{IV} centres and other high oxidation state metal complexes and the use of both Pd and Ir in the metal mediated formation of a ^{18}F–C bond for use in PET tracers. The advancement in asymmetric fluorinations has been extensive with the opening of expoxides and α-fluorination of carbonyl compounds providing enantioenriched fluorides with high levels of selectivity.

Though huge steps have been made in the insertion of a C–F bond with a transition metal, development is still needed to have methodologies that require less or no preactivation of the substrate, are tolerant of all functionalities and capable of imparting high levels of enantiomeric excess. The ultimate goal is to provide practising chemists with a toolbox of reactions that would enable one to label any organic molecule selectively at any position with either ^{19}F or radioactive ^{18}F, as solving this fundamental problem could have far reaching impact. It is anticipated that this field will continue to rapidly develop over the next five to ten years.

Acknowledgements

We thank EPSRC for financial support and Dr J. M. Brown F.R.S. for useful discussions.

References

1. D. O'Hagan, *Chem. Soc. Rev.*, 2008, **37**, 308.
2. S. Purser, P. R. Moore, S. Swallow and V. Gouverneur, *Chem. Soc. Rev.*, 2008, **37**, 320.
3. W. K. Hagmann, *J. Med. Chem.*, 2008, **51**, 4359.
4. P. Jeschke, *Chembiochem*, 2004, **5**, 570.
5. V. Gouverneur and K. Muller, (eds.), *Fluorine in Pharmaceutical and Medicinal Chemistry*, Imperial College Press, London, 2012.
6. P. W. Miller, N. J. Long, R. Vilar and A. D. Gee, *Angew. Chem., Int. Ed.*, 2008, **47**, 8998.
7. H. Suschitzky, *Adv. Fluorine Chem.*, 1965, **4**, 1.
8. J. H. Clark, C. H. Clark and D. Wails, (eds.), *Aromatic Fluoriantion*, CRC Press, 1996.
9. D. J. Adams and J. H. Clark, *Chem. Soc. Rev.*, 1999, **28**, 225.
10. J. F. Hartwig, *Acc. Res. Chem.*, 1998, **31**, 852.
11. J. F. Hartwig, *Inorg. Chem.*, 2007, **46**, 1936.
12. A. H. Roy and J. F. Hartwig, *J. Am. Chem. Soc.*, 2001, **123**, 1232.
13. S. L. Fraser, M. Y. Antipin, V. N. Khroustalyov and V. V. Grushin, *J. Am. Chem. Soc.*, 1997, **119**, 4769.
14. V. V. Grushin, *Chem.–Eur. J.*, 2002, **8**, 1006.
15. V. V. Grushin and W. J. Marshall, *J. Am. Chem. Soc.*, 2006, **128**, 12644.
16. V. V. Grushin and W. J. Marshall, *J. Am. Chem. Soc.*, 2006, **128**, 4632.
17. D. V. Yandulov and N. T. Tran, *J. Am. Chem. Soc.*, 2007, **129**, 1342.
18. K. L. Hull, W. Q. Anani and M. S. Sanford, *J. Am. Chem. Soc.*, 2006, **128**, 7134.
19. A. W. Kaspi, A. Yahav-Levi, I. Goldberg and A. Vigalok, *Inorg. Chem.*, 2008, **47**, 5.
20. X. Wang, T.-S. Mei and J.-Q. Yu, *J. Am. Chem. Soc.*, 2009, **131**, 7520.
21. K. S. L. Chan, M. Wasa, X. Wang and J.-Q. Yu, *Angew. Chem., Int. Ed.*, 2011, **50**, 9081.

22. N. D. Ball and M. S. Sanford, *J. Am. Chem. Soc.*, 2009, **131**, 3796.
23. T. Furuya, H. M. Kaiser and T. Ritter, *Angew. Chem., Int. Ed.*, 2008, **47**, 5993.
24. A. J. Canty, *Acc. Chem. Res.*, 1992, **25**, 83.
25. A. J. Canty, H. Jin, A. S. Roberts, B. W. Skelton, P. R. Traill and A. H. White, *Organometallics*, 1995, **14**, 199.
26. A. J. Canty, M. C. Denney, G. van Koten, B. W. Skelton and A. H. White, *Organometallics*, 2004, **23**, 5432.
27. J. Cámpora, P. Palma, D. del Rio, J. A. López, E. Álvarez and N. G. Connelly, *Organometallics*, 2005, **24**, 3624.
28. T. Furuya and T. Ritter, *J. Am. Chem. Soc.*, 2008, **130**, 10060.
29. T. Furuya, D. Benitez, E. Tkatchouk, A. E. Strom, P. Tang, W. A. Goddard III and T. Ritter, *J. Am. Chem. Soc.*, 2010, **132**, 3793.
30. T. Furuya and T. Ritter, *Org. Lett.*, 2009, **11**, 2860.
31. T. Furuya, A. E. Strom and T. Ritter, *J. Am. Chem. Soc.*, 2009, **131**, 1662.
32. P. Tang, T. Furuya and T. Ritter, *J. Am. Chem. Soc.*, 2010, **132**, 12150.
33. P. Tang and T. Ritter, *Tetrahedron*, 2011, **67**, 4449.
34. D. A. Watson, M. Su, G. Teverovskiy, Y. Zhang, J. García-Fortanet, T. Kinzel and S. L. Buchwald, *Science*, 2009, **325**, 1661.
35. B. P. Fors, D. A. Watson, M. R. Biscoe and S. L. Buchwald, *J. Am. Chem. Soc.*, 2008, **130**, 13552.
36. T. J. Maimone, P. J. Milner, T. Kinzel, Y. Zhang, M. K. Takase and S. L. Buchwald, *J. Am. Chem. Soc.*, 2011, **133**, 18106.
37. E. García-Toraño, V. P. Medina and M. R. Ibarra, *Appl. Radiation Isotopes*, 2010, **68**, 1561.
38. T. Nöel, T. J. Maimone and S. L. Buchwald, *Angew. Chem., Int. Ed.*, 2011, **50**, 8900.
39. E. Lee, A. S. Kamlet, D. C. Powers, C. N. Neumann, G. B. Boursalian, T. Furuya, D. C. Choi, J. M. Hooker and T. Ritter, *Science*, 2011, **334**, 639.
40. M. A. Subramanian and L. E. Manzer, *Science*, 2002, **297**, 1665.
41. K. M. Janmanchi and W. R. Dolbier, Jr., *Org. Process Res. Dev.*, 2008, **12**, 349.
42. V. Grushin, *Process for Preparing Fluoroarenes from Haloarenes*, U.S. Patent 7 202 388, 2007.
43. A. Casitas, M. Canta, M. Solà, M. Costas and X. Ribas, *J. Am. Chem. Soc.*, 2011, **133**, 19386.
44. J. Qian, Y. Liu, J. Zhu, B. Jiang and Z. Xu, *Org. Lett.*, 2011, **13**, 4220.
45. D. S. Laitar, P. Müller, T. G. Gray and J. P. Sadighi, *Organometallics*, 2005, **24**, 4503.
46. J. A. Akana, K. X. Bhattacharyya, P. Müller and J. P. Sadighi, *J. Am. Chem. Soc.*, 2007, **129**, 7736.
47. B. C. Gorske, C. T. Mbofana and S. J. Miller, *Org. Lett.*, 2009, **11**, 4318.
48. M. Schuler, F. Silva, C. Bobbio, A. Tessier and V. Gouverneur, *Angew. Chem., Int. Ed.*, 2008, **47**, 7927.

49. T. de Haro and C. Nevado, *Chem. Commun.*, 2011, **47**, 248.
50. M. N. Hopkinson, G. T. Giuffredi, A. D. Gee and V. Gouverneur, *Synlett*, 2010, 2737.
51. H. Peng and G. Liu, *Org. Lett.*, 2011, **13**, 772.
52. M. A. Tius and J. K. Kawakami, *Synth. Commun.*, 1992, **22**, 1461.
53. M. A. Tius and J. K. Kawakami, *Synlett*, 1993, 207.
54. M. A. Tius and J. K. Kawakami, *Tetrahedron*, 1995, **51**, 3997.
55. C. Cazorla, E. Métay, B. Andrioletti and M. Lemaire, *Tetrahedron Lett.*, 2009, **50**, 3936.
56. T. Xu, X. Mu, H. Peng and G. Liu, *Angew. Chem., Int. Ed.*, 2011, **50**, 8176.
57. V. V. Grushin, *Angew. Chem., Int. Ed.*, 1998, **37**, 994.
58. W. W. Dukat, J. H. Holloway, E. G. Hope, M. R. Rieland, P. J. Townson and R. L. Powell, *J. Chem. Soc., Chem. Commun.*, 1993, 1429.
59. J. E. Veltheer, P. Burger and R. G. Bergman, *J. Am. Chem. Soc.*, 1995, **117**, 12478.
60. P. Barthazy, R. M. Stoop, M. Wörle, A. Togni and A. Mezzetti, *Organometallics*, 2000, **19**, 2844.
61. P. Barthazy, A. Togni and A. Mezzetti, *Organometallics*, 2001, **20**, 3472.
62. T. Wu, G. Yin and G. Liu, *J. Am. Chem. Soc.*, 2009, **131**, 16354.
63. S. Qiu, T. Xu, J. Zhou, Y. Guo and G. Liu, *J. Am. Chem. Soc.*, 2010, **132**, 2856.
64. T. Xu, S. Qiu and G. Liu, *Chin. J. Chem.*, 2011, **29**, 2785.
65. S.-B. Zhao, J. J. Becker and M. R. Gagné, *Organometallics*, 2011, **30**, 3926.
66. A. W. Kaspi, I. Goldberg and A. Vigalok, *J. Am. Chem. Soc.*, 2010, **132**, 10626.
67. N. P. Mankad and F. D. Toste, *Chem. Sci.*, 2012, **3**, 72.
68. J. M. Racowski, J. B. Gray and M. S. Sanford, *Angew. Chem., Int. Ed.*, 2012, **51**, 3414.
69. A. Yahav, I. Goldberg and A. Vigalok, *J. Am. Chem. Soc.*, 2003, **125**, 13634.
70. A. Yahav, I. Goldberg and A. Vigalok, *Inorg. Chem.*, 2005, **44**, 1547.
71. K. B. McMurtrey, J. M. Racowski and M. S. Sanford, *Org. Lett.*, 2012, **14**, 4094.
72. W. J. Middleton, *J. Org. Chem.*, 1975, **40**, 574.
73. D. M. Gree, C. J. M. Kermarrec, J. T. Martelli, R. L. Gree, J. P. Lellouche and L. J. Toupet, *J. Org. Chem.*, 1996, **61**, 1918.
74. B. Greedy, J-M. Paris, T. Vidal and V. Gouverneur, *Angew. Chem., Int. Ed.*, 2003, **42**, 3291.
75. P. Barthazy, L. Hintermann, R. M. Stoop, M. Wörle, A. Mezzetti and A. Togni, *Helv. Chim. Acta*, 1999, **82**, 2448.
76. P. Barthazy, R. M. Stoop, M. Wörle, A. Togni and A. Mezzetti, *Organometallics*, 2000, **19**, 2844.
77. B. M. Trost and D. L. van Vranken, *Chem. Rev.*, 1996, **96**, 395.
78. B. M. Trost and M. L. Crawley, *Chem. Rev.*, 2003, **103**, 2921.
79. H. Amii and K. Uneyama, *Chem. Rev.*, 2009, **109**, 2119.

80. U. Burckhardt, M. Baumann and A. Togni, *Tetrahedron: Asymmetry*, 1997, **8**, 155.
81. E. M. Stang and M. C. White, *Nat. Chem.*, 2009, **1**, 547.
82. L. Hintermann, F. Läng, P. Maire and A. Togni, *Eur. J. Inorg. Chem.*, 2006, 1397.
83. A. Hazari, V. Gouverneur and J. M. Brown, *Angew., Chem. Int. Ed.*, 2009, **48**, 1296.
84. C. Hollingworth, A. Hazari, M. N. Hopkinson, M. Tredwell, E. Benedetto, M. Huiban, A. D. Gee, J. M. Brown and V. Gouverneur, *Angew. Chem., Int. Ed.*, 2011, **50**, 2613.
85. D. W. Kim, H.-J. Jeong, S. T. Lim and M.-H. Sohn, *Angew. Chem., Int. Ed.*, 2008, **47**, 8404.
86. D. W. Kim, D.-S. Ahn, Y.-H. Oh, S. Lee, H. S. Kil, S. J. Oh, S. J. Lee, J. S. Kim, J. S. Ryu, D. H. Moon and D. Y. Chi, *J. Am. Chem. Soc.*, 2006, **128**, 16394.
87. E. Lee and D. V. Yandulov, *J. Fluorine Chem.*, 2009, **130**, 474.
88. B. M. Trost, D. L. van Vranken and C. Bingel, *J. Am. Chem. Soc.*, 1992, **114**, 9327.
89. M. H. Katcher and A. G. Doyle, *J. Am. Chem. Soc.*, 2010, **132**, 17402.
90. M. H. Katcher, A. Sha and A. G. Doyle, *J. Am. Chem. Soc.*, 2011, **133**, 15902.
91. J. J. Topczewski, T. J. Tewson and H. M. Nguyen, *J. Am. Chem. Soc.*, 2011, **133**, 19318.
92. S. Bruns and G. Haufe, *J. Fluorine Chem.*, 2000, **104**, 247.
93. G. Haufe, S. Bruns and M. Runge, *J. Fluorine Chem.*, 2001, **112**, 55.
94. G. Haufe and S. Bruns, *Adv. Synth. Catal.*, 2002, **344**, 165.
95. J. A. Kalow and A. G. Doyle, *J. Am. Chem. Soc.*, 2011, **132**, 3268.
96. J. A. Kalow and A. G. Doyle, *J. Am. Chem. Soc.*, 2011, **133**, 16001.
97. D. Enders and M. R. M. Hüttl, *Synlett*, 2005, 991.
98. M. Marigo, D. Fielenbach, A. Braunton, A. Kjærsgaard and K. A. Jørgensen, *Angew. Chem., Int. Ed.*, 2005, **44**, 3703.
99. D. D. Steiner, N. Mase and C. F. Barbas III, *Angew. Chem., Int. Ed.*, 2005, **44**, 3706.
100. T. D. Beeson and D. W. C. MacMillan, *J. Am. Chem. Soc.*, 2005, **127**, 8826.
101. For a review see M. Marigo and K. A. Jørgensen, *Chem. Commun.* 2006, 2001 and references therein.
102. L. Hintermann and A. Togni, *Angew. Chem., Int. Ed.*, 2000, **39**, 4359.
103. M. Perseghini, M. Massaccesi, Y. Liu and A. Togni, *Tetrahedron*, 2006, **62**, 7180.
104. R. Frantz, L. Hintermann, M. Perseghini, D. Broggini and A. Togni, *Org. Lett.*, 2003, **5**, 1709.
105. D. P. Huber, K. Stanek and A. Togni, *Tetrahedron: Asymmetry*, 2006, **17**, 658.
106. J. X. Gao, T. Ikariya and R. Noyori, *Organometallics*, 1996, **15**, 1087.

107. C. Bonaccorsi, M. Althaus, C. Becker, A. Togni and A. Mezzetti, *Pure Appl. Chem.*, 2006, **78**, 391.
108. M. Althaus, C. Becker, A. Togni and A. Mezzetti, *Organometallics*, 2007, **26**, 5902.
109. Y. Hamashima, K. Yagi, H. Takano, L. Tamás and M. Sodeoka, *J. Am. Chem. Soc.*, 2002, **124**, 14530.
110. Y. Hamashima, H. Takano, D. Hotta and M. Sodeoka, *Org. Lett.*, 2003, **5**, 3225.
111. Y. Hamashima, T. Suzuki, Y. Shimura, T. Shimizu, N. Umebayashi, T. Tamura, N. Sasamoto and M. Sodeoka, *Tetrahedron Lett.*, 2005, **46**, 1447.
112. Y. Hamashima, T. Suzuki, H. Takano, Y. Shimura, Y. Tsuchiya, K. Moriya, T. Goto and M. Sodeoka, *Tetrahedron*, 2006, **62**, 7168.
113. Y. Hamashima, T. Suzuki, H. Takano, Y. Shimura and M. Sodeoka, *J. Am. Chem. Soc.*, 2005, **127**, 10164.
114. T. Suzuki, T. Goto, Y. Hamashima and M. Sodeoka, *J. Org. Chem.*, 2007, **72**, 246.
115. S. Suzuki, Y. Kitamura, S. Lectard, Y. Hamashima and M. Sodeoka, *Angew. Chem., Int. Ed.*, 2012, **51**, 4581.
116. S. M. Kim, H. R. Kim and D. Y. Kim, *Org. Lett.*, 2005, **7**, 2309.
117. H. R. Kim and D. Y. Kim, *Tetrahedron Lett.*, 2005, **46**, 3115.
118. Y. K. Kang, M. J. Cho, S. M. Kim and D. Y. Kim, *Synlett*, 2007, 1135.
119. N. Shibata, T. Ishimaru, T. Nagai, J. Kohno and T. Toru, *Synlett*, 2004, 1703.
120. N. Shibata, J. Kohno, K. Takai, T. Ishimaru, S. Nakamura, T. Toru and S. Kanemasa, *Angew. Chem. Int. Ed.*, 2005, **44**, 4204.
121. For a full dicussion of non-linear effect seen see S. Kanemasa, Y. Oderaotoshi, S. Sakaguchi, H. Yamamoto, J. Tanaka, E. Wada and D. P. Curran, *J. Am. Chem. Soc.*, 1998, **120**, 3074.
122. J.-A. Ma and D. Cahard, *J. Fluorine Chem.*, 2004, **125**, 1357.
123. J.-A. Ma and D. Cahard, *Tetrahedron: Asymmetry*, 2004, **15**, 1007.
124. D. A. Evans, T. Rovis, M. C. Kozlowski and J. S. Tedrow, *J. Am. Chem. Soc.*, 1999, **121**, 1994.
125. S. Suzuki, H. Furuno, Y. Yokoyama and J. Inanaga, *Tetrahedron: Asymmetry*, 2006, **17**, 504.
126. D. S. Reddy, N. Shibata, J. Nagai, S. Nakamura, T. Toru and S. Kanemasa, *Angew. Chem. Int. Ed.*, 2008, **47**, 164.
127. T. Suzuki, Y. Hamashima and M. Sodeoka, *Angew. Chem., Int. Ed.*, 2007, **46**, 5435.
128. D. H. Paull, M. T. Scerba, E. Alden-Danforth, L. R. Widger and T. Lectka, *J. Am. Chem. Soc.*, 2008, **130**, 17260.
129. J. Erb, D. H. Paull, T. Dudding, L. Belding and T. Lectka, *J. Am. Chem. Soc.*, 2011, **133**, 7536.
130. M. Althaus, A. Togni and A. Mezzetti, *J. Fluorine Chem.*, 2009, **130**, 702.
131. N. Ahlsten and B. Martín-Matute, *Chem. Commun.*, 2011, **47**, 8331.

CHAPTER 8

Coordination-Directed Metallation Strategy for C–H Functionalization

LUCIANO CUESTA AND ESTEBAN P. URRIOLABEITIA*

Instituto de Síntesis Química y Catálisis Homogénea, CSIC – Universidad de Zaragoza, Pedro Cerbuna 12, 50009 Zaragoza, Spain
*Email: esteban@unizar.es

8.1 Introduction

The formation of new C–C and/or C–heteroatom bonds is considered the most relevant process in synthetic organic chemistry for the build up of new compounds and structures. The cross-coupling processes are probably one of the most useful tools to achieve this task[1,2] and there are, at least, four different approximations to this problem, exemplified in Figure 8.1. Path (a) illustrates the use of prefunctionalized materials, where X can be a halide, a pseudohalide, a carboxylate or, in general, a good leaving group, while R is, in most cases, a nucleophilic organometallic reagent or, in general, a nucleophile. The use of such prefunctionalized substrates has provided new and versatile synthetic tools to the chemists, achieving the selective preparation of novel molecules and unleashing a fast and relevant progress. They are the well-known cross-couplings of Suzuki, Sonogashira, Stille, Kumada, Hiyama or Negishi, among others. The use of pre-functionalized substrates[3,4] has been prevalent for many decades over the use of non-functionalized precursors, due to the higher

RSC Catalysis Series No. 11
C–H and C–X Bond Functionalization: Transition Metal Mediation
Edited by Xavi Ribas

Published by the Royal Society of Chemistry, www.rsc.org

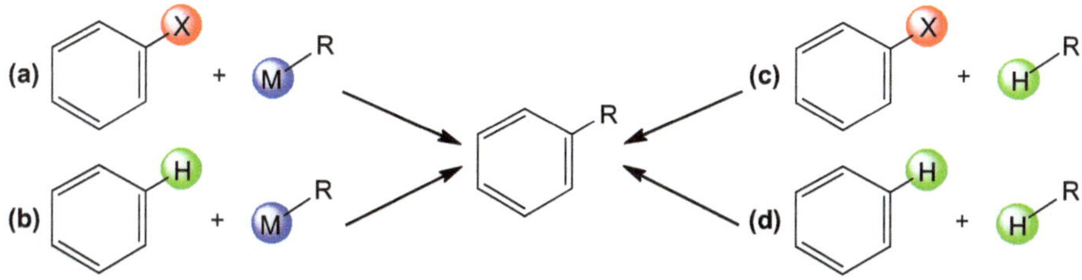

Figure 8.1 Different approximations to the functionalization of an arylic substrate as a function of the type of starting compound.

reactivity and selectivity of the formers. The importance of these processes has been recognized with the award of the Nobel Prize in Chemistry 2010 to Richard F. Heck, Ei-ichi Negishi and Akira Suzuki.[5]

However, the price we have to pay to achieve such degree of selectivity in these reactions is the use of expensive and pre-elaborated molecules. In fact, we have to pre-functionalize and isolate two independent species in order to perform the reaction, generating, in addition, stoichiometric amounts of waste materials (XM in path a). Due to these reasons, the efforts of the chemists during the last years have been directed to the search of new processes aiming to obtain identical final compounds, with the same selectivity, but in a greener and more sustainable way. In this respect, probably the best alternative fulfilling the mentioned requirements is the activation of C–H bonds. That is, the substitution of the reactivity of the C–X and C–M bonds by that of the C–H bonds. Why should we use C–H bonds?

The C–H bonds are ubiquitous in organic species, from natural feedstocks to the more elaborated natural products. Therefore, the use of C–H bonds as reactive points of a molecule should be very advantageous, since it will not be necessary to synthesise prefunctionalized substrates, usually prepared in tedious multi-step procedures, with the subsequent saving of resources, energy, time and, finally, money. In addition, the waste material produced during the C–C and/or C–X coupling will be reduced considerably. From a preparative point of view, several possibilities can be envisaged. It should be possible to perform the C–H activation on the electrophilic reagent, avoiding the use of halide or pseudohalide compounds (path b, Figure 8.1),[6–12] on the nucleophilic reagent (disregrading the organometallic compound, path c),[6,7,10–16] or even on the two cases, which is by far the most interesting option, the so-called dehydrogenative cross-coupling (path d).[6,7,16–20] All the four possibilities are known, and we can find excellent reviews for all of them.[6–20]

However, in spite of all the advantages, the use of C–H bonds as targets has two main drawbacks. The first one is the well-known fact that the C–H bond is relatively inert, since its cleavage involves energies of dissociation about 100 kcal·mol^{-1}, depending on the nature of the C–H bond (Csp^3, Csp^2 or Csp). This problem was solved many years ago by using transition metals (Pd, Pt, Rh, Ru, Cu, Fe, among others), which are able to produce the activation (relaxation) and cleavage of the C–H bond, very often under mild reaction conditions. Therefore, all functionalizations described in this chapter, either catalytic or stoichiometric, will be mediated by transition metals. A more difficult task arises when the selectivity (chemo-, regio-, enantio-) of the reaction is considered. The own ubiquity of the C–H bond can be a problem if several C–H bonds are amenable to activation at the same time. This fact results, unavoidably, in obtaining of different isomers, or even in the multi-functionalization of the starting compound.

There are different strategies to overcome the problem of the selectivity. There is, in principle, a natural reactivity based on steric or electronic biases, which direct the C–H functionalization to specific positions without any other external guidance. The functionalization of heterocycles or substituted arenes, where the selectivity is achieved due to the different distribution of the electron density, is a paradigm of this "innate" reactivity.[21,22] Some representative examples are shown in Figure 8.2, and a vast amount of fully regioselective chemical processes are based on this natural reactivity.

That said, it seems clear that the functionalization of the other positions of the same molecules, for instance, the 3 position of a pyrrole or the 2 position of an indole, have to be directed (or "guided") if a high degree of selectivity has to be achieved,[21] because these positions are not accessible using the same inherent preferences. Obviously, this directed reactivity must be seen as complementary to that obtained in natural positions, since in this way all targeted positions of a given substrate could be functionalized. There are several strategies to achieve a directed functionalization, but probably the most extended one lies on the use of functional groups with potential donor atoms such as N, O, P, S, or others; that is, atoms able to be coordinated, even weakly,[23] to the transition metal, as represented in Figure 8.3. This interaction results in the formation of a coordination complex, where the substrate to be functionalized is one of the ligands. In this way, once the substrate is bonded to

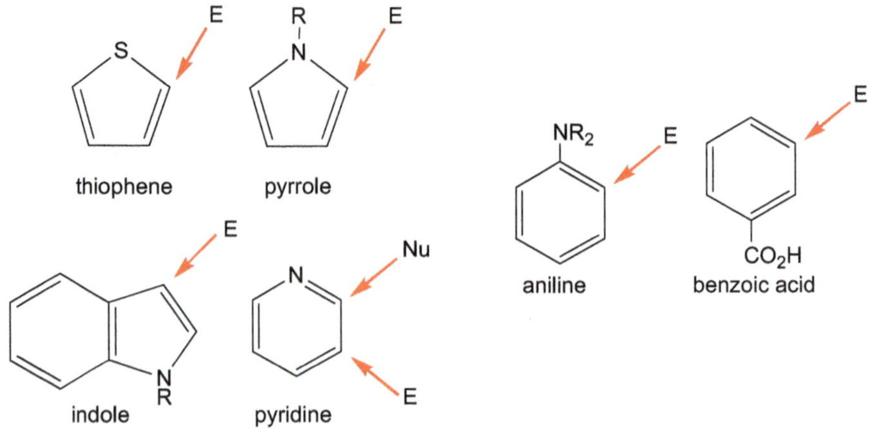

Figure 8.2 Innate[21] C–H functionalizations. E = electrophile; Nu = nucleophile.

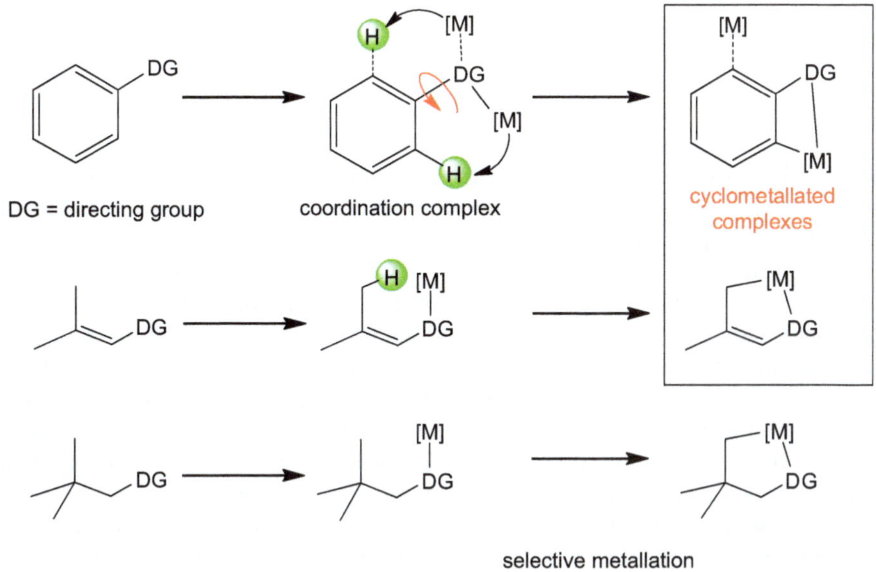

Figure 8.3 Strategy of the coordination-directed metallation using directing groups.

the metal, only certain positions fall in the vicinity of the metal and, therefore, only selected C–H bonds can be activated.

Due to these facts, these functional groups are properly called "directing groups" (DG), since they direct the metal to a specific position and, hence, the general strategy is termed "coordination-directed metallation". This method drastically reduces the number of available positions, and affords products with a high degree of selectivity. Following with the definitions, the metallic systems resulting from the bonding of a donor atom and subsequent intramolecular C–H bond activation, with concomitant C–M bond formation, are called cyclometallated complexes.[24]

In this chapter, we will analyze the use of these complexes as synthetic tools for the selective modification and transformation of organic entities. Because of the ease and specificity attained in the incorporation of the metal to the organic scaffold, and its versatility for further modification, these species have attracted considerable interest in recent times. After a general overview of relevant directing groups, we will focus on the most impacting results reported during the last 3–4 years for directed functionalization using organo-complexes of Pd, Ru, Rh and Cu, because the vast amount of results published up to 2009 is adequately covered in diverse reviews,[15–20,25–30] and because these metals cover more than 90% of the literature published in this subject. For the sake of clarity the results are clustered, within each metal, as a function of the type of reaction: C–C bond formation (arylation, olefination, carbonylation, acylation, and others) and C–heteroatom (B, O, S, N) bond formation.

8.2 Variety of Functional Groups Directing Arene C–H Functionalization

In principle, the number of functional groups amenable to be used as directing groups is almost unlimited, since the only requisite is to contain a heteroatom able to bind to the metal center. These heteroatoms are most usually N, O, P, S, As and Se, but examples containing other heteroatoms are also known. Typical directing groups, which should be seen just as a representative selection, are those shown in Figure 8.4. In addition to the specific electronic and steric properties of each heteroatom on each particular functional group, which result in a tailored tuning of the metal requirements, the directing groups can store additional information, for instance stereogenic (chiral groups), hemilabile behaviour, thermomorphic properties or optical applications (liquid crystals).

Even if a wide variety of directing groups is available, some limitations appear when a retrosynthetic organic process is designed. For instance, it is envisageable that typical good orienting groups such as pyridine, or other chelating fragments, are no longer necessary after the selective C–H bond activation. However, sometimes these efficient moieties can not be removed easily, or they are not prone to undergo further functionalization. Moreover, the need for greener reactants and reaction conditions, providing a high level of atom economy and minimizing wastes, does not fit with the presence of non-convertible functional groups. Due to these facts, an extensive research in multi-functional directing groups is being developed nowadays. That is, directing groups that can perform extra tasks, in addition to their directing purpose.

A clear example is the use of oxidizing directing groups (see Figure 8.5).[31–34] In these cases, the directing group contains a covalent bond which is the responsible of the oxidation of the metal, avoiding the use of external oxidants which usually generates waste subproducts. While, in principle, several bonds are amenable to act as oxidants, in all cases shown in Figure 8.5 the N–O bond plays this role. In spite of their similarity, mechanistic differences for the oxidation step have been observed for these groups.

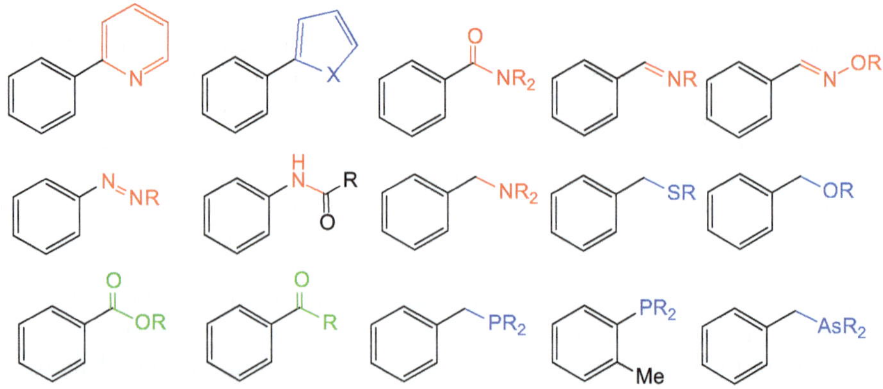

Figure 8.4 Selected functional groups displaying potential behaviour as directing groups.

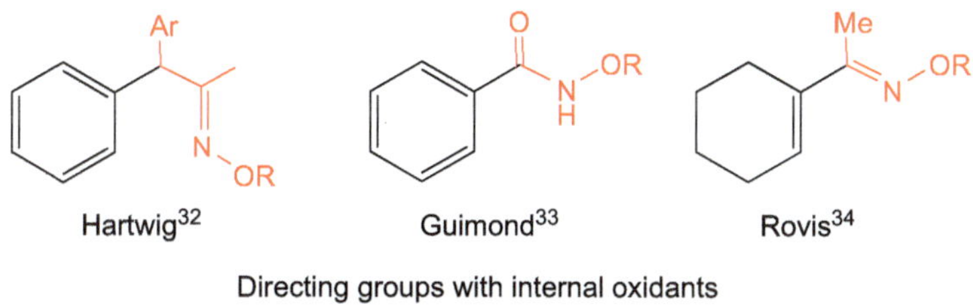

Figure 8.5 Directing groups with internal oxidants.

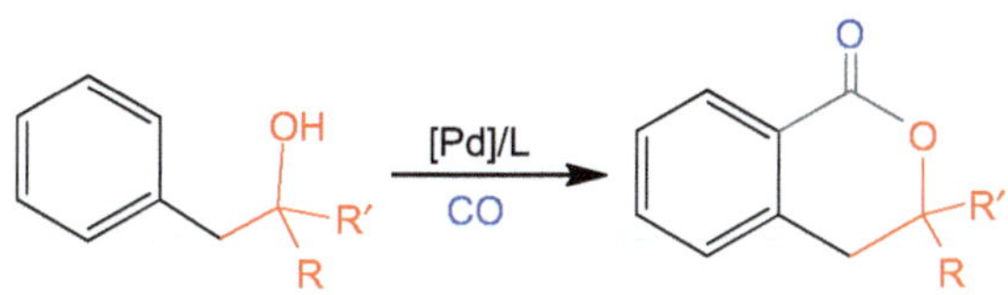

Figure 8.6 Incorporation of the directing group to the final target molecule.

The optimized performance of the capabilities of the directing groups can be shown properly when they are designed to be incorporated to the final molecule at the end of the reaction. This is one of the most challenging concepts currently, since it represents the best example of atom economy in retro-synthetic terms (nothing or almost nothing is lost during the reaction). It has been developed by several authors, and in Figure 8.6 we show an application of Yu *et al.*[35] Although we will find similar examples throughout this chapter, this really interesting and useful methodology is scarcely developed.

Another way to plan the multi-task character of the directing groups is the use of "traceless" functional groups. In this case the basic idea is that, after proper orientation of the reaction has been achieved, the functional group (or part of it) is removed by reaction with an external reagent. In this respect, cutting edge advances have been obtained through the use of silanols, 2-pyridylsulfonyl or carboxylates as traceless directing groups,[36–38] as shown in Figure 8.7. The reaction can always be considered as composed by two steps. In the first step the directing group orientates the reaction to the ortho position. In the second step, the directing group is transformed into a C–H bond or into another simpler functional group. It is remarkable that sometimes the reaction can be performed in a one-pot fashion, while in other cases the semi-one-pot procedure[37] (sequential addition of the reagents without isolation of the intermediates) is used in order to achieve the best efficiency.

The carboxylate, or the carboxylic acid functional group, is almost the paradigm of what a multi-purpose directing group could be. Although it coordinates weakly to the metal center,[23] it can behave as a classical directing group, selecting the position where the metal has to be incorporated and, typically, forming part of the target molecule by further functionalization after the C–H activation step. This two-step approach undoubtedly has a huge

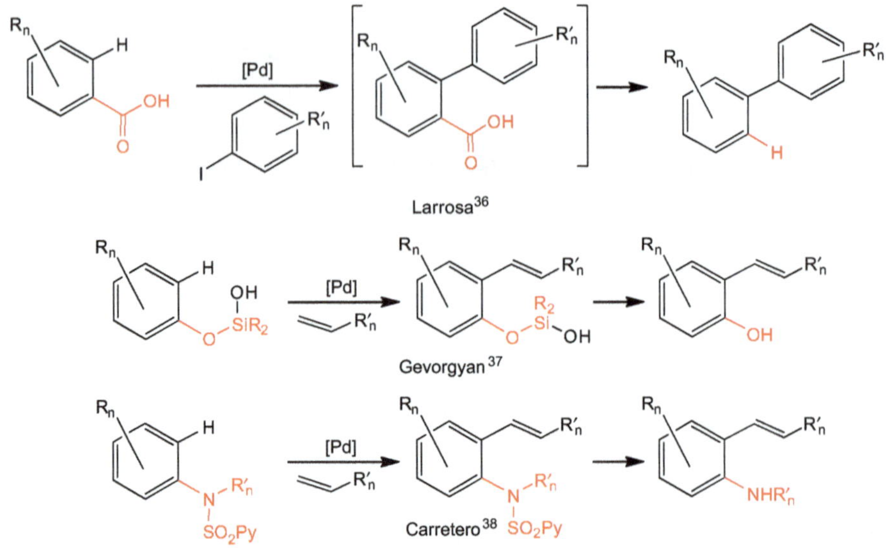

Figure 8.7 Some examples of traceless directing groups.

synthetic utility, and recent examples are presented in Figure 8.8,[39–42] where applications to the arylation or olefination of aromatic and aliphatic substrates are shown. It is remarkable, for instance, that the synthesis of common precursors or fragments of natural products is possible from accessible substrates such as 2-arylacetic acid derivatives.

In addition to this "normal" behaviour, the carboxylate groups can also acts as a traceless directing group, orienting the reaction in a first step and then decarboxylating, as we have described in Figure 8.7. The use of carboxylic acid derivatives as synthons of specific reactions (for instance, arene carboxylic acids as arylic coupling partners in the Heck reaction) was already known, but it has been only very recently that carboxylic acids has been employed as traceless directing groups. Additional examples are presented in Figure 8.9, and show the versatility of this approximation because it gives access to useful carbo- and heterocycles from common and accessible sources.[43–46]

A third option of the carboxylic fragment is its behaviour as carrier of other functional groups, for instance, the aryl moiety, as shown in the last example of Figure 8.9.[47]

Additional examples of the multi-functionality of the directing groups are those related to the modulation of the electronic and steric requirements of the C–H bond activation step. For instance, while the cyclopalladation of aromatic aldehydes is not favoured due to the high electron withdrawing character of the carbonyl moiety, and usually requires harsh conditions, that of the corresponding imines occurs under milder conditions and even at room temperature,[48] because imines are less electron-withdrawing and the aromatic ring bears more electron density. Thus, the functionalization of the imines will be more tolerant with hypothetical functional groups on the starting substrate than

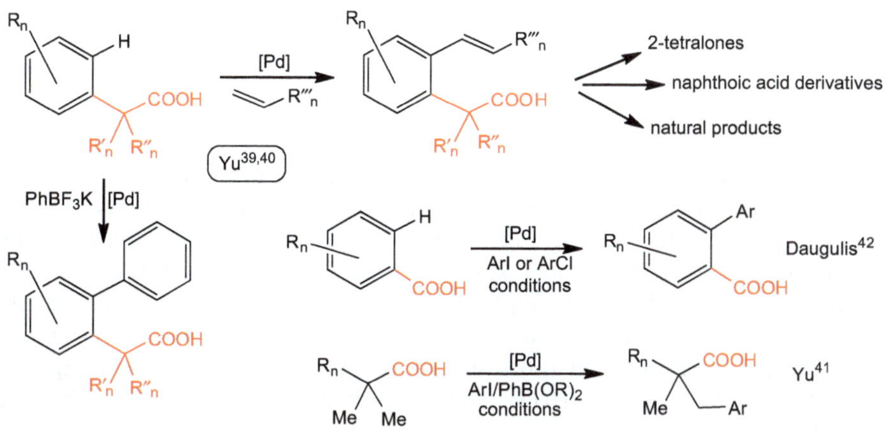

Figure 8.8 Carboxylic acids as directing groups and further functionalization.

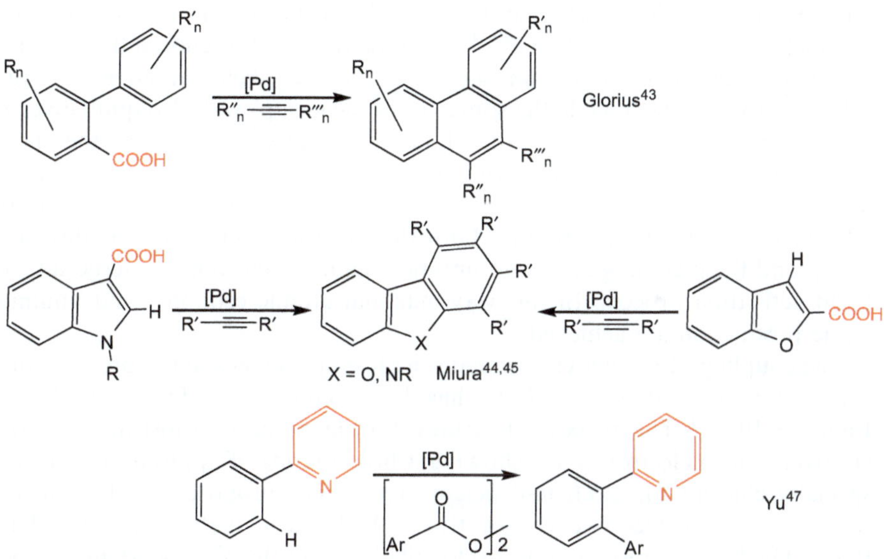

Figure 8.9 The carboxylate as a traceless directing ligand and carrier of the aryl group.

that of aldehydes, always keeping in mind that the imine is a masked form of an aldehyde.[48]

In this introductory section about directing groups we have detailed some aspects of their influence, but there are others. For the sake of clarity, these examples will be treated on each particular metal and on each specific reaction. In the following paragraphs we will detail the most relevant and recent facts on the catalytic functionalization promoted by palladium complexes and, in a second section, processes catalysed by copper, rhodium and ruthenium derivatives.

8.3 Palladium Catalysis Based on Coordination-Directed Metallation

8.3.1 Arylation

The occurrence and importance of the biaryl moiety in natural products and in many compounds with biological and pharmaceutical activity are the main reasons for the extensive research developed in arylation reactions. This is probably one of the most studied processes among the functionalizations, due to their deep implications in the type and nature of the accessible compounds. Previous works were always based on cross-couplings between pre-functionalized substrates (route A in Figure 8.1), including very useful processes such as Heck, Suzuki, Sonogashira, Kumada or Negishi couplings.

In contrast, scarce activity was observed in the field of oxidative C–C coupling. Pioneering work in Pd-catalysed directed arylations involved the reaction of robust substrates (phenylpyridine, benzylamines, benzamides, among others) with a proper source of aryl moieties in the presence of stoichiometric amounts of oxidants.[49–53] The oxidant could be chosen from a large variety of compounds (iodonium salts,[49] oxone,[50] O_2, benzoquinone[53] or Ag^+ salts,[50,52] for instance) while the source of arylic groups was also quite diverse, for example aryl halides,[51] diaryliodonium salts,[49] or even the non-prefunctionalized aromatic molecule itself.[16,17,50,52] It is evident that the most interesting option is the last one, also called "tandem directed arylation" or "2-fold C–H activation", because both the substrate containing the directing group and the incoming arene are unfunctionalized, and both undergo a C–H bond activation process. In this way maximal atomic economy and minimal waste production are achieved.[17]

The coupling of *O*-phenylcarbamates with simple arenes in the presence of a very common oxidant, $Na_2S_2O_8$, has been reported by Dong *et al.* (see Figure 8.10a).[54] The reaction tolerates a wide range of substituents, from electron-rich to electron-poor, showing a broad scope of application, but it is specially efficient for electron-deficient arenes. The proposed mechanism for this reaction involves two consecutive C–H bond activations promoted by Pd(II). The first one results in the cyclopalladation of the phenylcarbamate, and the second one is produced at the arene by S_EAr. The reductive elimination by C–C coupling generates the expected product and the reoxidation of the Pd(0) to Pd(II) induced by $Na_2S_2O_8$ closes the catalytic cycle. The method has been extended to other substrates such as phenylacetamides, benzamides and anilides (Figure 8.10b),[55] although in this case only with electron-rich arenes. An intramolecular variant gives access to phenanthridinones, an interesting class of compounds (Figure 8.10c).[55] However, in the latter cases (b, c), the mechanism of the reaction seems to involve Pd(III) or even Pd(IV) intermediates, showing that different directing groups can change dramatically the reaction mechanism.[56–62] See Chapters 4 and 5 for detailed information on Pd(III) and Pd(IV) compounds.

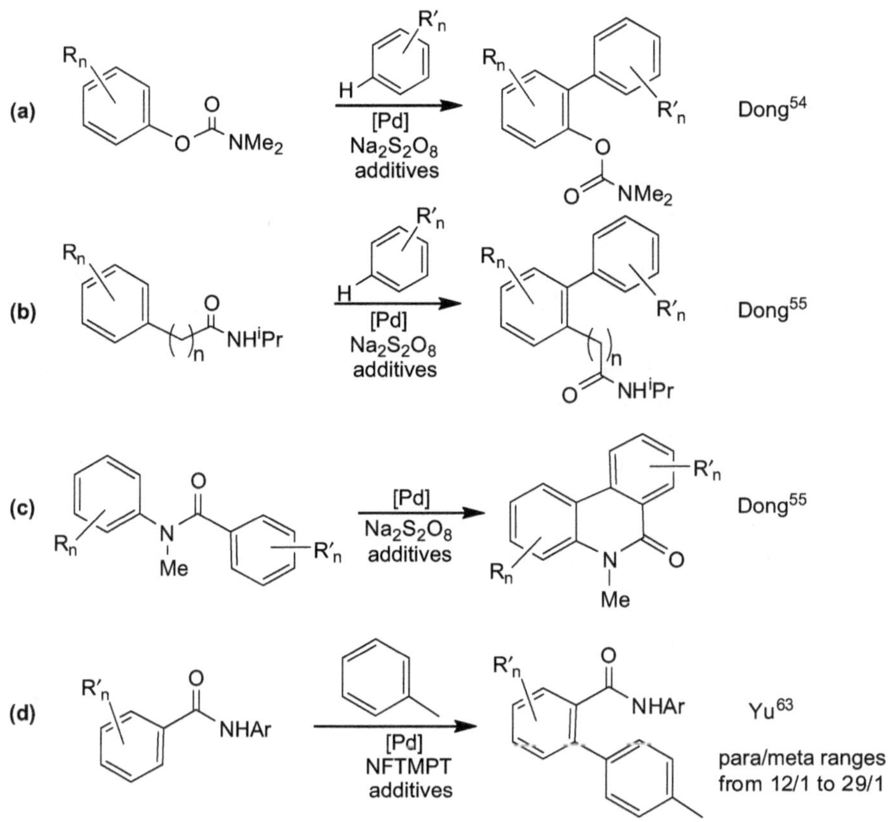

Figure 8.10 Oxidative coupling of arenes through double C–H bond activation.

An additional problem frequently encountered in arylation processes through unfunctionalized arenes is the selectivity of the position of the arene at which C–H bond activation takes place, for instance in mono-substituted arenes. While in the substrate to be functionalized this is achieved by using the directing group, there is not a general strategy for the arene, which behaves as the coupling partner. An outstanding contribution of Yu's group shows that the judicious choice of the oxidant allows a high selectivity in the arylation of benzamides with toluene, reaching values of selectivity *para*/*meta* up to 29/1 (Figure 8.10d).[63] The best oxidants in terms of yield and selectivity were NFTMPT (1-fluoro-2,4,6-trimethylpyridinium triflate) and NFSI (N-fluorobenzenesulfonimide), both of them acting as sources of electrophilic F^+. The authors have shown that, in fact, the presence of F^+ is crucial to attain high values of *para* selectivity. The method is valid not only for toluene, but also for other mono-substituted arenes, containing alkyl, alkoxy and halides as substituents.

In the same line, the selectivity in the functionalization of benzo[*h*]quinoline (bzq) with 1,3-dimethoxybenzene, promoted by Pd(II) complexes, has been studied.[64] In principle, the three isomers, A, B and C shown in Figure 8.11, can be expected, but the reaction is strongly sensitive to the changes in the reaction conditions.

A detailed mechanistic study, based on previous results,[65] shows that the modification of the selectivity is related to the change of the rate-determining step, which could be the quinone complexation or the C–H bond activation. It must be noted that a few homocoupling processes have been reported,[66] although they are scarcely developed.

The 2-fold C–H bond activation reactions are not the only synthetic pathways described for arylation of organic substrates. One of the most common methods deals with the use of aryl halides as the source of the aryl moiety. The synthesis of phenanthridinone derivatives can be accomplished either by tandem directed activation (see Figure 8.12a, right)[67] or by coupling of the corresponding benzamide with a variety of aryl halides (Figure 8.12a, left).[68] Notably, in both cases, the directing groups not only orientate the reaction, but also are part of the target structures. In the case of the 2-fold C–H activation the reaction proceeds with good yields for a large variety of substituents, both at the benzamide and the arene partners, while in the case of

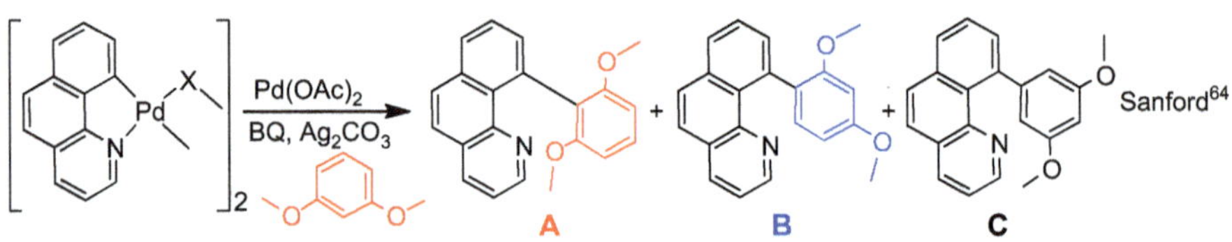

Figure 8.11 Functionalization of benzo[*h*]quinoline with 1,3-dimethoxybenzene.

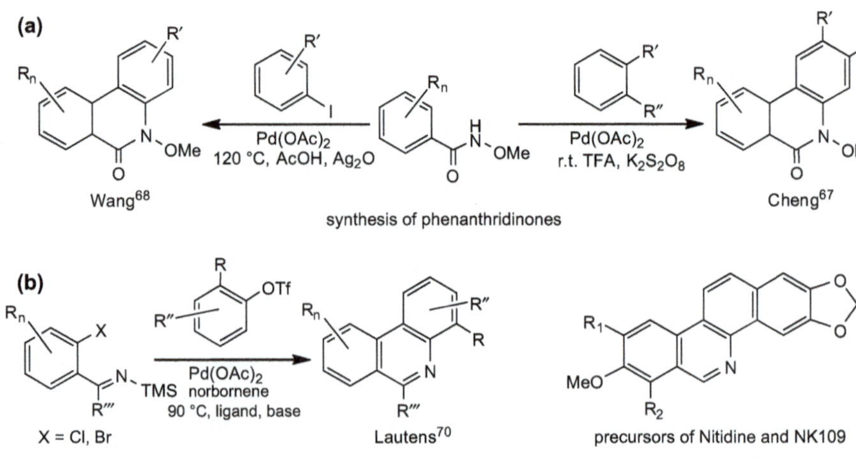

Figure 8.12 Synhtesis of phenanthridinones and phenanthridines, and some applications.

the aryl halides only the iodides proved to be efficient, bromides or chlorides being unreactive.

Another interesting class of compounds, such as phenanthridines, can also be obtained following this strategy, starting from anilides and aryl iodides.[69] An alternative synthesis involves the use of *o*-substituted halo-imines and *o*-substituted aryl triflates, as shown in Figure 8.12b.[70] This reaction, based on the Catellani coupling mediated by norbornene,[71] is particularly useful because it provides access to products with well-established biological activity. The use of aryl triflates instead of aryl iodides is key for the reaction.[70] It seems that the only requisite of the aryl triflate to work properly is to have a substituent *ortho* to the triflate group, because the *ortho* substituents prevent unwanted reductive eliminations and drive the whole reaction through the formation of Csp^2–Csp^2 bonds.[71] In this respect, it is worth noting two recent contributions from the groups of Catellani, Malacria and Lacôte, which show that the *ortho* effect has exceptions, and that Csp^2–Csp^3 couplings can occur, leading to interesting derivatives based on the dihydrophenanthrene (Figure 8.13a)[72] or 5*H*-dibenzo[*b*,*f*]azepines cores (Figure 8.13b).[73]

Although we have stated that the most interesting reagents are, in principle, those which do not contain functional groups at the reactive positions, the most common situation is the use of functionalized molecules as aryl carriers, and some examples are presented in Figure 8.14a–e. In these cases aryl radicals,[74] silanes,[75,76] iodonium salts[77,78] or arylboronic acids or salts[48,79–81] are attractive choices, in addition to the aryl halides.[68–73,82,83] For instance, free oximes can be arylated using aryl radicals (example a in Figure 8.14), modified cyclic enamides are easily obtained from arylsilanes (b), phenol esters from iodonium salts (c, compare with the method of Dong *et al.* in Figure 8.10), functionalized cyclopropanes can be prepared stereoselectively from boronic derivatives (d), and even nucleosides such as purine are adequately transformed using aryl iodides (e).

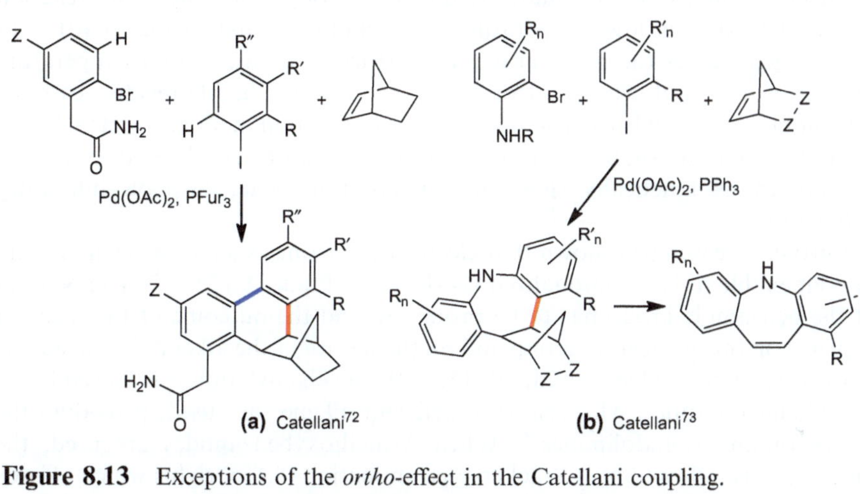

Figure 8.13 Exceptions of the *ortho*-effect in the Catellani coupling.

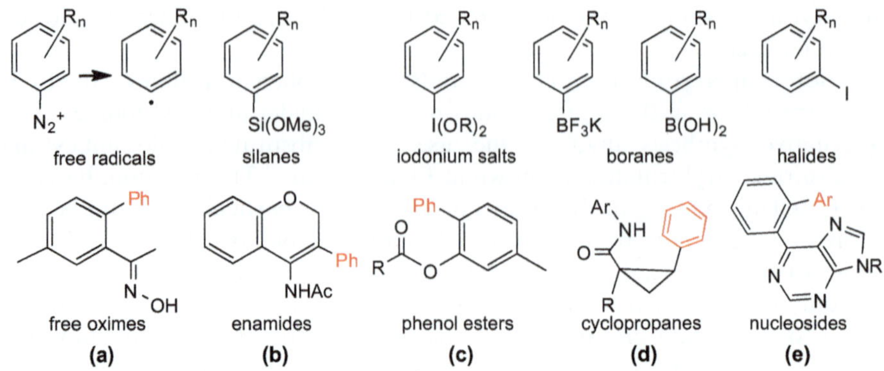

Figure 8.14 Different sources of aryl moieties and derived products.

8.3.2 Alkenylation

The Fujiwara–Moritani reaction[84] is related to the arylation process and possesses an outstanding importance from a synthetic point of view, because, in contrast with the genuine Mizoroki–Heck reaction, unfunctionalized substrates are demanded. In spite of this practical importance, general routes achieving Pd-catalyzed C–H olefination are still in their infancy because of several synthetic restraints.[39] Among them, we have to remark the lack of selectivity in multi-substituted arenes, the need of electron-rich substrates, or the use of harsh reaction conditions in most of the cases. In this respect, several improvements have been produced, as shown in Figure 8.15a–e.

The use of the traceless silanol or *N*,*N*-dimethylaminomethyl directing groups provides easily removable moieties, keeping the full regiocontrol of the reaction and, moreover, affording synthetically useful products from the monoinsertion of alkenes (Figures 8.15a and b).[85–87] In addition, the use of electron-rich ureas or acetanilides as directing groups allows for mild reaction conditions, such as low temperatures (c)[88] and/or benign oxidants (d).[89] In some specific cases the reaction can be performed even at room temperature (d)[89] in water,[90] but using electron-rich substrates and highly reactive cationic Pd complexes as catalysts. The use of water as reaction medium is remarkable, since the reaction usually needs acids as additives to be accelerated, otherwise the process is kinetically slow, and in addition water displays inhibitory behaviour.[90]

Further research has shown that electron-poor rings such as benzamides are also amenable to be functionalized, as shown in Figure 8.15e.[91–93] The N atom of the benzamide group has to be protected, and the outcome of the reaction depends on this protecting group and on the nature of the alkene. For instance, in the case of N-tosylbenzamides (8.15e, left) the alkenylation is followed by an aza-Michael addition when electron-deficient alkenes are used, providing the corresponding isoindolinones.[91] When *N*-methoxybenzamides are used, the olefination is followed by a Wacker-type amination (8.15e, right) which affords

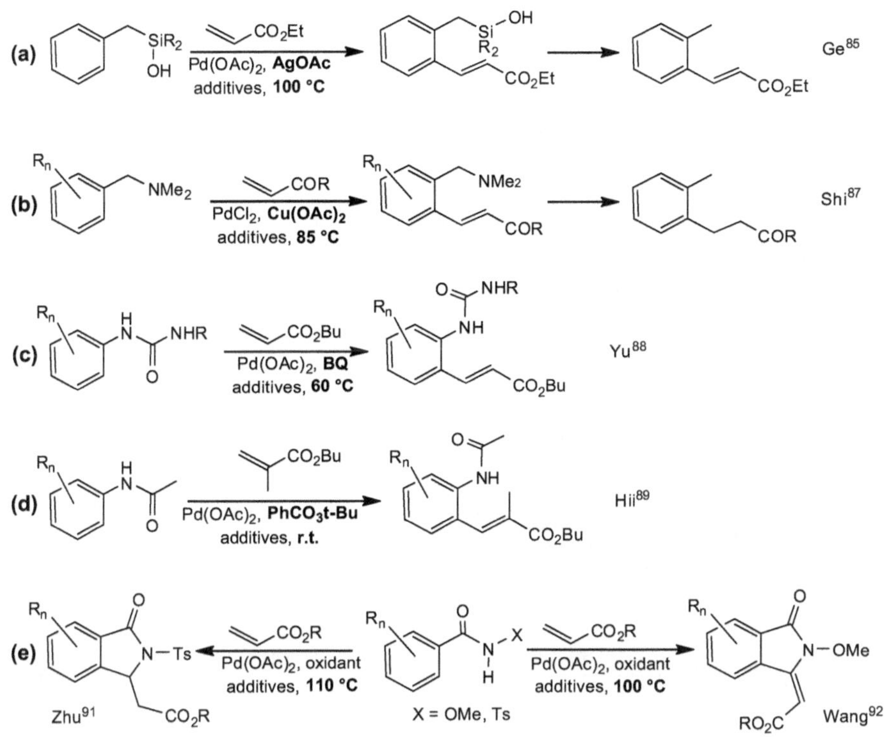

Figure 8.15 Recent improvements in the Fujiwara–Moritani reaction.

regioselectively the *E*-exocyclic C=C bond.[92] Using the same protocol a new synthesis of phenanthridines has been reported.[93]

Undoubtedly, the Yu's group is the most active in the field of the directed olefination of arenes currently. Remarkable examples are presented in Figure 8.16, such as the enantioselective olefination of diphenylacetic acids with styrenes and acrylates using protected α-amino acids as chiral auxiliary ligands.[94] The desymmetrized products are obtained with *ee* close to 100% in some cases (typically in the range 70–90%), although harsh reaction conditions are used (90 °C, 48 h). The reaction is tolerant to a wide array of functional groups on the diphenylacetic acid and the styrene (Figure 8.16a).

The olefination of alkyl derivatives has also been reported.[95] Figure 8.16b shows how to transform *N*-arylpivalamides into γ-lactams by olefination of the pivalamide with acrylates and, in a second step, 1,4-addition of the amide to the acrylate. The method is valid not only for compounds with quaternary atoms, but also for species with α-H atoms, just by changing the X group at the N atom. The yields are moderate to good and the conditions are not as exceptional as anticipated (DMF, 120 °C).

The use of amino acids as auxiliary ligands has provided unexpected results in the modification of weakly coordinating substrates such as primary,

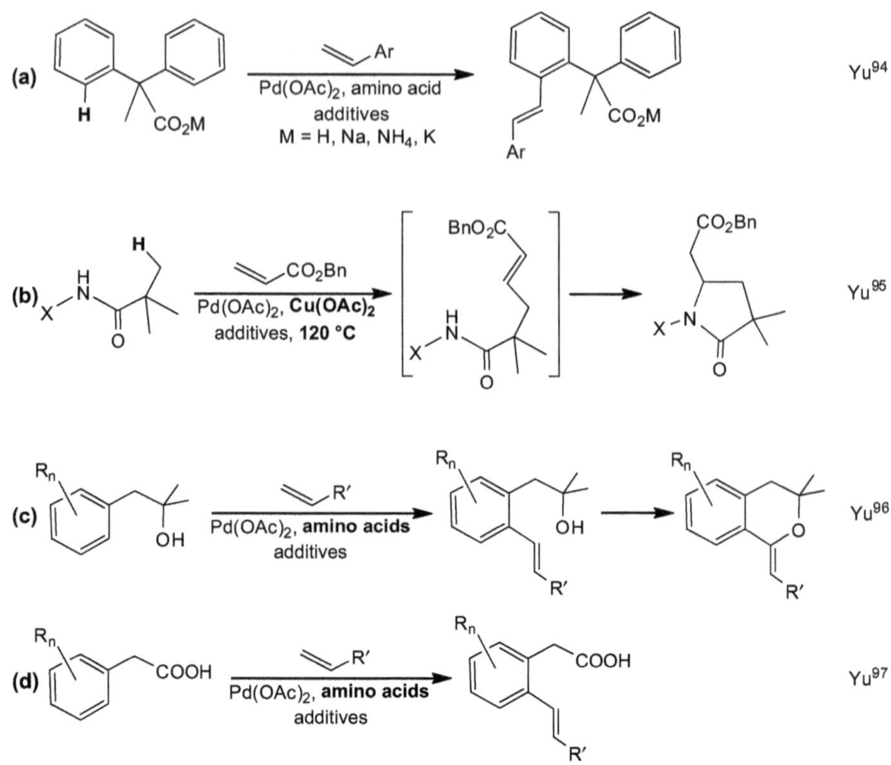

Figure 8.16 Representative examples of alkenylations performed in Yu's group.

secondary and tertiary alcohols (Figure 8.16c).[96] In the absence of amino acids the olefination occurs with low or no conversion, while in the presence of a variety of amino acids the reaction reaches a 90% yield. The process works efficiently with diverse substituents at the aryl ring, but better results were obtained with electron-releasing groups and, while the best yields are achieved with tertiary alcohols, primary and secondary alcohols also affords notable conversions. More interestingly, the type of final products depends on the type of olefin: the simple stilbenes shown in Figure 8.16c are obtained when R′ is an aryl group, but the final pyran derivatives are obtained when an acrylate is used as starting compound. Intriguingly, a vague mention of the precise role of the amino acid is made. These studies show how the coordination of the amino acid promotes a notable acceleration of the reaction, and that the origin of this phenomenon is a plausible change of the mechanism of the C–H bond activation step from the classical S_EAr when the amino acid is not present to a CMD mechanism (concerted metallation deprotonation) when the amino acid is coordinated. The application of this methodology to the synthesis of natural products and compounds with pharmacological activity has been successful.[97]

8.3.3 Reactivity with Alkynes

Conversely, while the dehydrogenative coupling of arenes and alkenes has proved to be a very fruitful reaction, the coupling of arenes and alkynes through Pd-promoted C–H bond activation is still far from being fully successful. Of course, this statement needs to be explained. There are numerous examples of *stoichiometric* reactions between palladacycles and internal alkynes, processes which afford a variety of carbo- and heterocycles.[98] There is also a huge amount of work, mostly developed by the group of Larock,[99] but also by other groups, which uses *halo-derivatives* as starting materials. In these cases the processes are catalytic, but the substrates need to be prefunctionalized. Recent work of M. Miura and N. Jiao reports the synthesis of a variety of heterocycles by reaction of indole derivatives with internal alkynes, but in these examples a *heterocyclic position* is involved in the cyclization, therefore they are out of the scope of this chapter.[100,101] We can find genuine examples of a directed arylic C–H bond activation, followed by the insertion of an alkyne into the M–C bond *in other metals* (see the section devoted to the rhodium and ruthenium), but not in palladium.

The only example found where an orthopalladation, produced through C–H bond activation, is followed by a migratory insertion to afford organic species (in this case isoquinolines and related species) has been very recently reported by Huang *et al.*[102] This process is shown in Figure 8.17a and, as it can be shown, it is not strictly a genuine orthopalladation since two bond activations have occurred (C–H and N–H) and the orthometallated ligand behaves as a dianion. There are two additional examples, shown in Figure 8.17b and c, which display a formal analogy with the expected processes. The first one accounts for the synthesis of indoles from anilines and alkynes (Figure 8.17b), where the amino moiety assumes the role of directing group. However, the analysis of the reaction mechanism shows that the initial step is the hydroamination of the alkyne by the aniline, and that the subsequent C–C coupling on the vinyl-aryl amine intermediate forms the final indol.[103] In principle, it seems that similar considerations could apply for the example shown in Figure 8.17c. The assumption that pyridine acts as directing group, due to its good coordinating ability, is not clear because it seems to be not necessary in these cases, although a detailed mechanism is not provided.[104]

In contrast, the alkynylation of arylic derivatives using alkynyl halides proceeds very efficiently, as shown by Chatani's group (Figure 8.17d). The incorporation of alkynyl groups at specific positions has been carried out by reaction of substituted benzamides or anilides anchored to different directing groups.[105] It seems that a pre-requisite for the proper behaviour of the reaction is that the directing group acts as a chelating fragment.

8.3.4 Carbonylation

Carbon monoxide, as well as alkenes and alkynes and other unsaturated molecules, is able to react with orthopalladated complexes through insertion

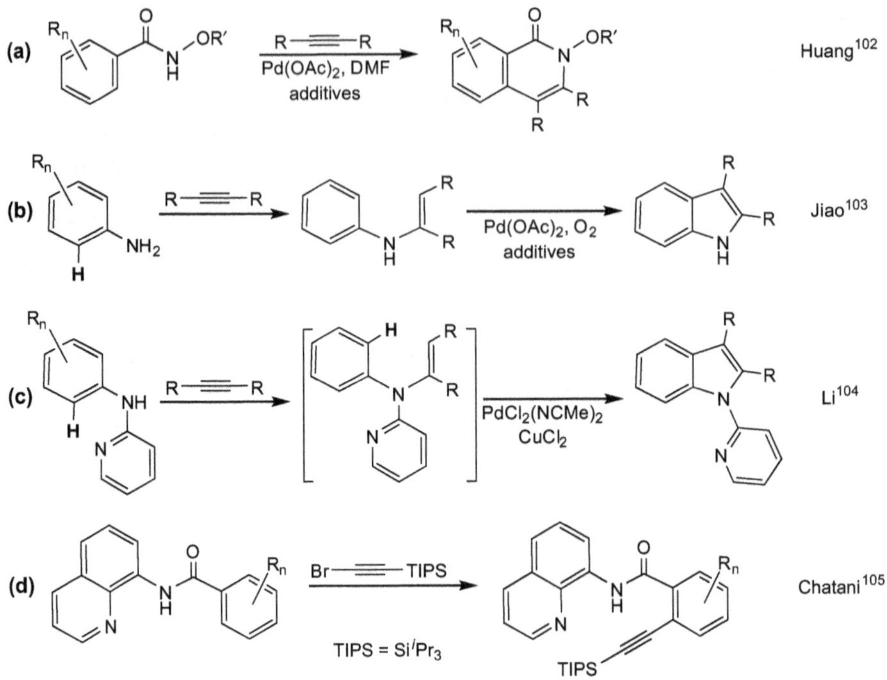

Figure 8.17 Diverse functionalizations by reactions involving alkynes.

into the Pd–C bond.[106] Orito has reported the carbonylation of *N*-alkyl-ω-arylalkylamines, catalyzed by $Pd(OAc)_2$, which affords the corresponding benzolactams under mild conditions. Both five- and six-membered heterocycles can be obtained, but the benzylic amines cyclize much faster than the phenethylamines, as shown in Figure 8.18a.[107] The presence of substituents with different electronic effects is not a critical parameter for the orientation of the reaction. The method works efficiently for secondary amines, while, if primary amines are used, ureas are obtained instead.[108] The further work of Gaunt's group also describes the necessity of a substituent at the N atom in order to achieve the catalytic functionalization of β-arylethylamines (Figure 8.18b),[109] and a recent report of Lloyd-Jones *et al.* about carbonylation of alkoxybenzamides points out in the same direction.[110]

Interestingly, in spite of these studies, the catalytic carbonylation of *N*-unprotected quaternary amino esters derived from phenylalanine to give benzolactams has been achieved by the group of Granell *et al.*[111] The process is shown in Figure 8.18c. The quaternary nature of the Cα of the amino ester is critical in order to achieve the best yield and the optimal ratio benzolactam/acetamide. On the other hand, the reaction is quite sensitive to the size of the benzolactam formed. Thus, when quaternary esters of phenylalanine are used, good yields of six-membered benzolactams are obtained, while no lactams were detected under the same catalytic conditions but starting from the closely related esters of phenylglycine, this result being in clear contrast to the process

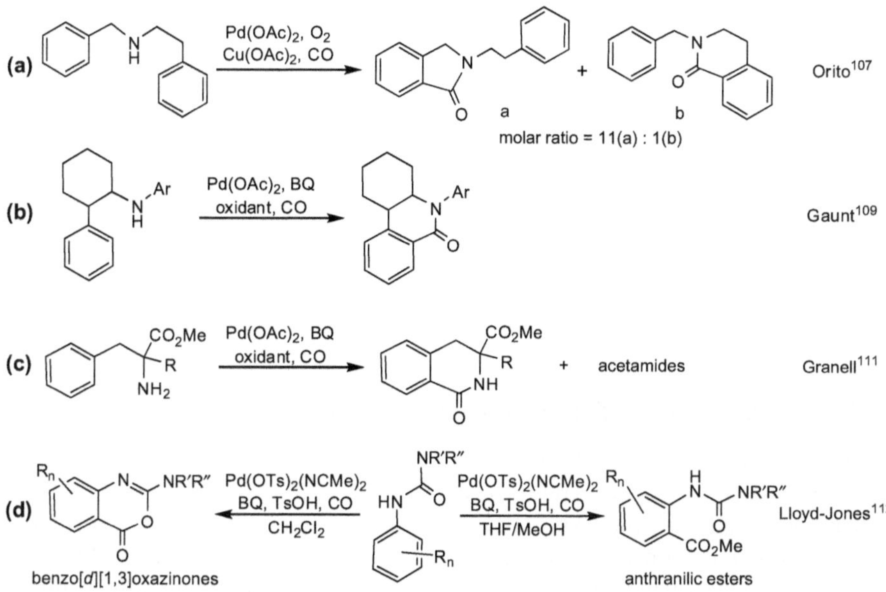

Figure 8.18 Catalytic carbonylation processes using different directing groups.

reported by Orito (Figure 8.18a). A fast equilibration between 5- and 6-membered orthopalladated intermediates and a higher reactivity of the latter towards CO has been invoked to explain these observations. *O*-Substituted ligands can also undergo carbonylation processes. The treatment of *m*-toluidine ureas with CO, catalyzed by Pd(II) complexes, gives anthranilic esters or cyclic imidates as a function of the reaction solvent,[112] as shown in Figure 8.18d. However carbonylation of *N*-aryl-alkylamides proceed with Csp^3-H activation and formation of the corresponding succinimides, which can be converted into useful 1,4-dicarbonyl derivatives.[113]

In addition to the direct reaction with CO gas there are other methods to produce the directed carbonylation of a given substrate, but avoiding the use of this very toxic reagent. Usually the source of CO is an aldehyde, but it can be another reagent. The acylation of acetanilides has been carried out by treatment with aromatic aldehydes in presence of TBHP (*tert*-butyl hydroperoxide), catalyzed by Pd(TFA)$_2$, as shown in Figure 8.19 (left).[114] The same type of compounds can be obtained through a slightly different pathway, shown in Figure 8.19 (right), by reaction of the acetanilides with the corresponding substituted benzyl alcohols, under milder reaction conditions.[115] This last reaction can be considered formally analogous to the former one, since it is proposed that the alcohol is oxidized to aldehyde before it is incorporated to the catalytic cycle. Similar results have been reported by other authors using phenylpyridine as substrate.[116,117] A different approach has been described by Shi,[118] who reports the use of strained oxaziridines for the ethoxycarbonylation of phenylpyridine derivatives.

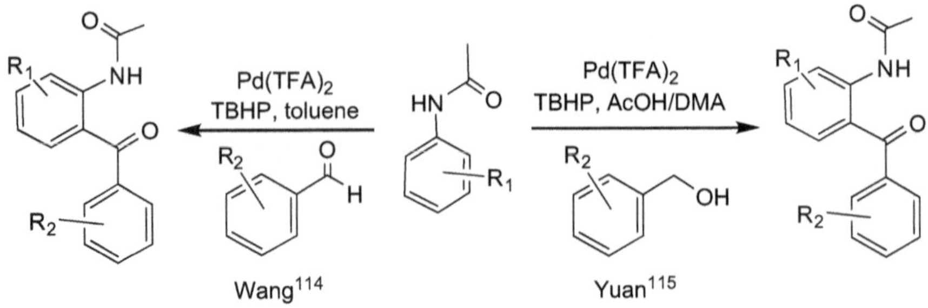

Figure 8.19 Carbonylation using aldehydes or alcohols as CO sources.

8.3.5 Alkylation and Perfluoroalkylation

In the preceding paragraphs we have stated the difficulty of functionalizing Csp^3–H bonds in the substrate to be modified (see, for instance, Figure 8.16b), when compared with the Csp^2–H bonds.[119] The higher inertia of the former is usually translated into harsher reaction conditions and/or lower yields. The same applies when we try to introduce a Csp^3-based fragment into a target molecule, for instance when alkylation of aryl rings is attempted.[119] In this case, it is extremely difficult to promote the coupling of the two molecules through C–H bond activation and, usually, one of the species has to be activated in the form of halides or borates, for instance. Representative examples are shown in Figure 8.20a–b.

The alkylation of a series of picolinamide-protected benzylamine derivatives has been accomplished by their reaction with a large variety of alkyl halides in presence of a halide scavenger (NaOTf) under strong reaction conditions (pure O_2 atmosphere, 135 °C).[120] The use of the picolinamide-protecting group seems to be critical for the correct development of the reaction, and can be easily removed. The products, shown in Figure 8.20a, are interesting precursors for valuable products, such as tetrahydroisoquinolines. On the other hand, aryl derivatives of the easily removable directing group 2-pyridylsulfinyl (see a related removable group in Figure 8.7) can be methylated by reaction with boron compounds in the presence of *p*-benzoquinone, as presented in Figure 8.20b.[121] In the same context, the introduction of trifluoromethyl groups into organic scaffolds is a hot area of research in the synthesis of pharmaceuticals for the unique properties that it confers to the molecule (widely discussed in Chapter 8.7). Recent advances in this area use reagents able to generate the highly reactive "CF_3^+" moiety, such as I(III) or sulfonium salts, as shown in Figure 8.20 (c–d). Phenylpyridine can be trifluoromethylated very efficiently by reaction with sulfonium salts (Figure 8.20c),[122] although under harsh reaction conditions ($ClCH_2CH_2Cl$, CF_3CO_2H, 110 °C, 48 h). Heterocycles such as pyrimidines, imidazoles or thiazoles also behave as adequate directing groups.[122] Interestingly, the use of a similar ligand, benzo[*h*]quinoline, but in different reaction conditions and using different CF_3 sources, allows for the isolation and characterization of a mononuclear Pd(IV) derivative

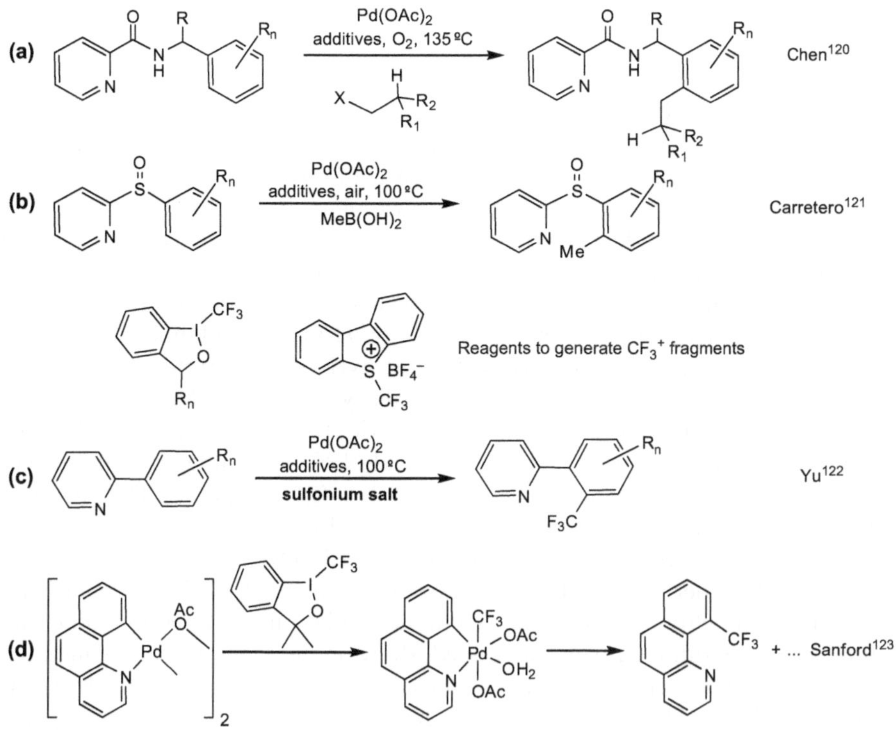

Figure 8.20 Reactions of alkylation and trifluoromethylation.

(Figure 8.20d).[123] This Pd(IV) compound is a true intermediate in the catalytic reaction, which is remarkable because related dinuclear acetate-bridging Pd(II)–Pd(II) systems studied by Ritter *et al.* allow for the synthesis of the corresponding dinuclear Pd(III)–Pd(III) derivatives, also kinetically competent, in the presence of iodonium salts.[56,57,124] In this particular case, several reductive eliminations can occur (C–C *versus* C–O coupling) from the Pd(IV) intermediate, but a judicious choice of the solvent drives the reaction almost totally to the desired C–CF_3 coupling.

8.3.6 C–B, C–O, C–S and C–N Bond Formation

Although the directed formation of C–C bonds, catalyzed by Pd species, is by far the most developed topic in this area, the formation of C–X bonds, where X represents a heteroatom (usually B, O, N or halogen), has also undergone an spectacular development during the last years.[125]

Aryl boronic acids or aryl borates are key intermediates in relevant cross-coupling processes (Suzuki reactions, for instance). Therefore, the borylation of arene substrates through C–H functionalization is a fundamental reaction and numerous efforts have been made in this direction, mainly by the group of Hartwig using Ir complexes.[126,127] The use of Pd complexes to achieve this task is limited to a few cases.[127] Very recently, the group of Yu has reported the oxidative borylation of *N*-arylbenzamides in MeCN using $B_2(pin)_2$ (pin = pinacol); a process catalyzed by $Pd(OAc)_2$ *via* Pd(II)/Pd(0) redox steps, in the presence of $K_2S_2O_8$ as oxidant, NaOTs as base and a dba-based ligand (dba = dibenzyleneacetone).[128] A variety of benzamides have been borylated at the gram scale, even with strongly electron-withdrawing substituents.

The oxidation of C–H bonds with concomitant C–O bond formation by directed oxidative coupling is now a well established topic.[16,30] Recent outstanding contributions cover the use of benign "green" oxidants, traceless directing groups, delicate multifunctional starting materials or unexpected cyclizations.

Vedernikov has reported the use of O_2 or H_2O_2 as oxidants for the acetoxylation or hydroxylation of benzylic and pyridinic substrates, as shown in Figure 8.21a and b, using hemilabile ancillary ligands.[129] These ligands are substituted di(2-pyridyl)sulfonates or di(2-pyridyl)ketones, and their role is the stabilization of Pd(IV) intermediates by facial N,N,O-bonding, and the further promotion of the reductive elimination by C,O-coupling. The benzylic derivatives are more reactive, and can be oxidized by molecular O_2, while pyridinic derivatives are more resistant and need H_2O_2 to be converted into the reactive Pd(IV) derivatives. The processes shown in Figure 8.21 represent stoichiometric functionalizations, but catalytic oxidations of 8-methylquinolines and 2-benzylpyridines are also described, showing the generality of the method.

The use of traceless directing groups is an attractive synthetic strategy. Recently, the use of silanols as traceless directing groups has been applied to the synthesis of industrially important catechols, process shown in Figure 8.21c. The method developed by Gevorgyan *et al.* allows the synthesis of a broad scope of substituted catechols, including those with electron-deficient substituents.[130] Chelating protecting groups such as picolinamides or aminoquinolines have also been used in the highly efficient acetoxylation of different benzyl or benzamide functional groups, as shown in Figure 8.21d.[131] The key feature of these bidentate systems seems to be the formation of two-fused 5-membered rings at the time of the cyclopalladation, and the stability of this arrangement. Wang and coworkers have achieved the alkoxylation of *N*-methoxy-benzamides (Figure 8.21d, right) after intensive screening of different *N*-substituents.[132] It is remarkable that in this case the *N*-methoxy group does not behave as an internal oxidant (see, for instance, Figure 8.5). This methodology can also be applied, under certain conditions, to C–O bond formation when dealing with delicate substrates, for instance those containing multiple functional groups or chiral centers. In this context, the acetoxylation and alkoxylation of several C6 and C2-arylpurines, very relevant from the point of view of the medicinal chemistry, has been reported by the group of Qu.[133] On the other hand, the acetoxylation of phenethylamines has been described by Yu *et al.*[134] These reactions use *t*-butylperoxyacetate as a mild oxidant, MeCN as additive, and can be applied to substrates such as aminoesters of phenylalanine, as shown in Figure 8.21e. Furthermore, the resulting acetylated derivative is still reactive enough to undergo an intramolecular C–H amination furnishing the corresponding indoline.

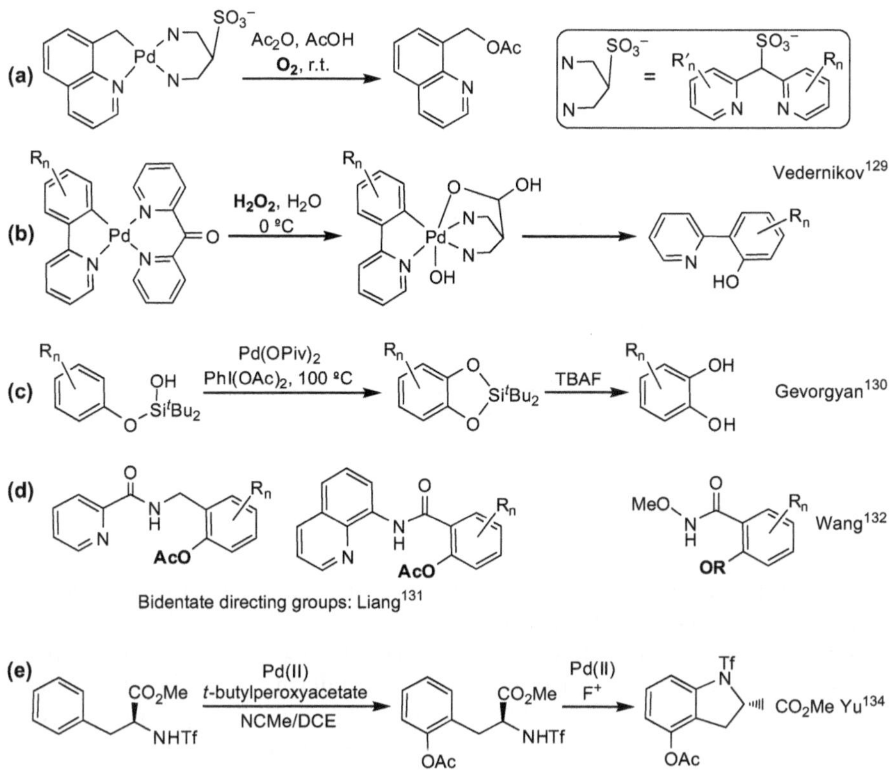

Figure 8.21 Different intermolecular C–O bond formations.

The C–O bond coupling can also occur intramolecularly, affording very interesting oxygen-based heterocycles. In these cases, the directing groups remain in the target molecules, and have to be designed accordingly. The intramolecular cyclization of arylsubstituted tertiary alcohols affords dihydrobenzofuranes, as shown in Figure 8.22a.[135] The use of alcohol functional groups as directing moieties is still scarce, but it is clear that they perform efficiently this task (see also Figure 8.16c). The choice of the oxidant is critical, and only $PhI(OAc)_2$ gives good results. Under these conditions, alcohols with both electron-withdrawing and electron-releasing groups at the aryl ring have been cyclized, and the method has been further applied successfully to the synthesis of natural products. Along the same lines, the cyclization of arylphenols and other phenol derivatives afford dibenzofuranes, as shown by two different groups at almost the same time. The reactions are presented in Figure 8.22b.

The work reported by Liu uses air as the simple oxidant,[136] which is very convenient as a green reagent, but also uses quite specific ligands such as the *N*-heterocyclic carbene IPr and diazafluorenone, in addition to base and another additives to promote the C–O reductive coupling. It is remarkable that

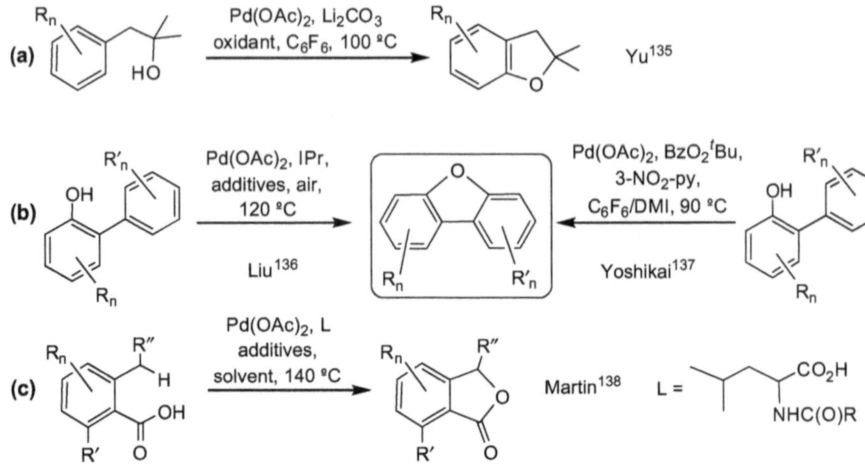

Figure 8.22 Intramolecular cyclizations through C–O bond formation.

the reaction is totally regiospecific, when several activations are possible, and that this specificity is controlled by steric factors. On the other hand, the work reported by Yoshikai[137] describes an alternative access to the same type of compounds (see Figure 8.22b). In spite of the similarity of the two processes it seems that the mechanisms are different: the C–H activation step appears as the limiting step in the Yoshikai work, while that for the mechanism proposed by Liu is the C–O bond formation. The cyclization of organic species by the tandem process C–H activation/C–O bond coupling can involve functional groups others than alcohols. The carboxylic acid is a paradigm of this multi-task behaviour, as described previously. The intramolecular coupling of a carboxylic acid and an ortho-alkyl group gives the corresponding benzolactones, as shown in Figure 8.22c.[138] It is remarkable that the reaction implies a Csp^3–H bond activation. A critical parameter for the success of the reaction is the presence of a *N*-protected amino acid, in concordance with the results reported by Yu *et al.* for Csp^2–H bond activation and alkenylation (see Figure 8.16). The results were particularly good when *N*-acetyl-leucine was used (R = Me), although small changes of the R group dramatically alter the reaction yield and the product distribution. This protocol tolerates a variety of substituents and provides access to benzolactones not available by conventional routes.

In contrast with the well-developed C–O coupling process, the Pd-mediated directed C–S coupling is still very scarcely developed. The group of Dong has reported the only example of intermolecular C–S bond formation through C–H bond activation, not only with Pd, but with any transition metal.[139] The reaction of arylsulfonyl chlorides with substituted phenylpyridines is catalyzed by $PdCl_2(NCMe)_2$, giving the corresponding aryl-sulfonated materials. However, the reaction is sensitive to the solvent and some additives, and alternative elimination pathways can occur when the reaction conditions are changed. Further studies of the same group have been centered on the C–S bond forming step by reductive elimination, showing that the C–S coupling occurs from a Pd(IV) species.[140] The electron-rich complex $[Pd(phpy)_2]$ (phpy = 2-phenyl-pyridine) reacts with sulfonyl chlorides RSO_2–Cl resulting in the formation of stable Pd(IV) derivatives by oxidative addition of the Cl–S bond to the Pd(II) center. Although a clear mechanistic picture is missing, heating of these complexes promotes the C–S coupling by reductive elimination to Pd(II), showing that this elimination is favoured over C–C, C–Cl or even desulfitative C–C coupling, mainly in presence of Ag^+ salts.[140]

Recently there have been improvements in the still difficult field of intermolecular C–N bond formations. Directed C–H amination has been achieved by the group of Yu.[141] Treatment of substituted benzamides with secondary amines and an oxidant (typically a peroxide) or, alternatively, with *O*-benzoylhydroxylamines, gives the corresponding 2-aminobenzamides (see Figure 8.23a). This is worthy to note because only catalytic amidation of C–H bonds were reported until this contribution, with the amination only achieved through the Buchwald–Hartwig protocol from functionalized precursors. Moreover, the reaction is tolerant to a broad scope of substituents in the aryl ring and, although it is limited to secondary amines, notably improves previous results. The mechanism of this reaction is not clear, and proposals based on the Pd(0)/Pd(II) and Pd(II)/Pd(IV) manifolds are reported. On the other hand, a broad scope amidation of Csp^3–H bonds has been achieved by Muñiz *et al.*[142] In this respect, 8-methyl-quinolines, *tert*-butylpyridines and 2-methylarylethers have been successfully amidated by reaction with $FN(SO_2Ar)_2$ (Figure 8.23b); the reaction catalyzed by $Pd(hfacac)_2$ (hfacac = hexafluoro-acetylacetonate) in the case of the quinolines and pyridines, and by a mixture of $Pd(OAc)_2$ and bathocuproine in the case of the ethers. Mechanistic studies showed that the limiting step is the C–H bond activation process, that $FN(SO_2Ar)_2$ behaves as a fluorinating reagent for palladium (instead of amidating), forming Pd(IV)

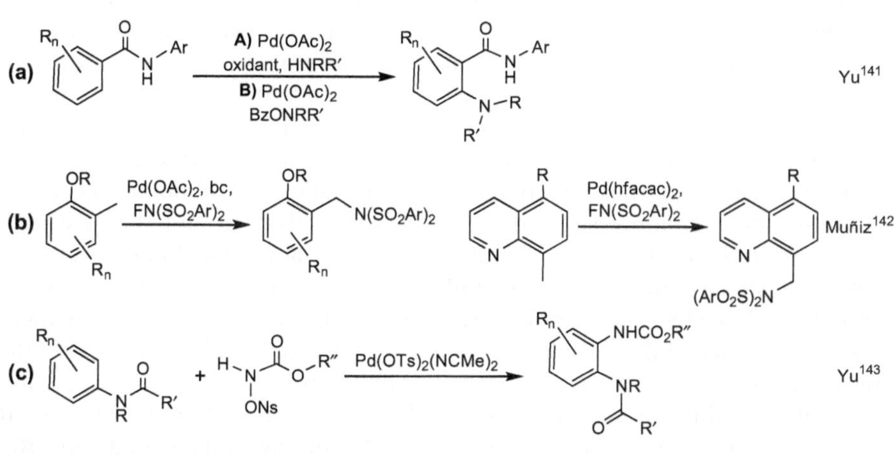

Figure 8.23 Directed intermolecular amidations catalyzed by palladium.

intermediates, and that the C–N coupling occurs by nucleophilic attack of the imide anion to the palladated carbon (nucleophilic substitution).[142]

The intermolecular *ortho*-amidation of a variety of anilines has also been carried out (Figure 8.23c) using a different strategy, as is the generation of nitrenes from *N*-nosyloxycarbamates.[143] In a first step, the anilide is ortho-palladated, and the resulting complex is allowed to react with different *N*-nosyloxycarbamates. In the reaction conditions, it is likely that the decomposition of the *N*-nosyloxycarbamates with concomitant formation of nitrenes takes place. Further reaction of the nitrenes with the orthopalladated Pd–C bond in different ways affords the final ortho-bisamido derivatives, which are very interesting from a pharmacological point of view.

The intramolecular version of the C–N coupling is useful in most cases for the synthesis of interesting heterocycles, as we have shown in some examples throughout this chapter (see, for instance, Figures 8.17 and 8.18). In this context, there are new synthetic concepts worth citing here. The seminal work of Buchwald about the synthesis of carbazoles by intramolecular C–N coupling on 2-aminobiphenyl derivatives substituted with *N*-electrowithdrawing groups has been expanded and improved.[144,145] Buchwald himself reports a method tolerant with all types of functional groups, just replacing the oxidant $Cu(OAc)_2$ by DMSO (see Figure 8.24a), although the reaction only works under harsh conditions.[144]

On the other hand, the Gaunt's process works at room temperature using electron-donating *N*-substituents and a large variety of substituents at both the anilinic and arylic parts, being, therefore, more versatile.[145] An analogous and very general reaction, but starting from *N*-tosyl-2-arylanilines, has been developed by Youn.[146] The process also has the advantage that milder reaction conditions (25 °C) are required, and the use of inexpensive and benign oxidants (Oxone®), compared with the Buchwald method, and also displays a wide range of applicability. An interesting approximation for the one-pot synthesis of carbazoles from very simple precursors has been describe by Wu.[147] There, a tandem C–H activation/C–C coupling, followed by a C–H activation/C–N coupling is reported. However, mixtures of arylated and cyclizated products are obtained in most the cases. Indoles have also been obtained by cyclization of oximes, as shown in Figure 8.24b (see also Figure 8.5).[32] The reaction uses a functional group (*N*-acetate oxime) that works simultaneously as oxidant and directing group, therefore no external oxidant is needed. Interestingly, this process starts with the oxidative addition of the N–O bond to the Pd(0) catalyst, and this unprecedented species has been isolated and characterized. On the other hand, 1*H*-benzo[*d*]imidazoles have been prepared in two steps.[148] The first step is the reaction of anilines with nitriles, affording the corresponding *N*-arylbenzimidamides, this being followed by the subsequent cyclization under Pd catalysis, promoted by tetramethylthiourea, to give the targeted imidazoles. This second step is resumed in Figure 8.24c. The reaction lacks of generality, because the presence of strong electrowithdrawing groups at either the nitrile moiety (R'_n in Figure 8.24c) or the aniline fragment (R_n) completely stops the cyclization, and NH moieties do not tolerate the presence

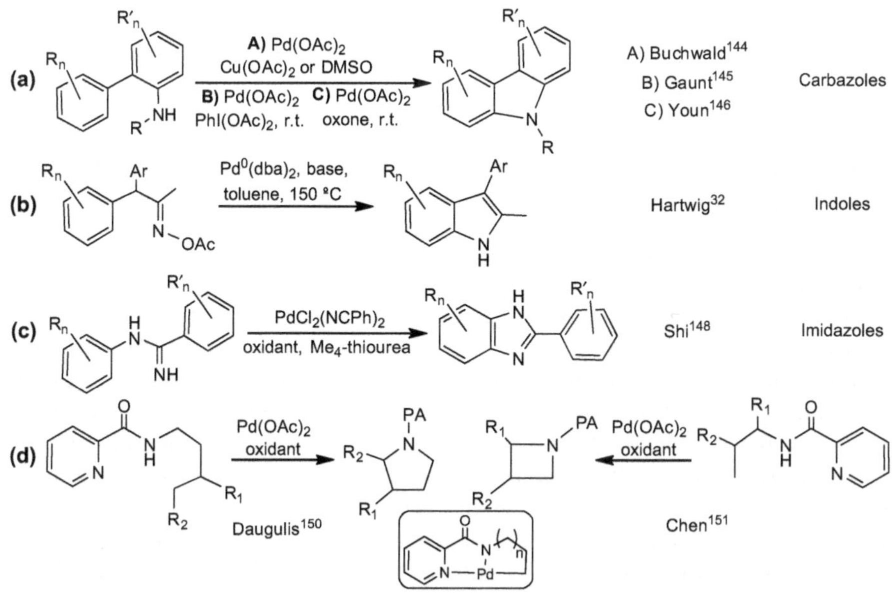

Figure 8.24 Intramolecular C–N couplings affording different types of heterocycles.

of substituents either. The authors provide evidence of the orthopalladation of the benzimidamide-type starting materials and their reactivity towards Me_4-thiourea (two X-ray structures) and how the mononuclear adduct could be the true catalyst of this transformation.[148] A conceptually similar reaction, but with a wider scope of achievable compounds, has been reported recently describing the synthesis of 1-aryl-1*H*-benzotriazoles from aryltriazene derivatives.[149] Two recent reports of Daugulis[150] and Chen[151] describe the use of the picolinamide directing group for the synthesis of heterocycles, as shown in Figure 8.24d. In the case of the Daugulis work,[150] pyrrolidines, indolines and isoindolines were prepared by smooth amination of non-activated alkane Csp^3–H bonds, while in the work of Chen[151] strained azetidines and other heterocycles can be obtained by intramolecular cyclization of alkylic amines and/or amino acids.

Due to its importance, the coordination directed formation of C–halogen bonds is widely discussed in Chapter 7.

8.4 Ru-, Rh- and Cu-Catalyzed Functionalizations of Arene Groups

Although the Pd-catalyzed processes are the most representative for the coordination-directed formation of C–C and C–X bonds, the use of other metals to perform the same tasks is really interesting because, among other reasons, they provide additional patterns of reactivity, which are often complementary to the palladium ones. In the following sections we will present recent achievements with ruthenium, rhodium and copper that complement the chemistry described with palladium.

8.4.1 Rh- and Ru-Catalyzed Annulation Reactions Involving Alkynes

One of the most outstanding results obtained during the last years is the catalytic annulation of arene fragments with internal alkynes, promoted by both ruthenium and rhodium complexes. A vast amount of work has been performed using Pd (section 8.3.3), but all processes involving C–H activation in arene moieties are stoichiometric, and more than one equivalent of alkyne were incorporated following non-predictable rules. Conversely, the reactions involving Ru- and Rh-orthometallated intermediates with internal alkynes are often catalytic and regularly (up to now) incorporate only one alkyne per arene or heteroarene unit. In addition, in all cases the directing group belongs to the target molecule, being good examples of what we called 'multi-purpose' groups.

The synthesis of *N*-heterocycles and related annulated derivatives, catalyzed by Ru- and Rh-complexes, attracts a considerable interest due to their ubiquitous presence in biologically relevant molecules. The group of Ackermann has reported the synthesis of 3,4-disubstituted isoquinolin-1(2*H*)-one cores by oxidative annulation of *N*- and aryl-substituted benzamides with internal alkynes. The reaction is conducted in *t*-AmOH at 100 °C, and is catalyzed by Ru species (Figure 8.25a).[152] The catalytic process tolerates

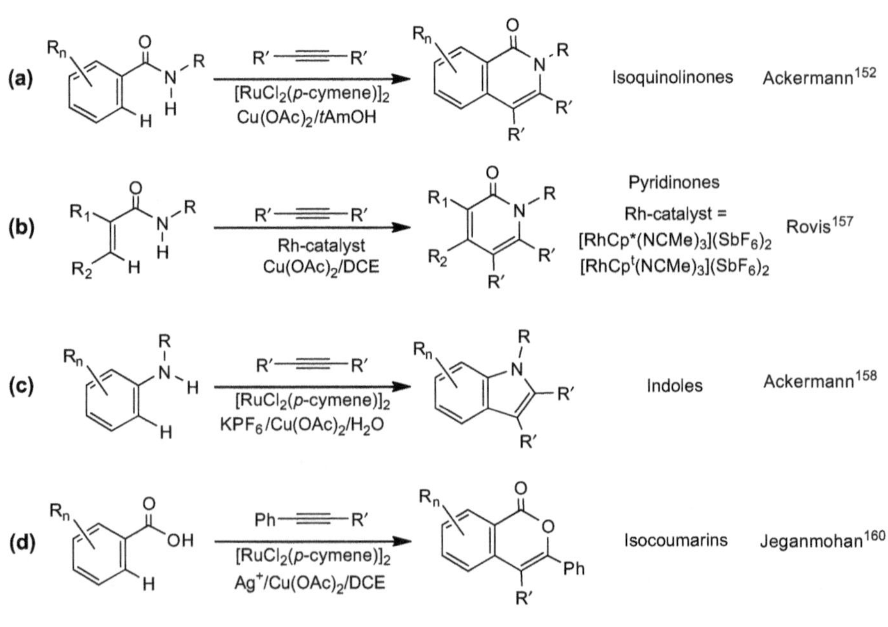

Figure 8.25 Synthesis of diverse heterocycles by annulation reactions involving alkynes.

diverse substituents at the aryl ring and the N atom, but is mostly centered on the use of diphenylacetylene as the alkyne. Nevertheless, alkynes with alkylic and alkenylic substituents also afford the targeted annulated products. This procedure was further improved and extended to a wider variety of alkynes using an internal oxidants, starting from *N*-methoxybenzamides, and performing the reactions in water.[153] The synthesis of the same or related isoquinolin-1(2*H*)-ones has also been achieved using Rh-catalysts, mainly by the groups of Fagnou,[154] Miura[155] and Rovis.[156] In spite of the different metal centers used, similar experimental parameters are reported for the above examples. At this point it is necessary to remark that the use of the cheap ruthenium catalysts provide similar efficiency and activity to that of the considerably more expensive rhodium derivatives.

Along the same lines, the group of Rovis has reported the Rh-catalysed synthesis of a large variety of pyridin-2(1*H*)-ones, carried out by reaction of substituted *N*-acrylamides with internal alkynes, as shown in Figure 8.25b.[157] The catalyst is based on the RhCp unit, and fine tuning of the Cp moiety (C_5Me_5 *vs.* $C_5H_3{}^tBu_2$) results in a spectacular improvement of the regioselectivity when the insertion of unsymmetrical alkynes is being considered.

Other heterocycles are amenable to be prepared using this strategy. For instance, a wide range of 2,3-modified indoles are obtained by dehydrogenative coupling of anilines and internal alkynes in water, catalyzed for the first time by Ru-complexes, as represented in Figure 8.25c.[158] A high degree of regioselectivity is observed for unsymmetrical alkynes, which always places the alkylic moiety distal to the N atom. An even more versatile process, reported by Fagnou, uses enynes and is catalyzed by rhodium complexes.[159] This method displays full regioselectivity when unsymmetrical alkynes are used, and provides single regioisomers with the alkenyl fragment at the C2 positions. It can also be used for the synthesis of 2,3-disubstituted pyrroles, starting from the corresponding acrylamide.[159] In addition to *N*-heterocycles, *O*-heterocycles have also been prepared by cyclization reactions. For instance, a family of isocoumarins and pyrones, very important cores in natural products, have been obtained by Jeganmohan, by reaction of many different aromatic or acrylic acids with diverse symmetric- and unsymmetrically substituted alkynes, this process being catalyzed by Ru compounds (Figure 8.25d).[160] Interestingly, the presence of silver salts (*e.g.* $AgSbF_6$) allows for a complete control of the regioselectivity of the process when aryl-alkynes such as Ph–C≡C–R′ are involved. In these cases the C bearing the Ph group binds directly with the oxygen of the carboxylate, regardless the type of aromatic acid or R′ group. Moreover, the silver salt also suppresses the plausible decarboxylative pathway, which should afford naphthalene derivatives after double alkyne insertion. A contribution of Ackermann reporting similar experimental findings appeared almost simultaneously.[161] Obviously, the synthesis of isocoumarins, pyrones and related O-heterocycles has also been achieved by coupling of aromatic acids and alkynes using Rh derivatives, and has been reviewed elsewhere.[162]

Despite all these significant contributions, the synthesis of isoquinolines is the most extensively studied process due to the industrial interest of these compounds. The pioneering work of Fagnou and Miura (Figure 8.26a) showed that it is possible to produce the dehydrogenative cyclization of imines and internal alkynes through C–H bond activation under Rh-catalysis. The method is quite general and overcomes the prefunctionalization of the substrates needed by the Pd-mediated methods.[163,164] As far as we know, the synthesis of isoquinolines by oxidative coupling of imines and alkynes, catalysed by ruthenium complexes, is still unknown. Concerning the Rh-catalysed processes, further improvements of have been produced through the use of internal oxidants, as reported by the groups of Chiba[165] and Li,[166] which are schematically shown in Figure 8.26b. The starting substrates are arylketone *O*-acetyloximes ($R_2 = Ac$) and arylketone oximes ($R_2 = H$), respectively. In both cases the reaction occurs under mild conditions (MeOH, 60 °C) and very low catalysts loadings (1–2.5%), and tolerates different substituents at the aryl ring and the alkyne. However, the change of the leaving group at the N atom significantly alters the formation of the target isoquinoline.[165] In addition, the method has been extended to a wide variety of starting substrates, allowing for the production of different target species. For instance, the reaction of α,β-unsaturated oximes with internal alkynes, catalysed by $[Cp^tRh(\mu\text{-}Cl)Cl]_2$, affords a wide range of pyridines substituted in all positions of the aromatic ring (Figure 8.26c).[34] Interestingly, this catalyst provides an excellent regioselectivity when unsymmetrical alkynes are used, but with opposite

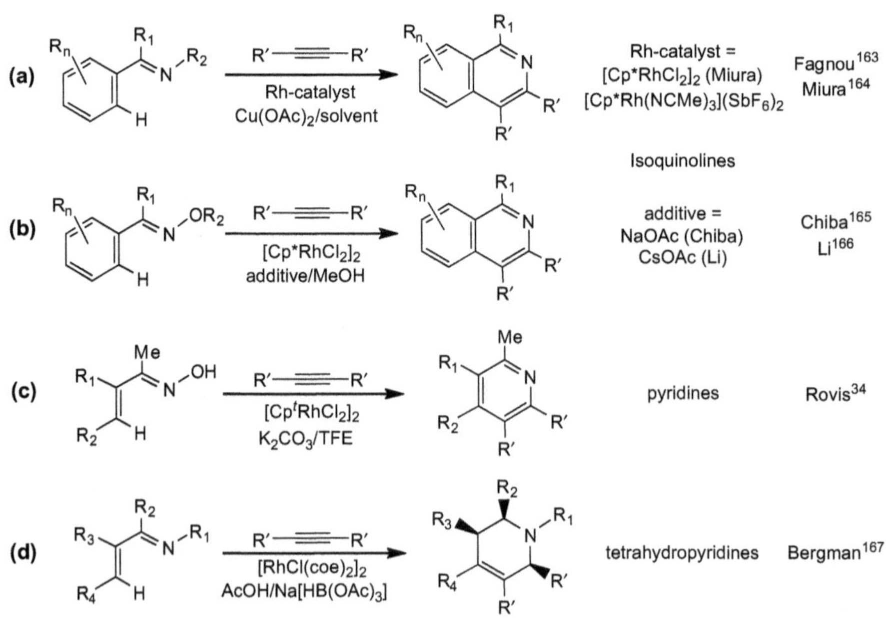

Figure 8.26 Synthesis of isoquinolines and related species by Rh-catalysed annulation reactions involving alkynes.

orientation with respect to the results obtained with pyridones (see preceding paragraphs).

1,2,3,6-Tetrahydropyridines can be obtained in a cascade procedure by reaction of α,β-unsaturated imines with internal alkynes, followed by electrocyclization, acid treatment and final reduction of the dihydropyridine intermediate with $Na[HB(OAc)_3]$.[167] This process is catalysed by the Rh(I) complex $[Rh(\mu\text{-}Cl)(coe)_2]_2$ (coe = cyclooctene), as shown in Figure 8.26d, under redox-neutral conditions. While different substituents on the imines and alkynes are well tolerated by the method, the choice of the acid and the reducing agent is critical, with the couple $AcOH/Na[HB(OAc)_3]$ displaying the best results. It is noteworthy that the hexasubstituted products are obtained with excellent diastereoselectivity.

8.4.2 Alkenylation, Carbonylation and Arylation under Rh- or Ru-Catalysis

The coupling of alkenes and other small unsaturated substrates with aryl fragments is also catalysed by Ru- and Rh-derivatives. In this field it is still difficult to compete with the extensively developed Pd-catalysed Fujiwara–Moritani reaction (see section 8.3.2), but a growing number of contributions dealing with Ru- and Rh-complexes as catalysts have appeared in recent years, showing the interest in the area. As described for Pd, the nature of the final product depends on the directing group of the starting material, the alkene substituents and the metallic complex used as catalyst.

Benzoic acids react with alkenes through a dehydrogenative coupling, in presence of $Cu(OAc)_2$ as oxidant and water as solvent. This step is usually followed by an oxa-Michael reaction, producing the corresponding annulated lactones, as shown in Figure 8.27a.[168] The reaction is restricted to activated alkenes (esters of the acrylic acid; acrylonitrile) although a wide number of substituents, in different positions, is tolerated at the aryl ring. However, the alkenylation of a variety of benzoic and heterocyclic esters, under rhodium catalysis, affords the *ortho*-vinyl derivatives without further cyclization (Figure 8.27b).[169] The procedure is general for activated alkenes and even for the less reactive styrenes. Similarly, the chelation-assisted alkenylation of substituted aromatic ketones[170] and aldehydes,[171] catalysed by Ru complexes in both cases, affords the corresponding *ortho*-vinylated compounds without subsequent formation of cyclic species, following processes analogous to those depicted in Figure 8.27b. In this respect, the selective functionalization of the aldehydes is quite remarkable, since this is a scarcely used directing group and because, moreover, in this case its modification is carried out in presence of an oxidant. Coming back to the synthesis of cyclic derivatives, it seems that the presence of acidic H atoms at the directing group is required to have an additional cyclization process between the alkenyl unit and the directing moiety.

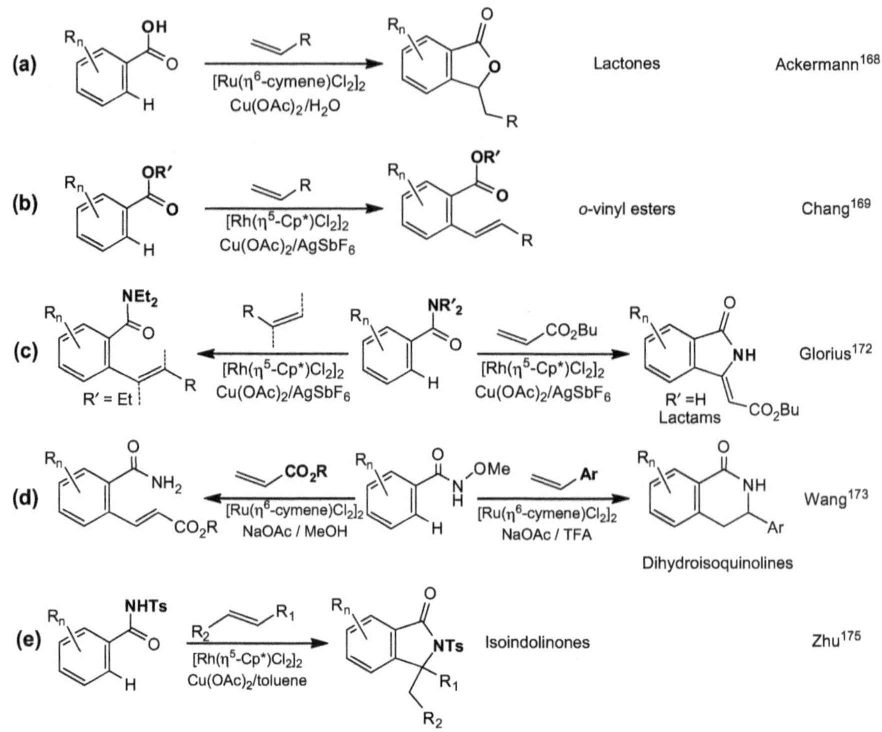

Figure 8.27 Alkenylation of aromatic substrates catalyzed by Ru and Rh complexes.

Notably, when benzamides are used as starting materials two different products can be prepared from the same type of precursors, as reported by Glorius *et al.* (processes shown in Figure 8.27c).[172] When a tertiary benzamide is the starting material, the expected alkenylated compounds are obtained, this process being efficient for a wide diversity of benzamides and even for the synthesis of highly substituted olefins (Figure 8.27c, left). However, when primary benzamides were reacted with highly activated acrylates, the cyclized γ-butyrolactams are formed as the main reaction products (Figure 8.27c, right). In the same contribution, the olefination of aryl ketones with styrene is also described, affording the expected stilbenes.[172]

The reactivity of *N*-substituted benzamides can be even more intricate. The use of *N*-alkoxybenzamides as starting compounds allows the reaction to occur without the need of externals oxidants. A very recent contribution of Wang *et al.* reports that two different products can be obtained starting from *N*-methoxybenzamides and alkenes, as shown in Figure 8.27d.[173] The reaction with activated alkenes (acrylates) affords a large variety of aromatic and heteroaromatic Heck-type products, and in this case no cyclization has been observed (Figure 8.27d, left). Surprisingly, when substituted styrene or norbornadiene derivatives were used, 3,4-dihydroisoquinolinones were obtained, as shown in Figure 8.27d (right), under apparently the same reaction conditions. Therefore, a very interesting six-membered heterocycle has been formed,[173] instead of the five-membered lactam formed in previous[172] or related examples.[174] It seems that both the presence of the *N*-alkoxy group and its specific nature are also directly related with the final outcome of the reaction. On the other hand, while terminal olefins are extensively investigated as precursors in this type of reactions, internal alkenes (which result in highly substituted olefins) are scarcely represented. A recent report of Zhu *et al.* describes the general synthesis of 3,3-disubstituted isoindolinones, by reaction of *N*-benzoylsulfonamides with diethylfumarate and other internal olefins, catalysed by Rh, as represented in Figure 8.27e.[175] In addition, Bergman, Ellman *et al.* have reported a method to oxidatively couple oximes and the typically unreactive aliphatic alkenes.[176]

After this short survey about the possibilities of the olefination, it is necessary to remark that the general method (regioselective ortho-monolefination) has been extended to other substrates, for which representative examples are collected in Figure 8.28.[177–180]

The catalytic carbonylation of organic substrates by CO insertion into C–H bonds is also a powerful and widely used synthetic strategy (see Section 8.3.4). While carbonylation of Csp^3–H bonds is a well-known process in Pd-catalysis, it is scarcely developed in Rh and Ru. The most important results, up to 2009, about chelation-assisted carbonylation catalyzed by Rh or Ru, appear nicely reviewed in a recent paper by Chatani *et al.*[181]

Besides that, it is worth noting the work of the Chatani group in the carbonylation of inert Csp^3–H bonds, using $Ru_3(CO)_{12}$ as catalyst and a pyridine-based bidentate auxiliary, as shown in Figure 8.29a.[182] In this process aliphatic amides were regioselectively transformed into succinimides by their reaction with CO (10 atm) and ethylene (7 atm), under harsh conditions (160 °C, 3 days). The selectivity in the activation of methyl groups in preference of methylene units is outstanding, always affording five-membered rings. Also noteworthy is the fact that the pyridinic fragment has to chelate the metal forming five-membered metallacycles, otherwise the carbonylation does not take place.

The harsh conditions and the need of a chelating auxiliary inherent to the ruthenium catalysis can be avoided if rhodium is used as catalytic metal. The group of Rovis has reported a simple protocol for the synthesis of phthalimides, shown in Figure 8.29b,[183] based on previous work of the same

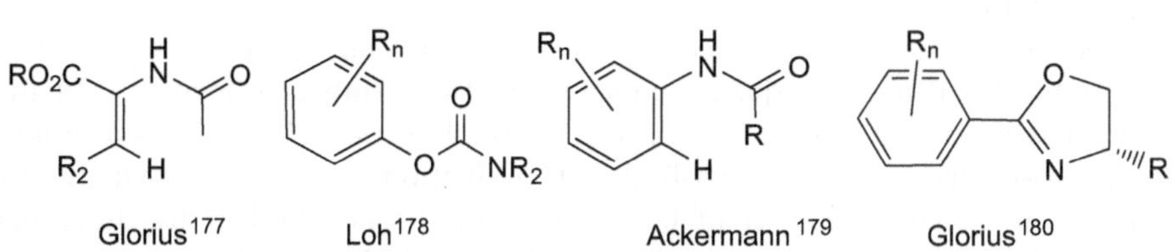

Figure 8.28 Substrates amenable to olefination reactions.

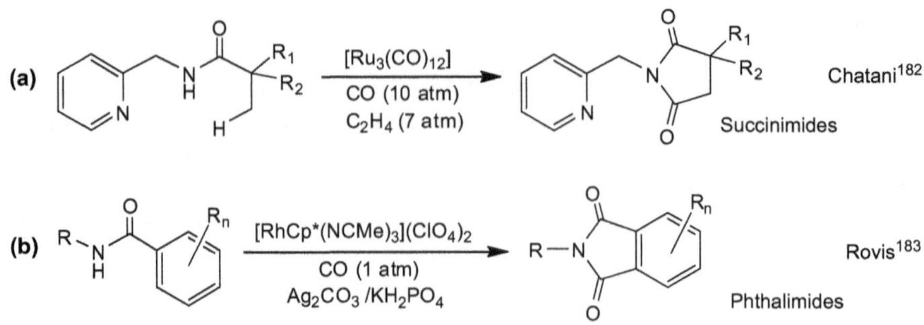

Figure 8.29 Carbonylations catalysed by Ru- and Rh-derivatives.

group on alkynes, herein just replacing the alkynes by CO (compare with the process in Figure 8.25b). This protocol works under 1 atm of CO, without chelating ligands, and using simple additives such as Ag_2CO_3 and KH_2PO_4. Under these conditions, a wide range of benzamides are efficiently transformed into the target phthalimides, regardless of the substituent nature at the aryl ring or the N atom.

Carbonylated derivatives can also be prepared avoiding the use of toxic CO and using, instead, aldehydes as carbonyl sources (see also section 8.3.4). A large variety of diaryl ketones can be prepared by reaction of *N*,*N*-dialkylbenzamides with aromatic aldehydes, the process catalysed by cationic complexes of Rh(III) prepared *in situ*, as shown in Figure 8.30a,[184] while when secondary *N*-alkyl benzamides are used as starting materials the outcome of the reaction is different, and the cyclic 3-hydroxyisoindolin-1-one species, shown in Figure 8.30b, are obtained.[185]

The importance of the directing group has been highlighted by Shi, by reporting the synthesis of bis-aryl methanols by arylic C–H bond addition to aryl aldehydes, catalysed by Rh(III) derivatives (Figure 8.30c).[186] A similar result for the Grignard arylation of aldehydes, catalysed by Rh(III) complexes, has been reported by Li *et al.*[187] Along the same lines, the group of Ellman has described the arylic C–H bond addition of benzamides to aryl imines, which results in the synthesis of branched amines (Figure 8.30d).[188]

Concerning the arylation process, although in its infancy under Ru- and Rh-catalysis compared to Pd- catalysis, the promising results obtained in recent years have prompted further developments this area. The most representative examples of arylation reactions catalysed by either Ru- or Rh-complexes are shown in Figure 8.31.

The only example reported up to now of dehydrogenative arylation promoted by Rh(III) complexes comes from the group of Glorius.[189] In this contribution, the regioselective aryl–aryl coupling of *N*,*N*-dialkylbenzamides with substituted arenes, by two-fold C–H activation, is studied. The method is efficient for a large number of benzamides and arenes, and many functional groups are tolerated even under harsh reaction conditions. However, most of the reported cases of aryl–aryl coupling require the use of haloaryl derivatives,

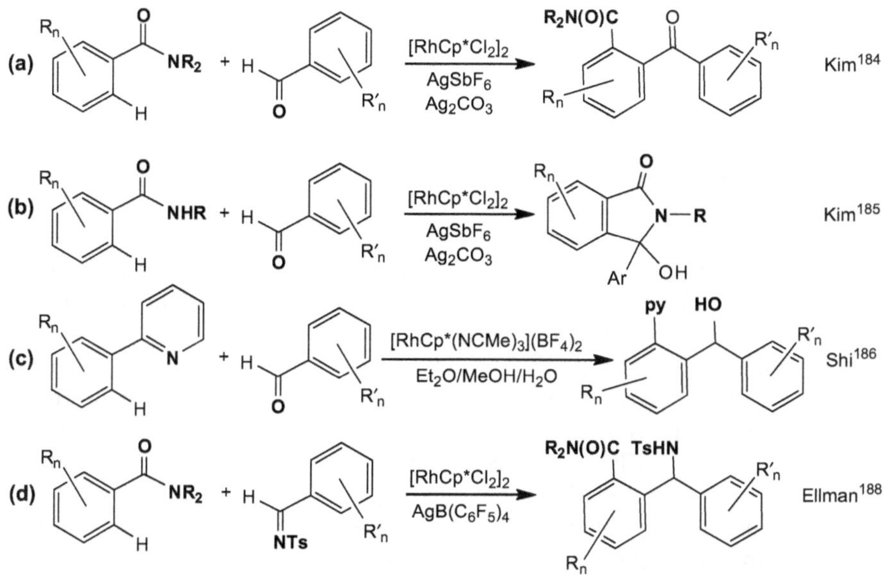

Figure 8.30 Carbonylation processes using aldehydes as CO source: different outcomes.

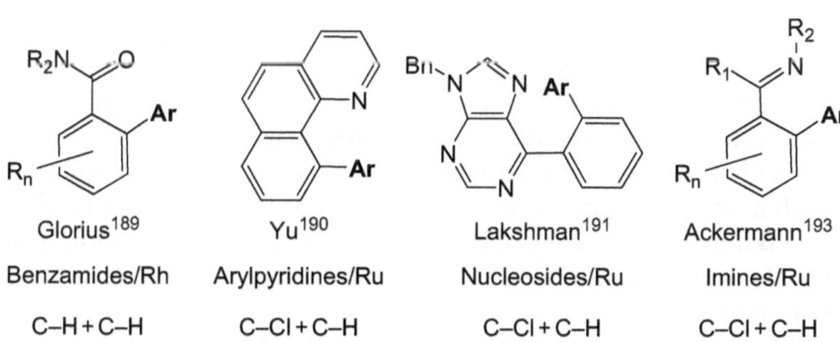

Figure 8.31 Representative examples of arylated substrates, catalyst and method.

at least in one of the counterparts, and examples are found in the arylation of simple compounds, such as benzo[*h*]quinolines,[190] or of delicate nucleosides.[191] The mechanism for the arylation of these substrates is believed to occur in three main steps, being the first one the oxidative addition of the aryl halide to the Ru(II) center, which is followed by the C–H bond activation step at the generated Ru(IV) intermediate,[190,191] and closing of the cycle by the reductive elimination of the functionalized species.

A very interesting example of phenol as carrier of the electrophilic arylic moiety (instead of the aryl halide) has been reported by Ackermann.[192] The C–O and C–H bond activations, as well as the further C–C coupling, are promoted by the Ru(II) complex [Ru(η^6-cymene)(O_2CMes)$_2$], which has emerged as a very efficient arylation catalyst even in water. The same author has studied the excellent performance of this catalyst in the arylation of imines (Figure 8.31), phenylpyridines, oxazolines and related precursors.[193] Based on experimental data, it is suggested that the mechanism of the arylation starts with the reversible carboxylate-assisted C–H bond activation, promoted by Ru(II), affording a cycloruthenated complex. This is followed by the oxidative addition of the aryl halide to the orthoruthenated derivative, which is the rate determining step. The final reductive elimination results in the formation of the target species and regeneration of the catalyst. This mechanism differs notably of that proposed by other authors.[190,191] In a recent contribution, Jutand *et al.* have disclosed the intimate mechanism of the arylation catalysed by $[Ru(\eta^6\text{-cymene})(O_2CMe)_2]$.[194]

Just as a final remark, many other coordination-directed functionalizations, catalysed by rhodium or ruthenium complexes, have been reported. Processes involving CN,[195,196] CB[197] and CS[198] couplings, similar or complementary to those described with palladium, have been published only some months ago, giving an idea of the potential and the interest of this area. It is certain that more exciting developments are waiting in the near future.

8.4.3 Cu-Catalyzed Functionalizations of Aryl C–H Bonds

In clear contrast with the extensive C–C and C–heteroatom functionalization of arenes developed with Pd-, Rh- and Ru-catalysts, fewer examples have been described using copper as catalyst. The advantages of copper are related to its very low cost, low toxicity and biological relevance. Due to these reasons, research into new processes and new applications using copper as catalyst is a continuous challenge. Some aspects of copper chemistry are treated along this book, such as the Ullmann reaction (Chapter 1), the mechanism of the C_{aryl}–X bond formation catalysed by Cu (Chapter 2), or the oxidation reactions (Chapter 11). In this section we will focus only in aspects related with the arene C–H bond activation.

Concerning the reactivity of C–H bonds, probably the most common functionalization pathway catalysed by copper complexes is the carbene or nitrene insertion into Csp^3–H bonds, a topic which has been recently reviewed,[199] while the coordination-directed strategy is still relatively recent and underdeveloped. Most of the work carried out up to now, dealing with directed C–C bond couplings, involves arylic species of the type Ar–H and Ar–X (X = halide), where Ar–H is usually a heterocycle (either electron-rich or electron-deficient) or an aryl fragment with at least two strong electron-withdrawing substituents. This chemistry has also been recently reviewed.[14,200] There are remarkable contributions in the *ortho*-directed coupling of aryl moieties with O- and N-nucleophiles. Some recent representative Cu-catalysed processes are found in Figure 8.32. As it is possible to see, many of them resemble examples already presented with Pd, Rh or Ru. However, this similarity is only limited, since the intimate mechanism of the Cu-catalysts differ considerably from those determined for the other noble metals (see Chapter 2).

Acetoxylation, halogenation and dimerization of phenylpyridine has been reported, among other reactions, by the group of Yu in a seminal contribution.[201] Examples are shown in Figure 8.32a, but have the drawbacks of high loads of catalysts (up to 20%) and lack of selectivity. For instance, the acetoxylation process results in the formation of the mono- and diacetoxylated products. The aryloxy group can be incorporated to a variety of phenylpyridine derivatives by reaction with acyl chlorides as shown in Figure 8.32b (left).[202] An intramolecular version of the hydroxylation results in the synthesis of dibenzofuranes, as represented in Figure 8.32 (right; compare with Figure 8.22).[203] All reactions afford the expected products in good yields, and are quite general because a wide number of functional groups are tolerated. Another advantage is that almost all reactions can be carried out in air (or in O_2 atmosphere) avoiding the use of other oxidants. However, in general, the comparison with the corresponding Pd- or Ru-catalysed reactions shows that Cu-mediated processes occur under much harsher conditions (*T* is usually higher than 120 °C) and need longer reaction times (typically more than 24 h), counterbalancing partially the lower cost of the copper complexes.

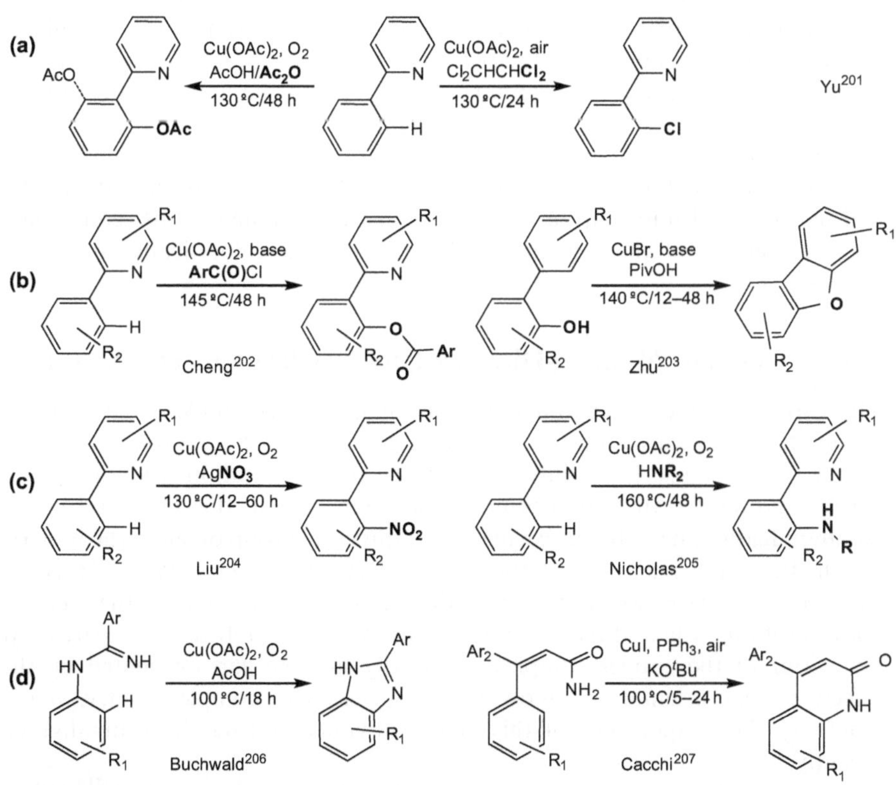

Figure 8.32 Selected copper catalysed C_{aryl}–O and C_{aryl}–N bond couplings.

A similar situation is found in the C_{aryl}–N coupling. The nitration of 2-phenylpyridine derivatives has been achieved using substoichiometric amounts of $Cu(OAc)_2$ (50%) and 1.5 equivalents of $AgNO_3$ as source of the NO_2 unit,[204] as seen in Figure 8.32c (left), while the amination of the same type of substrates (Figure 8.32c, right) has been reported by Nicholas *et al.* under really strong reaction conditions (160 °C, 48 h).[205] It seems difficult to make these processes compatible with other more fragile substrates, or with the presence of reactive functional groups on the starting materials.

Intramolecular versions of the C–N coupling have been reported by different authors. Two representative examples from Buchwald[206] and Cacchi[207] are shown in Figure 8.32d left and right, respectively. The oxidative cyclization of amidines, reported by Buchwald,[206] affords benzimidazoles and *N*-methylbenzimizadoles in very good yields using reaction conditions softer in comparison to those described in previous examples. Moreover, the method improves previous synthetic procedures because it is more general. On the other hand, 4-aryl-2-quinolones can be obtained by intramolecular cyclization of 3,3-diarylacrylamides, catalyzed by CuI in presence of PPh_3 and base. A fine tuning of the substituents in the Ar_1 and Ar_2 rings allows for the regioselective functionalization of only one of the arylic moities.[207]

Notably, in all the above papers[201–207] the behaviour of Cu(III) complexes as key species in the reaction mechanism is suggested, and sometimes recognized, but not fully understood. A full description of the role of Cu(III) species in the mechanism of Cu-catalysed reactions is presented in Chapter 2, and it will not be discussed here. But it is worth mentioning that the relevance of the high oxidation states in Cu- and Pd-catalysed processes has been highlighted very recently[208] and that in Chapters 4 [Pd(IV)] and 5 [Pd(III)] of this book the reader will find detailed information about the most outstanding results concerning the involvement of these highly oxidized centers as catalytically competent species.

8.4.4 Cu- and Pd- *meta*-Directed Functionalization: The New Tool

The coordination-directed functionalization using Cu-complexes still has an ace up its sleeve, consisting of a behaviour completely different to those shown by the other metals: the ability to orientate aryl substitutions *meta* to a pre-existing group.[209] All examples up to now reported in this chapter share a common feature, namely the existence of a directing group on an arylic moiety, which drives the reactivity of the metal selectively to the *ortho* C–H bond. The sequence activation–functionalization can not thus occur at the *meta* or *para* positions when directing groups are used. Therefore, this particular behaviour of the copper complexes partially fills the gap originated by the selective *ortho*-orientation, allowing for a selective, complementary, *meta*-reactivity. The importance of this fact is obvious, and has been highlighted recently.[210,211]

The original work of Phipps and Gaunt is shown in Figure 8.33.[212] When substituted anilides, either at the aryl ring and the keto group, are treated with

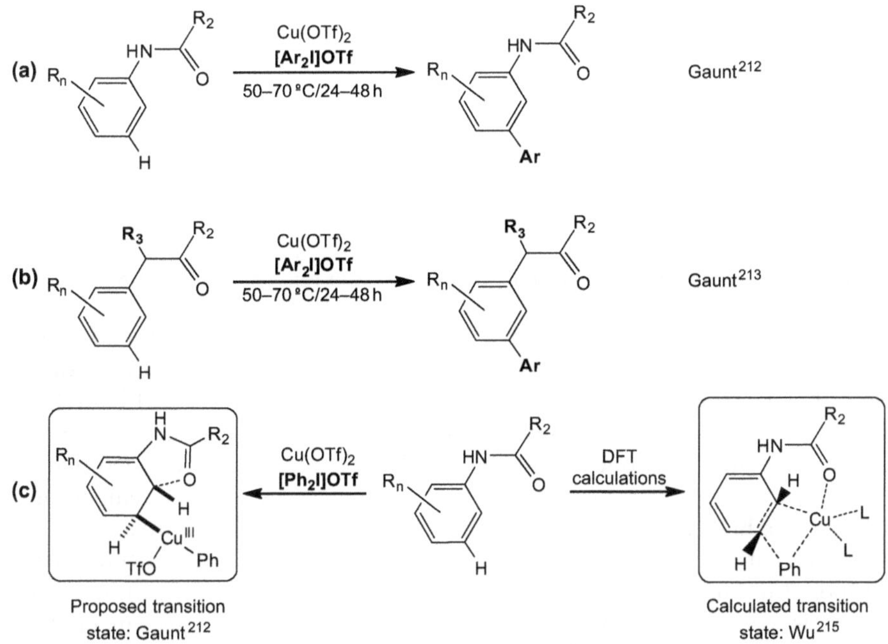

Figure 8.33 Recent achievements in meta-oriented functionalizations catalysed by Cu.

oxidant I(III) salts [Ar_2I]A (Ar = aryl; A = OTf or BF_4) under mild reaction conditions, the incorporation of the aryl group is produced selectively at the *meta* position with respect to the anilido group. This is quite remarkable, since the anilido group orientates *ortho* electronically and, moreover, the coordination to the carbonyl group also orientates *ortho*. Compare this situation with, for instance, the processes shown in Figures 8.15, 8.18, 8.19 and 8.23. The reaction proceeds using softer conditions compared with previous processes, and is quite general with respect to the substituents. The variation of the aryl group on the I(III) salt, the aryl ring of the anilide and/or the carbonyl group promote yield fluctuations, but do not limit the method. In this way, tri- and tetra-substituted anilides, difficult to obtain by classical pathways, can be synthesized. The method is particularly tolerant to the presence of halogen groups, therefore it is suitable for further orthogonal functionalization *via* Heck, Suzuki, or other couplings. Furthermore, the method has been extended to other α-aryl carbonyl substrates,[213] as shown in Figure 8.33b. The initial hypothesis for this expansion lies in the consideration of the carbonyl group as truly responsible for the *meta* orientation and, apparently, this is a sensible proposal. Following this assumption, a variety of α-aryl carbonyl compounds (α-arylacetamides, esters of α-arylcarboxylic acids, α-arylketones or *N*-methoxy-α-arylacetamides) have been *meta*-functionalized using I(III) salts as aryl carriers and oxidants, with the reactions proceeding with good yields and under very mild reaction conditions. Additional advantages are that chiral products can be functionalized without loss of the chiral information.

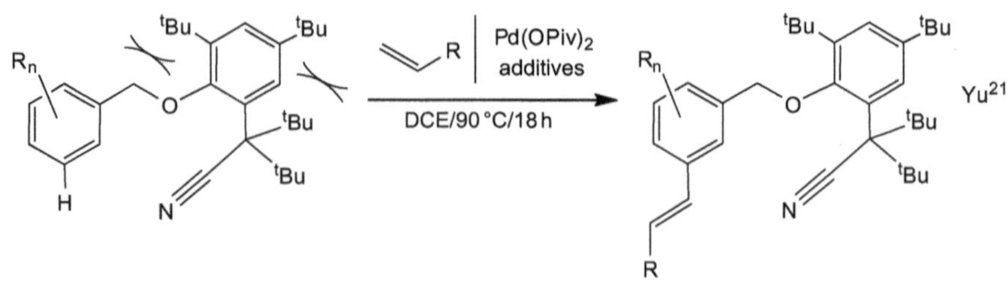

Figure 8.34 Template-assisted functionalization at a remote *meta*-position.

While this *meta*-directed reaction is already being used as synthetic tool,[214] its intimate mechanism is poorly understood and data gathered are controversial. The group of Gaunt made an initial proposal, shown in Figure 8.33c (left), in which an anti-oxocupration involving a Cu(III)-aryl species is involved.[212] In a subsequent contribution the same group has reported that the reaction works without metal,[213] albeit in a very narrow range of temperatures (80–90 °C), higher than the temperature of the metal-catalysed process (70 °C). Very recent theoretical calculations (DFT methods) have determined a proposed mechanism based on a Cu(I)-Cu(III) cycle, in which the attack of the metal is produced at the *ortho*-position through a concerted metalation-deprotonation scheme, followed by a Heck-like four-membered transition state, also shown in Figure 8.33c, as the key species.[215] The same calculations show that the original proposal of Gaunt is extremely unfavourable energetically, due to the loss of aromaticity, while the proposed Heck-like mechanism keeps some degree of aromaticity, and shows lower activation barriers. Although it seems that this new proposal is more likely than the initial one of Gaunt, additional experiments are necessary. As expected, and in order to cover all possibilities, it worth mentioning that the group of Gaunt has reported the *para*-selective Cu-catalysed arylation of several substrates.[216]

We are now almost ready to close the circle: a very recent work of Yu *et al.* has reported that Pd can also be incorporated at remote positions of arylic rings, more specifically at *meta* positions, using a nitrile-containing template.[217] Therefore, the ring to be modified is tethered to the nitrile-containing template, decorated with very bulky groups as shown in Figure 8.34, forming a cyclophane-like structure which directs the incorporation of the metal to its *meta* position. This very new concept works for the olefination of aryl rings with electron-donating and -attracting substituents, and using electro-deficient olefins or even highly congested alkenes.

8.5 Conclusions and Outlook

In summary, the development of the directed catalytic functionalization of organic substrates has undergone an impressive advance in the last 4–5 years, as has been shown in the presented examples. A large number of difficult challenges have been overcome, and in our present position we can state that most organic procedures, usually performed without the assistance of transition metals, have an alternative pathway using organometallic complexes, and that the latter reactions occur with high efficiency (catalytically) and selectivity (directed).

While many advances have been achieved recently, it is also clear that much remains to be done. Considering the incalculable importance of aryl-based organic moieties, as well as the burgeoning literature devoted to improve cyclometallation-based methods, we predict that this area of research will grow faster still in the near future. It is hoped that the present review will help stimulate this growth.

Acknowledgment

Funding by the Ministerio de Economia y Competitividad (Spain, Project CTQ2011-22589) and Gobierno de Aragón (Spain, group E97) is gratefully acknowledged. L. C. thanks MICINN and Consejo Superior de Investigaciones Científicas (Spain) for a Juan de la Cierva contract.

References

1. K. Tamao and N. Miyaura, *Top. Curr. Chem.*, 2002, **219**, 1.
2. A. de Meijere and F. Diederich, ed., *Metal-Catalyzed Cross Couplig Reactions*, Springer Verlag, Berlin, 2010.
3. V. F. Slagt, A. H. M. de Vries, J. G. de Vries and R. M. Kellogg, *Org. Process Res. Dev.*, 2010, **14**, 30.
4. D.-G. Yu, B.-J. Li and Z.-J. Shi, *Acc. Chem. Res.*, 2010, **43**, 1486.
5. The Nobel Prize in Chemistry 2010, Nobelprize.org. See the complete scientific background at: http://www.nobelprize.org/nobel_prizes/chemistry/laureates/2010/. See also C. C. C. Johansson Seechurn, M. O. Kitching, T. J. Colacot and V. Snieckus, *Angew. Chem., Int. Ed.*, 2012, **51**, 5062.
6. L. Ackermann, R. Vicente and A. R. Kapdi, *Angew. Chem., Int. Ed.*, 2009, **48**, 9792.
7. J. Wencel-Delord, T. Dröge, F. Liu and F. Glorius, *Chem. Soc. Rev.*, 2011, **40**, 4740.
8. C.-L. Sun, B.-J. Li and Z.-J. Shi, *Chem. Commun.*, 2010, **46**, 677.
9. X. Chen, K. M. Engle, D.-H. Wang and J.-Q. Yu, *Angew. Chem., Int. Ed.*, 2009, **48**, 5094.
10. R. Jazzar, J. Hitce, A. Renaudat, J. Sofack-Kreutzer and O. Baudoin, *Chem. Eur. J.*, 2010, **16**, 2654.
11. W. Shi, C. Liu and A. Lei, *Chem. Soc. Rev.*, 2011, **40**, 2761.
12. C. Liu, H. Zhang, W. Shi and A. Lei, *Chem. Rev.*, 2011, **111**, 1780.
13. B. Liégault, I. Petrov, S. I. Gorelsky and K. Fagnou, *J. Org. Chem.*, 2010, **75**, 1047.
14. O. Daugulis, H.-Q. Do and D. Shabashov, *Acc. Chem. Res.*, 2009, **42**, 1074.
15. B. Liégault, D. Lapointe, L. Caron, A. Vlassova and K. Fagnou, *J. Org. Chem.*, 2009, **74**, 1826.

16. T. W. Lyons and M. S. Sanford, *Chem. Rev.*, 2010, **110**, 1147.
17. C. S. Yeung and V. M. Dong, *Chem. Rev.*, 2011, **111**, 1215.
18. S.-L. You and J.-B. Xia, *Top. Curr. Chem.*, 2010, **219**, 165.
19. W.-J. Yoo and C.-J. Li, *Top. Curr. Chem.*, 2010, **219**, 281.
20. J. Le Bras and J. Muzart, *Chem. Rev.*, 2011, **111**, 1170.
21. T. Bruckl, R. D. Baxter, Y. Ishihara and P. S. Baran, *Acc. Chem. Res.*, 2012, **45**, 826.
22. T. Newhouse and P. S. Baran, *Angew. Chem., Int. Ed.*, 2011, **50**, 3362.
23. K. M. Engle, T.-S. Mei, M. Wasa and J.-Q. Yu, *Acc. Chem. Res.*, 2012, **45**, 788.
24. M. Albrecht, *Chem. Rev.*, 2010, **110**, 576.
25. K. Godula and D. Sames, *Science*, 2006, **312**, 67.
26. J.-Q. Yu, R. Giri and X. Chen, *Org. Biomol. Chem.*, 2006, **4**, 4041.
27. D. Kalyani and M. S. Sanford, *Top. Organomet. Chem.*, 2007, **24**, 85.
28. J. Spencer in *Palladacycles, Synthesis, Characterization and Applications*, ed. M. Pfeffer and J. Dupont, Wiley-VCH Verlag GmbH & Co. KGaA, Weinheim, Germany, 2008, p. 109.
29. D. A. Colby, A. S. Tsai, R. G. Bergman and J. A. Ellman, *Acc. Chem. Res.*, 2012, **45**, 814, and references contained therein.
30. D. Aguilar, L. Cuesta, S. Nieto, E. Serrano and E. P. Urriolabeitia, *Curr. Org. Chem.*, 2011, **15**, 3441.
31. F. W. Patureau and F. Glorius, *Angew. Chem., Int. Ed.*, 2011, **50**, 1977.
32. Y. Tan and J. F. Hartwig, *J. Am. Chem. Soc.*, 2010, **132**, 3676.
33. N. Guimond, C. Gouliaras and K. Fagnou, *J. Am. Chem. Soc.*, 2010, **132**, 6908.
34. T. K. Hyster and T. Rovis, *Chem. Commun.*, 2011, **47**, 11846.
35. Y. Lu, D. Leow, X. Wang, K. M. Engle and J.-Q. Yu, *Chem. Sci.*, 2011, **2**, 967.
36. J. Cornella, M. Righi and I. Larrosa, *Angew. Chem., Int. Ed.*, 2011, **50**, 9429.
37. C. Huang, B. Chattopadhyay and V. Gevorgyan, *J. Am. Chem. Soc.*, 2011, **133**, 12406.
38. A. García-Rubia, B. Urones, R. Gómez Arrayás and J. C. Carretero, *Angew. Chem., Int. Ed.*, 2011, **50**, 10927.
39. D.-H. Wang, K. M. Engle, B.-F. Shi and J.-Q. Yu, *Science*, 2010, **327**, 315.
40. K. M. Engle, P. S. Thuy-Boun, M. Dang and J.-Q. Yu, *J. Am. Chem. Soc.*, 2011, **133**, 18183.
41. R. Giri, N. L. Maugel, J.-J. Li, D.-H. Wang, S. P. Breazzano, L. B. Saunders and J.-Q. Yu, *J. Am. Chem. Soc.*, 2007, **129**, 3510.
42. H. A. Chiong, Q. N. Pham and O. Daugulis, *J. Am. Chem. Soc.*, 2007, **129**, 9879.
43. C. Wang, S. Rakshit and F. Glorius, *J. Am. Chem. Soc.*, 2010, **132**, 14006.
44. M. Yamashita, K. Hirano, T. Satoh and M. Miura, *Org. Lett.*, 2009, **11**, 2337.

45. M. Yamashita, H. Horiguchi, K. Hirano, T. Satoh and M. Miura, *J. Org. Chem.*, 2009, **74**, 7481.
46. R. Shang, Z.-W. Yang, Y. Wang, S.-L. Zhang and L. Liu, *J. Am. Chem. Soc.*, 2010, **132**, 14391.
47. W.-Y. Yu, W. N. Sit, Z. Zhou and A. S.-C. Chan, *Org. Lett.*, 2009, **11**, 3174.
48. M. T. Tredwell, M. Gulias, N. Gaunt Bremeyer, C. C. C. Johansson, B. S. L. Collins and M. J. Gaunt, *Angew. Chem., Int. Ed.*, 2011, **50**, 1076.
49. D. Kalyani, N. R. Deprez, L. V. Desai and M. S. Sanford, *J. Am. Chem. Soc.*, 2005, **127**, 7330.
50. K. L. Hull, E. L. Lanni and M. S. Sanford, *J. Am. Chem. Soc.*, 2006, **128**, 14047.
51. A. Lazareva and O. Daugulis, *Org. Lett.*, 2006, **8**, 5211.
52. D. Shabashov and O. Daugulis, *Org. Lett.*, 2006, **8**, 4947.
53. K. L. Hull and M. S. Sanford, *J. Am. Chem. Soc.*, 2007, **129**, 11904.
54. X. Zhao, C. S. Yeung and V. M. Dong, *J. Am. Chem. Soc.*, 2010, **132**, 5837.
55. C. S. Yeung, X. Zhao, N. Borduas and V. M. Dong, *Chem. Sci.*, 2010, **1**, 331.
56. D. C. Powers and T. Ritter, *Top. Organomet. Chem.*, 2011, **35**, 129.
57. D. C. Powers and T. Ritter, *Acc. Chem. Res.*, 2012, **45**, 840.
58. P. Sehnal, R. J. K. Taylor and I. J. S. Fairlamb, *Chem. Rev.*, 2010, **110**, 824.
59. L.-M. Xu, B.-J. Li, Z. Yang and Z.-J. Shi, *Chem. Soc. Rev.*, 2010, **39**, 712.
60. K. Muñiz, *Angew. Chem., Int. Ed.*, 2009, **48**, 9412.
61. A. J. Canty, *Dalton Trans.*, 2009, **47**, 10409.
62. J. M. Racowski, N. D. Ball and M. S. Sanford, *J. Am. Chem. Soc.*, 2011, **133**, 18022.
63. X. Wang, D. Leow and J.-Q. Yu, *J. Am. Chem. Soc.*, 2011, **133**, 13864.
64. T. W. Lyons, K. L. Hull and M. S. Sanford, *J. Am. Chem. Soc.*, 2011, **133**, 4455.
65. K. L. Hull and M. S. Sanford, *J. Am. Chem. Soc.*, 2009, **131**, 9651.
66. D. G. Pintori and M. F. Greaney, *Org. Lett.*, 2011, **13**, 5713.
67. J. Karthikeyan and C.-H. Cheng, *Angew. Chem., Int. Ed.*, 2011, **50**, 9880.
68. G.-W. Wang, T.-T. Yuan and D.-D. Li, *Angew. Chem., Int. Ed.*, 2011, **50**, 1380.
69. D. Shabashov and O. Daugulis, *J. Org. Chem.*, 2007, **72**, 7720.
70. M. Blanchot, D. A. Candito, F. Larnaud and M. Lautens, *Org. Lett.*, 2011, **13**, 1486.
71. G. Maestri, E. Motti, N. Della Ca', M. Malacria, E. Derat and M. Catellani, *J. Am. Chem. Soc.*, 2011, **133**, 8574.
72. M.-H. Larraufie, G. Maestri, A. Beaume, E. Derat, C. Ollivier, L. Fensterbank, C. Courillon, E. Lacôte, M. Catellani and M. Malacria, *Angew. Chem., Int. Ed.*, 2011, **50**, 12253.
73. N. Della Ca', G. Maestri, M. Malacria, E. Derat and M. Catellani, *Angew. Chem., Int. Ed.*, 2011, **50**, 12257.

74. D. Kalyani, K. B. McMurtrey, S. R. Neufeldt and M. S. Sanford, *J. Am. Chem. Soc.*, 2011, **133**, 18566.
75. H. Zhou, Y.-H. Xu, W.-J. Chung and T.-P. Loh, *Angew. Chem., Int. Ed.*, 2009, **48**, 5355.
76. W. Li, Z. Yin, X. Jiang and P. Sun, *J. Org. Chem.*, 2011, **76**, 8543.
77. N. R. Deprez and M. S. Sanford, *J. Am. Chem. Soc.*, 2009, **131**, 11234.
78. B. Xiao, Y. Fu, J. Xu, T.-J. Gong, J.-J. Dai, J. Yi and L. Liu, *J. Am. Chem. Soc.*, 2010, **132**, 468.
79. M. Wasa, K. M. Engle, D. W. Lin, E. J. Yoo and J.-Q. Yu, *J. Am. Chem. Soc.*, 2011, **133**, 19598.
80. T. Nishikata, A. R. Abela, S. Huang and B. H. Lipshutz, *J. Am. Chem. Soc.*, 2010, **132**, 4978.
81. C.-L. Sun, N. Liu, B.-J. Li, D.-G. Yu, Y. Wang and Z.-J. Shi, *Org. Lett.*, 2010, **12**, 184.
82. H.-M. Guo, L.-L. Jiang, H.-Y. Niu, W.-H. Rao, L. Liang, R.-Z. Mao, D.-Y. Li and G. R. Qu, *Org. Lett.*, 2011, **13**, 2008.
83. M. Wasa, K. M. Engle and J.-Q. Yu, *J. Am. Chem. Soc.*, 2009, **131**, 9886.
84. C. Jia, T. Kitamura and Y. Fujiwara, *Acc. Chem. Res.*, 2001, **34**, 633.
85. C. Wang and H. Ge, *Chem.–Eur. J.*, 2011, **17**, 14371.
86. M. Mewald, J. A. Schiffner and M. Oestreich, *Angew. Chem., Int. Ed.*, 2012, **51**, 1763.
87. G. Cai, Y. Fu, Y. Li, X. Wan and Z.-J. Shi, *J. Am. Chem. Soc.*, 2007, **129**, 7666.
88. L. Wang, S. Liu, Z. Li and Y. Yu, *Org. Lett.*, 2011, **13**, 6137.
89. X. Liu and K. K. Hii, *J. Org. Chem.*, 2011, **76**, 8022.
90. T. Nishikata and B. H. Lipshutz, *Org. Lett.*, 2010, **12**, 1972.
91. C. Zhu and J. R. Falck, *Org. Lett.*, 2011, **13**, 1214.
92. D.-D. Li, T.-T. Yuan and G.-W. Wang, *Chem. Commun.*, 2011, **47**, 12789.
93. Y.-Y. Liu, R.-J. Song, C.-Y. Wu, L.-B. Gong, M. Hu, Z.-Q. Wang, Y.-X. Xie and J.-H. Li, *Adv. Synth. Catal.*, 2012, **354**, 347.
94. B.-F. Shi, Y.-H. Zhang, J. K. Lam, D.-H. Wang and J.-Q. Yu, *J. Am. Chem. Soc.*, 2010, **132**, 460.
95. M. Wasa, K. M. Engle and J.-Q. Yu, *J. Am. Chem. Soc.*, 2010, **132**, 3680.
96. Y. Lu, D.-H. Wang, K. M. Engle and J.-Q. Yu, *J. Am. Chem. Soc.*, 2010, **132**, 5916.
97. D.-H. Wang and J.-Q. Yu, *J. Am. Chem. Soc.*, 2011, **133**, 5767.
98. M. Pfeffer, *Recl. Trav. Chim. Pays-Bas*, 1990, **109**, 567.
99. G. Zeni and R. C. Larock, *Chem. Rev.*, 2006, **106**, 4644.
100. M. Yamashita, K. Hirano, T. Satoh and M. Miura, *Org. Lett.*, 2009, **11**, 2337.
101. S. Ding, Z. Shi and N. Jiao, *Org. Lett.*, 2010, **12**, 1540.
102. H. Zhong, D. Yang, S. Wang and J. Huang, *Chem. Commun.*, 2012, **48**, 3236.
103. Z. Shi, C. Zhang, S. Li, D. Pan, S. Ding, Y. Cui and N. Jiao, *Angew. Chem., Int. Ed.*, 2009, **48**, 4572.

104. J. Chen, Q. Pang, Y. Sun and X. Li, *J. Org. Chem.*, 2011, **76**, 3523.
105. Y. Ano, M. Tobisu and N. Chatani, *Org. Lett.*, 2012, **14**, 354, and references therein.
106. *Palladium Reagents and Catalysis, New Perspectives for the 21st Century*, ed. J. Tsuji, John Wiley & Sons, Chichester, 2004.
107. K. Orito, A. Horibata, T. Nakamura, H. Ushito, H. Nagasaki, M. Yuguchi, S. Yamashita and M. Tokuda, *J. Am. Chem. Soc.*, 2004, **126**, 14342.
108. K. Orito, M. Miyazawa, T. Nakamura, A. Horibata, H. Ushito, H. Nagasaki, M. Yuguchi, S. Yamashita, T. Yamazaki and M. Tokuda, *J. Org. Chem.*, 2006, **71**, 5951.
109. B. Haffemayer, M. Gulias and M. J. Gaunt, *Chem. Sci.*, 2011, **2**, 312.
110. J. W. Wrigglesworth, B. Cox, G. C. Lloyd-Jones and K. I. Booker-Milburn, *Org. Lett.*, 2011, **13**, 5326.
111. B. López, A. Rodriguez, D. Santos, J. Albert, X. Ariza, J. Garcia and J. Granell, *Chem. Commun.*, 2011, **47**, 1054.
112. C. E. Houlden, M. Hutchby, C. D. Bailey, J. G. Ford, S. N. G. Tyler, M. R. Gagné, G. C. Lloyd-Jones and K. I. Booker-Milburn, *Angew. Chem., Int. Ed.*, 2009, **48**, 1830.
113. E. J. Yoo, M. Wasa and J.-Q. Yu, *J. Am. Chem. Soc.*, 2010, **132**, 17378.
114. C. Li, L. Wang, P. Li and W. Zhou, *Chem.–Eur. J.*, 2011, **17**, 10208.
115. Y. Yuan, D. Chen and X. Wang, *Adv. Synth. Catal.*, 2011, **353**, 3373.
116. O. Baslé, J. Bidange, Q. Shuai and C.-J. Li, *Adv. Synth. Catal.*, 2010, **352**, 1145.
117. F. Xiao, Q. Shuai, F. Zhao, O. Baslé, G. Deng and C.-J. Li, *Org. Lett.*, 2011, **13**, 1614.
118. X. Peng, Y. Zhu, T. A. Ramirez, B. Zhao and Y. Shi, *Org. Lett.*, 2011, **13**, 5244.
119. E. M. Simmons and J. F. Hartwig, *Nature*, 2012, **483**, 70. For additional information of direct alkylations see: L. Ackermann, *Chem. Commun.*, 2010, **46**, 4866.
120. Y. Zhao and G. Chen, *Org. Lett.*, 2011, **13**, 4850.
121. J. A. Romero-Revilla, A. García-Rubia, R. Gómez-Arrayás, M. A. Fernández-Ibáñez and J. C. Carretero, *J. Org. Chem.*, 2011, **76**, 9525.
122. X. Wang, L. Truesdale and J.-Q. Yu, *J. Am. Chem. Soc.*, 2010, **132**, 3648.
123. Y. Ye, N. D. Ball, J. W. Kampf and M. S. Sanford, *J. Am. Chem. Soc.*, 2010, **132**, 14682.
124. D. C. Powers, D. Y. Xiao, M. A. L. Geibel and T. Ritter, *J. Am. Chem. Soc.*, 2010, **132**, 14530.
125. J. F. Hartwig, *Nature*, 2008, **455**, 314.
126. J. F. Hartwig, *Acc. Chem. Res.*, 2012, **45**, 864.
127. I. A. I. Mkhalid, J. H. Barnard, T. B. Marder, J. M. Murphy and J. F. Hartwig, *Chem. Rev.*, 2010, **110**, 890.
128. H.-X. Dai and J.-Q. Yu, *J. Am. Chem. Soc.*, 2012, **134**, 134.

129. A. N. Vedernikov, *Acc. Chem. Res.*, 2012, **45**, 803.
130. C. Huang, N. Ghavtadze, B. Chattopadhyay and V. Gevorgyan, *J. Am. Chem. Soc.*, 2011, **133**, 17630.
131. F.-R. Gou, X.-C. Wuang, P.-F. Huo, H.-P. Bi, Z.-H. Guan and Y.-M. Liang, *Org. Lett.*, 2009, **11**, 5276.
132. G.-W. Wang and T.-T. Yuan, *J. Org. Chem.*, 2010, **75**, 476.
133. H.-M. Guo, W.-H. Rao, H.-Y. Niu, L.-L. Yiang, G. Meng, J.-J. Jin, X.-N. Yang and G.-R. Qu, *Chem. Commun.*, 2011, **47**, 5608.
134. C. J. Vickers, T.-S. Mei and J.-Q. Yu, *Org. Lett.*, 2010, **12**, 2511.
135. X. Wang, Y. Lu, H.-X. Dai and J.-Q. Yu, *J. Am. Chem. Soc.*, 2010, **132**, 12203.
136. B. Xiao, T.-J. Gong, Z.-J. Liu, J.-H. Liu, D.-F. Luo, J. Xu and L. Liu, *J. Am. Chem. Soc.*, 2011, **133**, 9250.
137. Y. Wei and N. Yoshikai, *Org. Lett.*, 2011, **13**, 5504.
138. P. Novák, A. Correa, J. Gallardo-Donaire and R. Martín, *Angew. Chem., Int. Ed.*, 2011, **50**, 12236.
139. X. Zhao, E. Dimitrijevic and V. M. Dong, *J. Am. Chem. Soc.*, 2009, **131**, 3466.
140. X. Zhao and V. M. Dong, *Angew. Chem. Int. Ed.*, 2011, **50**, 932.
141. E. J. Yoo, S. Ma, T.-S. Mei, K. S. L. Chan and J.-Q. Yu, *J. Am. Chem. Soc.*, 2011, **133**, 7652.
142. A. Iglesias, R. Álvarez, A. R. de Lera and K. Muñiz, *Angew. Chem., Int. Ed.*, 2012, **51**, 2225.
143. K.-H. Ng, A. S. C. Chan and W.-Y. Yu, *J. Am. Chem. Soc.*, 2010, **132**, 12862.
144. W. C. P. Tsang, R. H. Munday, G. Brasche, N. Zheng and S. L. Buchwald, *J. Org. Chem.*, 2008, **73**, 7603.
145. J. A. Jordan-Hore, C. C. C. Johansson, M. Gulias, E. M. Beck and M. J. Gaunt, *J. Am. Chem. Soc.*, 2008, **130**, 16184.
146. S. W. Youn, J. H. Bihn and B. S. Kim, *Org. Lett.*, 2011, **13**, 3738.
147. J.-H. Chu, P.-S. Lin, Y.-M. Lee, W.-T. Shen and M.-J. Wu, *Chem.–Eur. J.*, 2011, **17**, 13613.
148. Q. Xiao, W.-H. Wang, G. Liu, F.-K. Meng, J.-H. Chen, Z. Yang and Z.-J. Shi, *Chem.–Eur. J.*, 2009, **15**, 7292.
149. R. K. Kumar, M. A. Ali and T. Punniyamurthy, *Org. Lett.*, 2011, **13**, 2102.
150. E. T. Nadres and O. Daugulis, *J. Am. Chem. Soc.*, 2012, **134**, 7.
151. G. He, Y. Zhao, S. Zhang, C. Lu and G. Chen, *J. Am. Chem. Soc.*, 2012, **134**, 3.
152. L. Ackermann, A. V. Lygin and N. Hofmann, *Angew. Chem., Int. Ed.*, 2011, **50**, 6379.
153. L. Ackermann and S. Fenner, *Org. Lett.*, 2011, **13**, 6548.
154. N. Guimond, C. Gouliaras and K. Fagnou, *J. Am. Chem. Soc.*, 2010, **132**, 6908.
155. S. Mochida, N. Umeda, K. Hirano, T. Satoh and M. Miura, *Chem. Lett.*, 2010, **39**, 744.

156. T. K. Hyster and T. Rovis, *J. Am. Chem. Soc.*, 2010, **132**, 10565.
157. T. K. Hyster and T. Rovis, *Chem. Sci.*, 2011, **2**, 1606.
158. L. Ackermann and A. V. Lygin, *Org. Lett.*, 2012, **14**, 764.
159. M. P. Huestis, L. Chan, D. R. Stuart and K. Fagnou, *Angew. Chem., Int. Ed.*, 2011, **50**, 1338.
160. R. K. Chinnagolla and M. Jeganmohan, *Chem. Commun.*, 2012, **48**, 2030.
161. L. Ackermann, J. Pospech, K. Graczyk and K. Rausch, *Org. Lett.*, 2012, **14**, 930.
162. T. Satoh and M. Miura, *Chem.–Eur. J.*, 2010, **16**, 11212.
163. N. Guimond and K. Fagnou, *J. Am. Chem. Soc.*, 2009, **131**, 12050.
164. T. Fukutani, N. Umeda, K. Hirano, T. Satoh and M. Miura, *Chem. Commun.*, 2009, 5141.
165. P. C. Too, Y.-F. Wang and S. Chiba, *Org. Lett.*, 2010, **12**, 5688.
166. X. Zhang, D. Chen, M. Zhao, J. Zhao, A. Jia and X. Li, *Adv. Synth.Catal.*, 2011, **353**, 719.
167. S. Duttwyler, C. Lu, A. L. Rheingold, R. G. Bergman and J. A. Ellman, *J. Am. Chem. Soc.*, 2012, **134**, 4064.
168. L. Ackermann and J. Pospech, *Org. Lett.*, 2011, **13**, 4153.
169. S. H. Park, J. Y. Kim and S. Chang, *Org. Lett.*, 2011, **13**, 2372.
170. K. Padala and M. Jeganmohan, *Org. Lett.*, 2011, **13**, 6144.
171. K. Padala and M. Jeganmohan, *Org. Lett.*, 2011, **13**, 1134.
172. F. W. Patureau, T. Besset and F. Glorius, *Angew. Chem., Int. Ed.*, 2011, **50**, 1064.
173. B. Li, J. Ma, N. Wang, H. Feng, S. Xu and B. Wang, *Org. Lett.*, 2012, **14**, 736.
174. X. Li, X. Gong, M. Zhao, G. Song, J. Deng and X. Li, *Org. Lett.*, 2011, **13**, 5808.
175. C. Zhu and J. R. Falck, *Chem. Commun.*, 2012, **48**, 1674.
176. A. S. Tsai, M. Brasse, R. G. Bergman and J. A. Ellman, *Org. Lett.*, 2011, **13**, 540.
177. T. Besset, N. Kuhl, F. W. Patureau and F. Glorius, *Chem.–Eur. J.*, 2011, **17**, 7167.
178. C. Feng and T.-P. Loh, *Chem. Commun.*, 2011, **47**, 10458.
179. L. Ackermann, L. Wang, R. Wolfram and A. V. Lygin, *Org. Lett.*, 2012, **14**, 728.
180. N. Schröder, T. Besset and F. Glorius, *Adv. Synth. Catal.*, 2012, **354**, 579.
181. N. Chatani, S. Inoue, K. Yokota, H. Tatamidani and Y. Fukumoto, *Pure Appl. Chem.*, 2010, **82**, 1443.
182. N. Hasegawa, V. Charra, S. Inoue, Y. Fukumoto and N. Chatani, *J. Am. Chem. Soc.*, 2011, **133**, 8070.
183. Y. Du, T. K. Hyster and T. Rovis, *Chem. Commun.*, 2011, **47**, 12074.
184. J. Park, E. Park, A. Kim, Y. Lee, K.-W. Chi, J. H. Kwak, Y. H. Jung and I. S. Kim, *Org. Lett.*, 2011, **13**, 4390.
185. S. Sharma, E. Park, J. Park and I. S. Kim, *Org. Lett.*, 2012, **14**, 906.

186. Y. Li, X.-S. Zhang, K. Chen, K.-H. He, F. Pan, B.-J. Li and Z.-J. Shi, *Org. Lett.*, 2012, **14**, 636.
187. L. Yang, C. A. Correia and C.-J. Li, *Adv. Synth. Catal.*, 2011, **353**, 1269.
188. K. D. Hesp, R. G. Bergman and J. A. Ellman, *Org. Lett.*, 2012, **14**, 2304.
189. J. Wencel-Delord, C. Nimphius, F. W. Patureau and F. Glorius, *Angew. Chem., Int. Ed.*, 2012, **51**, 2247.
190. N. Luo and Z. Yu, *Chem.–Eur. J.*, 2010, **16**, 787.
191. M. K. Lakshman, A. C. Deb, R. R. Chamala, P. Pradhan and R. Pratap, *Angew. Chem., Int. Ed.*, 2011, **50**, 11400.
192. L. Ackermann, J. Pospech and H. K. Potukuchi, *Org. Lett.*, 2012, **14**, 2146.
193. L. Ackermann, R. Vicente, H. K. Potukuchi and V. Pirovano, *Org. Lett.*, 2010, **12**, 5032.
194. E. Ferrer Flegeau, C. Bruneau, P. H. Dixneuf and A. Jutand, *J. Am. Chem. Soc.*, 2011, **133**, 10161.
195. K.-H. Ng, Z. Zhou and W.-Y. Yu, *Org. Lett.*, 2012, **14**, 272.
196. C. Grohmann, H. Wang and F. Glorius, *Org. Lett.*, 2012, **14**, 656.
197. S. Kawamorita, T. Miyazaki, H. Ohmiya, T. Iwai and M. Sawamura, *J. Am. Chem. Soc.*, 2011, **133**, 19310.
198. O. Saidi, J. Marafie, A. E. W. Ledger, P. M. Liu, M. F. Mahon, G. Kociok-Köhn, M. K. Whittlesey and C. G. Frost, *J. Am. Chem. Soc.*, 2011, **133**, 19298.
199. M. M. Díaz-Requejo and P. J. Pérez, *Chem. Rev.*, 2008, **108**, 3379.
200. O. Daugulis, *Top. Curr. Chem.*, 2010, **292**, 57.
201. X. Chen, X.-S. Hao, C. E. Goodhue and J.-Q. Yu, *J. Am. Chem. Soc.*, 2006, **128**, 6790.
202. W. Wang, C. Pan, F. Chen and J. Cheng, *Chem. Commun.*, 2011, **47**, 3978.
203. J. Zhao, Y. Wang, Y. He, L. Liu and Q. Zhu, *Org. Lett.*, 2012, **14**, 1078.
204. L. Zhang, Z. Liu, H. Li, G. Fang, B.-D. Barry, T. A. Belay, X. Bi and Q. Liu, *Org. Lett.*, 2011, **13**, 6536.
205. A. John and K. M. Nicholas, *J. Org. Chem.*, 2011, **76**, 4158.
206. G. Brasche and S. L. Buchwald, *Angew. Chem., Int. Ed.*, 2008, **47**, 1932.
207. R. Berrino, S. Cacchi, G. Fabrizi and A. Goggiamani, *J. Org. Chem.*, 2012, **77**, 2537.
208. A. J. Hickman and M. S. Sanford, *Nature*, 2012, **484**, 177.
209. There is another contribution of Yu *et al.* (Y.-H. Zhang, B.-F. Shi and J.-Q. Yu, *J. Am. Chem. Soc.*, 2009, **131**, 5072) where the *meta* olefination of electron-deficient arenes, catalysed by Pd(II) is described. However, the work of Yu is based on the steric hindrance of ancillary bulky ligands, which ultimately are not functionalized, while the work of Gaunt[212] is based on the electronic effects of the substrate to be functionalized, where the directing group plays an active role.
210. R. E. Maleczka Jr., *Science*, 2009, **323**, 1572.
211. Y. Zhou, J. Zhao and L. Liu, *Angew. Chem., Int. Ed.*, 2009, **48**, 7126.
212. R. J. Phipps and M. J. Gaunt, *Science*, 2009, **323**, 1593.

213. H. A. Duong, R. E. Gilligan, M. L. Cooke, R. J. Phipps and M. J. Gaunt, *Angew. Chem., Int. Ed.*, 2011, **50**, 463.
214. E. Y. Lee and J. Park, *ChemCatChem*, 2011, **3**, 1127.
215. B. Chen, X.-L. Hou, Y.-X. Li and Y.-D. Wu, *J. Am. Chem. Soc.*, 2011, **133**, 7668.
216. C.-L. Ciana, R. J. Phipps, J. R. Brandt, F.-M. Meyer and M. J. Gaunt, *Angew. Chem., Int. Ed.*, 2011, **50**, 458.
217. D. Leow, G. Li, T.-S. Mei and J.-Q. Yu, *Nature*, 2012, **486**, 518.

CHAPTER 9

Transition Metal-Catalysed Direct Arylation of Unactivated Arenes with Aryl Halides

AIWEN LEI* AND HUA ZHANG

The College of Chemistry and Molecular Sciences, Wuhan University, 430072, Wuhan, Hubei, P. R. China
*Email: aiwenlei@whu.edu.cn

9.1 Introduction

The biaryl structural motif constitutes the core of many pharmaceutically relevant and biologically active compounds and is extensively applied in the material sciences.[1–3] Although there exist several methods for the construction of aryl–aryl bonds, the most common route is transition metal-catalysed cross-coupling reactions, such as Stille,[4,5] Suzuki–Miyaura,[6,7] Negishi,[8–10] Hiyama[11,12] and Kumada[13–15] couplings (Scheme 9.1). Typically, these reactions involve either the coupling of an aryl halide or pseudohalide with an organometallic reagent, which often originates from a corresponding simple arene.

Recently, direct arylation of arenes with aryl halides *via* C–H bond activation has received significant attention, as prefunctionalization to form organometallic reagents from arenes was avoided and the corresponding metal wastes were reduced (Scheme 9.2).[16–25] Direct arylation of arenes bearing directing groups can be regarded as one category (Scheme 9.2A).[20–22,26–29] Heterocycles were directly arylated successfully in many cases and could be regarded in another category (Scheme 9.2B).[28] The other type of direct arylation

RSC Catalysis Series No. 11
C–H and C–X Bond Functionalization: Transition Metal Mediation
Edited by Xavi Ribas

Published by the Royal Society of Chemistry, www.rsc.org

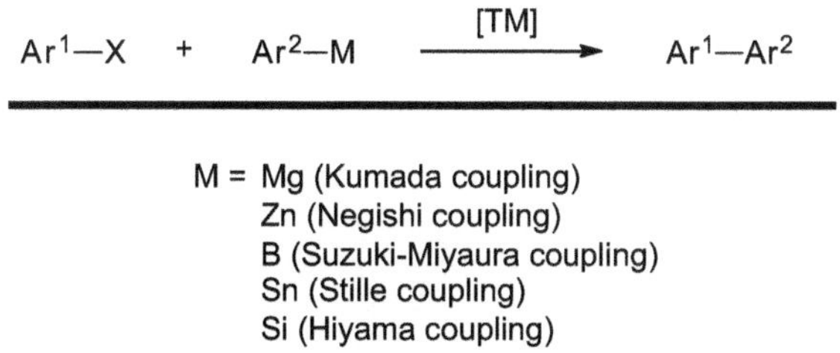

Scheme 9.1 Transition metal catalysed cross-coupling reactions.

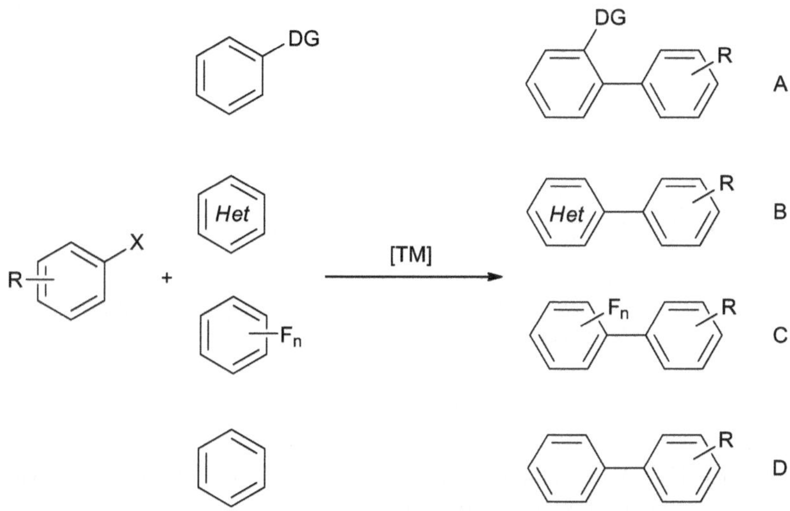

Scheme 9.2 Transition metal catalysed direct arylation reactions with aryl halides.

involves the cascade deprotonation/metallation of electron-deficient arenes (Scheme 9.2C).[17]

Unactivated arenes (unactivated arenes refer to benzene, naphthalene, toluene, anisole, xylene and mesitylene in this text) have been among the most salient components in the chemical industry and are produced from petrochemicals in ton-scale quantities. However, the direct arylation of unactivated arenes is one of the most important and challenging topics in directly utilizing C–H bonds because of the low acidity of aromatic C–H bonds and the lack of directing groups to assist the aromatic C–H bond cleavage (Scheme 9.2D).[22–24] To date, very limited examples have been successfully reported on the direct arylation of unactivated arenes using the Ar–X as the electrophiles, and are summarised in this chapter.

9.2 Iridium and Rhodium Catalysed Direct Arylation

The first direct arylation of unactivated benzene with aryl iodides catalysed by a noble Ir complex in the presence of potassium *tert*-butoxide (KO^{t}Bu) was

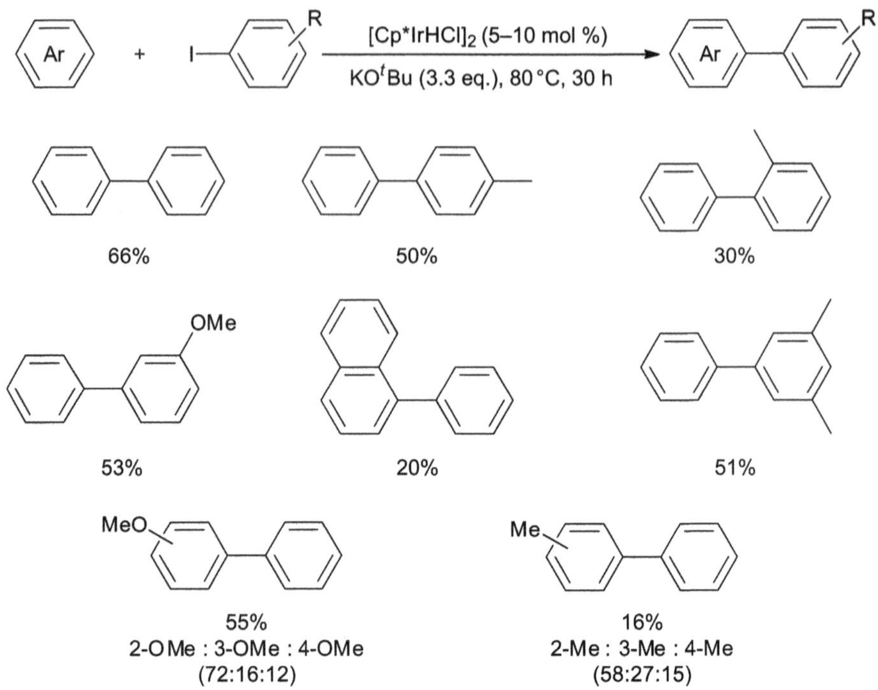

Scheme 9.3 Ir-catalysed direct arylation of unactivated benzene.

reported by Fujita in 2004 (Scheme 9.3).[30] $[Cp^*Ir(H)Cl]_2$ ($Cp^* = C_5Me_5$) as catalyst gave the best yields and benzene was used as solvent in this transformation. The regioselectivities in the reaction of anisole and toluene with iodobenzene are very similar to those for the radical aromatic substitution reactions reported by Davies,[31,32] suggesting that radical intermediates would be involved in this reaction. When the reaction was performed in the presence of a radical scavenger (2,6-di-*tert*-butylphenol), the yield of biaryl product considerably decreased.

The first Rh-catalysed direct C–H arylation of benzene derivatives was reported by Itami *et al.* in 2006.[33,34] The direct arylation of anisole and 1,3-dimethoxybenzene with *p*-nitrophenyl iodide worked well under a catalytic amount of Rh catalyst bearing a strongly π-accepting ligand and silver carbonate, furnishing the corresponding biaryls in moderate to good yields (Scheme 9.4). Notably, the clear *ortho-para* selectivity in this transformation is consistent with the C–H bond cleavage based on electrophilic metalation but not directed *ortho*-metalation and/or C–H oxidative addition.

Kempe *et al.* later reported a $[Rh^I]^+[Rh^{III}]^-$ catalyst for the direct arylation of unactivated arenes with aryl iodides, bromides and chlorides (Scheme 9.5).[35] Benzene is arylated by iodobenzene under milder conditions with a TON of 780 in a shorter period of time compared to the former Ir catalytic system. The regioselectivities in arylation of toluene with iodobenzene and the results of a

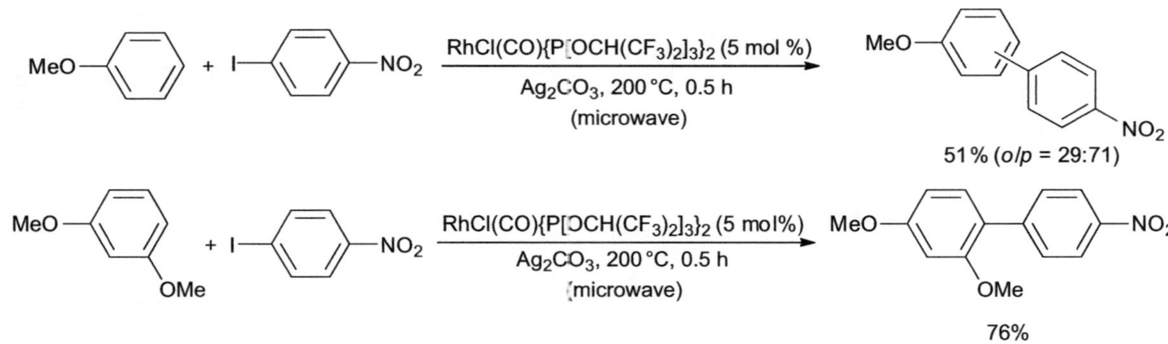

Scheme 9.4 Rh-catalysed direct arylation of benzene derivatives.

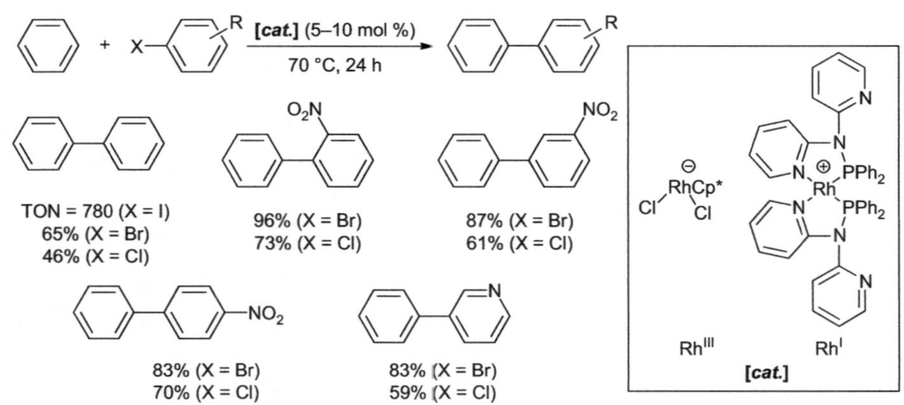

Scheme 9.5 Bimetallic Rh-catalysed direct arylation of unactivated arenes.

Hammett correlation indicated the participation of radical intermediates in this reaction.

9.3 Palladium Catalysed Direct Arylation

The Pd-catalysed regioselective arylation of polar hydrocarbon azulene with iodobenzene at the electron-rich 1-position was first reported by Dyker *et al.* in 2000 (Scheme 9.6).[36] Although azulene is one of the most reactive aromatic hydrocarbons, it suggested the possibility for the direct C–H arylation of less reactive aromatic arenes, such as benzene and toluene.

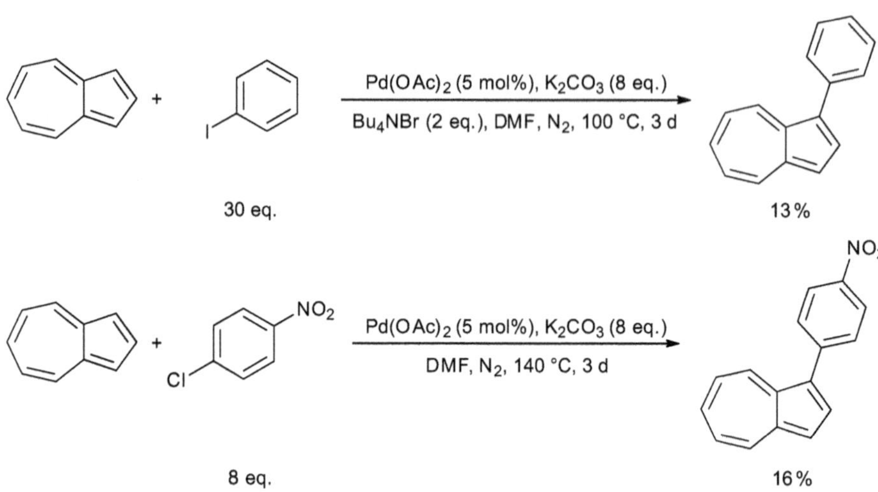

Scheme 9.6 Pd-catalysed C–H arylation of azulene.

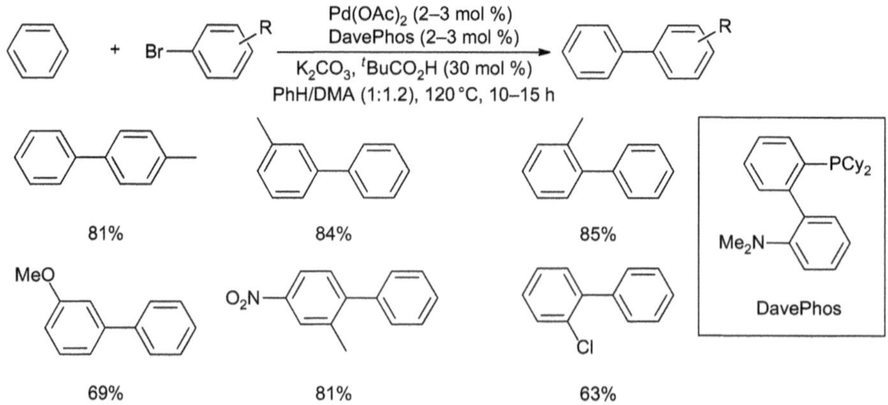

Scheme 9.7 Pd-catalysed direct arylation of benzene with aryl bromides.

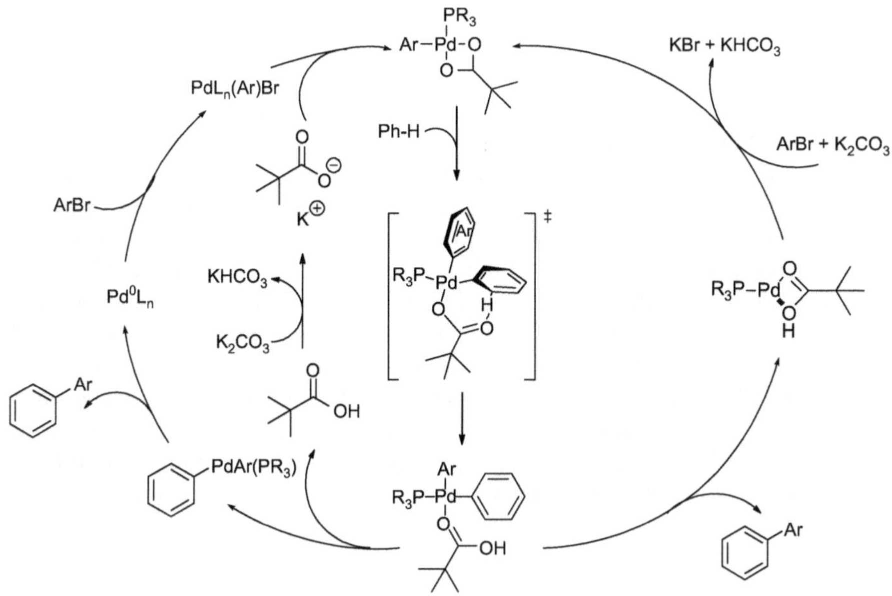

Scheme 9.8 Proposed mechanism for Pd-catalysed direct arylation of benzene with aryl bromides.

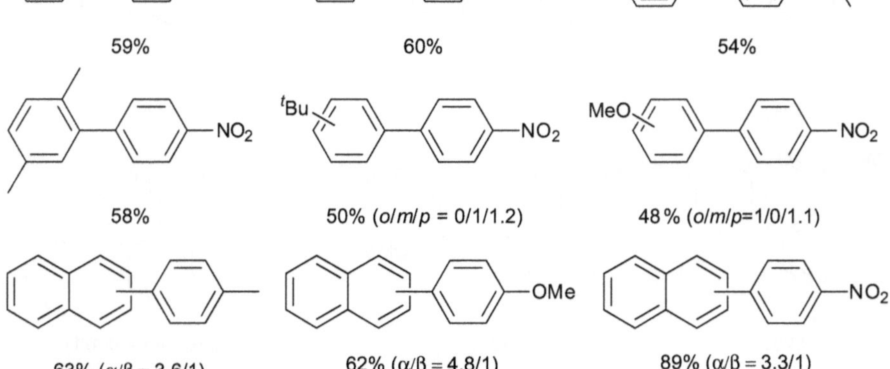

Scheme 9.9 Phosphine-free Pd-catalysed direct arylation.

Later on, Fagnou and coworkers reported the first Pd-catalysed direct arylation of benzene using Pd^{II}/DavePhos [2-Dicyclohexylphosphino-2′-(*N*,*N*-dimethylamino)biphenyl] as the catalyst (Scheme 9.7).[37] The substrate scope was limited to aryl bromides, while more reactive aryl iodides and inactive aryl chlorides showed less efficiency. The choice of base and additive was crucial to the transformation. Experimental and computational evidence indicated that the pivalate anion is a key component in C–H bond cleavage, which lowers the energy of C–H bond cleavage and acts as a catalytic proton shuttle from benzene to the stoichiometric carbonate base (Scheme 9.8).

A phosphine-free Pd-catalysed arylation of naphthalene and benzene with aryl iodides to give biaryls in moderate to good yields were reported by Lu *et al.* (Scheme 9.9).[38] A $Pd^{II/IV}$ catalytic cycle was postulated for this reaction.

In 2010, Charette *et al.* disclosed an efficient Pd-catalysed direct arylation of unactivated arenes with aryl bromides containing an α-directing group (Scheme 9.10).[39] The reaction requires only a catalytic amount of $Pd(OAc)_2$ and a substoichiometric amount of silver salts, without any external base or ligand.

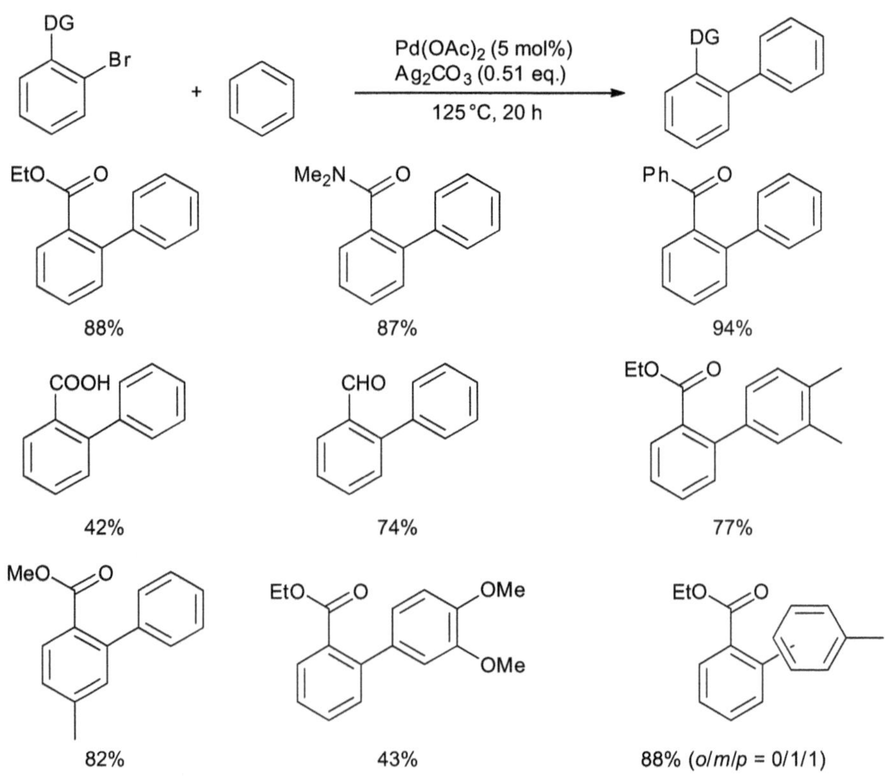

Scheme 9.10 Pd-catalysed direct arylation with aryl bromides bearing directing group.

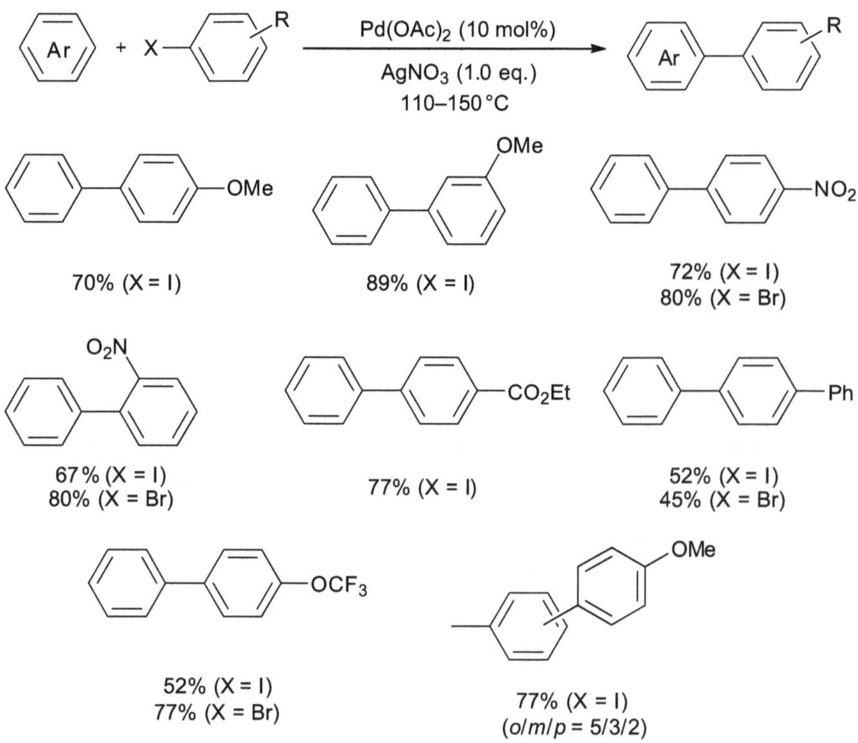

Scheme 9.11 Direct arylation catalysed by $Pd(OAc)_2/AgNO_3$ system.

Very recently, a $Pd(OAc)_2/AgNO_3$ catalyst system was employed to achieve the direct C–H arylation of unactivated arenes (Scheme 9.11).[40]

9.4 Iron, Cobalt and Nickel Catalysed Direct Arylation

Nickel-catalysed direct arylation of unactivated benzene and naphthalene was first developed by Yamakawa (Scheme 9.12).[41] The arylation was successfully catalysed by Cp_2Ni in the presence of KO^tBu and a catalytic amount of BEt_3. All the components have been found to be indispensable for the arylation.

In 2010, Lei and coworkers successfully developed the first iron-catalysed direct arylation of unactivated arenes with a broad range of aryl halides through C–H bond activation to prepare birayl compounds (Scheme 9.13).[42] The choice of base and ligand was essential to the success of this transformation. Optimizations on the reaction parameters showed that 15 mol% $FeCl_3$ combined with 30 mol% N^1,N^2-dimethylethane-1,2-diamine (DMEDA) afforded the highest yield of direct arylated products in the presence of 90 equiv. of benzene and 2 equiv. of lithium bis(trimethylsilyl)amide (LiHMDS).

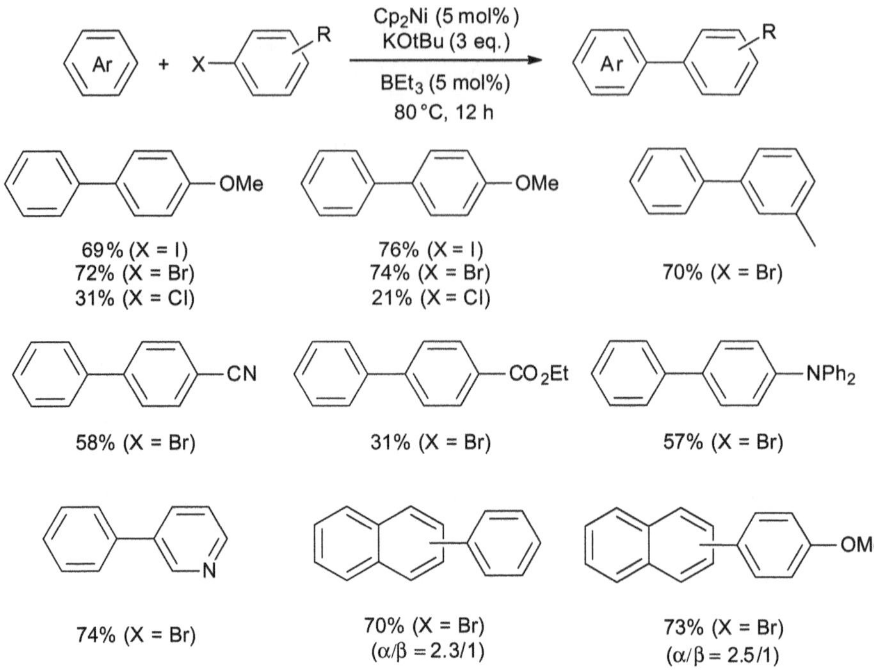

Scheme 9.12 Cp_2Ni-KO^{t}Bu catalysed direct arylation.

Various aryl halides including aryl iodides, bromides and chlorides underwent the direct arylation smoothly with unactivated benzene. Toluene and naphthalene were also investigated and afforded mixtures of direct arylation products. It's noteworthy that ferrocene could be successfully arylated with bromobenzene to give an acceptable 54% yield of the corresponding monoarylated product.

A benzyne mechanism was ruled out, as the reactions of various substituted aryl halides with benzene all produced the sole corresponding direct arylation products. The mechanism involving base-assisted metallation and subsequent cross-coupling was also ruled out because of the high C–H bond energy (110 kcal mol^{-1}) and pK_a value (44.7) of the unactivated benzene. No conversion of aryl halide was observed in the presence of radical scavenger TEMPO (1.0 equiv.) and a H/D kinetic isotope effect (KIE) of 1.7 was obtained in labeling experiments. The radical trapping experiment suggests that a radical pathway was possibly involved, since the yield of coupling product decreased from 81% to 16%, and the coupling with the radical trap afforded 57% yield (Scheme 9.14).

Charette simultaneously reported a similar version of direct arylation of unactivated benzene and other simple arenes catalysed by $Fe(OAc)_2$ and bathophenanthroline in the presence of potassium *tert*-butoxide (KO^{t}Bu) (Scheme 9.15).[43] The substrate scope of aryl halides was limited to aryl iodides.

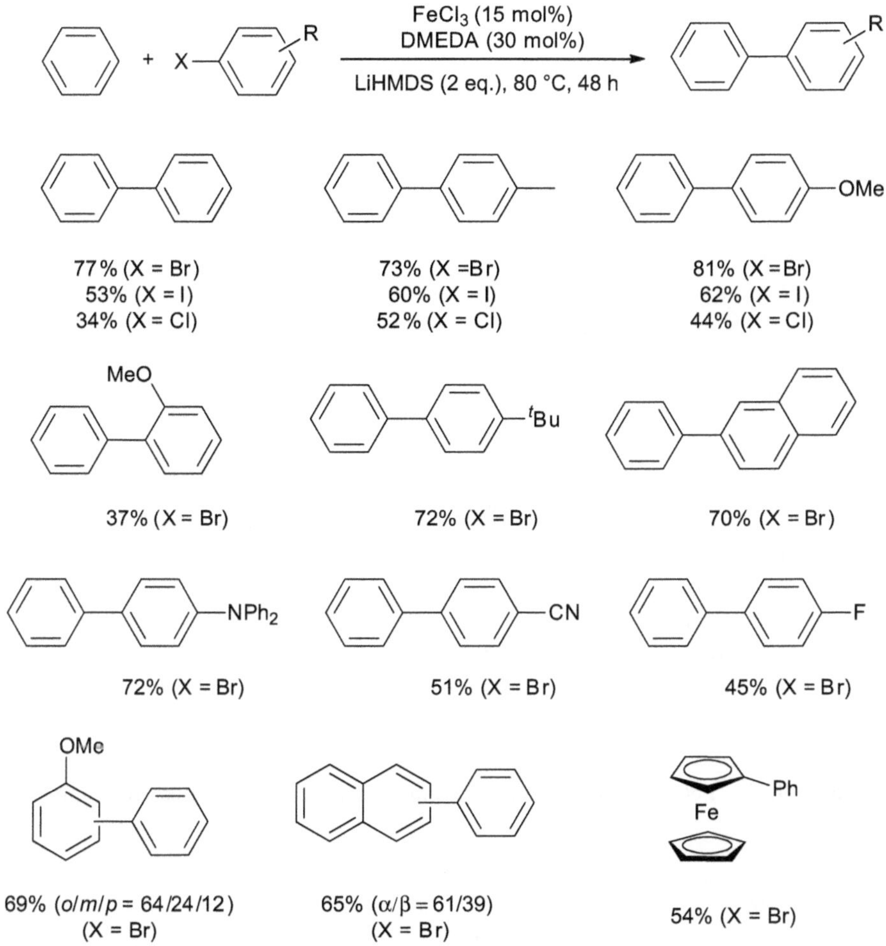

Scheme 9.13 $FeCl_3$-DMEDA-LiHMDS catalysed direct arylation of unactivated arenes with aryl halides.

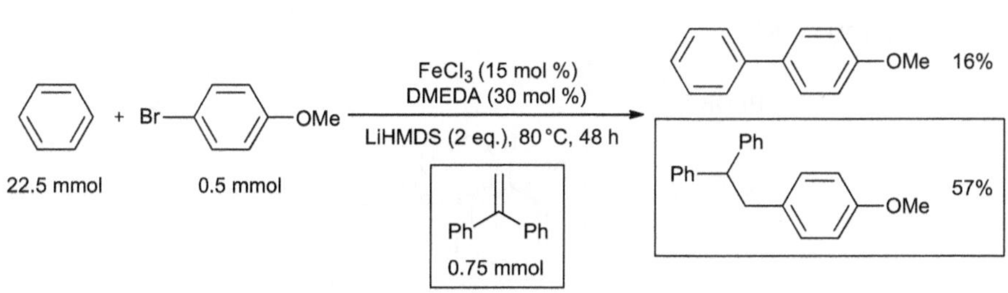

Scheme 9.14 Aryl radical trap experiment in $FeCl_3$-DMEDA-LiHMDS catalysed direct arylation.

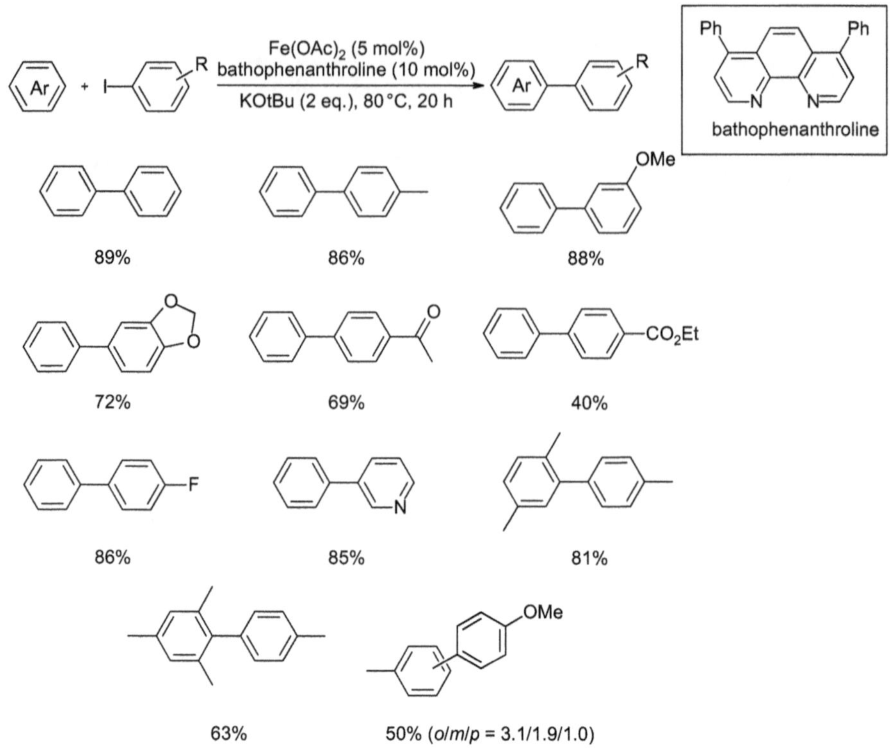

Scheme 9.15 $Fe(OAc)_2$-bathophenanthroline-KO^tBu catalysed direct arylation.

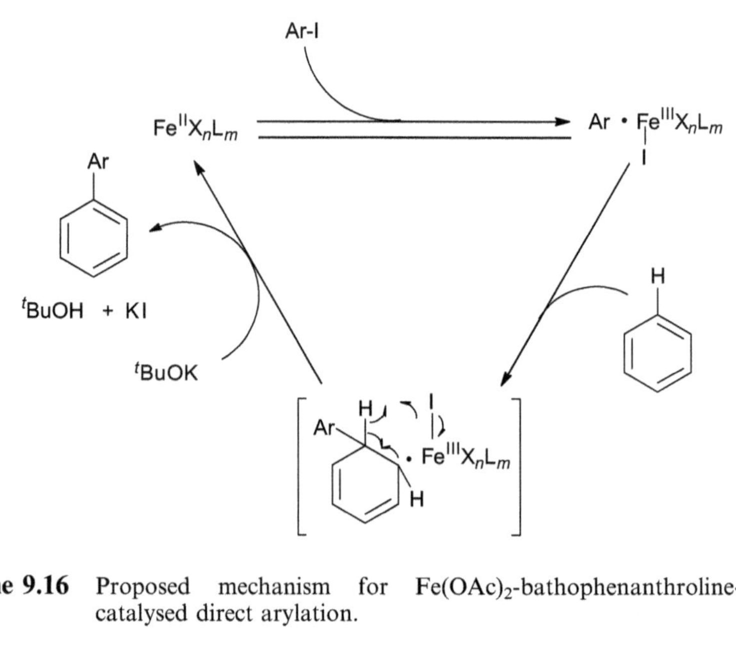

Scheme 9.16 Proposed mechanism for $Fe(OAc)_2$-bathophenanthroline-KO^tBu catalysed direct arylation.

Mechanistic evidence suggested that the transformation proceeded through an Fe-catalysed radical process (Scheme 9.16). The activation of the aryl–halogen bond *via* an iodine atom transfer formed an initiating radical species and an oxidized metallo-intermediate. The biaryl product was produced *via* radical addition onto an arene and proximal abstraction of a halogen atom from the Fe^{III} complex. The active Fe catalyst was regenerated as well as HI that could be quenched by KO^{t}Bu.

Lei *et al.* further reported an efficient cobalt catalysed direct arylation of unactivated arenes, such as benzene, with a broad range of aryl halides, including aryl iodides, aryl bromides, and aryl chlorides at a relatively mild reaction temperature (Scheme 9.17).[44] [Co(acac)$_3$] alone gave the best results in the reaction of aryl iodides and bromides while $CoBr_2$ combined with DMEDA showed higher efficiency when aryl chlorides were employed as substrates (Scheme 9.18). Compared with iron catalysts developed before, the inexpensive

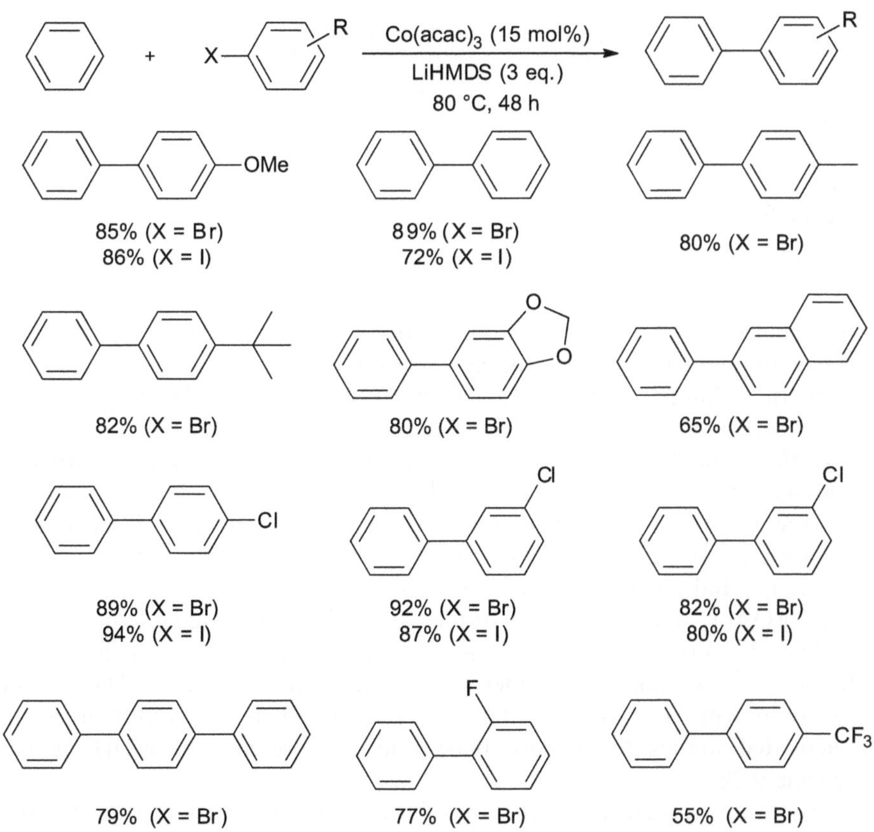

Scheme 9.17 Co(acac)$_3$-catalysed direct arylation of unactivated arenes with aryl iodides and bromides.

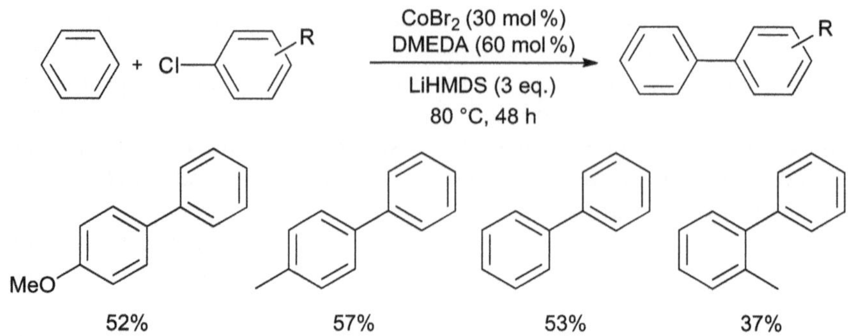

Scheme 9.18 $CoBr_2$-catalysed direct arylation of unactivated arenes with aryl chlorides.

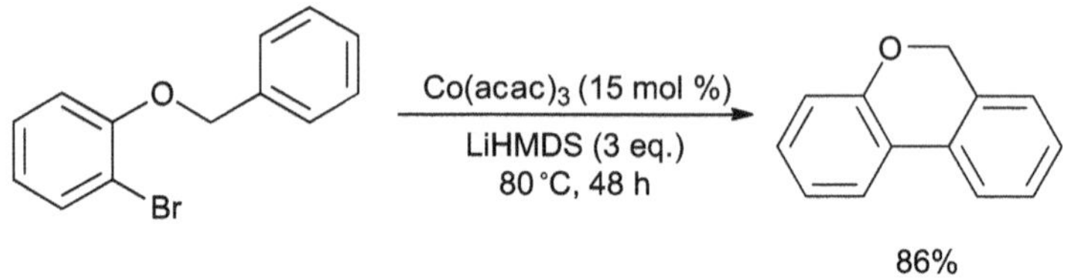

Scheme 9.19 Co-catalysed intramolecular direct arylation.

and easy to handle [$Co(acac)_3$] catalyst showed higher efficiency, in most cases even without extra ligands, and led to various biaryl products in excellent yields. In addition, an intramolecular direct arylation was also achieved using simple cobalt catalysis at 80 °C (Scheme 9.19).

At the same time, Shi *et al.* developed the direct C–H transformation catalysed by various transition metal catalysts based on Nb, Mo and Co (Scheme 9.20).[45] The construction of biaryls by cheaper and more easily available, and sometimes lower or nontoxic, transition metal complexes showed advantages and great potential for industry.

A $Ni(OAc)_2 \cdot 4H_2O$-KO*t*Bu-1,10-phenanthroline catalytic system is able to catalyze the direct arylation of unactivated arenes with aryl halides in the absence of any additives under mild conditions (Scheme 9.21).[46] Co(II) porphyrin can also catalyze the cross-coupling between aryl halides and unactivated arenes to prepare biaryls in the presence of KOH as base (Scheme 9.22).[47]

Very recently, a cheap magnetic Fe-Ni alloy has been applied to the direct arylation of unactivated arenes with aryl halides (Scheme 9.23).[48] The catalyst exhibits excellent activity and can be reused at least eight times without a significant loss of catalytic activity.

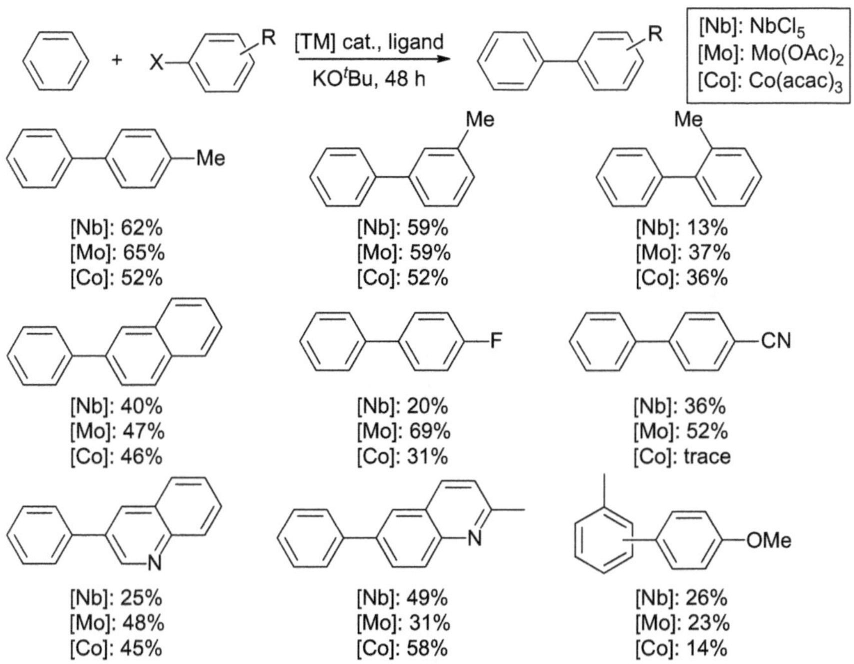

Scheme 9.20 $NbCl_5$, $Mo(OAc)_2$ and $Co(acac)_3$ catalysed direct arylation.

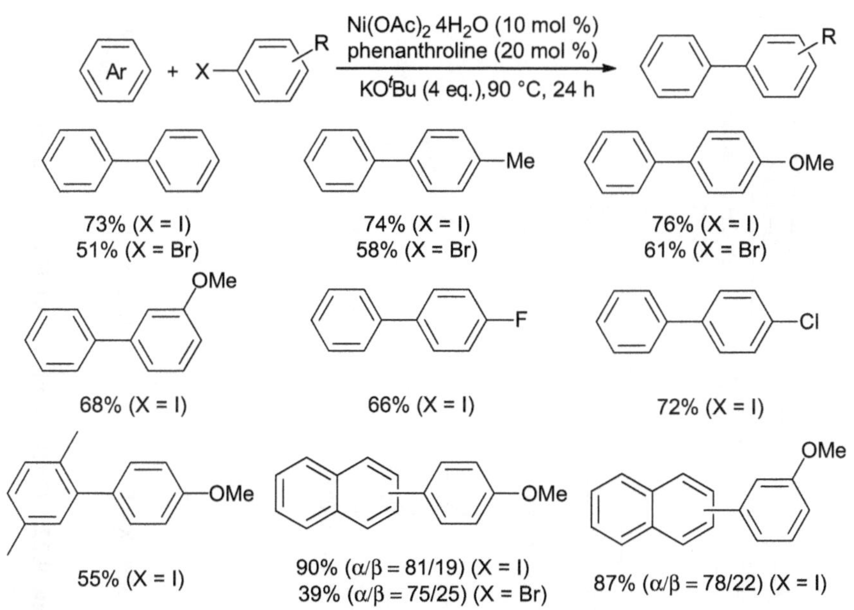

Scheme 9.21 $Ni(OAc)_2 \cdot 4H_2O$ catalysed direct arylation.

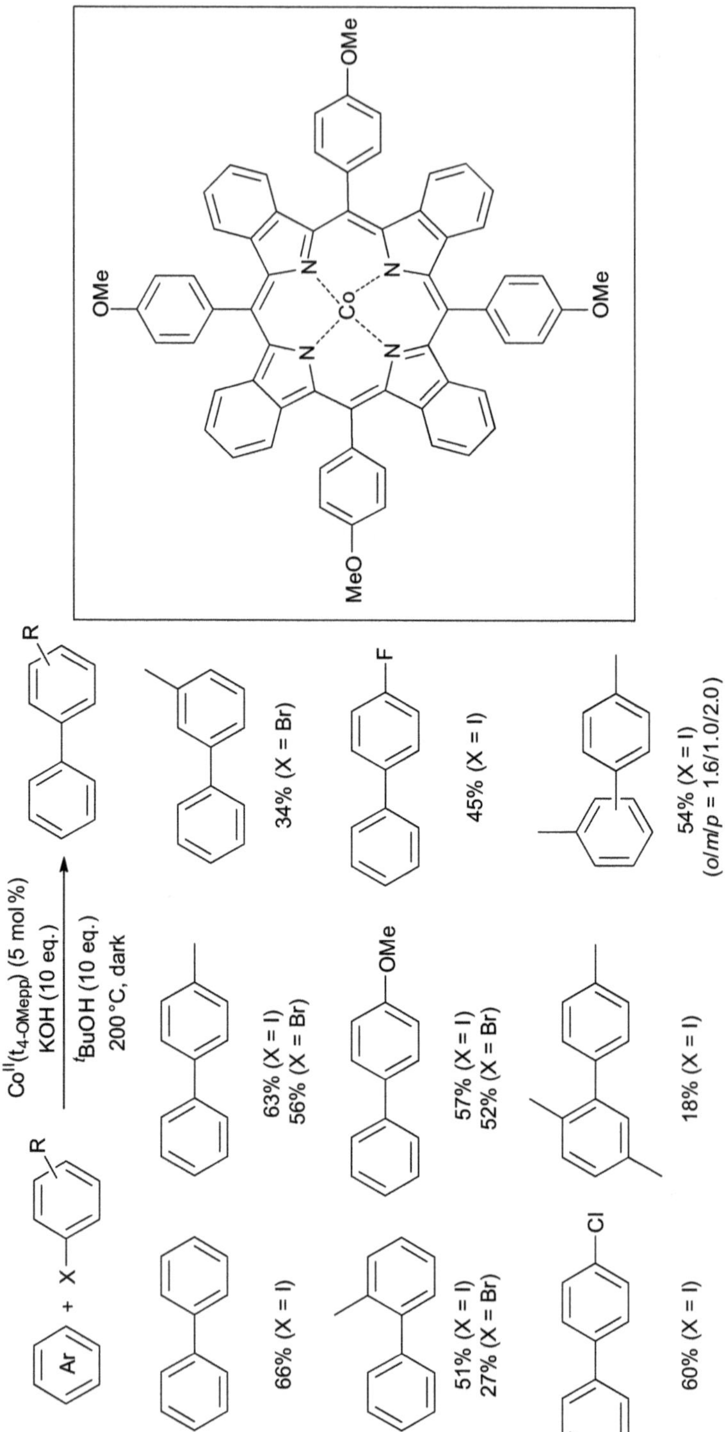

Scheme 9.22 Co(II) porphyrin-KOH catalysed direct arylation.

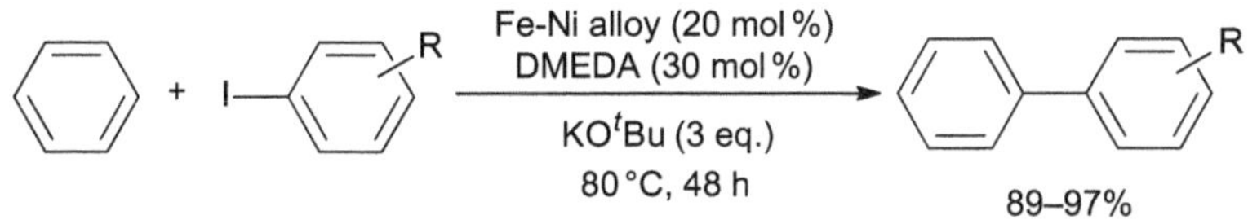

Scheme 9.23 Magnetic Fe-Ni alloy catalysed direct arylation.

9.5 Conclusion

During the past four decades, transition metal-catalysed cross-coupling reactions involving organohalides as electrophiles and organometallic reagents as nucleophiles have revolutionized chemical syntheses and become one of the most popular strategies to construct carbon–carbon bonds. A number of improvements have developed the former process into an industrially very useful and attractive method for the construction of aryl–aryl bonds. Obviously, the direct arylation of aromatic rings *via* C–H bond activation is a more economical and direct method towards the synthesis of biaryl compounds.

Most efforts have been devoted to the direct arylation of activated arenes such as arenes bearing directing groups, heterocycles and electron-deficient arenes in the past two decades. However, the direct arylation of unactivated aromatic rings like benzene is more challenging due to their inert C–H bonds. To date, only several elegant examples have been reported. Among most of them, strong bases like KO^{t}Bu, KOH and LiHMDS were essential for the success of this transformation. The corresponding mechanistic studies revealed the participation of radicals in many reported protocols. Although both noble and cheap metal catalytic systems have been developed, great challenges including regioselectivity for substituted unactivated arenes, substrate scope extension of aryl halides/pseudohalides and development of mild conditions still remain. Considering the significant importance of such a process, further breakthroughs are expected in the future.

References

1. F. Diederich and P. J. Stang, (ed.), *Metal-catalyzed Cross-coupling Reactions*, Wiley-VCH, New York, 1998.
2. J. Hassan, M. Sevignon, C. Gozzi, E. Schulz and M. Lemaire, *Chem. Rev.*, 2002, **102**, 1359.
3. D. A. Horton, G. T. Bourne and M. L. Smythe, *Chem. Rev.*, 2003, **103**, 893.
4. D. Milstein and J. K. Stille, *J. Am. Chem. Soc.*, 1978, **100**, 3636.
5. P. Espinet and A. M. Echavarren, *Angew. Chem., Int. Ed.*, 2004, **43**, 4704.
6. N. Miyaura and A. Suzuki, *Chem. Rev.*, 1995, **95**, 2457.
7. P. Lloyd-Williams and E. Giralt, *Chem. Soc. Rev.*, 2001, **30**, 145.
8. S. Baba and E. Negishi, *J. Am. Chem. Soc.*, 1976, **98**, 6729.

9. E. Negishi and S. Baba, *J. Chem. Soc., Chem. Commun.*, 1976, 596.
10. E. Negishi, *Acc. Chem. Res.*, 1982, **15**, 340.
11. K. Takahashi, T. Minami, Y. Ohara and T. Hiyama, *Tetrahedron Lett.*, 1993, **34**, 8263.
12. S. E. Denmark and R. F. Sweis, *Acc. Chem. Res.*, 2002, **35**, 835.
13. R. J. P. Corriu and J. P. Masse, *J. Chem. Soc., Chem. Commun.*, 1972, 144.
14. K. Tamao, Y. Kiso, K. Sumitani and M. Kumada, *J. Am. Chem. Soc.*, 1972, **94**, 9268.
15. K. Tamao, K. Sumitani and M. Kumada, *J. Am. Chem. Soc.*, 1972, **94**, 4374.
16. M. Catellani, E. Motti, N. Della Ca and R. Ferraccioli, *Eur. J. Org. Chem.*, 2007, 4153.
17. S. Pascual, P. de Mendoza and A. M. Echavarren, *Org. Biomol. Chem.*, 2007, **5**, 2727.
18. I. V. Seregin and V. Gevorgyan, *Chem. Soc. Rev.*, 2007, **36**, 1173.
19. A. Gunay and K. H. Theopold, *Chem. Rev.*, 2010, **110**, 1060.
20. L.-C. Campeau and K. Fagnou, *Chem. Commun.*, 2006, 1253.
21. O. Daugulis, H.-Q. Do and D. Shabashov, *Acc. Chem. Res.*, 2009, **42**, 1074.
22. L. Ackermann, R. Vicente and A. R. Kapdi, *Angew. Chem., Int. Ed.*, 2009, **48**, 9792.
23. B.-J. Li, S.-D. Yang and Z.-J. Shi, *Synlett*, 2008, 949.
24. G. P. McGlacken and L. M. Bateman, *Chem. Soc. Rev.*, 2009, **38**, 2447.
25. G. Dyker, *Handbook of C-H Transformations: Applications in Organic Synthesis*, Wiley-VCH, Weinheim, 2005.
26. V. Ritleng, C. Sirlin and M. Pfeffer, *Chem. Rev.*, 2002, **102**, 1731.
27. O. Daugulis, V. G. Zaitsev, D. Shabashov, Q.-N. Pham and A. Lazareva, *Synlett*, 2006, 3382.
28. D. Alberico, M. E. Scott and M. Lautens, *Chem. Rev.*, 2007, **107**, 174.
29. D. A. Colby, R. G. Bergman and J. A. Ellman, *Chem. Rev.*, 2010, **110**, 624.
30. K.-i. Fujita, M. Nonogawa and R. Yamaguchi, *Chem. Commun.*, 2004, 1926.
31. D. I. Davies, D. H. Hey and B. Summers, *J. Chem. Soc. C*, 1970, 2653.
32. D. I. Davies, D. H. Hey and B. Summers, *J. Chem. Soc. C*, 1971, 2681.
33. S. Yanagisawa, T. Sudo, R. Noyori and K. Itami, *J. Am. Chem. Soc.*, 2006, **128**, 11748.
34. S. Yanagisawa, T. Sudo, R. Noyori and K. Itami, *Tetrahedron*, 2008, **64**, 6073.
35. S. Proch and R. Kempe, *Angew. Chem., Int. Ed.*, 2007, **46**, 3135.
36. G. Dyker, S. Borowski, J. Heiermann, J. Korning, K. Opwis, G. Henkel and M. Kockerling, *J. Organomet. Chem.*, 2000, **606**, 108.
37. M. Lafrance and K. Fagnou, *J. Am. Chem. Soc.*, 2006, **128**, 16496.
38. C. Qin and W. Lu, *J. Org. Chem.*, 2008, **73**, 7424.
39. J. J. Mousseau, F. Vallee, M. M. Lorion and A. B. Charette, *J. Am. Chem. Soc.*, 2010, **132**, 14412.

40. W. Hong, Y. Qiu, Z. Yao, Z. Wang and S. Jiang, *Tetrahedron Lett.*, 2011, **52**, 4916.
41. O. Kobayashi, D. Uraguchi and T. Yamakawa, *Org. Lett.*, 2009, **11**, 2679.
42. W. Liu, H. Cao and A. Lei, *Angew. Chem., Int. Ed.*, 2010, **49**, 2004.
43. F. Vallee, J. J. Mousseau and A. B. Charette, *J. Am. Chem. Soc.*, 2010, **132**, 1514.
44. W. Liu, H. Cao, J. Xin, L. Jin and A. Lei, *Chem.–Eur. J.*, 2011, **17**, 3588.
45. H. Li, C.-L. Sun, M. Yu, D.-G. Yu, B.-J. Li and Z.-J. Shi, *Chem.–Eur. J.*, 2011, **17**, 3593.
46. G. Xie, T. Li, X. Qu and J. Mao, *J. Mol. Catal., A: Chem.*, 2011, **340**, 48.
47. C. T. To, T. L. Chan, B. Z. Li, Y. Y. Hui, T. Y. Kwok, S. Y. Lam and K. S. Chan, *Tetrahedron Lett.*, 2011, **52**, 1023.
48. R. Zhang, C.-X. Miao, S. Wang, C. Xia and W. Sun, *ChemCatChem*, 2012, **4**, 192.

CHAPTER 10

Double C–H Activation in Pd-Catalyzed Cross-Coupling Reactions of Non-Preactivated Arenes

WEIPING SU,* MIN ZHANG AND GE WU

State Key Laboratory of Structural Chemistry, Fujian Institute of Research on the Structure of Matter, Chinese Academy of Sciences, Fuzhou, Fujian 350002, China
*Email: wpsu@fjirsm.ac.cn

10.1 Introduction

The formation of (hetero)aryl–(hetero)aryl bonds constitutes one of the most active research areas in modern organic synthesis because these bonds are ubiquitous in natural products, biologically active parts of pharmaceutical and agrochemical specialties, and functional molecules in material science.[1] Their prevalence has promoted synthetic organic chemists to seek efficient methods for the construction of this kind of molecular framework.

Pd-catalyzed cross-coupling reactions of (hetero)aryl halides or pseudohalides with (hetero)aryl organometallic reagents (Scheme 10.1) provided the most powerful tools for (hetero)aryl–(hetero)aryl bond formation due to their generally high yield, excellent selectivity and good functional group tolerance.[2] These traditional Pd-catalyzed cross-coupling reactions are classified into several types depending on the arylmetal reagents used, such as the

RSC Catalysis Series No. 11
C–H and C–X Bond Functionalization: Transition Metal Mediation
Edited by Xavi Ribas

Published by the Royal Society of Chemistry, www.rsc.org

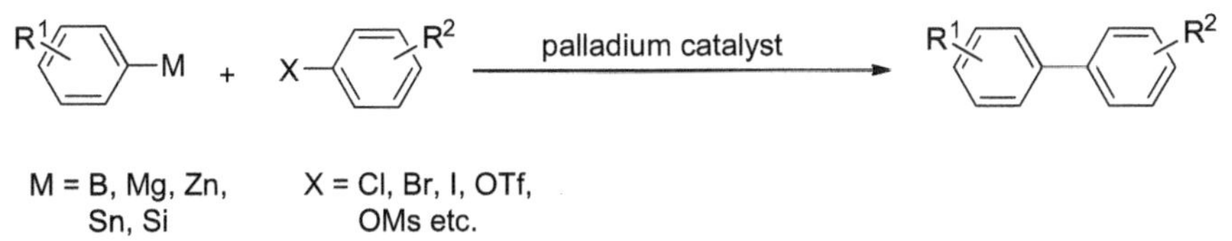

Scheme 10.1 The traditional Pd-catalyzed cross-coupling reactions for (hetero)aryl–(hetero)aryl bond formation.

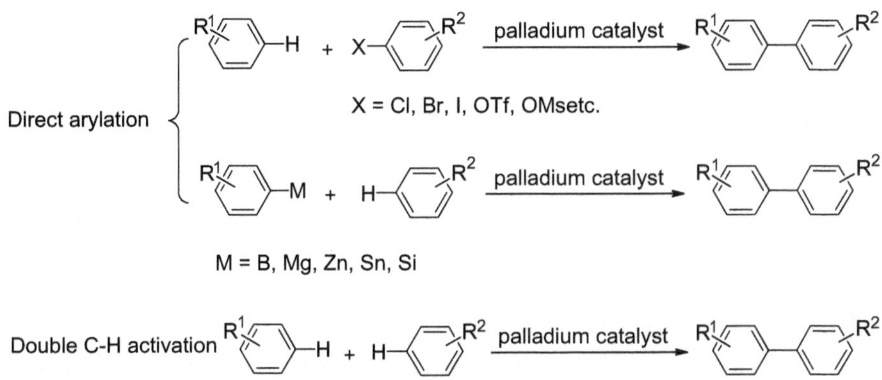

Scheme 10.2 Evolution of Pd-catalyzed cross-coupling reactions.

Suzuki–Miyaura reaction (arylboronic acids),[3] Neigishi reaction (arylzinc reagents),[4] Stille reaction (aryltin reagents)[5] and Hiyama reaction (arylsilicon reagents)[6] (see Chapter 3 of this book). Although greatly successful, these traditional methods still suffer from a lack of atom-economy[7,8] because of the requirement for the pre-activated starting materials as well as the release of metal salt waste that is harmful to the environment.

An ideal approach to bi(hetero)aryl motif is the direct cross-coupling between two non-preactivated arene partners *via* double C–H bond cleavages (Scheme 10.2) since such a oxidative C–H/C–H cross-coupling reaction eliminates the need for preactivation of starting materials, and therefore minimizes waste by-products. As illustrated by the possible mechanism of C–H/C–H cross-coupling (Scheme 10.3), achieving two different metallation steps in a tandem fashion is the most crucial prerequisite for this transformation and also poses a considerable challenge in terms of both reactivity and selectivity.[9]

In the past decade, palladium catalyst systems have played an important role in the development of C–H functionalization reaction, as described by a series of reviews.[10–14] The current chapter will focus specifically on the palladium-catalyzed C–H/C–H cross-coupling reactions for the synthesis of bi(hetero)aryl compounds,[15–21] while the applications of other metal catalysts such as Rh catalysts[22–24] and Cu catalysts[25,26] to such transformations will not be covered. Some C–H/C–H cross-coupling reactions using stoichiometric palladium will

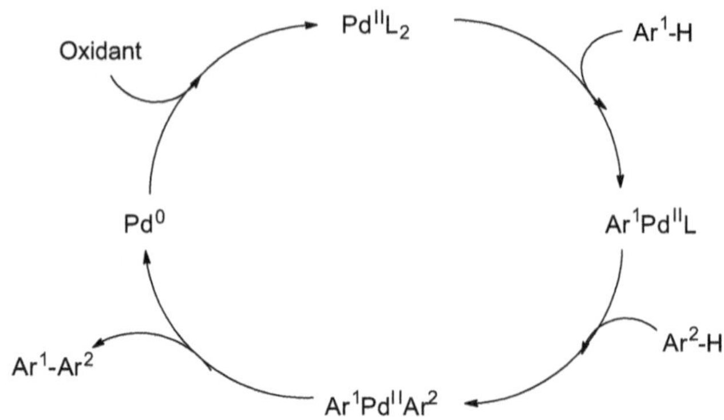

Scheme 10.3 Catalytic cycle of Pd-catalyzed oxidative C–H/C–H cross-coupling reactions.

also be included where appropriate. This chapter is organized by the type of substrate combinations.

10.2 Pd-Catalyzed Oxidative C–H/C–H Cross-Coupling Reaction Between Two Simple Arenes

Since van Helden and Verberg disclosed the formation of biphenyl from benzene by using a stoichiometric amount of $PdCl_2$ in 1965,[27] a great deal of effort has been devoted to develop a more synthetically useful method for the cross-coupling of two different arenes to form unsymmetrical biaryls *via* double C–H activation. Substantial progress toward this goal was made successfully in 2006 by Lu and co-workers, who achieved the formation of unsymmetrical biaryls from simple arenes using a $Pd(OAc)_2$/trifluoroacetic acid (TFA)/$K_2S_2O_8$ catalytic system merely by tuning the concentrations of arenes and TFA under mild conditions.[28] When the authors examined the $Pd(OAc)_2$-catalyzed cross-coupling of benzene with anisole, they found that both lower concentration of TFA and higher concentration of benzene enhanced efficiency of the selective cross-coupling reaction. With a high concentration of TFA (6.3 equiv) and excess benzene (benzene/anisole = 10 : 1), the major product of the reaction was biphenyl **1** (cross-coupling/homocoupling = 44 : 56). Interestingly, decreasing the amount of TFA to 0.63 equiv improved the selectivity for cross-coupling (cross-coupling/homocoupling = 79 : 21) under otherwise the same conditions. No homo-coupling products from anisole were detected in the above two processes. However, when the ratio of benzene to anisole was less than 10 : 1, the homocoupling of anisole occurred. Employing more active naphthalene in place of anisole almost suppressed the homo-coupling product completely, albeit the yield of cross-coupling product was moderate (Scheme 10.4). The selectivity of this cross-coupling reaction likely stemmed from the use of a large excess of one of coupling partners. However, the dependence of reaction

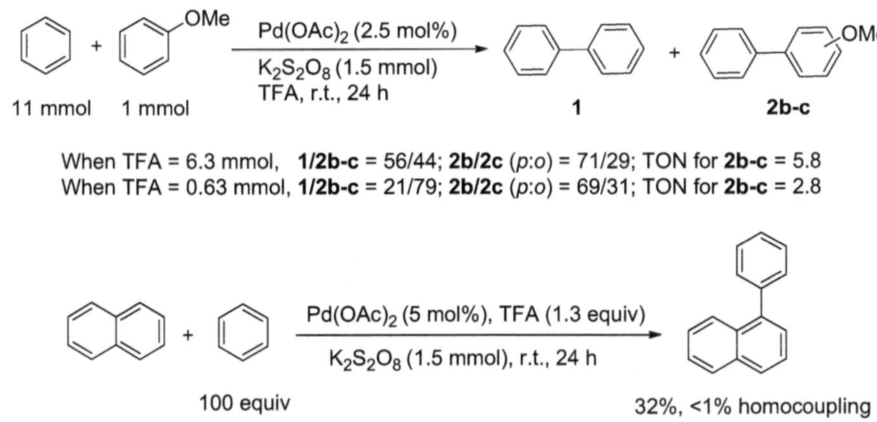

Scheme 10.4 Pd-catalyzed oxidative C–H/C–H cross-coupling of benzene with anisole or naphthalene.

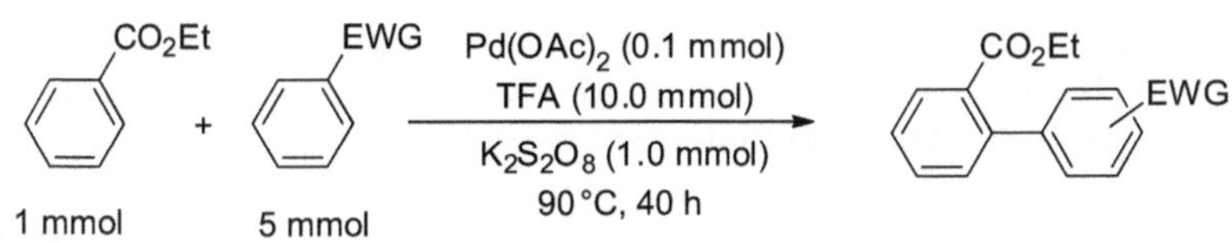

EWG	Selecitivity	Yield
NO_2	*m*:*p* = 4.4:1	55%
CF_3	*m*:*p* = 3.5:1	56%

Scheme 10.5 Pd-catalyzed oxidative C–H/C–H cross-coupling of two electron-deficient arenes.

selectivity on the amount of TFA could be explained by the effect of the TFA concentration on the activity of Pd catalyst. The high concentration of TFA led to more active Pd catalyst, as a result, homocoupling of excess benzene occurred as a main reaction. The Pd catalyst was less active at low concentration of TFA, and therefore reacted with the more active electron-rich arene in the first palladation step and then reacted with benzene favorably in the second palladation steps because of the use of excess benzene to generate the cross-coupling product. Although the authors proposed a mechanism for this reaction, the details remain to be clarified.

The report from the same group described the extension of cross-coupling reaction between two substituted benzenes. By tuning the molar ratio of two coupling partners and a TFA loading, two different benzenes bearing electron-withdrawing groups could be cross-coupled in synthetically useful yields (Scheme 10.5).[29]

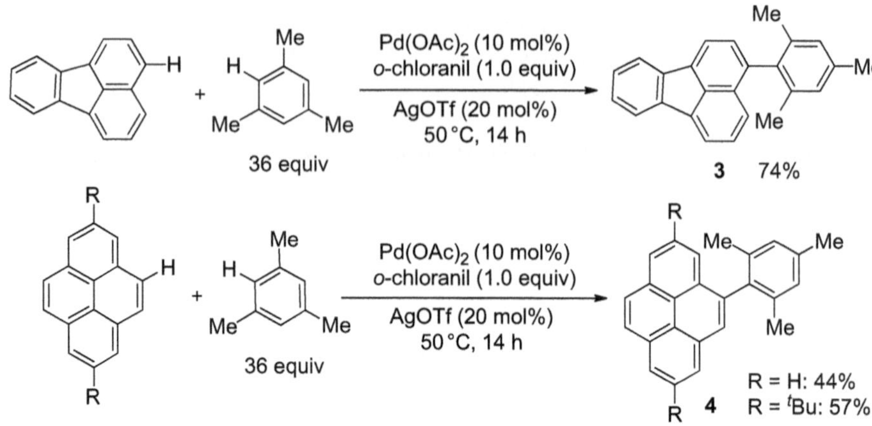

Scheme 10.6 Synthesis of polycyclic aromatic hydrocarbons (PAHs) *via* Pd-catalyzed oxidative C–H/C–H cross-coupling reactions.

Very recently, Itami and co-workers have applied Pd-catalyzed oxidative C–H/C–H cross-coupling reaction to the synthesis of polycyclic aromatic hydrocarbons (PAHs)(**3**–**4**) that are very useful in the field of material science and nanocarbon chemistry (Scheme 10.6).[30] Importantly, the C–H/C–H cross-coupling reactions not only offer an atom-economical approach to PAHs, but also produce PAH products that are difficult to access by traditional synthetic methods.

In view of the great importance of fluorobiphenyls in pharmaceutical and material applications, Su and co-workers developed the $Pd(OAc)_2$-catalyzed oxidative C–H/C–H cross-coupling of diverse electron-deficient polyfluoroarenes with substituted benzenes for the synthesis of fluorobiphenyls **5** in 2010 (Scheme 10.7).[31] Based on the difference between the reactivity of polyfluorobenzene and simple arenes toward C–H cleavage/palladation process, the sequence of two C–H cleavages in the catalytic cycle could be controlled to generate the Pd intermediate containing the two cross-coupling fragments $Pd(Ar^1)(Ar^2)$, and thus achieve selective cross-coupling reactions. This catalytic system displays a wide range of substrate scope with respect to both coupling partners under the optimal conditions (10 mol% $Pd(OAc)_2$, 2 equiv. of $Cu(OAc)_2$, 0.75 equiv. of Na_2CO_3, 1.5 equiv. of PivOH in DMA at 110 °C). Tetrafluoroarenes containing additional substituents, even sensitive unprotected aniline, provided fluorinated biaryls in moderate to good yields. Both electron-rich, -neutral, and -deficient arenes are suitable substrates for this transformation, and the steric factors of the substituents play a key role in controlling the reactivity and regioselectivity. A primary kinetic isotope effects (KIE) of 4.8–6.5 with respect to arenes indicated that the C–H bond cleavage of simple arenes is a significant event in the rate-limiting step.

Soon after, Shi and co-workers described a similar reaction of polyfluoroarenes with arenes by using stoichiometric amount of Ag_2CO_3 as the oxidant in

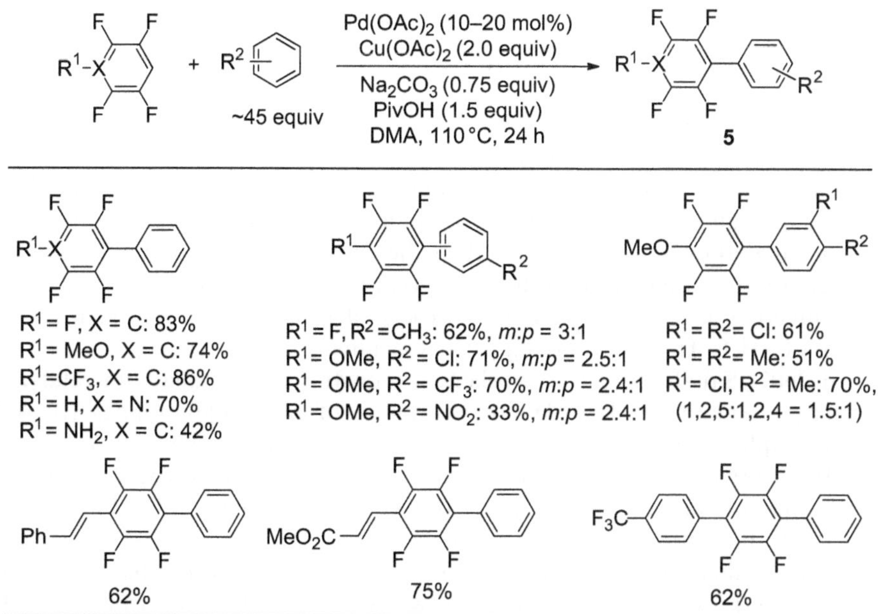

Scheme 10.7 Pd-catalyzed oxidative C–H/C–H cross-coupling of polyfluoroarenes with arenes.

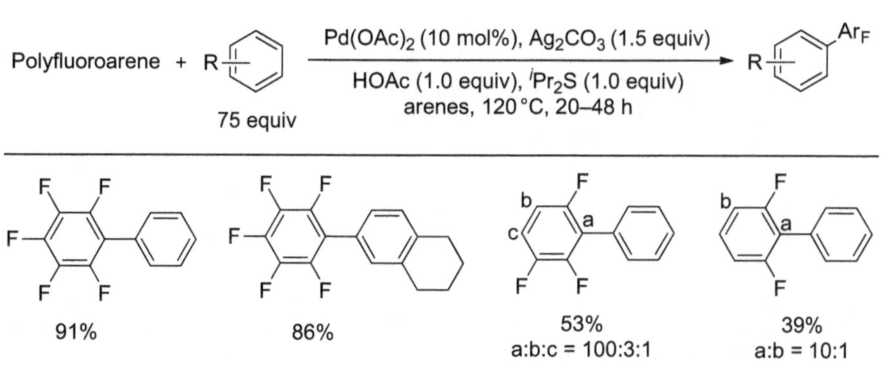

Scheme 10.8 Pd-catalyzed oxidative C–H/C–H cross-coupling of polyfluoroarenes with arenes in the presence of diisopropyl sulfide.

conjunction with diisopropyl sulfide as additive.[32] A quantitative yield was obtained for cross-coupling of pentafluorobenzene and benzene, various of polyfluoroarenes including trifluoro- and even some difluorobenzene could be selectively arylated with high efficiency (Scheme 10.8). Notably, the introduction of diisopropyl sulfide into reaction system in Shi's method improved the efficiency of Pd catalyst probably because diisopropyl sulfide could stabilize Pd catalyst.

10.3 Pd-Catalyzed Oxidative C–H/C–H Cross-Coupling Reaction Between a Directing-Group-Containing Arene and an Arene

One useful strategy to achieve efficient oxidative C–H/C–H cross-coupling reactions is to employ the arene substrate bearing directing group. The directing group can coordinate to the palladium center to facilitate the first C–H palladation regioselectively, then the formed arylpalladium species can attack another arene, leading to the second C–H palladation. In many cases, the directing groups can undergo a variety of transformations, which makes such a C–H/C–H cross-coupling very appealing.

In 2007, Sandford and co-workers demonstrated the palladium-catalyzed selective oxidative C–H/C–H cross-coupling of benzo[h]quinolines (Bzq) **6** with simple arenes in the presence of Ag_2CO_3 oxidant and benzoquinone (BQ) additive (Scheme 10.9).[33] The coordination of nitrogen atom of benzo[h]quinoline to the palladium catalyst promoted the first C–H activation on the C10-position, while the regioselectivity of the second C–H activation was predominantly controlled by steric factor of arenes rather than electronic effect. For 1,2-disubstituted arenes, in which the two different aromatic C–H bonds are electronically similar but have very different steric environments, the reaction proceeded with high selectivity (>10 : 1) at the less hindered C4-position of arenes. Sterically encumbered *p*-xylene only afforded the desired

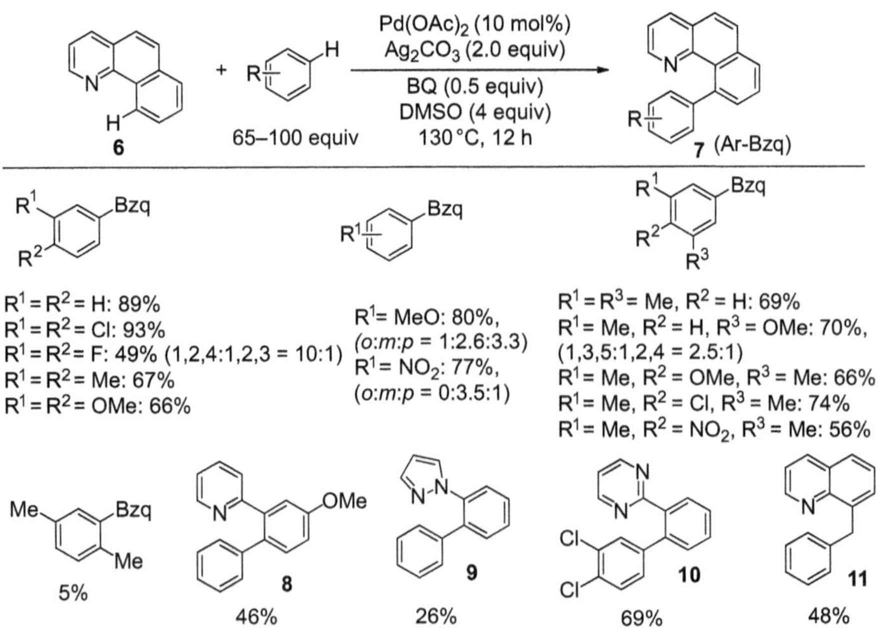

Scheme 10.9 Pd-catalyzed oxidative C–H/C–H cross-coupling of benzo[h]quinoline (or related directing-group-containing arenes) with simple arenes.

cross-coupling product in 5% yield. In the case of 1,3-di- and 1,2,3-trisubstituted arenes, the C–H activation mainly or solely occurred at the less hindered C5-position of arenes. Under identical conditions, other classes of nitrogen-containing heterocycles such as 2-arylpyridine, 1-arylpyrazole, 2-arylpyrimidine derivatives and 8-methylquinoline (**8**–**11**), were also suitable substrates in the cross-coupling with arenes. To collect the information about the role of BQ in this reaction, the author examined the dehydrogenation cross-coupling of benzo[h]quinoline with anisole in the presence of a series of methyl-substituted BQ derivatives and found that the structure of BQ additives has a profound effect on the ratio between *ortho*, *meta* and *para* isomers. They initially proposed that BQ may act as a ligand during the C–H activation. However, further studies indicated that BQ promotes reductive elimination of the diarylpalladium(II) species in the catalytic cycle.[34,35]

At the same time, You and co-workers disclosed the oxidative C–H/C–H cross-coupling of ferrocenyl oxazoline **12** and simple arenes that occurred in the presence of a stoichiometric amount of $Pd(OAc)_2$ or a catalytic amount of $Pd(OAc)_2$ along with excess oxidant such as $Cu(OAc)_2$. The oxazoline served as the directing group to allow the reaction to occur preferentially at the position *ortho* to the oxazoline ring (Scheme 10.10).[36] A better result was obtained if the base K_2CO_3 was added after refluxing ferrocenyl oxazoline **12** and $Pd(OAc)_2$ in benzene for 3 h to produce the dimer **13** intermediate initially. The oxidative cross-coupling of ferrocenyl oxazoline **12** and benzene under a stoichiometric amount of $Pd(OAc)_2$ afforded the desired product **14** in 70% yield and other

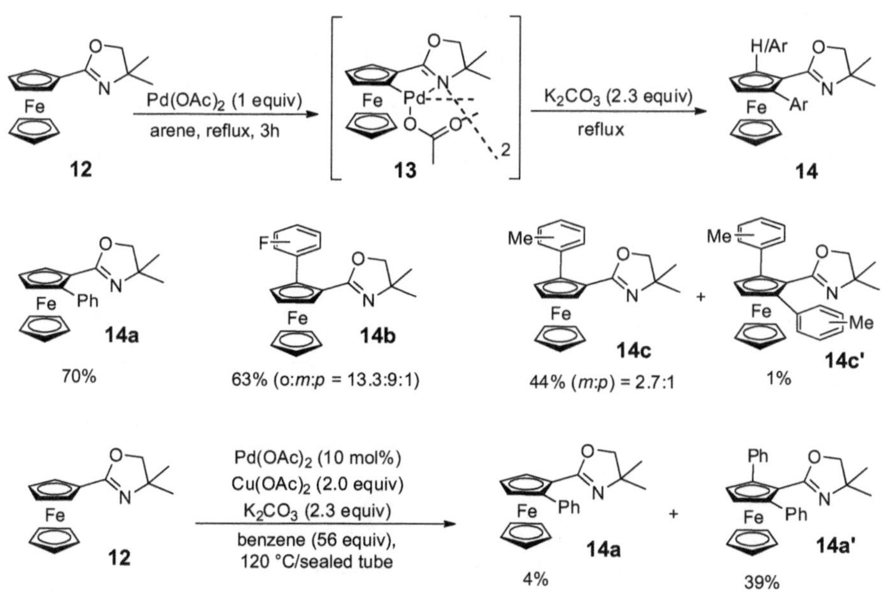

Scheme 10.10 Pd-catalyzed oxidative C–H/C–H cross-coupling of ferrocenyl oxazoline with arenes.

substituted benzenes such as toluene, fluorobenzene proved to be successful (**14a–c**) under identical reaction conditions. However, the palladium-catalytic version of this transformation gave a mixture of mono- or di-arylated ferrocene (**14a** and **14a′**) in a moderate yield (43%). An excellent diastereoselectivity was obtained in the cross-coupling reactions when chiral ferrocenyl oxazolines were used, providing a useful method for the syntheses of planar chiral aryl-substituted ferrocenyl oxazolines. An intermolecular competition reaction between equimolar benzene and benzene-d_6 shows a KIE of 1.8. Based on this KIE value, authors suggested that the C–H activation of arene may be involved in the rate-limiting step.

In 2008, Shi and co-workers developed the palladium-catalyzed *ortho*-arylation of acetanilides with substituted benzenes by employing 0.1–1.0 equiv. $Cu(OTf)_2$ as the co-oxidant, O_2 (1 atm) as the terminal oxidant, and $EtCO_2H$ as the co-solvent (Scheme 10.11).[37] Acetamino acted as a directing group to control the regioselectivity of acetanilides, while the regioselectivity of simple arenes mainly stems from steric hindrance. Under the standard conditions, an array of derivatives of *N*-acetyl-1,2,3,4-tetrahydroquinoline **15**, regardless of substitution at the aliphatic or aromatic ring, were suitable substrates for this transformation, and both electron-rich, -neutral, and -deficient arenes underwent the reaction in moderate to good yields with the high regioselectivity at the less-hindered *meta*- and *para*-positions. In combination with the C–H amination reported by Buchwald,[38] this C–H/C–H cross-coupling reaction can

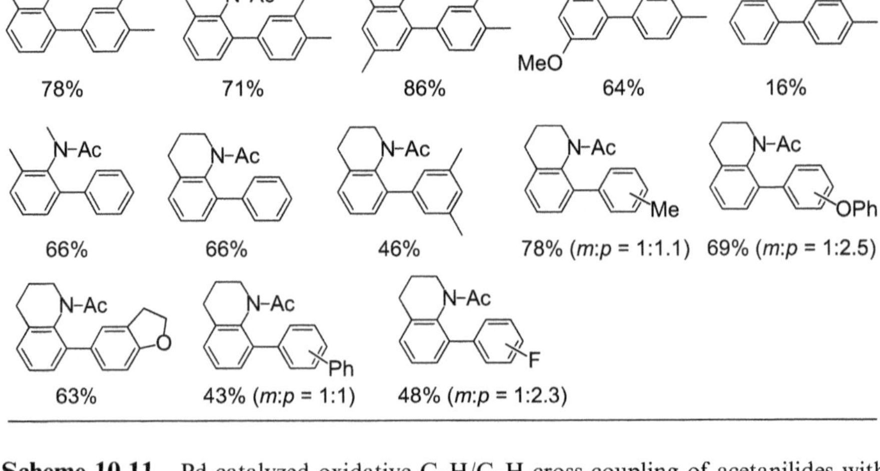

Scheme 10.11 Pd-catalyzed oxidative C–H/C–H cross-coupling of acetanilides with arenes.

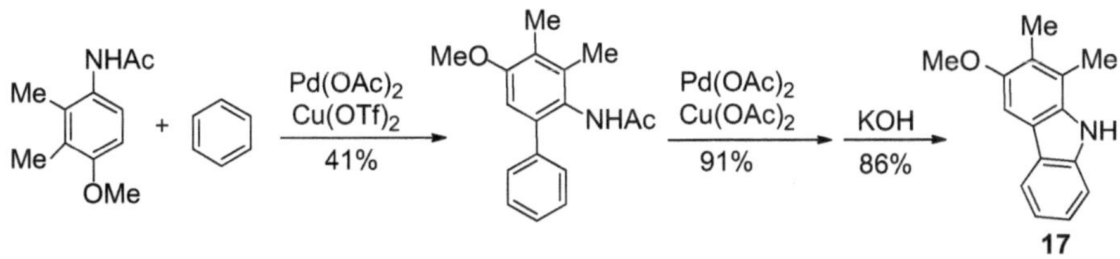

Scheme 10.12 Synthesis of 4-deoxycarbazomycin B.

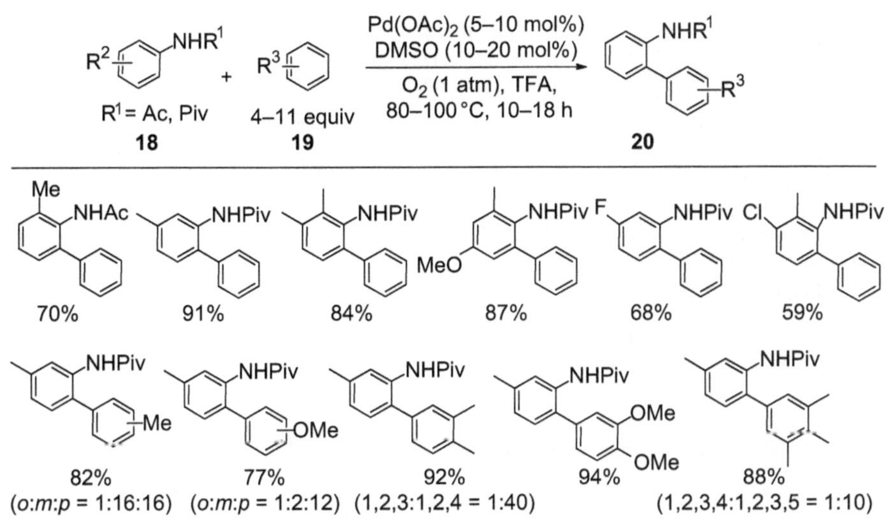

Scheme 10.13 Pd-catalyzed oxidative C–H/C–H cross-coupling of pivalanilldes with arenes.

be used for the facile syntheses of functionalized carbazoles from simple staring materials. For example, the 4-deoxycarbazomycin B **17**, a degradation product of the natural product carbazomycin B, was synthesized in three steps (Scheme 10.12).[37]

Shortly thereafter, Buchwald and co-workers reported a catalytic method similar to Shi's for the *ortho*-arylation of pivalanilides **18** with simple arenes **19**.[39] Buchwald's method used O_2 as the sole oxidant and eliminated the need for Cu salt as co-oxidant. According to the authors, introducing 10–20 mol% DMSO into the reaction system avoided the agglomeration of palladium catalyst, and thus increased the catalyst lifetime. Unlike other oxidative C–H/C–H cross-coupling reactions in which a large excess of a simple arene substrate (often as solvent) was required because of the low reactivity of simple benzene, this cross-coupling reaction worked well in TFA with a wide range of substrates using only 4–11 equiv. of simple benzene **19** coupling partner (Scheme 10.13).

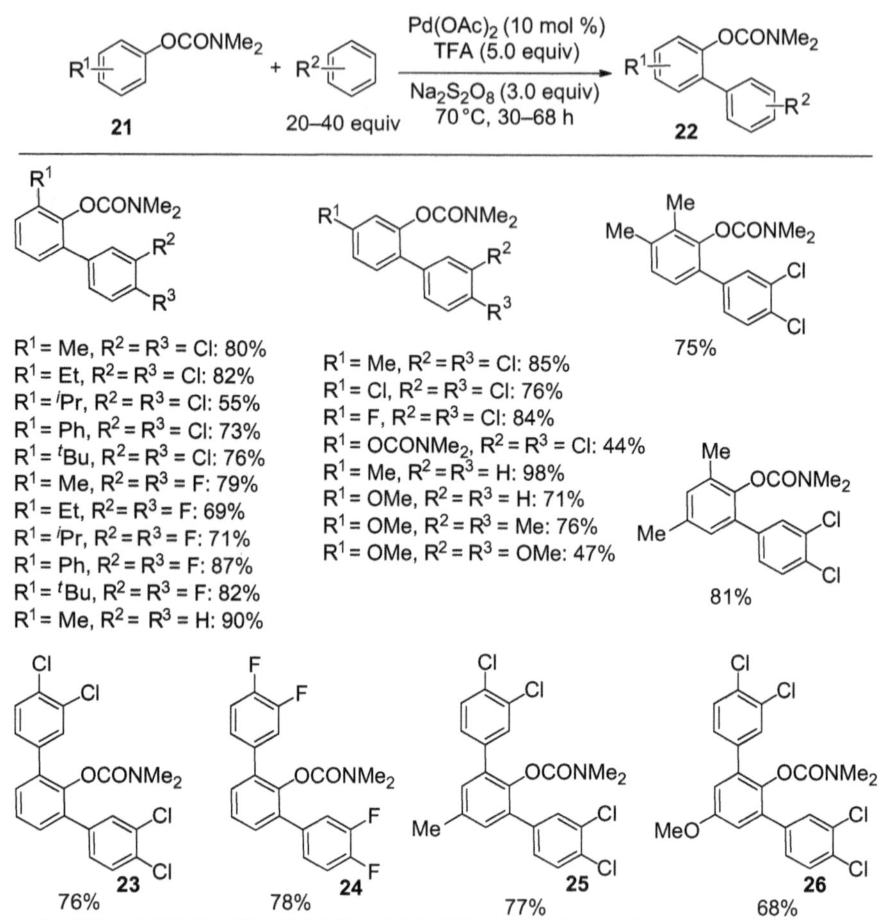

Scheme 10.14 Pd-catalyzed oxidative C–H/C–H cross-coupling of *O*-phenylcarbamates with arenes.

In 2010, Dong and co-workers reported a Pd-catalyzed method for the oxidative C–H/C–H cross-coupling of *O*-phenylcarbamates **21** with arenes using inexpensive and environmentally friendly $Na_2S_2O_8$ as the oxidant. In this reaction, the *O*-carbamate group proved to be a suitable directing group to control the *ortho*-selectivity of *O*-phenylcarbamates (Scheme 10.14).[40] *O*-Phenylcarbamates bearing substitution at the *ortho* or *meta* positions underwent selective monoarylation to form products **22** in excellent reaction efficiency and selectivity. In the case of *O*-phenylcarbamates bearing two reactive C–H bond, selective diarylation can be achieved *via* 4-fold C–H bond functionalization with unprecedented efficiency (**23**–**26**). A range of electron-rich, electron-neutral, and electron-deficient arenes are suitable coupling partners for this process, exclusive functionalization was observed at the C4-position for simple *o*-disubstituted arenes. To obtain insight into the

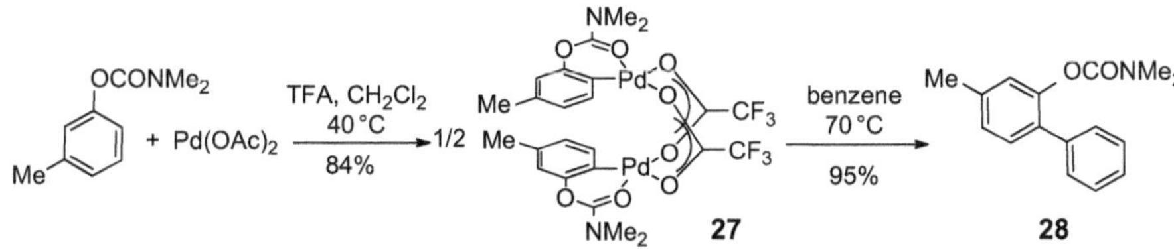

Scheme 10.15 Direct phenylation of palladium complex **27**.

mechanism, the palladacycle intermediate **27** derived from an *O*-phenylcarbamate was synthesized and characterized by NMR and X-ray crystallography spectroscopy. Once subjected to benzene in the absence of any external oxidant, the palladium complex **27** reacted with benzene to produce the *ortho*-arylation products **28** in excellent yield (Scheme 10.15), which suggested that a Pd^0/Pd^{II} mechanism may be involved in the catalytic cycle. Competition experiments between benzene and *o*-dimethoxybenzene, as well as between benzene and *o*-dichlorobenzene revealed that more electron-rich arenes reacted with the palladacycle intermediate **27** faster than less electron-rich arenes, indicating that the second C–H palladation occurred *via* electrophilic substitution. Based on these mechanistic investigations, the authors proposed a mechanism whereby two C–H bond activations occurred *via* cyclopalladation and electrophilic palladation, respectively.

Shortly after, Dong and co-workers further expanded their reaction to three types of substrates **29** bearing amide directing groups, that is, phenylacetamides, benzamides and anilides. This method featured mild reaction conditions, good yield and high chemo- or regioselectivity. (Scheme 10.16).[41] Different from *O*-phenylcarbamate substrates in their previous report, electron-deficient arenes were ineffective in this catalytic system. Mechanism studies indicated that the analogous palladacycle intermediate **31** and **33** derived from anilide substrates gave the expected *ortho*-arylation products **32** and **34** only when the oxidant $Na_2S_2O_8$ was added to the system (Scheme 10.17). Based on the high redox potential of $Na_2S_2O_8$, an oxidation of the palladacycle intermediate **31** and **33** to either bimetallic Pd^{III} or Pd^{IV} complexes may be involved. These results suggested that the reaction mechanism may differ greatly depending on the directing group as well as the oxidant chosen. Some examples of the palladium-catalyzed oxidative C–H/C–H cross-coupling of simple arenes with other substrates bearing coordination nitrogen atom as the directing groups, such as benzaldimines **35**, (phenylsulfonyl)pyridines **37** were reported by Gaunt and co-workers (Scheme 10.18)[42] and Zhang and co-workers (Scheme 10.19),[43] respectively.

A common problem in those oxidative C–H/C–H cross-coupling reactions is the formation of intractable mixture of regioisomers especially when a simple monosubstituted arene is used as the coupling partner. In 2011, Yu and co-workers described a Pd-catalyzed method for the selective C–H/C–H cross-coupling reactions between benzamides **39** and monosubstituted arenes **40**

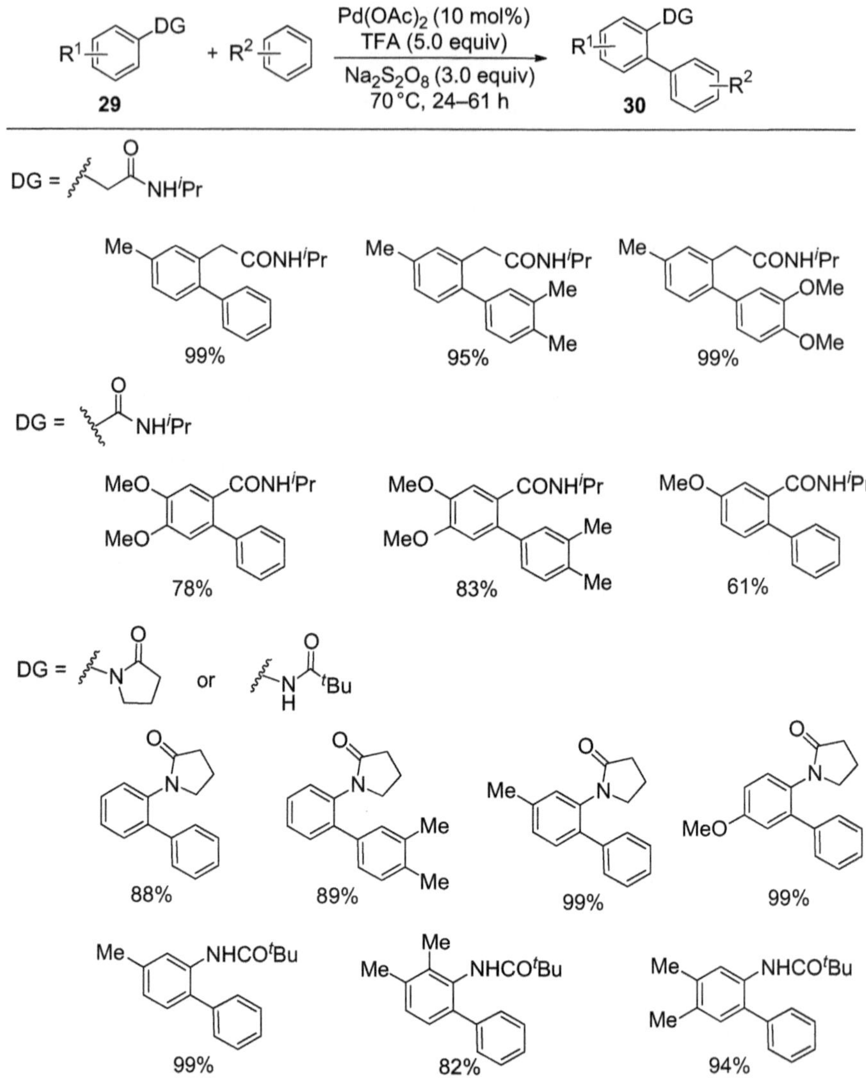

Scheme 10.16 Pd-catalyzed *ortho*-arylation of phenylacetamides, benzamides and anilides *via* double C–H activation.

(Scheme 10.20).[44] An exceedingly high *para*-selective for arenes was achieved in the presence of F^+ reagent NFSI (*N*-fluorobenzenesulfonimide) as oxidant. The regioselectivity of the first C–H bond activation was predominantly controlled by an acidic amide directing group. The *para*-selectivity of arenes observed for such oxidative C–H/C–H cross-coupling reaction seems to depend on both the directing group and the F^+ reagent. An array of synthetically useful benzamides, including those containing electron-withdrawing groups such as

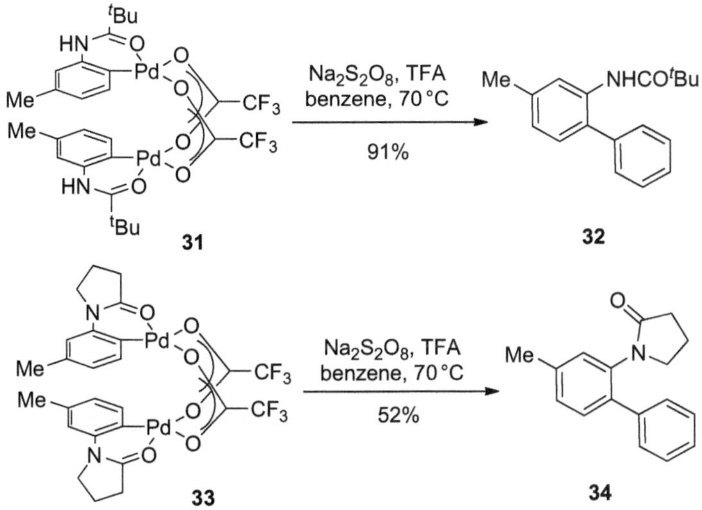

Scheme 10.17 Direct phenylation of Pd complex **31** and **33**.

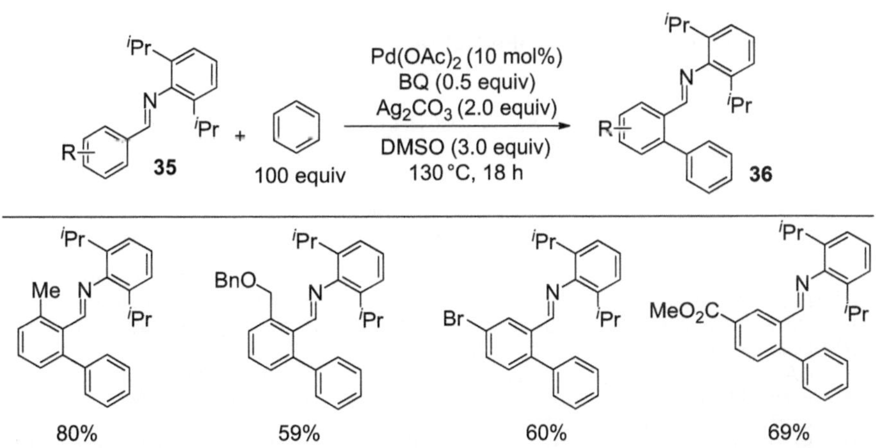

Scheme 10.18 Pd-catalyzed *ortho*-arylation of benzaldimines *via* double C–H activation.

trifluoromethyl, ketone, and cyano groups are compatible with this catalytic system, and both alkyl, methoxy, and halogenated arene underwent the reaction smoothly to give the corresponding biaryl products **41** with similarly high levels of *para*-selectivity. A oxidation of Pd^{II} to Pd^{IV} species by F^+ reagent may be involved in the catalytic cycle and the formed [$ArPd^{IV}F$] species seems to partially responsible for selective *para*-C–H cleavage.

In addition to the aforementioned two-fold C–H activation reaction, by using CONHOMe as the directing group, Cheng and co-workers realized a Pd-catalyzed dehydrogenative cyclization of *N*-methoxybenzamides **42** with

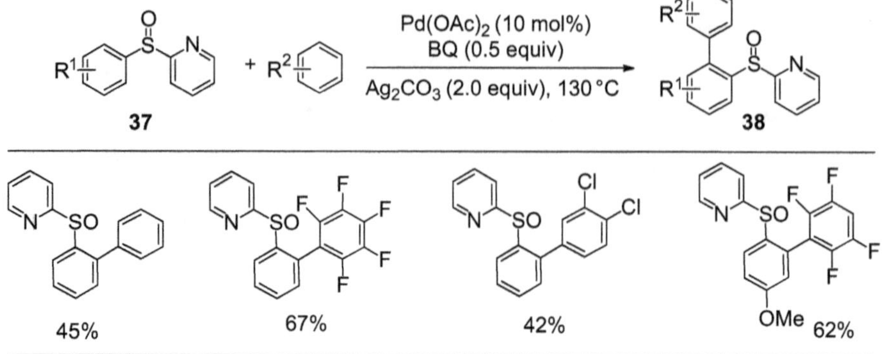

Scheme 10.19 Pd-catalyzed *ortho*-arylation of (phenylsulfonyl)pyridines *via* double C–H activation.

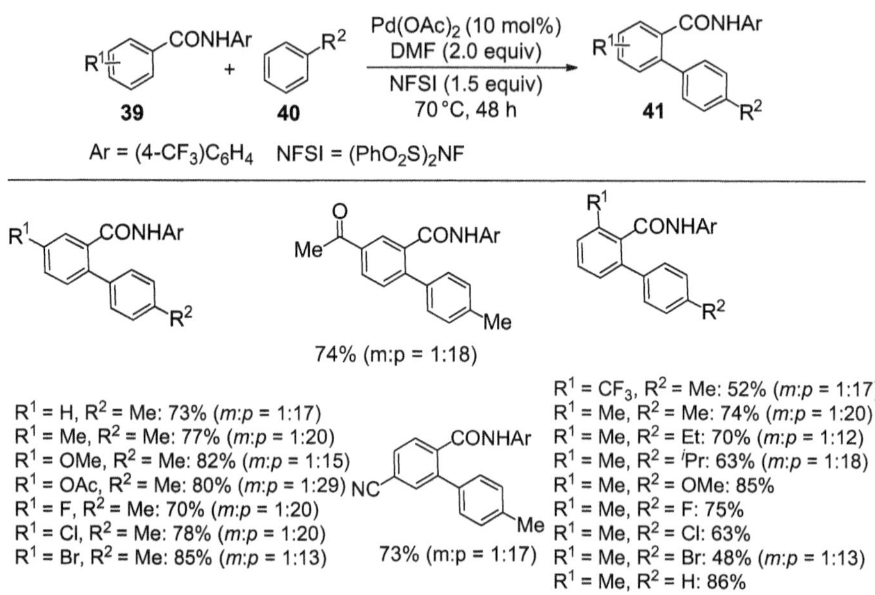

Scheme 10.20 Pd-catalyzed selective oxidative C–H/C–H cross-coupling of benzamides with monosubstituted arenes

arenes to synthesize biologically active phenanthridinones **43** through multiple oxidative C–H activation and C–C/C–N formation steps in one pot under mild conditions (Scheme 10.21).[45] The reaction has excellent functional group tolerance and is highly selective with respect to both coupling partners. For monosubstituted arenes, the transformation is highly site-selective at the C4-position and then at C3-position of arenes. Using the reported method, Crinasiadine **45**, a natural product which has been isolated from

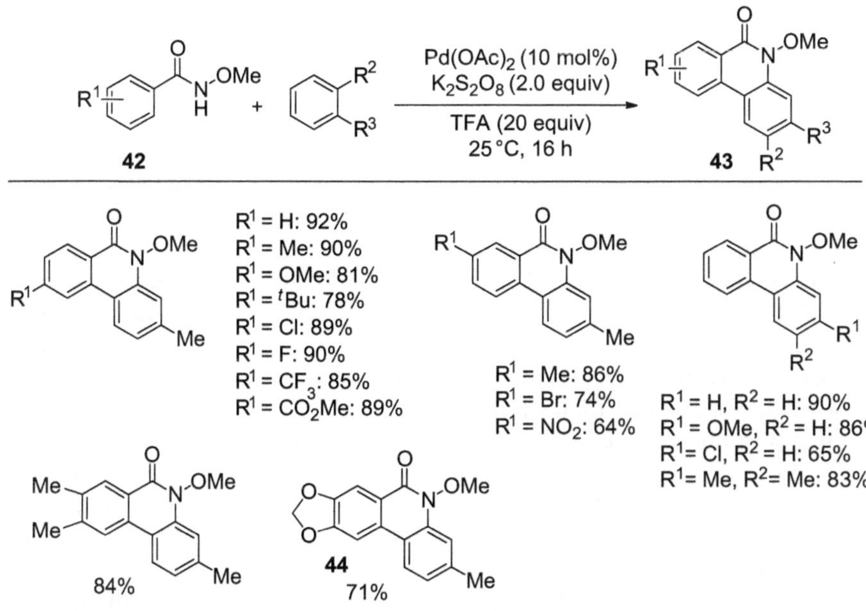

Scheme 10.21 Pd-catalyzed dehydrogenative cyclization of *N*-methoxybenzamides with arenes through multiple oxidative C–H activation.

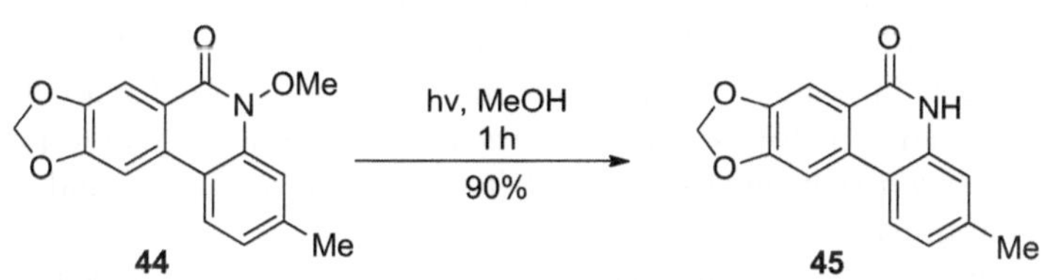

Scheme 10.22 Synthesis of Crinasiadine **45** from **44**.

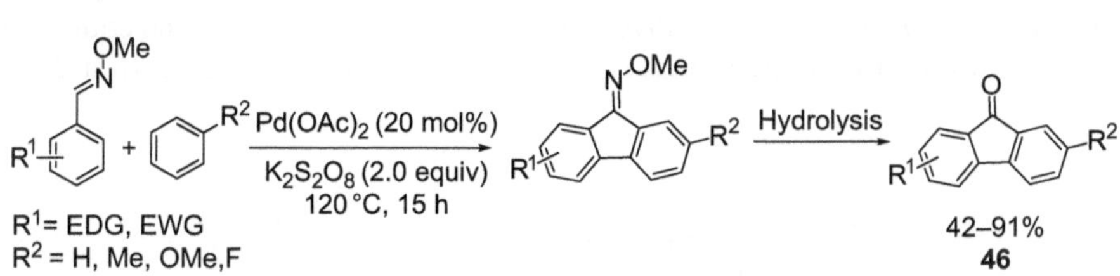

Scheme 10.23 Pd-catalyzed dehydrogenative cyclization of aromatic aldoxmine ethers with arenes.

Crinum asiaticum plant, can be very conveniently obtained in 90% yield (Scheme 10.22). Afterwards, the same group also described a similar dehydrogenative cyclization process which employs aldoxime ether as the directing group to synthesize fluorenone derivatives **46** (Scheme 10.23).[46]

10.4 Pd-Catalyzed Oxidative C–H/C–H Cross-Coupling Reaction Between a Heteroarene and an Arene

Given that the Pd-catalyzed oxidative C–H/C–H cross-coupling reactions between two arenes suffers from the problem of low activity and regioselectivity in both coupling partners, replacing one of the arene partner by a heteroarene substantially improves the activity and selectivity.

In 1981, Itahara and co-workers reported the palladium-mediated oxidative C–H/C–H cross-coupling of pyrroles and indoles with benzene in AcOH, refluxing l-acetylindole and 3-methyl-l-acetylindole with stoichiometric Pd(OAc) in AcOH-benzene gave the cross-coupling products **47** in 22% and 48% yield, respectively. (Scheme 10.24).[47]

An important advance was made by Fagnou and co-workers in 2007.[9] Using $Pd(TFA)_2$ as a catalyst, $Cu(OAc)_2$ as an oxidant in combination with 3-nitropyridine and CsOPiv as additives, this group achieved a highly efficient C3 arylation of *N*-acetylindoles **48** with simple arenes as arylating reagents (Scheme 10.25). Although the addition of the two additives was not crucial to achieve catalytic turnover, they were observed to greatly improve the reaction efficiency. 3-Nitropyridine presumably acted as a ligand to stabilize the palladium(0) species formed after reductive elimination, whereas CsOPiv may interact with $Pd(TFA)_2$ to generate palladium pivalate early in the reaction. The substituents on indole nitrogen atom also dramatically influenced the reactivity of indoles. Free N–H indole did not react at all while *N*-methylindole produced self-dimerization product predominantly. The reaction worked well for *N*-acetylindoles with different electronic properties, and both electron-rich, -neutral and -deficient simple arenes all underwent the reaction smoothly.

One of the most interesting aspects in this transformation is that the reacting site of indole can be inverse to C2 by using AgOAc instead of $Cu(OAc)_2$ as the oxidant.[48] For example, treatment of *N*-pivalyl indole with $Pd(TFA)_2$ (5 mol%), AgOAc (3.0 equiv.), PivOH (6.0 equiv.) and benzene ($\sim$60 equiv.) at 110 °C afforded the desired cross-coupling product in 84% yield with a 25 : 1

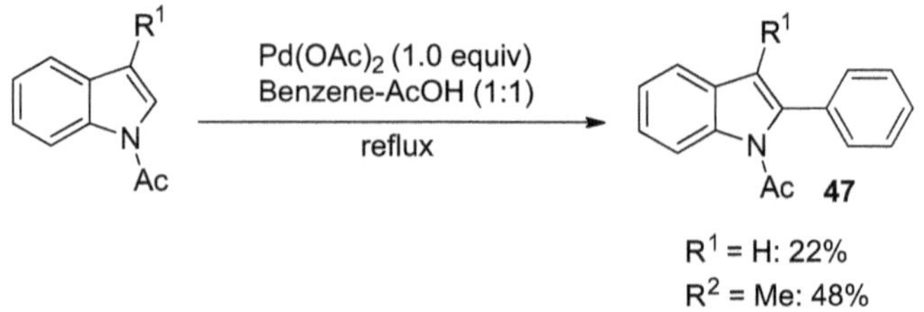

Scheme 10.24 Pd-mediated oxidative C–H/C–H cross-coupling of indoles with benzene.

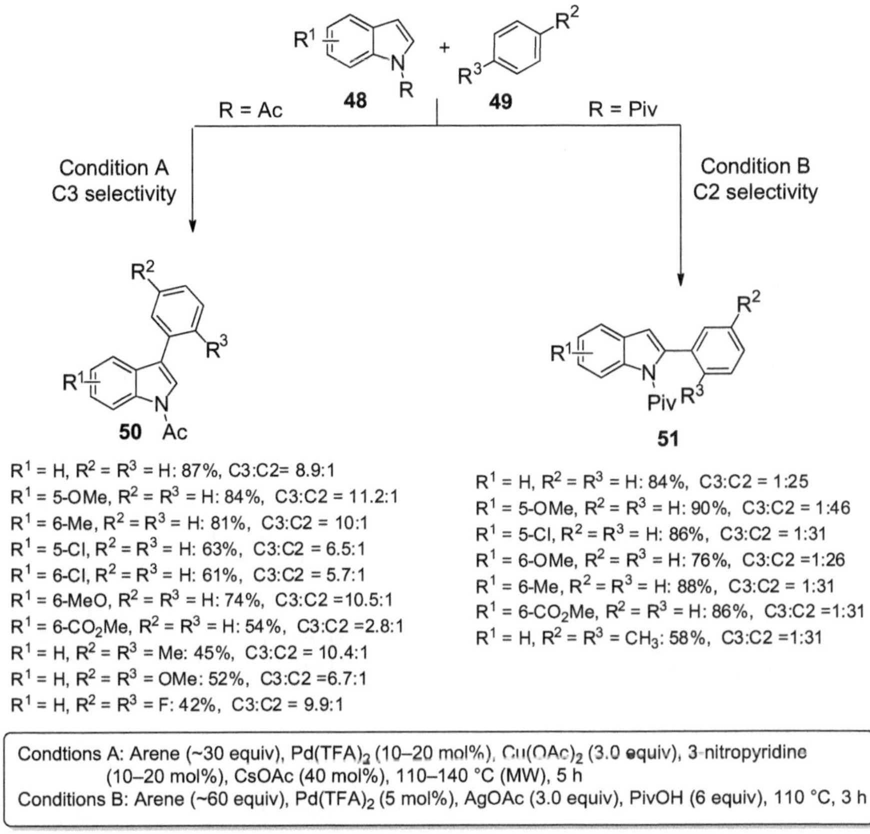

Scheme 10.25 Pd-catalyzed oxidative C–H/C–H cross-coupling of indoles with benzene.

ratio of C2 : C3 arylated products **51** (Scheme 10.25). Under the optimized conditions, pyrroles **52** can also successfully couple with arenes in good yield (Scheme 10.26). As to the reason behind the inversion in C2/C3 selectivity, further experiments showed that the C2/C3 selectivity of indoles was influenced by Pd concentration and acetate additive: high $Pd(TFA)_2$ loading led to a higher C3 selectivity while the additional introduction of CsOAc favored C2 selectivity. The authors proposed that when excessive $Cu(OAc)_2$ was added to a catalytic quantity of $Pd(TFA)_2$, mixed Pd-Cu clusters which favor the C3 selectivity may be formed. On the other hand, the acetate anion of AgOAc or CsOAc may induce the cleavage of higher-order Pd clusters and thus result in the formation of monomeric Pd catalyst which was assumed to be responsible for the C2 selectivity of indoles.

Almost at the same time, DeBoef and co-workers developed the Pd-catalyzed oxidative C–H/C–H cross-coupling of benzofuran **54** with benzene by using 3 atm of O_2 together with 10 mol% heteropolymolybdovanadic acid

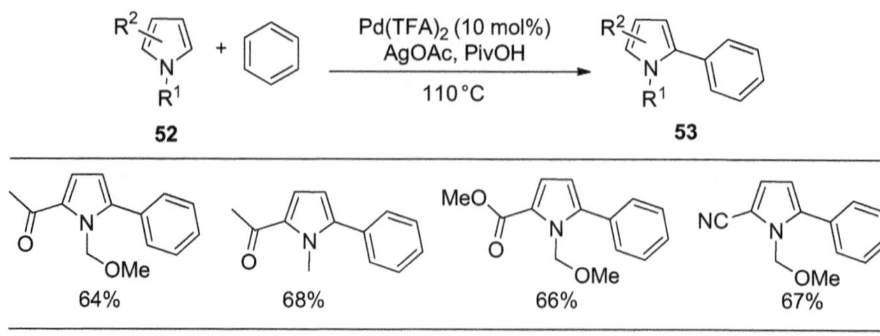

Scheme 10.26 Pd-catalyzed oxidative C–H/C–H cross-coupling of pyrroles with benzene.

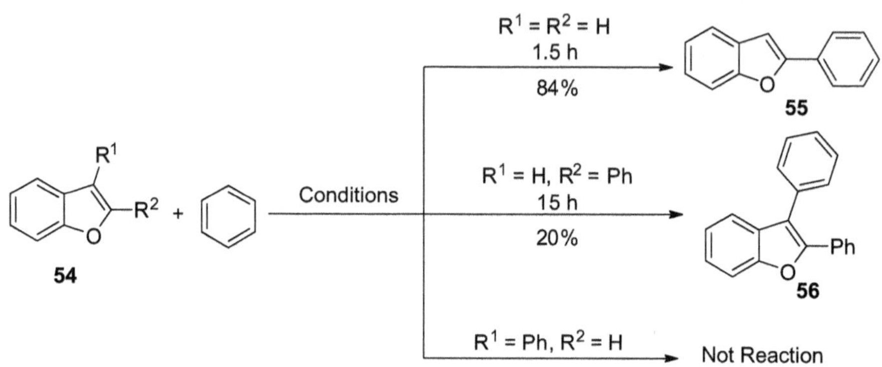

Conditions: $Pd(OAc)_2$ (10 mol%), HMPV (10 mol%), O_2 (3 atm), AcOH/PhH = 3/2 (v/v), 120 °C.

Scheme 10.27 Pd-catalyzed oxidative C–H/C–H cross-coupling of benzofurans with benzene.

$H_4PMO_{11}VO_{40}$ (HMPV) as the oxidant.[49] This reaction produced an exclusive formation of 2-phenylbenzofuran **55**, the homocoupling behavior of benzene was only observed after complete consumption of the benzofuran substrate. When 2-phenylbenzofuran was subjected to the reaction conditions, arylation at the C3-position to form **56** was observed, but the cross-coupling of 3-methylbenzofuran with benzene resulted in the recovery of the starting material (Scheme 10.27). These data suggested that arylation at the C2-position might result from an initial palladation at the C3-position, followed by a migration to the C2-position. In their later study on oxidative arylation of *N*-acetylindoles,[50] they also observed the effect of oxidant chosen on regioselectivity. The use of $Cu(OAc)_2$ as the oxidant produced selective arylation at the C3-position of indole while AgOAc produced C2 selectivity, which is in agreement with Fagnou's work.[9,48]

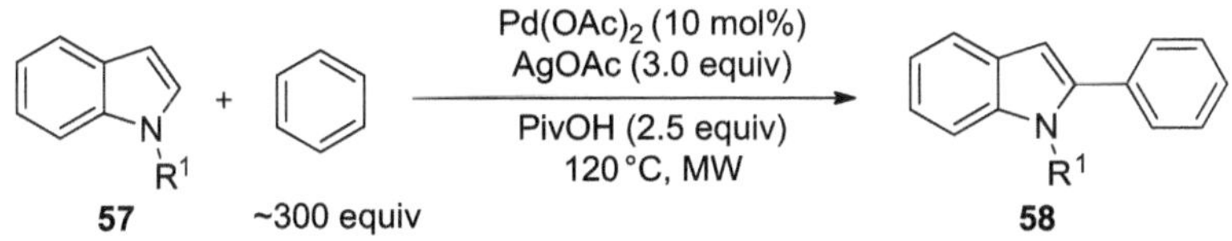

R^1 = SEM : 69%, C2:C3 = 3:1
R^1 = Bn : 57%, C2:C3 = 4:1
R^1 = MOM : 56%, C2:C3 = 4:1
R^1 = Me : 31%, C2:C3 = 4:1

Scheme 10.28 Pd-catalyzed oxidative C–H/C–H cross-coupling of *N*-alkyl indoles with benzene. SEM = 2-(trimethylsilyl)ethoxymethyl; Bn = benzyl; MOM = methoxymethyl.

When using the more electron-rich *N*-alkyl indole substrates, none of the conditions using HMPV, $Cu(OAc)_2$ or AgOAc as the oxidant, were effective because a rapid decomposition occurred under the acidic and oxidizing conditions. DeBoef and co-workers subsequently overcame this limitation by modulating the acidity in the presence of AgOAc as the oxidant and base. Under the buffered conditions containing nearly equal amount of PivOH and AgOAc in combination with 10 mol% $Pd(OAc)_2$, a wide range of *N*-alkyl-indoles **57** underwent the cross-coupling with good C2-regioselectivity and yield (Scheme 10.28).[51] Both experimental and computational data indicated that the C–H palladation of both the indoles and simple arenes proceeded *via* the concerted metalation–deprotonation (CMD) mechanism.

The regioselectivity of indoles in the oxidative C–H/C–H cross-coupling reactions was not only affected by the oxidant employed. Stahl and co-workers described in 2011 that regioselective aerobic oxidative arylation of indoles can also be achieved by proper selection of the anionic and neutral ligand for the Pd catalyst: $Pd(OPiv)_2$ and diazafluorenone ligand **60** favored the C2-functionalized product while $Pd(TFA)_2$ and ligand **61** favored the C3-functionalized product (Scheme 10.29).[52]

In 2008, Chang and co-workers described the palladium-catalyzed oxidative C–H/C–H cross-coupling reaction between pyridine N-oxides **64** and simple arenes.[53] Under the optimized conditions (10 mol% of $Pd(OAc)_2$, 2.2 equiv. of Ag_2CO_3 at 130 °C), a variety of N-oxides underwent the reaction smoothly with moderate to excellent selectivity for the formation of monoarylated products **65** (Scheme 10.30). It is worth mentioning that N-oxides of quinoline and benzo[h]quinoline were arylated selectively at the *ortho*-position in good yields (**65e** and **65f**). Since the directing group-assisted cross-coupling of benzo[h]quinoline was reported to proceed selectively at the C10-position,[33,54] this transformation offers a useful complementary route for the selective aryl introduction at the *ortho* position.

In 2010, the palladium-catalyzed oxidative C–H/C–H cross-coupling of aromatic heterocycles **67** with electron-deficient polyfluoroarenes **68** was developed by Zhang and co-workers (Scheme 10.31).[55] Both DMSO and silver

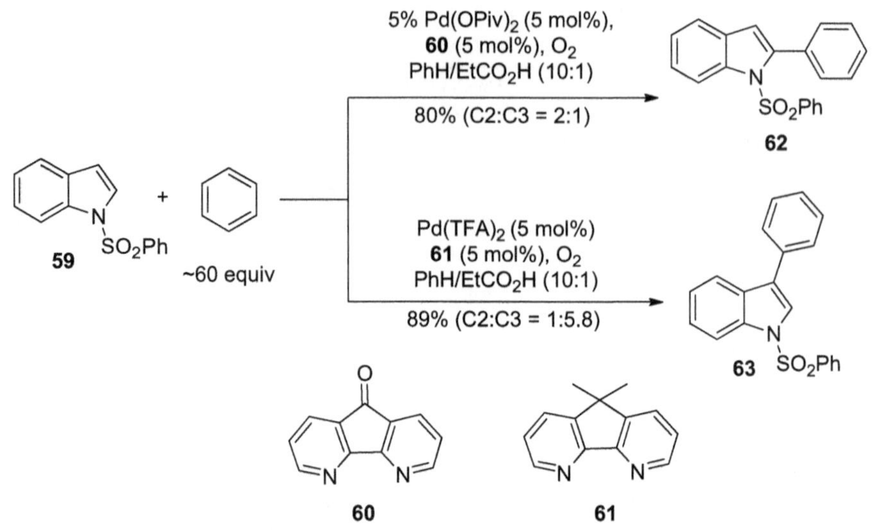

Scheme 10.29 Pd-catalyzed oxidative C–H/C–H cross-coupling of indole with benzene under O_2.

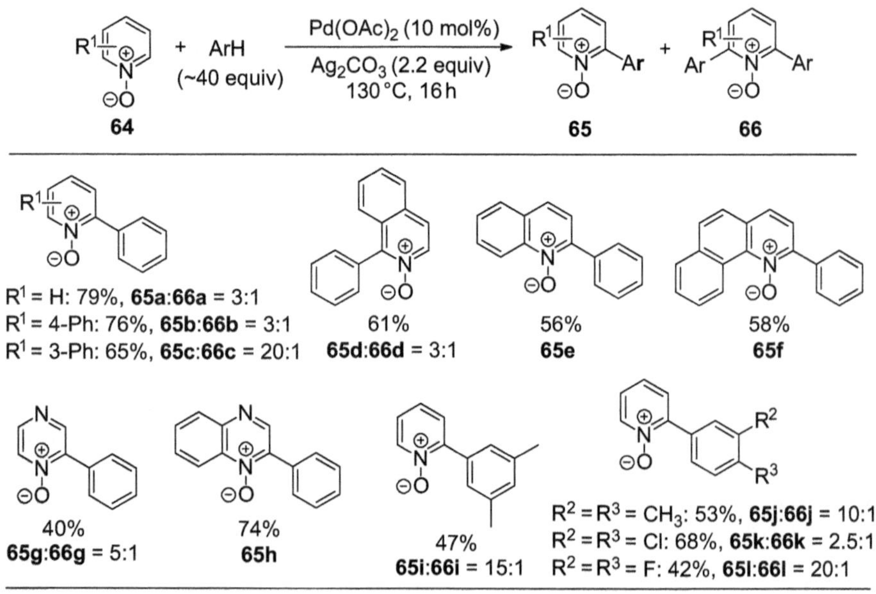

Scheme 10.30 Palladium-catalyzed oxidative C–H/C–H cross-coupling of pyridine N-oxides with arenes.

salt contributed to the reaction efficiency. It was proposed that DMSO might function as a ligand to activate the Pd catalyst and prevent the formation of inactive palladium black. The reaction is notable for the low catalyst loading (2.5 mol%), high reaction efficiency, good chemo- and regioselectivity, and

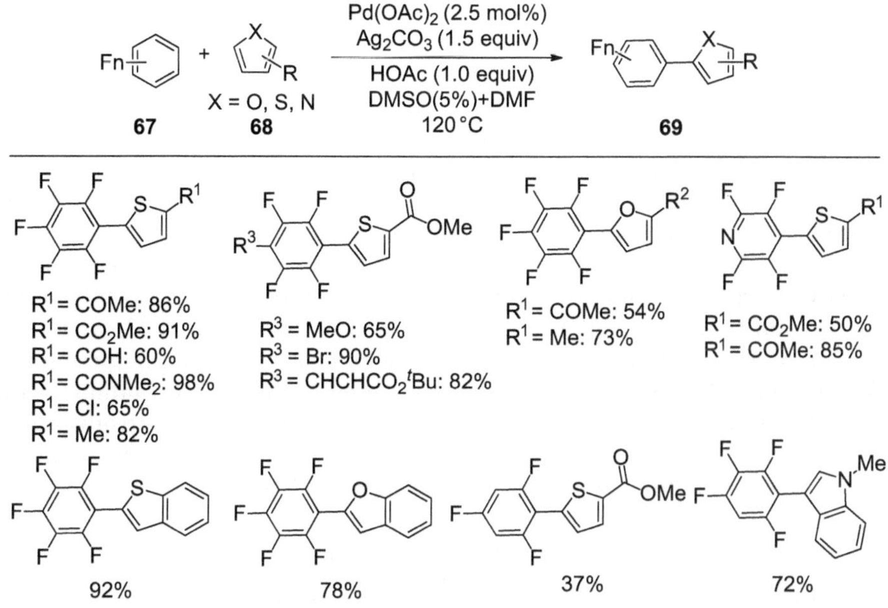

Scheme 10.31 Pd-catalyzed oxidative C–H/C–H cross-coupling of aromatic heterocycles with polyfluoroarenes.

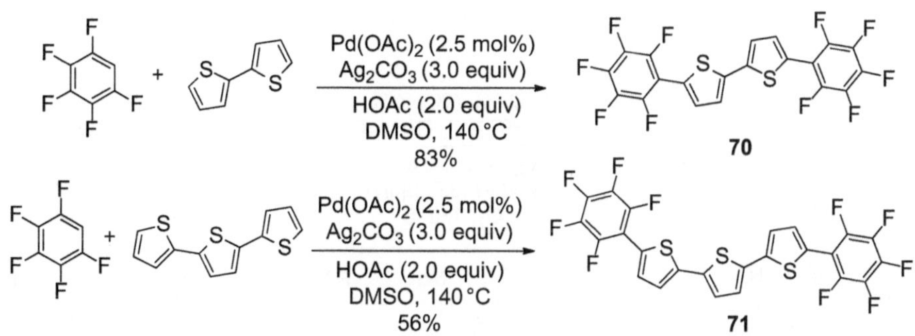

Scheme 10.32 Synthesis of n-type organic semiconductors **70** and **71**.

excellent functional-group compatibility. Under the optimum reaction conditions, a variety of aromatic heterocycles including simple thiophenes, benzo[b]thiophene, furans and indoles were suitable substrates for this transformation. For polyfluoroarenes bearing three or four fluorine atoms, which have more than one reaction site, reasonable yields of monoarylated products were still obtained. The n-type organic semiconductors **70** and **71** were easily prepared by using this procedure (Scheme 10.32).

The establishment of new methods that employ an inexpensive oxidant alternative to the expensive Ag_2CO_3 to effect efficient cross-coupling between thiophenes and polyfluoroarenes would make this reaction more practical. Mindful of this target, the same group further developed an improved procedure for oxidative C–H/C–H cross-coupling of polyfluoroarenes with thiophenes using O_2 as terminal oxidant (Scheme 10.33).[56] A KIE of 2.86 was observed for benzo[b]thiophene and its deuterated derivative, which indicated that C–H bond cleavage of benzo[b]thiophene is the rate-determining step in the overall catalytic process. Mechanistic studies revealed that the reactions were initiated by the C–H bond cleavage of polyfluoroarenes.

In 2011, Beifuss and co-workers demonstrated the Pd-catalyzed oxidative C–H/C–H cross-coupling of xanthines **72** with arenes utilizing a combination of Ag^I and O_2 as oxidants to synthesize C8 arylated xanthines **73**, which are known to act as selective antagonists of the human A_{2B} adenosine receptor.[57] The established method can be applied to various N-substituted xanthines with $Pd(PPh_3)_2Cl_2$ as catalyst, as well as various mono-, di-, and trisubstituted arenes with $Pd(OAc)_2$ as catalyst, and excellent regioselectivity was obtained with respect to both coupling partners (Scheme 10.34).

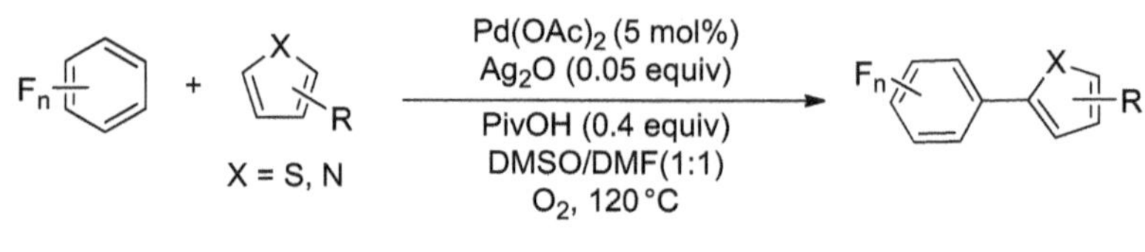

Scheme 10.33 Pd-catalyzed oxidative C–H/C–H cross-coupling of aromatic heterocycles with perfluoroarenes under O_2

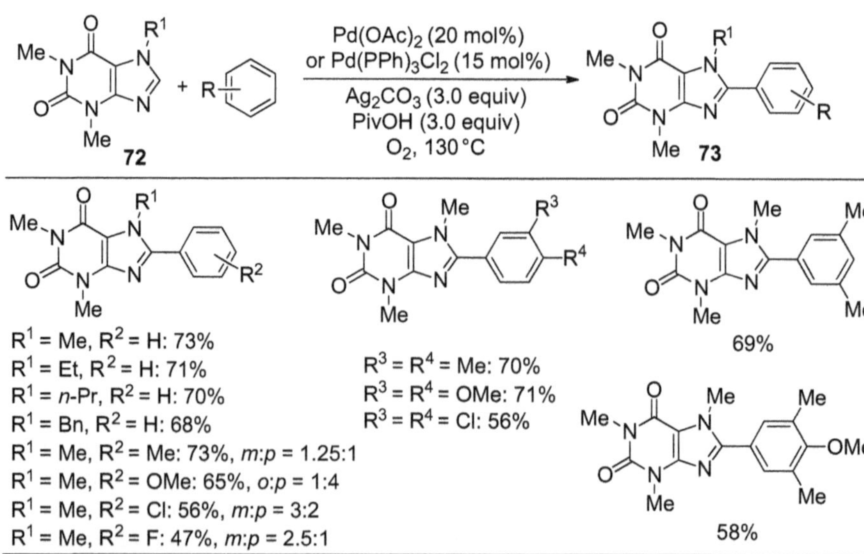

Scheme 10.34 Pd-catalyzed oxidative C–H/C–H cross-coupling of N-substituted xanthines with arenes.

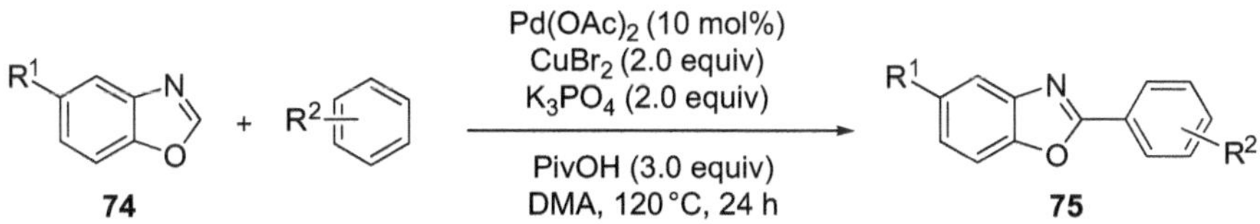

R^1 = H, R^2 = H: 79%
R^1 = Me, R^2 = H: 93%
R^1 = F, R^2 = H: 91%
R^1 = Cl, R^2 = H: 83%
R^1 = Br, R^2 = H: 52%
R^1 = MeO, R^2 = H: 74%
R^1 = COMe, R^2 = H: 65%
R^1 = CO_2Me, R^2 = H: 77%
R^1 = NO_2, R^2 = H: 63%
R^1 = CF_3, R^2 = H: 66%
R^1 = Ph, R^2 = H: 75%
R^1 = ^{t}Bu, R^2 = H: 94%

R^1 = Me, R^2 = F: 80%, *o*:*m*:*p* = 1:1.2:4
R^1 = Me, R^2 = Cl: 68%, *m*:*p* = 1.4:1
R^1 = Me, R^2 = Me: 74%, *o*:*m*:*p* = 1:1.8:1.6
R^1 = Me, R^2 = 1,4-2F: 76%
R^1 = Me, R^2 = 1,2-2F: 84%, 2,3,4:1,3,4 = 1:1.2
R^1 = Me, R^2 = 1,2-2Cl: 69%, 2,3,4:1,3,4 = 1.3:1
R^1 = Me, R^2 = 1,2-2Me: 65%, 2,3,4:1,3,4 = 1:5.4
R^1 = Me, R^2 = 1,3-2Me: 77%, 1,3,5:2,3,5 = 2.4:1
R^1 = Me, R^2 =1,3-2Cl: 78%, 1,3,5:2,3,5 = 17:1
R^1 = Me, R^2 = CO_2Me: 73%, *m*:*p* = 2:1
R^1 = Me, R^2 = 1,3-2Me-2-Cl: 52%,(1,2,3,5 exclusively)

Scheme 10.35 Pd-catalyzed oxidative C–H/C–H cross-coupling of benzoxazoles and arenes.

In 2012, Su and co-workers developed the Pd-catalyzed oxidative C–H/C–H cross-coupling of benzoxazoles **74** with arenes using $CuBr_2$ as an additive in combination with O_2 as the terminal oxidant.[58] Removal of $CuBr_2$ from the reaction system completely shut down the reaction, thus it was proposed that $CuBr_2$ may act as a Lewis acid to coordinate to benzoxazole, hence increasing the C–H acidity of benzoxazole and preventing catalyst poisoning caused by the coordination of 2-arylbenzoxazole to palladium center. The reaction displays a broad substrate scope with respect to both benzoxazoles and arenes, and the regioselectivity of arenes is mainly dependent on their steric hindrance (Scheme 10.35). The monosubstituted benzenes gave *ortho*, *meta* and *para* isomers, with less *ortho* isomer than either of the other two. In the case of multisubstituted benzenes, the reactions preferentially occurred at less hindered positions. For instance, in the case of 1,2,3-trisubstituted benzene, the cross-coupling reaction exclusively occurred at the less hindered C5-position. A KIE of 5.6 for benzene and benzene-d_6 indicated that the cleavage of the C–H bond of benzene is an important event in the rate-limiting step. The observation that the electron-deficient fluorobenzene was more active than benzene in the reaction with 5-methybenzoxazole implied that the C–H cleavage of arenes might process *via* the concerted metalation-deprotonation mechanism (CMD).

The palladium-catalyzed oxidative C–H/C–H cross-coupling of azole-4-carboxylates **76** with arenes have already been realized by Yao and co-workers in 2012.[59] Under the optimized conditions, a range of thiazoles and oxazoles smoothly coupled with a variety of arenes to afford the corresponding products **77** in good yields. The selectivity of arene C–H activation was mainly

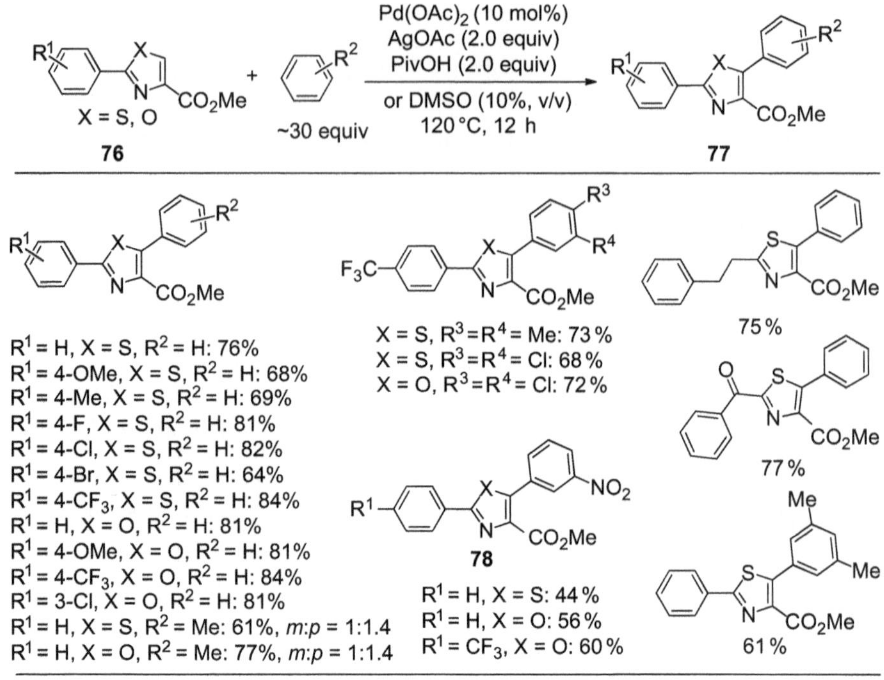

Scheme 10.36 Pd-catalyzed oxidative C–H/C–H cross-coupling of azole-4-carboxylates with arenes.

controlled by steric hindrance. Remarkably, nitrobenzene coupled with azoles to furnish 1,3-substituted products **78** in acceptable yields (Scheme 10.36).

10.5 Pd-Catalyzed Oxidative C–H/C–H Cross-Coupling Reaction Between Two Heteroarenes

To realize the Pd-catalyzed oxidative C–H/C–H cross-coupling of two different heteroarenes to build unsymmetrical biheteroaryl blocks, chemists must address the inherent difficulties that the heteroatom in both the substrates and products may strongly ligate the palladium center which will result in low reactivity and selectivity.

In 1976, Kozhevnikov observed the Pd-catalyzed cross-coupling between furan and thiophene to afford a high yield of furylthiophenes during their study on the oxidative homocoupling of furan or thiophene with $Cu(OAc)_2$ as the oxidant.[60]

A significant breakthrough in the Pd-catalyzed oxidative C–H/C–H cross-coupling between two different heteroarenes was made by You, Hu and co-workers. In 2010, they reported the $Pd(OAc)_2$-catalyzed oxidative C–H/C–H cross-coupling of various N-containing heteroarenes **79** including electron-rich xanthines, azoles, and indolizines and electron-poor pyridine

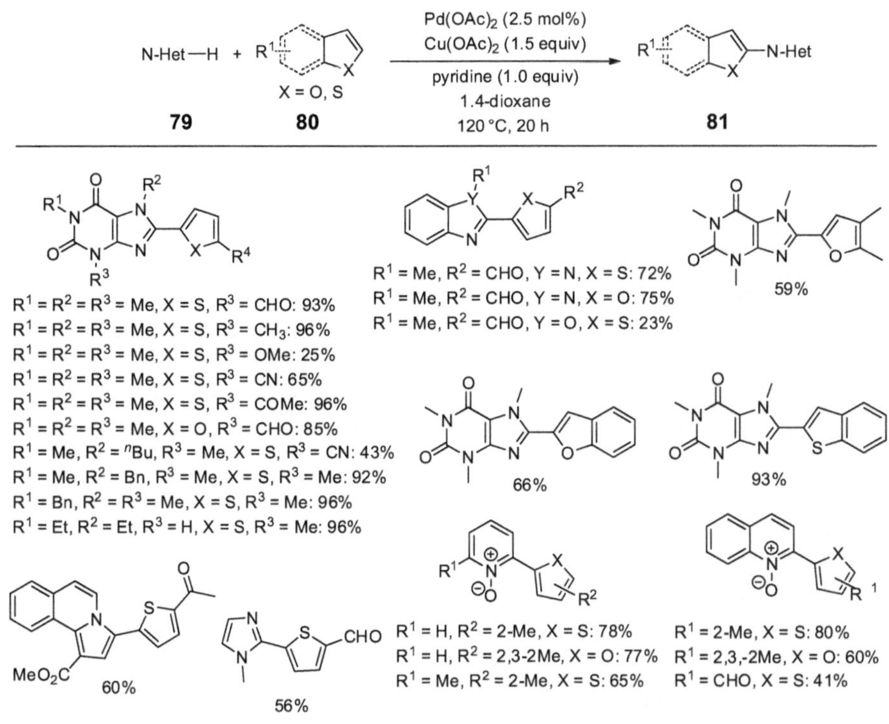

Scheme 10.37 Pd-catalyzed oxidative C–H/C–H cross-coupling of various N-containing heteroarenes with thiophenes or furans.

N-oxides, with π-electron-excessive five-membered heterocycles **80**, such as furans and thiophenes.[61] Even though up to 3–4 equiv. of five-membered heteroarene were used, a selective cross-coupling product **81** was obtained, which is contrary to the statistical distribution of products. The distinctly different electronic characteristics of the electron-rich five-membered heteroarene and the acidic heteroarene were assumed to be responsible for the excellent cross-coupling selectivity. The transformation was compatible with a broad range of substituents in good to excellent yields, and the C–H functionalization of furans and thiophene, including benzothiophene and benzofuran, took place at exclusively C2-position (Scheme 10.37).

A DFT calculation based on the reaction of *N*-methylimidazole and thiophene was performed to gain some insight into the reaction mechanism. It is proposed that the electron-rich thiophene was attacked by $Pd(OAc)_2$ through a regioselective electrophilic C–H substitution to generate (thien-2-yl-)Pd(OAc) intermediate in the first step. Although the subsequent reaction with a second molecule of thiophene is possible, calculations suggested that the catalytic system inverted its selectivity to preferentially react with *N*-methylimidazole *via* the concerted metalation–deprotonation (CMD) process, thus forming the key intermediate (thien-2-yl-)Pd(*N*-methylimidazol-2-yl).

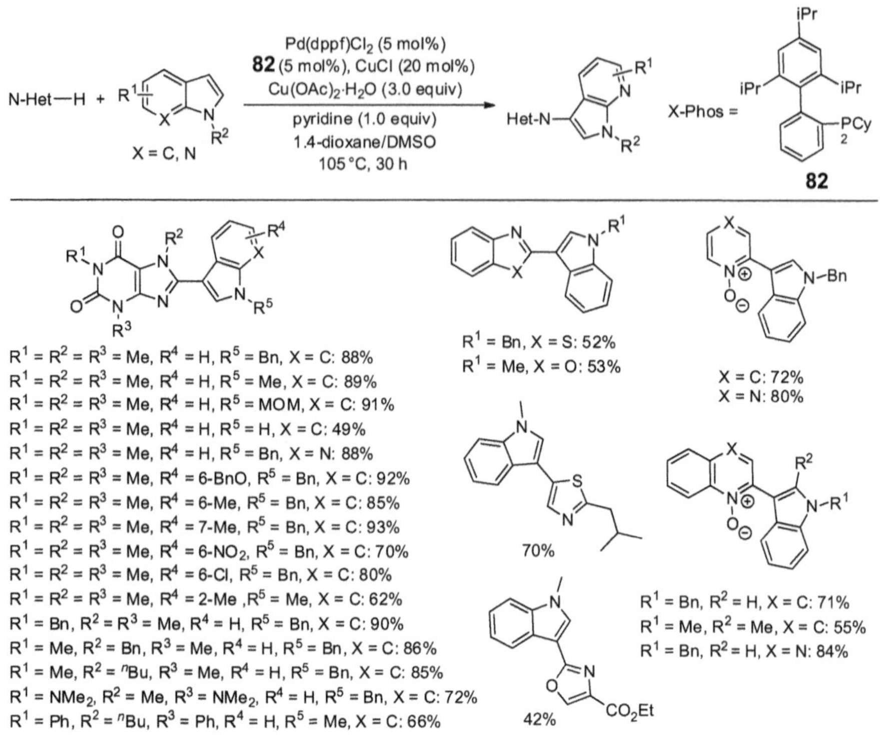

Scheme 10.38 Pd-catalyzed oxidative C–H/C–H cross-coupling of various N-containing heteroarenes with indoles.

In general, the oxidative C–H/C–H cross-coupling methods of indoles and pyrroles pose a greater challenge because the indole and pyrrole substrates, as well as their corresponding products, are susceptible to oxidative decomposition under the oxidative reaction conditions. In their later study, You and co-workers further overcame this limitation by introducing 2-dicyclohexylphosphino-2′,4′,6′-triisopropylbiphenyl (X-Phos) **82** as the ligand and CuCl as the cocatalyst (Scheme 10.38).[62] This transformation was suitable for a relatively broad range of indoles with complete C3 selectivity and good yields. Interestingly, subjecting pyrroles **83** to the established catalytic system gave only the C3 product **84**, which is different from other previously reported oxidative coupling reactions of pyrroles, thus providing an efficient and complementary route to form 3-heteroarylated pyrroles (Scheme 10.39).

Zhang, Li and co-workers also demonstrated an analogous palladium-catalyzed oxidative C–H/C–H cross-coupling between the C2-position of pyridine N-oxide **85** and the C3-position of indoles **86** using Ag_2CO_3 as the oxidant (Scheme 10.40).[63] A higher amount of pyridine (4 equiv.) and the introduction of tetrabutylammonium bromide (TBAB) as a co-additive proved to be beneficial for the reaction efficiency. Under the optimized conditions, a

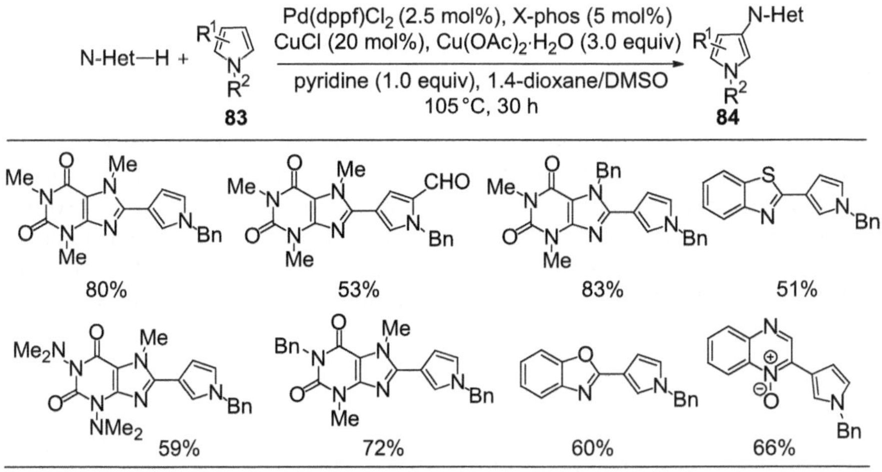

Scheme 10.39 Pd-catalyzed oxidative C–H/C–H cross-coupling of various N-containing heteroarenes with pyrroles.

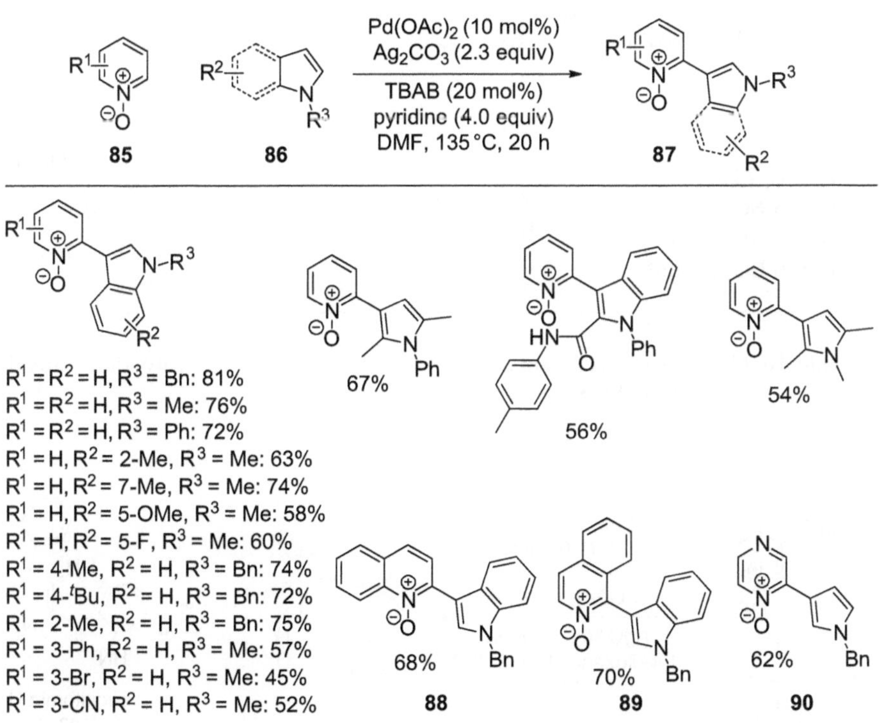

Scheme 10.40 Pd-catalyzed oxidative C–H/C–H cross-coupling of pyridine N-oxide with indoles.

variety of N-substituted indoles all underwent the reaction in moderate to good yields. In addition to pyridine N-oxide, the catalytic system can be simultaneously applied to N-oxides of quinoline, isoquinoline and pyrazine (**88–90**). A kinetic isotope effect of 3.6 was obtained for pyridine N-oxide and pyridine-d_5 N-oxide, which suggested that C–H bond cleavage of the pyridine N-oxides is involved in the rate-determining step. Concurrently, Itami and co-workers developed a similar process using MOM-protected indoles **91** as substrates, 2,6-lutidine as ligand and AgOAc as the oxidant (Scheme 10.41).[64] They applied their methods successfully for the synthesis of the marine indole alkaloid eudistomin U **93**, which possesses DNA-binding activity (Scheme 10.42).

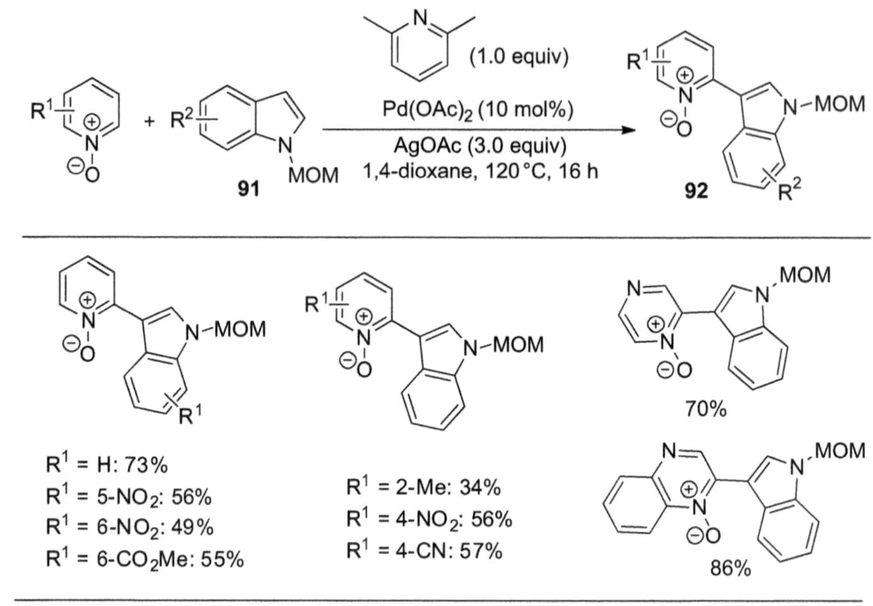

Scheme 10.41 Pd-catalyzed oxidative C–H/C–H cross-coupling of pyridine N-oxide with indoles.

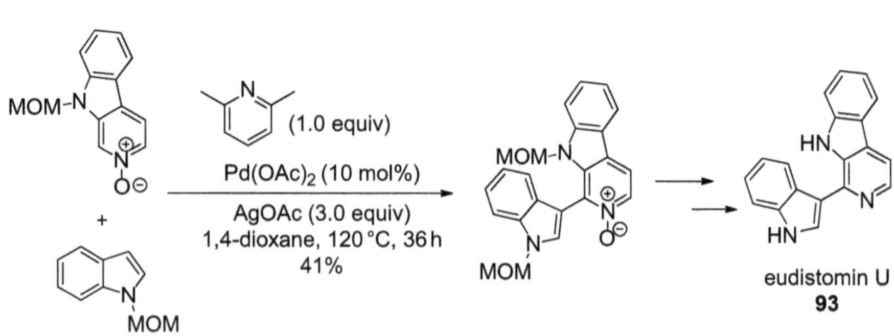

Scheme 10.42 Synthesis of eudistomin U **93** *via* oxidative C–H/C–H cross-coupling.

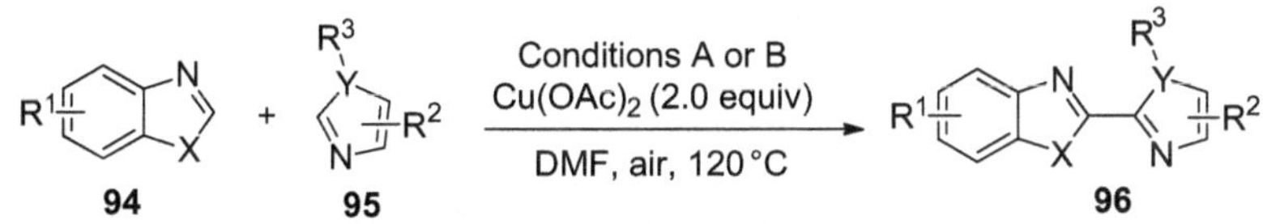

Conditions A: $Pd(OAc)_2$ (5 mol%), $KF/AgNO_3$ (3+1.5 equiv), 24 h
Conditions B: $Pd(OAc)_2$ (10 mol%), AgF (2.0 equiv), 48 h

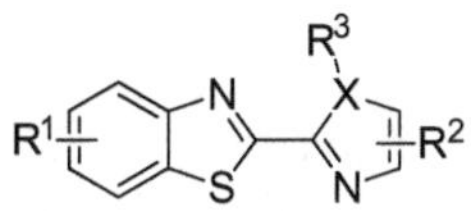

R^1 = H, R^2 = H, R^3 = Me, X = N, conditions A: 91%
R^1 = H, R^2 =5-Cl , R^3 = Me, X = N, conditions B: 95%
R^1 = H, R^2 = H, R^3 = H, X = O, conditions B: 86%
R^1 = H, R^2 = 5-(*p*-BrPh), R^3= H, X = O, conditions B: 71%
R^1 = H, R^2 = H, R^3 = H, X = S, conditions A: 65%
R^1 = H, R^2 = 4-Me, R^3 = H, X = S, conditions A: 67%
R^1 = H, R^2 = 4,5-2Me, R^3 = H, X = S, conditions A: 92%

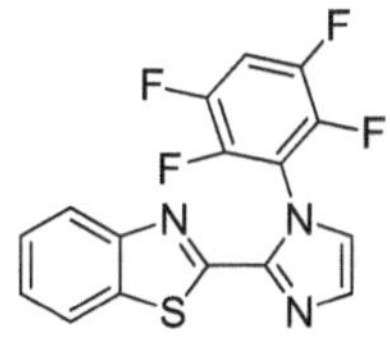

91%

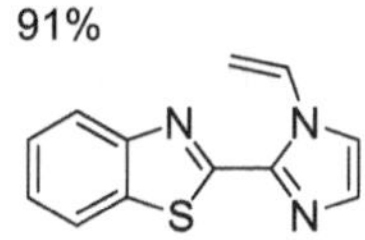

87%

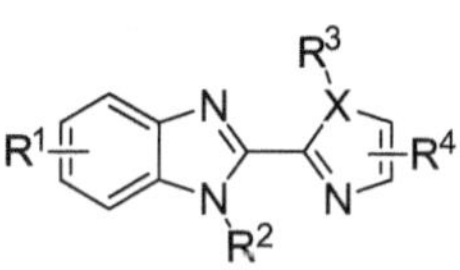

R^1 = H, R^2 = Me, R^3 = Me, R^4 = H, X = N,conditions B: 90%
R^1 = 5,6-2Me, R^2 = R^3 = Me, X = N, conditions B: 91%
R^1 = 5,6-2Me, R^2 = Et , R^3 = Me, X = N, conditions B: 88%
R^1 = H, R^2 = Bz, R^3 = Me, X = N, conditions B: 89%
R^1 = 5,6-2Me, R^2 = Bz, R^3 = Me, X = N, conditions B: 84%
R^1 = H, R^2 = Me, R^3 = Me, R^4 = 5-Cl, X = N, conditions B: 74%
R^1 = H, R^2 = Me, R^3 = H, R^4 = H, X = O, conditions B: 61%
R^1 = 5,6-2Me, R^2 = Me, R^3 = H, R^4 = H, X = O, conditions B: 71%
R^1 = H, R^2 = Me, R^3 = H, R^4 = 4,5-2Me, X = S, conditions B: 86%

Scheme 10.43 Pd-catalyzed oxidative C–H/C–H cross-coupling of benzothiazole and benzimidazole with azoles.

Ofial and co-workers independently developed efficient Pd-catalyzed oxidative C–H/C–H cross-coupling of benzothiazole and benzimidazole **94** with various N-, O-, and S-containing azoles **95** in the presence of $Cu(OAc)_2$ combined with either AgF or $AgNO_3/KF$. This transformation took place selectively at the C2 positions of azoles, thus offering an efficient access to various unsymmetrical 2,2′-bisheteroaryls **96** (Scheme 10.43).[65] X-Ray powder diffraction analysis indicated that Ag^I acted as terminal oxidants in these reactions. The role of the Cu^{II} ions is less clear at present; however, replacing $Cu(OAc)_2$ with $CuCl_2$, or $CuBr_2$, or $Cu(OTf)_2$ led to poor results, indicating that the presence of acetate ions is crucial, probably because Pd-bound acetate plays the role of a proton acceptor in the C–H bond cleavage. A set of mechanistic experiments were carried out to understand the parameters that determine the selectivity, which revealed that the presence of Ag^I suppressed the homocoupling products to a great extent.

Meanwhile, You and co-workers developed Pd-catalyzed oxidative C–H/C–H cross-coupling of two non-benzofused azoles **97** and **98** that bear a similar structure.[66] The challenge posed by such a reaction is to restrain the homocoupling and to efficiently promote cross-coupling. After a series of optimization studies, the conditions involving $PdCl_2$ (5 mol%) in combination with 10 mol% of PPh_3, 10 mol% of CuCl and 1.5 equiv. of $Cu(OAc)_2$ in 1,4-dioxane/DMSO were established. With the optimized conditions, a variety of oxazoles could couple with a diverse range of thiazoles with excellent functional group tolerance in satisfactory yields, and the C2 regioselective for azoles was observed. In addition, the established catalytic conditions can be applied to the reactions of oxazoles or thiazoles with imidazoles to afford the desired cross-coupling products, and the homocoupling of imidazoles was completely suppressed (Scheme 10.44). Mechanistic studies indicated that the difference in electronic nature between the two coupling partners might result in the expected selective cross-coupling, as the reaction of the thiazole **100** with the

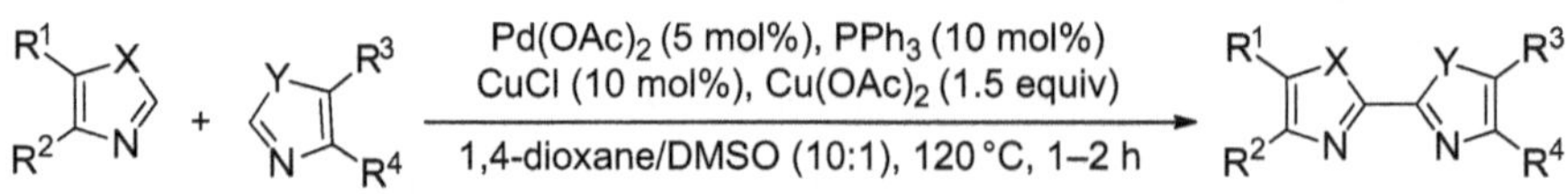

97 **98** **99**

R^1 = Ph, R^2 = H, X = O, R^3 = R^4 = Me, Y = S: 68%
R^1 =2-MeOPh, R^2 = H, X = O, R^3 = R^4 = Me, Y = S: 70%
R^1 =4-ClPh, R^2 = H, X = O, R^3 = R^4 = Me, Y = S: 67%
R^1 =2-Py, R^2 = H, X = O, R^3 = R^4 = Me, Y = S: 66%
R^1 =Ph, R^2 = H, X = O, R^3 = CH_2CH_2OH, R^4 = Me, Y = S: 70%
R^1 =Ph, R^2 = H, X = O, R^3 = CH_2CH_2OMe, R^4 = Me, Y = S: 68%
R^1 =Ph, R^2 = H, X = O, R^3 = CHO, R^4 = Me, Y = S: 68%
R^1 =Ph, R^2 = H, X = O, R^3 = CO_2Et, R^4 = H, Y = S: 66%
R^1 = CO_2Et, R^2 = Me, X = O, R^3 = R^4 = Me, Y = S: 90%
R^1 = CO_2Et, R^2 = Me, X = O, R^3 = CH_2CH_2OH, R^4 = Me, Y = S: 73%
R^1 = CO_2Et, R^2 = Me, X = O, R^3 = CH_2CH_2OMe, R^4 = Me, Y = S: 78%
R^1 = CO_2Et, R^2 = Me, X = O, R^3 = CHO, R^4 = Me, Y = S: 70%
R^1 = CO_2Et, R^2 = Me, X = O, R^3 = CO_2Et, R^4 = H, Y = S: 71%
R^1 = CO_2Et, R^2 = Me, X = O, R^3 = CO_2Et, R^4 = Me, Y = S: 73%
R^1 = CO_2Et, R^2 = Me, X = O, R^3 = R^4 = H, Y = S: 58%
R^1 = CO_2Et, R^2 = Me, X = O, R^3 = H, R^4 = H, Y = NMe: 67%
R^1 = CO_2Et, R^2 = Me, X = O, R^3 = H, R^4 = H, Y = $NCH{=}CH_2$: 60%
R^1 = CO_2Et, R^2 = Me, X = O, R^3 = R^4 = Me, Y = NMe: 68%
R^1 = CO_2Et, R^2 = Me, X = O, R^3 = H, R^4 = NO_2, Y = NMe: 66%
R^1 = CHO, R^2 = Me, X = S, R^3 = R^4 = H, Y = $NCH{=}CH_2$: 64%
R^1 = CHO, R^2 = Me, X = S, R^3 = R^4 = Me, Y = NMe: 71%
R^1 = CHO, R^2 = Me, X = S, R^3 = CH_2CH_2OAc, R^4 = Me, Y = S: 76%

Scheme 10.44 Pd-catalyzed oxidative C–H/C–H cross-coupling between two similar azoles.

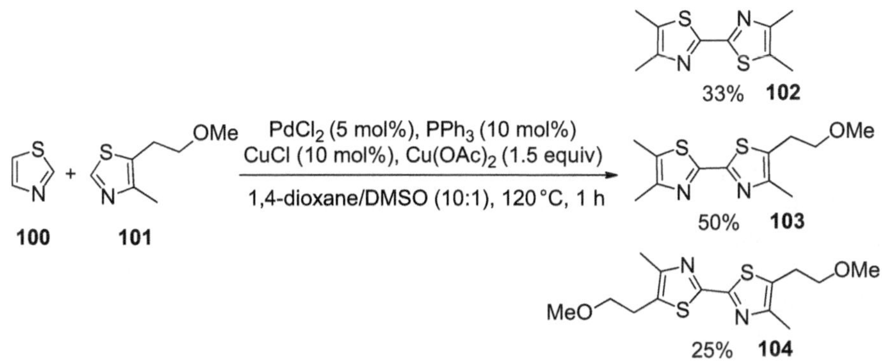

Scheme 10.45 Competition reactions of oxidative C–H/C–H cross-coupling between azoles.

electronically similar thiazole **101** with a ratio of 1 : 1 gave an approximate statistical distribution of products **102–104** (Scheme 10.45). Both acetate anion and Cu^{I} salt play critical roles in this transformation, and the replacement of $Cu(OAc)_2$ with other Cu^{II} salts or the removal of CuCl from the system greatly reduced the yield of reaction.

10.6 Summary and Outlook

The past ten years has witnessed the rapid development of Pd-catalyzed oxidative C–H/C–H cross-coupling to construct bi(hetero)aryl motifs. A variety of substrate combinations, such as simple arene/simple arene, simple arene/heteroarene and heteroarene/heteroarene, can be cross-coupled with good selectivity. The strategic solutions to chemoselectivity issues arising from the C–H/C–H cross-coupling reactions include: (i) choosing two electronically different arenes as coupling partners; (ii) using the directing-group-containing substrates, or (iii) the use of an oxidant capable of oxidizing Pd(II) to Pd(III) or Pd(IV), for example $K_2S_2O_8$ is promising for addressing the regioselectivity issue associated with the C–H/C–H cross-coupling of monosubstituted arenes.

The rapidly growing Pd-catalyzed C–H/C–H cross-coupling to bi(hetero)aryls is still full of many challenging problems that need to be solved in the coming years. Among them are the C–H activation of unactivated arene, high Pd catalyst loading, use of expensive terminal oxidants such as silver salts, and chemo- and regio-selectivity issues. The solutions to these problems will make this oxidative C–H/C–H cross-coupling more efficient and practical.

References

1. J. Hassan, M. Sévignon, C. Gozzi, E. Schulz and M. Lemaire, *Chem. Rev.*, 2002, **102**, 1359–1470.
2. K. C. Nicolaou, P. G. Bulger and D. Sarlah, *Angew. Chem., Int. Ed.*, 2005, **44**, 4442–4489.

3. A. Suzuki, *Angew. Chem. Int. Ed.*, 2011, **50**, 6722–6737.
4. S. Baba and E. Negishi, *J. Am. Chem. Soc.*, 1976, **98**, 6729–6731.
5. D. Milstein and J. K. Stille, *J. Am. Chem. Soc.*, 1978, **100**, 3636–3638.
6. Y. Hatanaka and T. Hiyama, *J.Org.Chem.*, 1988, **53**, 918–920.
7. B. Trost, *Science*, 1991, **254**, 1471–1477.
8. B. M. Trost, *Angew.Chem., Int. Ed.*, 1995, **34**, 259–281.
9. D. R. Stuart and K. Fagnou, *Science*, 2007, **316**, 1172–1175.
10. A. E. Shilov and G. B. Shul'pin, *Chem. Rev.*, 1997, **97**, 2879–2932.
11. D. Alberico, M. E. Scott and M. Lautens, *Chem. Rev.*, 2007, **107**, 174–238.
12. L. Ackermann, R. Vicente and A. R. Kapdi, *Angew. Chem., Int. Ed.*, 2009, **48**, 9792–9826.
13. X. Chen, K. M. Engle, D. H. Wang and J. Q. Yu, *Angew. Chem., Int. Ed.*, 2009, **48**, 5094–5115.
14. O. Daugulis, H.-Q. Do and D. Shabashov, *Acc. Chem. Res.*, 2009, **42**, 1074–1086.
15. J. A. Ashenhurst, *Chem. Soc. Rev.*, 2010, **39**, 540–548.
16. S. L. You and J. B. Xia, *Top. Curr. Chem.*, 2010, 165–194.
17. S. H. Cho, J. Y. Kim, J. Kwak and S. Chang, *Chem. Soc. Rev.*, 2011, **40**, 5068–5083.
18. C. Liu, H. Zhang, W. Shi and A. Lei, *Chem. Rev.*, 2011, **111**, 1780–1824.
19. W. Han and A. R. Ofial, *Synlett*, 2011, **2011**, 1951–1955.
20. C. S. Yeung and V. M. Dong, *Chem. Rev.*, 2011, **111**, 1215–1292.
21. D. Zhao, J. You and C. Hu, *Chem.–Eur. J.*, 2011, **17**, 5466–5492.
22. J. Wencel-Delord, C. Nimphius, F. W. Patureau and F. Glorius, *Angew. Chem., Int. Ed.*, 2012, **51**, 2247–2251.
23. N. Kuhl, M. N. Hopkinson and F. Glorius, *Angew. Chem., Int. Ed.*, 2012, **51**, 8230–8234.
24. F. W. Patureau, J. Wencel-Delord and F. Glorius, *Aldrichimica Acta*, 2012, **45**, 31–41.
25. M. Kitahara, N. Umeda, K. Hirano, T. Satoh and M. Miura, *J. Am. Chem. Soc.*, 2011, **133**, 2160–2162.
26. M. Nishino, K. Hirano, T. Satoh and M. Miura, *Angew. Chem., Int. Ed.*, 2012, **51**, 6993–6997.
27. R. van Helden and G. Verberg, *Recl. Trav. Chim. Pays-Bas*, 1965, **84**, 1263–1273.
28. R. Li, L. Jiang and W. Lu, *Organometallics*, 2006, **25**, 5973–5975.
29. L. Zhou and W. Lu, *Organometallics*, 2012, **31**, 2124–2127.
30. K. Kawasumi, K. Mochida, T. Kajino, Y. Segawa and K. Itami, *Org Lett*, 2012, **14**, 418–421.
31. Y. Wei and W. Su, *J. Am. Chem. Soc.*, 2010, **132**, 16377–16379.
32. H. Li, J. Liu, C. L. Sun, B. J. Li and Z. J. Shi, *Org Lett.*, 2011, **13**, 276–279.
33. K. L. Hull and M. S. Sanford, *J. Am. Chem. Soc.*, 2007, **129**, 11904–11905.
34. K. L. Hull and M. S. Sanford, *J. Am. Chem. Soc.*, 2009, **131**, 9651–9653.
35. T. W. Lyons, K. L. Hull and M. S. Sanford, *J. Am. Chem. Soc.*, 2011, **133**, 4455–4464.

36. J.-B. Xia and S.-L. You, *Organometallics*, 2007, **26**, 4869–4871.
37. B.-J. Li, S.-L. Tian, Z. Fang and Z.-J. Shi, *Angew. Chem., Int. Ed.*, 2008, **47**, 1115–1118.
38. W. C. P. Tsang, N. Zheng and S. L. Buchwald, *J. Am. Chem. Soc.*, 2005, **127**, 14560–14561.
39. G. Brasche, J. García-Fortanet and S. L. Buchwald, *Org. Lett.*, 2008, **10**, 2207–2210.
40. X. Zhao, C. S. Yeung and V. M. Dong, *J. Am. Chem. Soc.*, 2010, **132**, 5837–5844.
41. C. S. Yeung, X. Zhao, N. Borduas and V. M. Dong, *Chem. Sci.*, 2010, **1**, 331–336.
42. M. J. Tredwell, M. Gulias, N. G. Bremeyer, C. C. Johansson, B. S. Collins and M. J. Gaunt, *Angew. Chem. Int. Ed.*, 2011, **50**, 1076–1079.
43. M. Yu, Z. Liang, Y. Wang and Y. Zhang, *J. Org. Chem.*, 2011, **76**, 4987–4994.
44. X. Wang, D. Leow and J.-Q. Yu, *J. Am. Chem. Soc.*, 2011, **133**, 13864–13867.
45. J. Karthikeyan and C.-H. Cheng, *Angew. Chem., Int. Ed.*, 2011, **50**, 9880–9883.
46. V. S. Thirunavukkarasu and C.-H. Cheng, *Chem.–Eur. J.*, 2011, **17**, 14723–14726.
47. T. Itahara, *J. Chem. Soc., Chem. Commun.*, 1981, 254–255.
48. D. R. Stuart, E. Villemure and K. Fagnou, *J. Am. Chem. Soc.*, 2007, **129**, 12072–12073.
49. T. A. Dwight, N. R. Rue, D. Charyk, R. Josselyn and B. DeBoef, *Org. Lett.*, 2007, **9**, 3137–3139.
50. S. Potavathri, A. S. Dumas, T. A. Dwight, G. R. Naumiec, J. M. Hammann and B. DeBoef, *Tetrahedron Lett.*, 2008, **49**, 4050–4053.
51. S. Potavathri, K. C. Pereira, S. I. Gorelsky, A. Pike, A. P. LeBris and B. DeBoef, *J. Am. Chem. Soc.*, 2010, **132**, 14676–14681.
52. A. N. Campbell, E. B. Meyer and S. S. Stahl, *Chem. Commun.*, 2011, **47**, 10257–10259.
53. S. H. Cho, S. J. Hwang and S. Chang, *J. Am. Chem. Soc.*, 2008, **130**, 9254–9256.
54. D. Kalyani, N. R. Deprez, L. V. Desai and M. S. Sanford, *J. Am. Chem. Soc.*, 2005, **127**, 7330–7331.
55. C.-Y. He, S. Fan and X. Zhang, *J. Am. Chem. Soc.*, 2010, **132**, 12850–12852.
56. C.-Y. He, Q.-Q. Min and X. Zhang, *Organometallics*, 2011, **31**, 1335–1340.
57. C. C. Malakar, D. Schmidt, J. r. Conrad and U. Beifuss, *Org. Lett.*, 2011, **13**, 1378–1381.
58. G. Wu, J. Zhou, M. Zhang, P. Hu and W. Su, *Chem. Commun.*, 2012, **48**, 8964–8966.
59. Z. Li, L. Ma, J. Xu, L. Kong, X. Wu and H. Yao, *Chem. Commun.*, 2012, **48**, 3763–3765.

60. I. V. Kozhevnikov, *React. Kinet. Catal. Lett.*, 1976, **4**, 451–458.
61. P. Xi, F. Yang, S. Qin, D. Zhao, J. Lan, G. Gao, C. Hu and J. You, *J. Am. Chem. Soc.*, 2010, **132**, 1822–1824.
62. Z. Wang, K. Li, D. Zhao, J. Lan and J. You, *Angew. Chem., Int. Ed.*, 2011, **50**, 5365–5369.
63. X. Gong, G. Song, H. Zhang and X. Li, *Org. Lett.*, 2011, **13**, 1766–1769.
64. A. D. Yamaguchi, D. Mandal, J. Yamaguchi and K. Itami, *Chem. Lett.*, 2011, **40**, 555–557.
65. W. Han, P. Mayer and A. R. Ofial, *Angew. Chem., Int. Ed.*, 2011, **50**, 2178–2182.
66. J. Dong, Y. Huang, X. Qin, Y. Cheng, J. Hao, D. Wan, W. Li, X. Liu and J. You, *Chem.–Eur. J.*, 2012, **18**, 6158–6162.

CHAPTER 11

Dioxygen-Coupled Palladium and Copper-Catalyzed C_{sp2}–H Functionalization: Reactions and Mechanisms

XIN MU AND GUOSHENG LIU*

Shanghai Institute of Organic Chemistry, CAS, Lingling road No. 345, Shanghai, China
*Email: gliu@mail.sioc.ac.cn

11.1 Introduction

Oxidation reactions play essential roles for functional group manipulations in modern organic synthesis. The demand for the development of robust and selective oxidation process relies on discovery of more efficient and practical catalytic systems.[1,2] Among reported methods, Pd-catalyzed oxidation reactions represent a fast growing realm during the last several decades.[3]

Since then, significant improvements have been made in Pd-catalyzed oxidation reaction using ligands and additives to elevate the catalytic efficiency. Although many kinds of terminal oxidants have been applied (Cu^{II}, Ag^{I}, BQ, *etc.*), stoichiometric amounts of wastes and relatively low catalytic efficiency have encouraged the employment of environmental benign dioxygen as terminal oxidant.[3–7]

Among those reported examples, Pd-catalyzed alcohol oxidation reactions under aerobic conditions represent a major development in this area. Ligands

RSC Catalysis Series No. 11
C–H and C–X Bond Functionalization: Transition Metal Mediation
Edited by Xavi Ribas

Published by the Royal Society of Chemistry, www.rsc.org

have been proved beneficial to improve the catalyst stability and efficiency in co-oxidant free conditions (Scheme 11.1).[6,8–13]

The first example was provided by Peterson and Larock using $Pd(OAc)_2$/DMSO catalyst system to achieve oxidation of a wide variety of benzylic and allylic alcohols under aerobic atmosphere.[14,15] The mechanistic rationale of this catalytic system provided by Stahl and coworkers suggested the DMSO acted as an effective ligand to promote palladium oxidation by oxygen and prevent the active palladium catalyst from aggregating.[16,17]

Uemura and co-workers reported several Pd-catalyzed aerobic alcohol oxidations using pyridine as the ligand.[18–20] This catalytic system is quite efficient for oxidation of a large class of alcohols, including primary, secondary, benzylic and allylic alcohols. Detailed mechanistic studies of this oxidation system were later illustrated by Stahl and coworkers.[21–23]

Sheldon and co-workers reported $Pd(OAc)_2$/bathophenanthroline disulfonate catalytic system for the aerobic oxidation of alcohols in aqueous solution.[24–27] This advantageous catalytic system provided the oxidation process with operational safety and separation efficiency.

Stoltz[28] and Sigman[29–31] independently demonstrated $PdCl_2$/(–)-sparteine catalyst system which are capable of achieving aerobic oxidative kinetic resolution of secondary alcohols with high enantiomeric excess (Scheme 11.4). Mechanistic studies carried out by Sigman and co-workers elucidated that exogenous (–)-sparteine played an essential role acted as both ligand and base to promote the reaction process.[32]

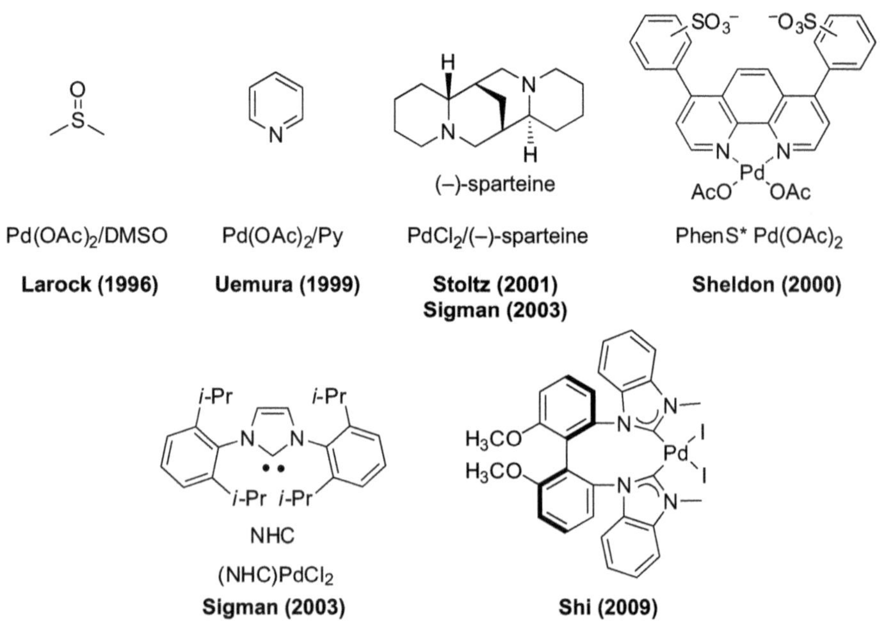

Scheme 11.1 Typical catalyst systems for aerobic alcohol oxidation reactions.

Sigman[33–35], Shi[36–38] and others[39–41] reported N-heterocyclic carbene (NHC) ligands were utilized to facilitate Pd-catalyzed aerobic alcohol oxidations due to their strong coordination capacity to stabilize Pd^0 and Pd^{II} center.

Pd-catalyzed oxidative functionalization of alkenes represents another type of strategy to install functional groups across double bonds, such as Wacker-type and aza-Wacker reactions to build C–O bond and C–N bond, and oxidative Heck reactions to form C–C bond.[42–46] Catalytic systems for aerobic oxidation of alcohols, which were mentioned above, have been extended to the related functionalization of alkenes. This chapter highlights recent progress on Pd-catalyzed aerobic oxidative functionalizations of alkenes, and detailed mechanistic studies involving the catalytic cycles, especially regeneration of the Pd^{II} catalysts by molecular oxygen are illustrated. In addition, copper-catalyzed aerobic C–H bond functionalization reactions have received much attention during last several decades. These catalytic systems greatly expanded the versatility of palladium-catalyze aerobic reactions.[47,48] Thus, a brief introduction to copper-catalyzed aerobic C–H bond functionalization reactions is provided.

11.2 Mechanistic studies on Pd^0 Reoxidation by Molecular Oxygen

Pd-catalyzed oxidations have been extensively studied, and one of the vital issues for aerobic catalytic cycle is the regeneration of Pd^{II} catalyst from reduced Pd^0 species by dioxygen.[49] Mechanistic studies of reaction between Pd^0 and dioxygen are essential for improvement of catalytic efficiency by preventing Pd^0 from aggregation to form inactive palladium black.

Recycling of palladium catalyst is commonly proposed to proceed through reductive elimination of the resultant Pd^{II}–hydride species and reoxidation of reduced Pd^0 species by oxygen (Scheme 2). Alternative possibilities can't be excluded. For instance, concomitant formation of a Pd^{II}–hydride species in the

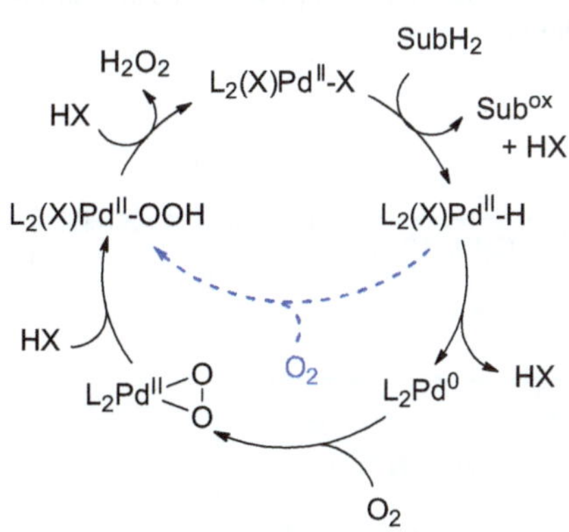

Scheme 11.2 Possible pathways for regeneration of Pd^{II} catalyst.

oxidation of alcohols process could directly react with oxygen to form a Pd^{II}–hydroperoxide intermediate, protonolysis of this intermediate will lead to the Pd^{II} catalyst (Scheme 11.2, dashed arrow).

An early example of a direct reaction between Pd^0 and dioxygen was reported by Wilke and co-workers in 1967.[50] Related η^2-peroxo complexes were also synthesized by Otsuka and other groups.[51–53] In 2001, Stahl and co-workers reported a bidentate nitrogen ligand coordinated η^2-peroxo palladium complex **2** synthesized from the related Pd^0 alkene complex **1**; treatment of **2** with acetic acid rapidly generated Pd^{II} catalyst **3** (Scheme 11.3). Upon isolation of the complex, kinetic studies of the ligand exchange pathway strongly supported an associative ligand exchange mechanism.[54]

To verify the alternative reaction pathway (Scheme 11.2, dashed arrow), Stahl and coworkers found that the exposure of Pd^{II}–hydride species **4** under O_2 atmosphere produced Pd^{II}-hydroperoxide complexes **5** in benzene at 40–50 °C.[55] This result indicated that the direct reaction of Pd^{II}–hydride species with O_2 is feasible (Scheme 11.4, respectively).

However, further mechanistic studies provided by Stahl suggested that the reaction pattern between **4** and dioxygen exhibit a first-order dependence on palladium catalyst and zero-order dependence on oxygen pressure. A radical scavenger exerted no influence on the reaction rate. A small kinetic isotopic effect ($k_{Pd\text{-}H}/k_{Pd\text{-}D} = 1.3$) was observed. These results strongly supported a HX reductive elimination pathway (Scheme 11.4).[56–58]

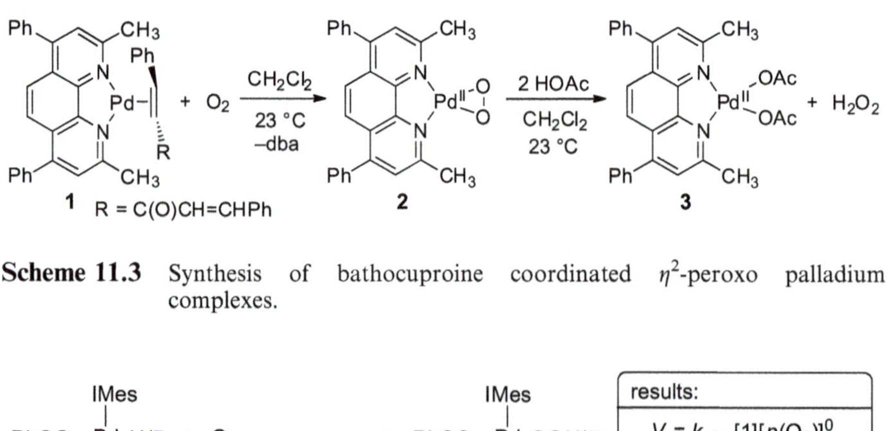

Scheme 11.3 Synthesis of bathocuproine coordinated η^2-peroxo palladium complexes.

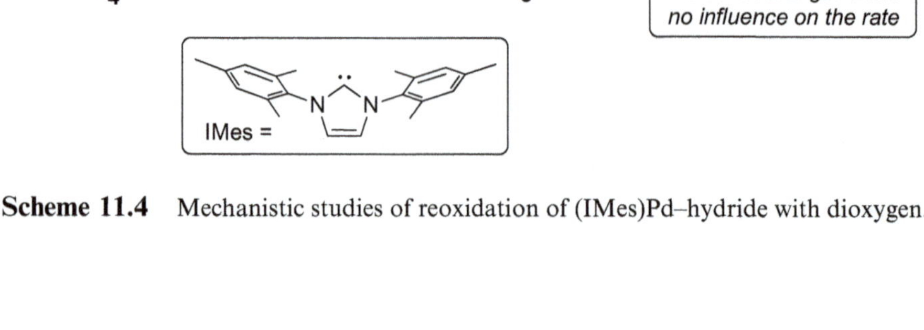

Scheme 11.4 Mechanistic studies of reoxidation of (IMes)Pd–hydride with dioxygen.

In the reaction of PCP–pincer Pd^{II}–hydride complex **6** with dioxygen to form hydroperoxide **7**, Goldberg and Goddard[59,60] provided experimental and computational evidence to support a hydrogen-atom abstraction mechanism (HAA). All observed kinetic data are consistent with proposed mechanism involving hydrogen-atom abstraction by dioxygen and subsequent rearrangement the of Pd^{I}–HOO intermediate to yield the Pd^{II}–hydroperoxide product (Scheme 11.5).

Meanwhile, Goddard carried out DFT calculations to investigate the detailed reaction process for Pd^{II}–hydride complex [((–)-sparteine)Pd(H)Cl] with dioxygen.[61] The computational results also provided support for a process involving a hydrogen-atom abstraction (HAA) mechanism. Stahl and coworkers recently reinvestigated the reaction of O_2 with Pd^{II}–hydride complex [((–)-sparteine)Pd(H)Cl].[62] However, contrary to previous report made by Goddard,[61] experimental studies suggested the reaction rate is independent of oxygen pressure, and DFT calculations identified an eminent "reductive elimination" mechanism in which chelating nitrogen ligand undergoes intramolecular deprotonation of the Pd^{II}–hydride species (Scheme 11.6). Finally, all catalyst systems with labile monodentate ligands[63] (pyridine, Et_3N, DMSO) exhibited a strong propensity to follow the same intramolecular deprotonation, reductive elimination reaction (HLRE) pathway.

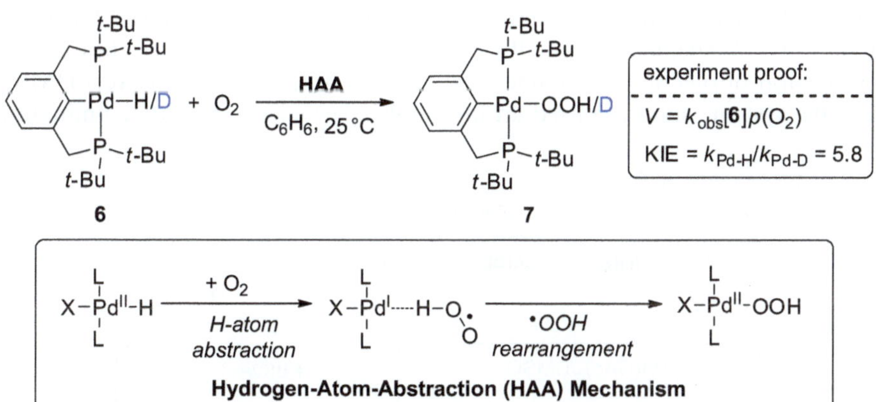

Scheme 11.5 Kinetic data for the reaction of (PCP)Pd^{II}–hydride complex with dioxygen and proposed hydrogen-atom abstraction (HAA) mechanism.

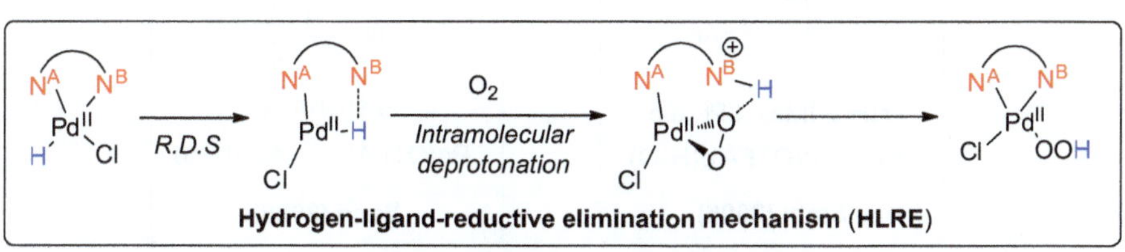

Scheme 11.6 Nitrogen-ligand assisted intramolecular deprotonation mechanism.

11.3 Pd-Catalyzed Aerobic Oxidative Functionalization of Alkenes

The traditional Wacker process employs water or hydroxide as nucleophiles to achieve aerobic oxidation of alkenes. Development of this classic transformation has been focused on the extension of different types of nucleophiles in view of synthetic versatility.[43,46,64–72] The mechanisms of these processes have received intensive attention to gain understandings on the detailed nucleopalladation pathways[42] and its stereochemical behaviors.

11.3.1 Pd-Catalyzed Aerobic Oxidative Amination of Alkenes

The intramolecular/intermolecular mode of the nucleopalladation to construct carbon–heteroatom bonds have been extensively studied by many research groups for broad applications in organic synthesis.[72–74] Initial nucleopalladation followed by β-hydride elimination of the alkyl–Pd^{II} intermediates provided the amination products.

11.3.1.1 Intramolecular Carbon–Nitrogen Bond Formation

A variety of Pd-catalyzed aerobic catalytic systems have been applied on intramolecular oxidative amination reactions (Scheme 11.7). Larock,[14] Bäckvall,[75,76] Hiemstra[77] and Broggini[78] have provided their studies on the cyclization of olefinic tosylamides and carboxamides under $Pd(OAc)_2$/DMSO aerobic system. Stahl reported an intramolecular aerobic oxidative amination

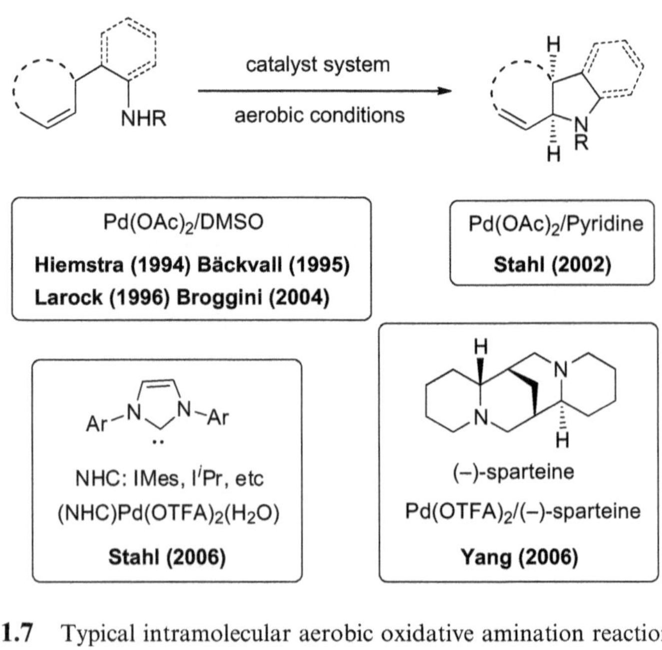

Scheme 11.7 Typical intramolecular aerobic oxidative amination reactions.

of unactivated alkenes under a $Pd(OAc)_2$/pyridine catalytic system.[79] And they also found that N-heterocyclic carbene Pd catalyst exhibited high reactivity towards the aerobic oxidative amination reactions, and a catalytic amount of acid additive is beneficial to promote the yields.[80] Yang and coworkers developed an enantioselective oxidative cyclization of **8** using $Pd(OTFA)_2$/(–)-sparteine catalytic system (Scheme 11.8).[81] Recently, Stahl and coworkers found that a variation of the $Pd(OAc)_2$/DMSO aerobic system to synthesize 1,2-diamine products could be achieved by Pd-catalyzed oxidative cyclization of allylic sulfamides (Scheme 11.9).[82]

Zhang and coworkers demonstrated that the employment of oxazoline-type ligands proved successful in achieving asymmetric oxidative amination reactions. Excellent catalytic activity and good enantioselectivity were obtained with quinoline oxazoline ligand **L1**.[83] Yang and coworkers reported a diastereoselective and enantioselective oxidative amination-coupled C–C bond forming reactions using *t*-butyl quinoline oxazoline ligand **L2**.[84] Stahl and coworkers also provided a asymmetric oxidative amination of unactivated alkenes by using pyridine oxazoline ligand **L3** to obtain very high enantioselectivity (Scheme 11.10).[85]

Yang and coworkers revealed that pyridine-type ligands also exhibited excellent activity to promote Pd-catalyzed aerobic oxidative amination reactions, which involve a cascade cyclization to achieve spiro-indolines with high diastereoselectivity (Scheme 11.11).[86] A *syn*-aminopalladation/electrophilic aromatic substitution sequence has been proposed to achieve the C–N and C–C bond formation under aerobic conditions by the same group using ethyl nicotinate ligand (Scheme 11.12).[87]

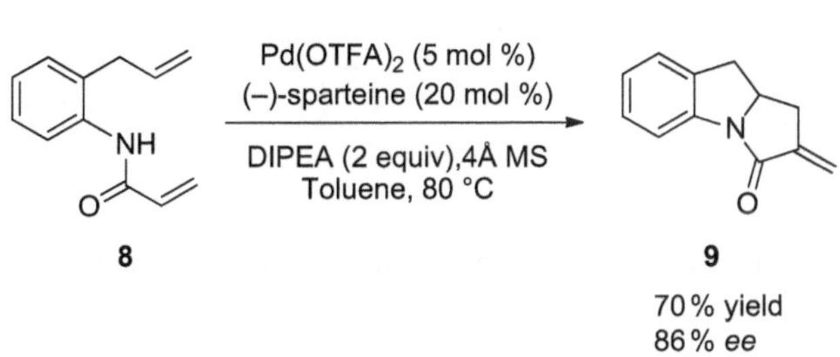

Scheme 11.8 Yang's intramolecular aerobic amination of alkenes.

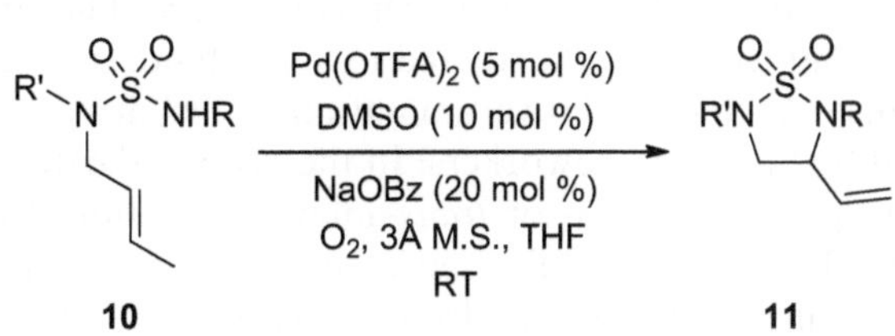

Scheme 11.9 Pd-catalyzed oxidative 1,2-diamination of allylic sulfamides.

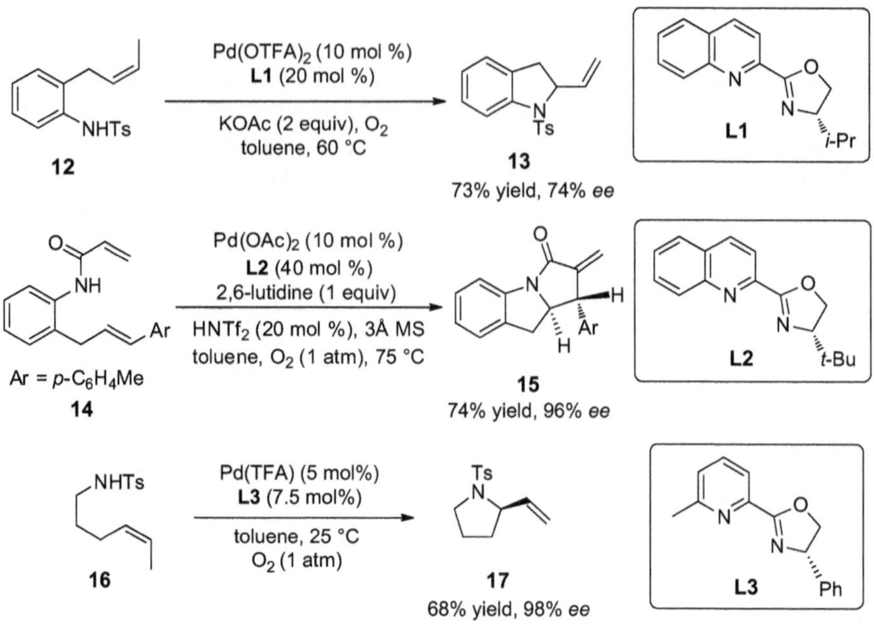

Scheme 11.10 Enantioselective oxidative amination of alkenes using oxazoline ligands.

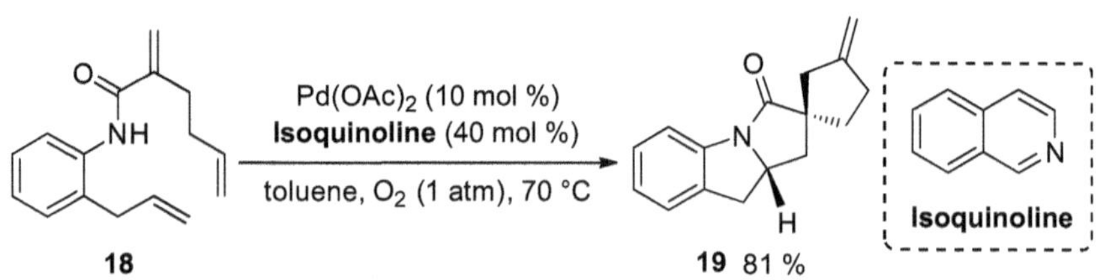

Scheme 11.11 Pd-catalyzed oxidative amination with isoquinoline ligands.

Zhang and coworkers reported a regioselective Pd-catalyzed aerobic aza-Wacker cyclization for preparation of isoindolinone heterocycles.[88] A switch of catalyst combination containing bidentate (phenanthroline) to monodentate ligands (triethylamine) led to 6-*indo* regioisomer **25.** The author indicated regioselectivity control may be attributed to steric hindrance (Scheme 11.13).

The mechanism of the oxidative amination under aerobic conditions has been illustrated by Stahl and coworkers. In the proposed catalytic cycle, initial cyclization may proceed *via cis*- or *trans*-aminopalladation. In 2006, Liu and Stahl systematically investigated five representative reported aerobic oxidation systems to differentiate the two cyclization pathways.[89] A stereo-embedded deuterium labeled substrate *trans*-**3**-*d*-**13** was subjected to these aerobic oxidation reactions to determine whether *cis*- or *trans*-aminopalladation are

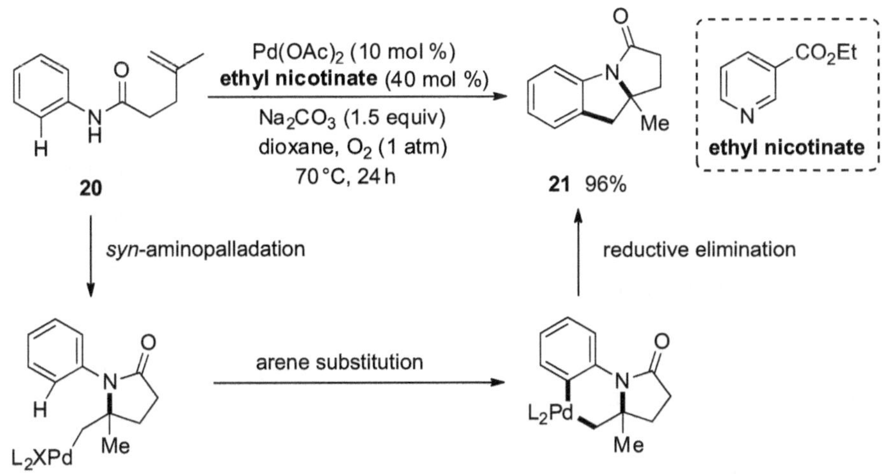

Scheme 11.12 Pd-catalyzed intramolecular aminoarylation of alkenes.

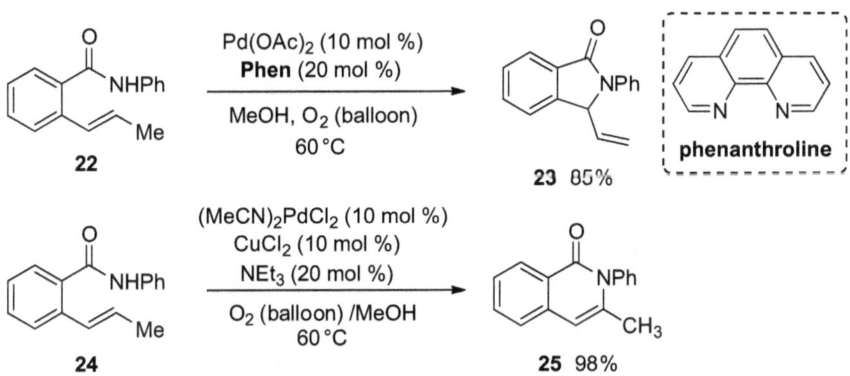

Scheme 11.13 Synthesis of isoindolinone *via* Pd-catalyzed aerobic cyclization.

operational. All of the possible *cis*- or *trans*-aminopalladation products are listed below (Scheme 11.14).

Experimental results revealed that four catalytic systems (**A**, **B**, **C** and **E**) supported exclusive *cis*-aminopalladation process, whereas both *cis*- and *trans*-aminopalladation occurred in **D** with the N-heterocyclic carbene catalytic system. If stoichiometric an amount of a Brønsted base (NaOAc, Na_2CO_3) was added to catalytic system **D**, only *cis*-aminopalladation is observed. The identity of nucleophiles also affected aminopalladation pathways; results ranged from exclusively *cis*- to *trans*-aminopalladation.

Upon stereochemical depiction of the aminopalladation step, full details of intramolecular aminocyclization[79] under $Pd(OAc)_2$/pyridine aerobic system were elucidated by Ye and Stahl recently (Scheme 11.15).[90]

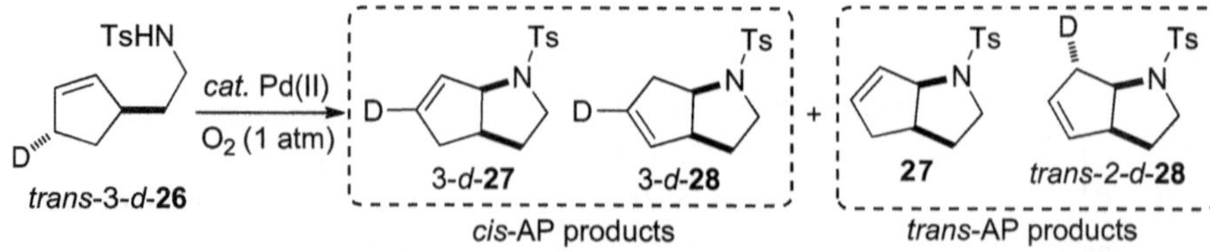

entry	Pd catalyst	yield (%)	product ratio 3-*d*-**27**/3-*d*-**28**	product ratio **27**/*trans*-2-*d*-**28**
1	$Pd(OAc)_2$/DMSO (**A**)	70	100:0	
2	$Pd(OAc)_2$/py (**B**)	84	98:2	
3	$Pd(O_2CCF_3)$/py (**C**)	85	88:12	
4	$Pd(IMes)(O_2CCF_3)_2$/BzOH (**D**)	60	43:8	37:12
5	$Pd(O_2CCF_3)_2$/sp (**E**)	37	59:41	

Scheme 11.14 Stereochemistry of Pd-catalyzed oxidative amination of *trans*-3-*d*-**26**.

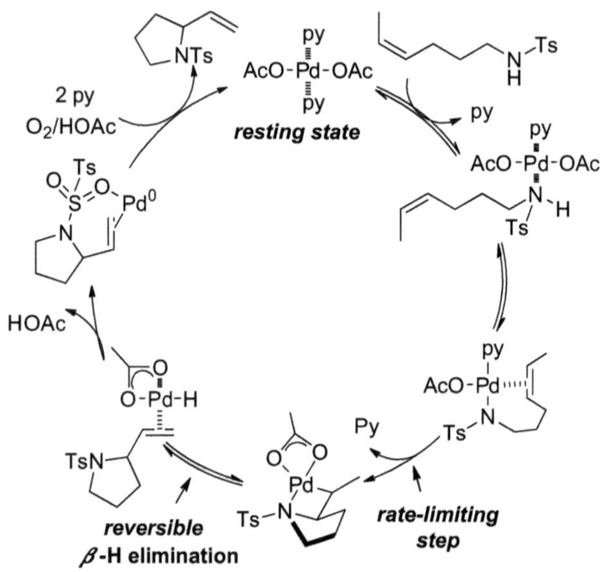

Scheme 11.15 Detailed mechanism of Pd-catalyzed cyclization of alkenoic tosylamide.

The *cis*-aminopalladation is an affirmative pathway in this reaction. The Hammett plot and kinetic measurement fits the proposed *turnover-limiting alkene insertion* into a Pd–N(amidate) bond. Kinetic isotope effects indicated *β*-hydride elimination of Pd–alkyl intermediate is a *reversible process*. According to the mechanistic elaboration, the full catalytic pattern is shown (Scheme 11.15). Furthermore, Stahl and coworkers recently disclosed direct evidence for *the reversible aminopalladation* by studying the stoichiometric reaction of a sulfonamidate–palladium complex (Scheme 11.16).[91]

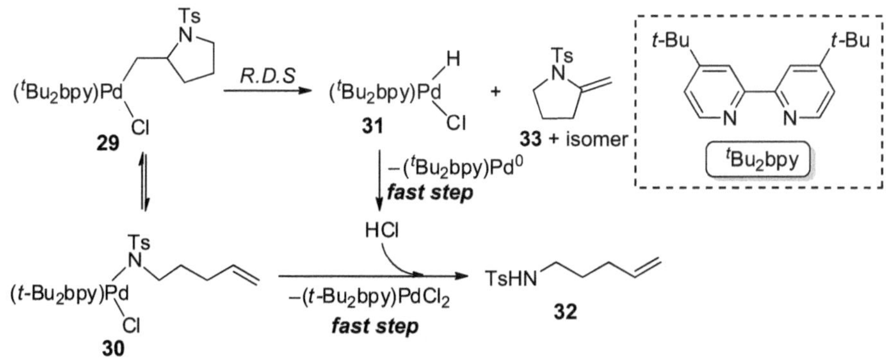

Scheme 11.16 Observation of reversible aminopalladation process.

11.3.1.2 Intermolecular Carbon–Nitrogen Bond Formation

A few examples have been provided of the intermolecular oxidative amination under aerobic conditions. Electron-deficient alkenes were considered as major candidates due to relative high reactivity in these reaction conditions. Hosokawa[92] and Cenini[93] have brought out early cases with different nitrogen nucleophiles.

A new procedure for the preparation of enamides has been developed by using Pd-catalyzed oxidative reactions between amides and conjugated olefins.[94] (*Z*)-Selective enamide **37** was obtained as the only regioisomer due to the six-membered ring transition state **36** (Scheme 11.17). The Pd-catalyzed 1,2-diamination of dienes using dioxygen or benzoquinone (BQ) as the terminal oxidants has been presented (Scheme 11.18).[95] The authors predict the activation of C(2) along with coordination of the unreacted olefin moiety help to stabilize alkyl-Pd species from *β*-hydride elimination (Scheme 11.19).

A broader spectrum of less activated alkenes was screened later by Stahl and coworkers. In 2003, they reported aerobic amination reaction of styrene substrates.[96] Employment of different palladium catalysts completely switched the regioselectivity of amination (Scheme 11.20). In the presence of catalyst **44**, Markovnikov product **46** was obtained. The role of triethylamine in catalyst **44** or added *in situ* served as a simple base to deprotonate the adduct after the nucleophile attack. The same regioselectivity was observed with the combination of catalyst **43** with exogenous anionic base. Kinetic investigations carried out by Stahl and coworkers provided insights on the role of Brønsted bases to affect the regioselectivity.[97]

A general method for the oxidative amination of unactivated alkyl olefins under co-catalyst free conditions was reported by the same group.[98] Addition of co-catalyst and stoichiometric oxidants instead of dioxygen diminished the yield. An excess of alkene substrates is essential to achieve good yields (Scheme 11.21). An improvement of this reaction was developed later using alkenes as the limiting reagent.[99] Different types of additives such as methyl

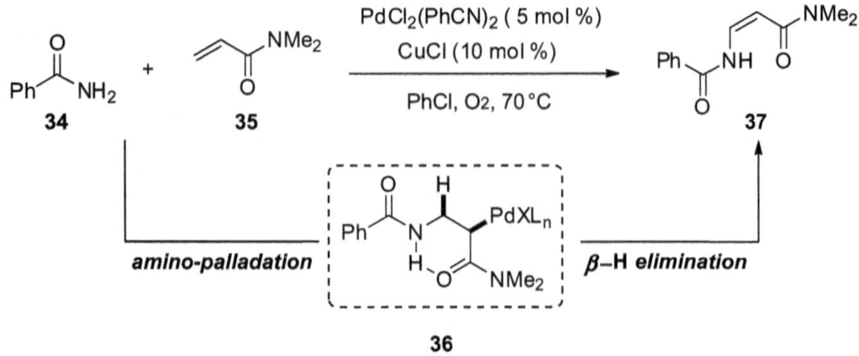

Scheme 11.17 Pd-catalyzed (*Z*)-selective oxidative amination to form enamides.

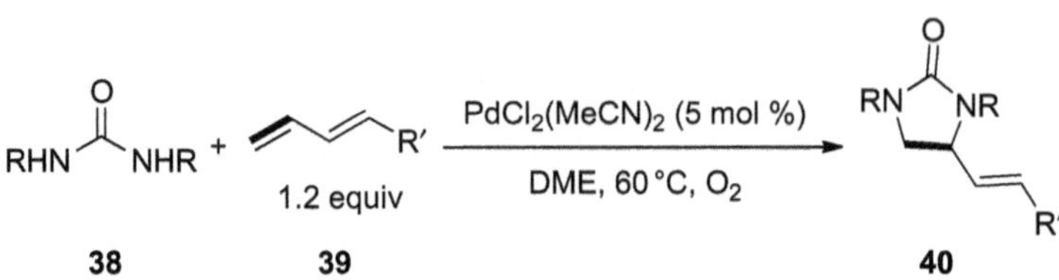

Scheme 11.18 Pd-catalyzed regioselective aerobic 1,2-diamination of dienes.

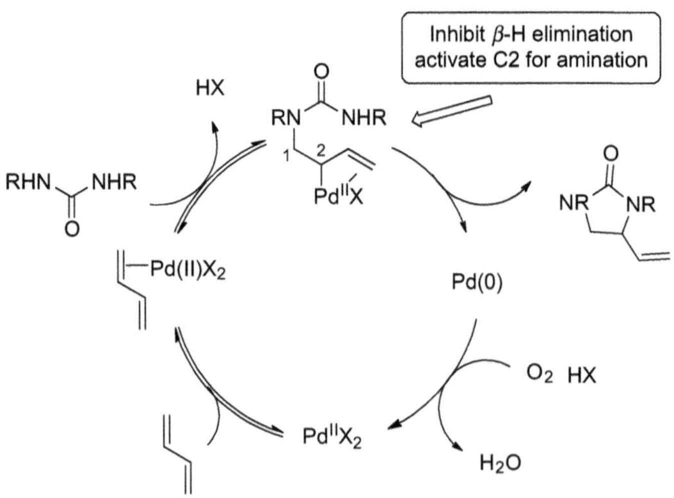

Scheme 11.19 Proposed path for Pd-catalyzed regioselective oxidative 1,2-diamination.

acrylate, COD (1,5-cyclooctadiene) and Brønsted acids were used to stabilize palladium catalyst. Elevated pressure also promoted the reaction efficiency by accelerating the reoxidation of reduced palladium catalyst.

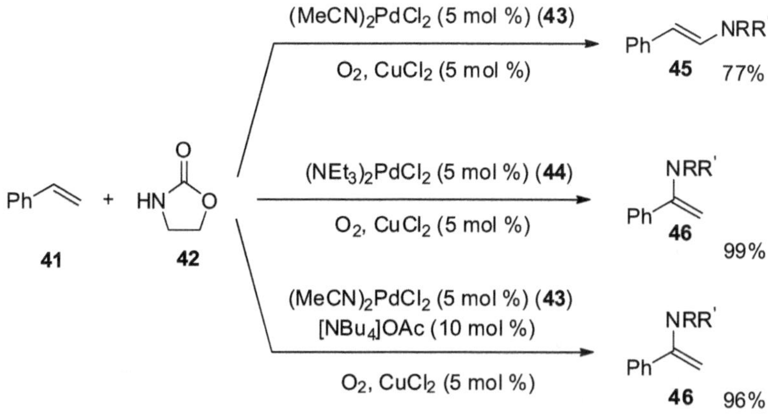

Scheme 11.20 Pd-catalyzed aerobic amination of styrene with controlled regioselectivity.

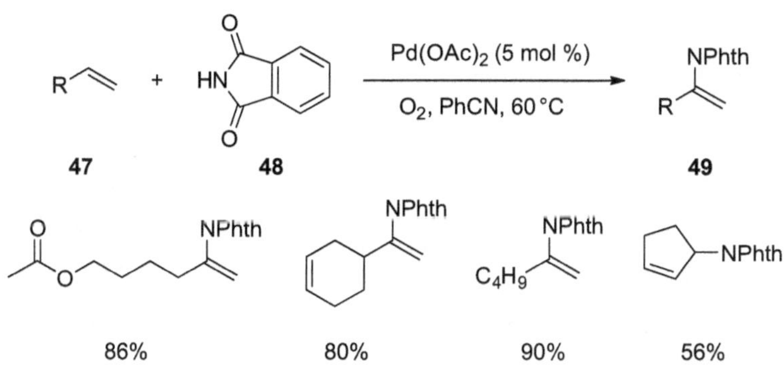

Scheme 11.21 Pd-catalyzed oxidative amination of unactivated olefins.

11.3.2 Pd-Catalyzed Oxidative Oxygenation of Alkenes

11.3.2.1 Intramolecular Carbon–Oxygen Bond Formation

Hiemstra[100] and Bäckvall[75] reported $Pd(OAc)_2$/DMSO catalyst systems to achieve oxidative cyclization of alkenoic alcohols **50**. In these cases, co-oxidants are unnecessary for highly efficient transformation (Scheme 11.22).

$Pd(OTFA)_2$/pyridine catalytic systems have been illustrated by Stoltz for oxidative oxygenation of a variety of oxygen nucleophiles including phenols, alcohols and carboxylic acids (Scheme 11.23).[101,102] Modification of the $Pd(OTFA)_2$/pyridine system to achieve high enantioselective induction has been developed by adding (–)-sparteine to the reaction system (Scheme 11.24).[101,102] Hosokawa[103–105] and Hayashi[106] reported application of optically active pinene–Pd complex and Boxax ligands to acquire high enantiomeric excess.

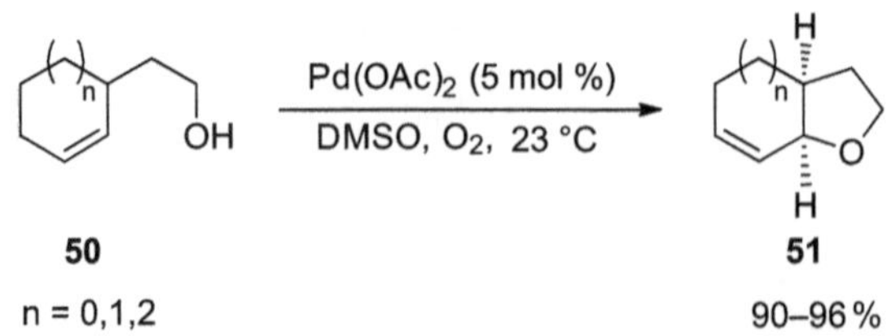

Scheme 11.22 Pd-catalyzed oxidative cyclization of alkenoic alcohols.

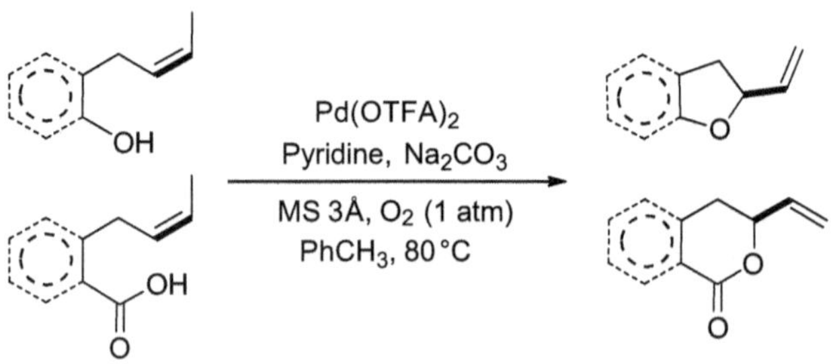

Scheme 11.23 Pd(OTFA)$_2$/pyridine catalytic systems for oxidative oxygenation.

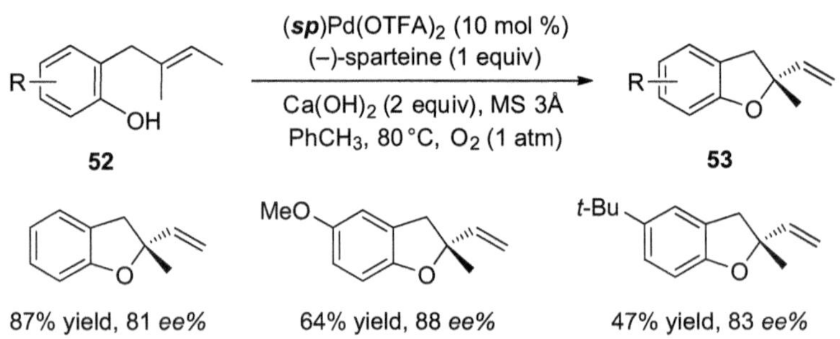

Scheme 11.24 Pd-catalyzed asymmetric oxidative oxygenation.

Intramolecular oxidative oxygenation reactions under ligand-free conditions were also demonstrated for different oxygen-containing substrates. Hosokawa[107,108] and Waegell[109] reported oxidative cyclization of phenols, alcohols and alkenoic carboxylic acids. Reiter and Gouverneur established novel protocols for preparation of diverse oxygenated heterocycles. Initial coordination of alkene to $PdCl_2$ activates the enone substrates toward intramolecular nucleophilic attack. Finally, β-hydride elimination provided the product **55** or **57**. The reduced Pd^0 species was reoxidized by O_2/CuCl to regenerate the catalyst (Scheme 11.25).[110,111]

Pd-catalyzed intramolecular acetalization of olefins represents another type of oxidative cyclization process. Hosokawa reported a Pd-catalyzed diastereoselective acetalization of hydroxyl alkenes **58** (Scheme 11.26).[112]

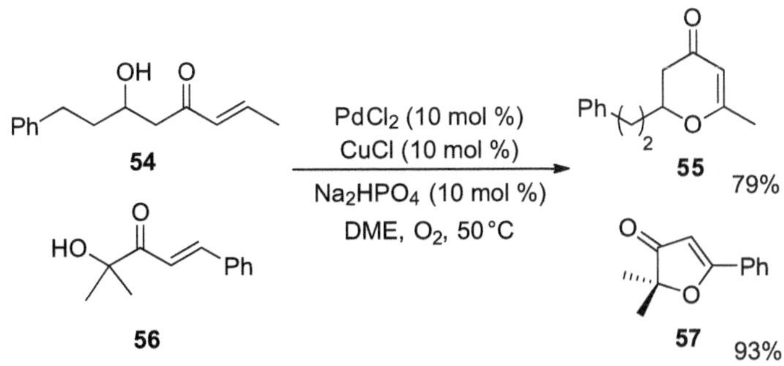

Scheme 11.25 Pd-catalyzed oxidative cyclization of β-hydroxyl enones.

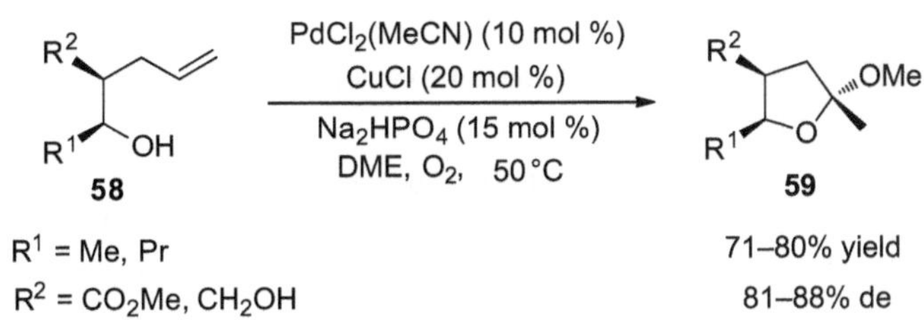

Scheme 11.26 Pd-catalyzed intramolecular acetalization of hydroxyl alkenes.

Various homo-allyl alcohols were regioselectively oxidized with molecular oxygen to afford α-alkoxy tetrahydrofurans.[113] A similar example was exhibited by Sharma on the Pd-catalyzed acetalization of terminal olefins hemiketals.[114]

Stereochemical processes of oxypalladation have been thoroughly investigated using deuterium-labeled substrates.[42,102] Stoltz and coworkers reported that Pd-catalyzed aerobic oxidative cyclization of primary alcohols under mono- (pyridine) or bidentate (bipyridine) ligated conditions occurs via similar *syn*-oxypalladation cyclization process (Scheme 11.27). On the contrary, *anti*-oxypalladation dominates carboxylic acid substrate cyclizations.

11.3.2.2 Intermolecular Carbon–Oxygen Bond Formation

Hosokawa reported a series of research papers on the Pd-catalyzed intermolecular oxidative acetalization reactions.[115–119] Terminal olefins bearing electron-withdrawing groups are regioselectively acetalized at the terminal carbon to form product **67** (Scheme 11.28). The authors substantiated that oxypalladation followed by β-hydride elimination and final intramolecular Michael addition provided the products. A Pd-catalyzed asymmetric Wacker-type coupling of allylic alcohols and vinyl ethers using oxygen as the terminal oxidant was reported by Hosokawa and coworkers (Scheme 11.29).[120]

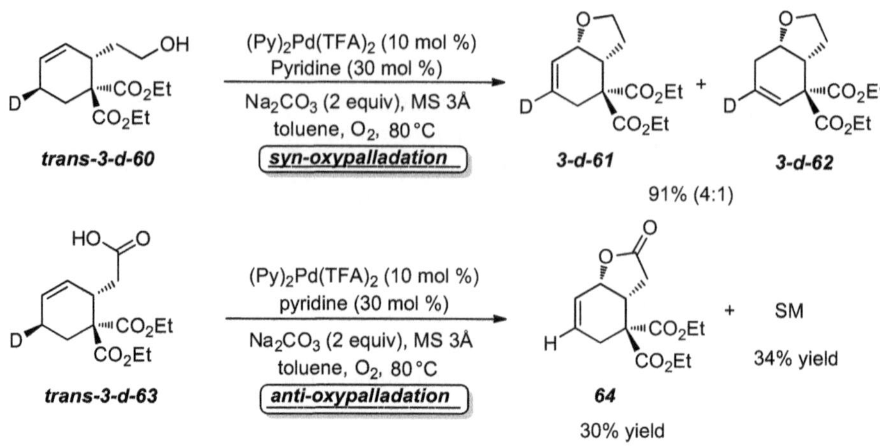

Scheme 11.27 Stereochemistry studies of oxypalladation process.

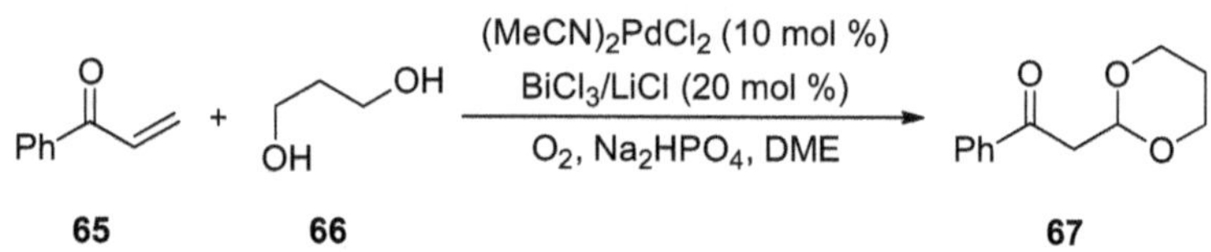

Scheme 11.28 Pd-catalyzed intermolecular oxidative acetalization reaction.

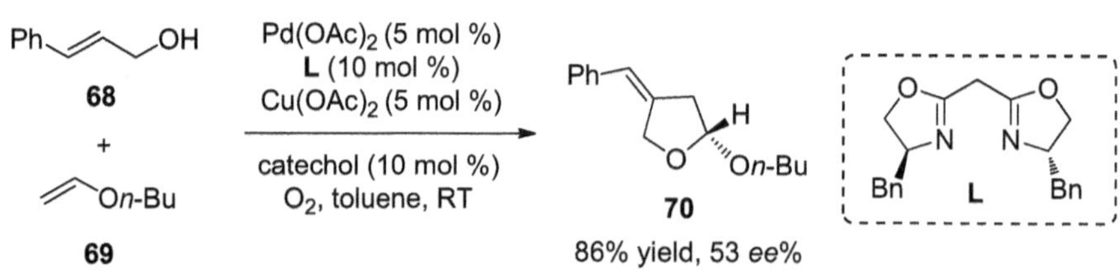

Scheme 11.29 Pd-catalyzed asymmetric coupling of allylic alcohols and vinyl ethers.

A novel advancement was reported by Sigman and coworkers in 2006.[121] The *o*-hydroxyl styrene was subjected to Pd-catalyzed aerobic conditions utilizing MeOH both as nucleophile and solvent to produce dialkoxylation products (Scheme 11.30). Control experiments suggested the phenol group on the benzene ring is essential to the success of this dialkoxylation process. The author proposed the mechanism included two sequential steps: (a) regio-selective oxypalladation of **A** *via* MeOH addition to the β-carbon of the styrene yielded **B** and (b) consecutive formation of a key intermediate *quinone methide species* **C** with concomitant reduction of palladium catalyst and addition of second equivalent of MeOH to **C**; final rearomatization led to the product.

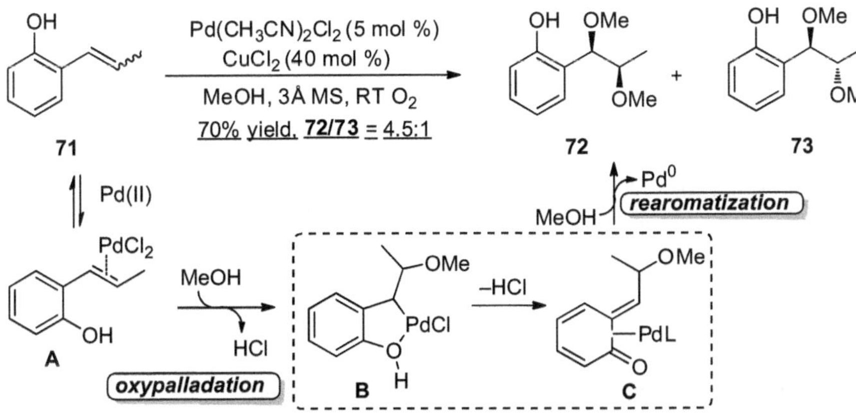

Scheme 11.30 Sigman's synthesis of dialkoxylation of styrene.

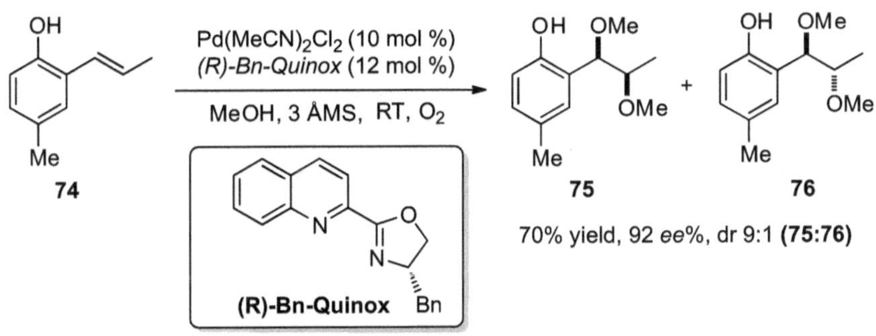

Scheme 11.31 Pd-catalyzed enantioselective dialkoxylation of styrene

Emergence of enantioselective reactions was developed by Sigman group using the asymmetric Quinox ligand to achieve high enantioselectivity.[122] For this approach, detrimental effects of additives such as copper salts and benthoquinone to disturb the coordination between Palladium catalyst and ligand should be avoided (Scheme 11.31). Further understanding of this enantioselective reaction suggested the absolute configuration is set by β-nucleopalladation. The same *palladium quinone methide* formation strategy has been realized on intramolecular enantioselective reactions with addition of two distinct nucleophiles across double bonds.[123,124]

As mentioned above, the C–Pd bond generally undergoes β-H elimination to give double bond and release Pd^0 rather than oxidative cleavage to afford a second C–X (X = heteroatom) bond. Recently, in 2009, Wang and Jiang provided the first example of Pd-catalyzed diacetoxylation of alkenes with dioxygen as the sole oxidant.[125] The *cis*-vicinal diol acetate **78** as products were achieved in HOAc with addition of a catalytic amount of KI under 8 atm pressure of oxygen (Scheme 11.32). This reaction was initiated by acetoxypalladation of alkene, oxidative cleavage of C–Pd bond to form a second

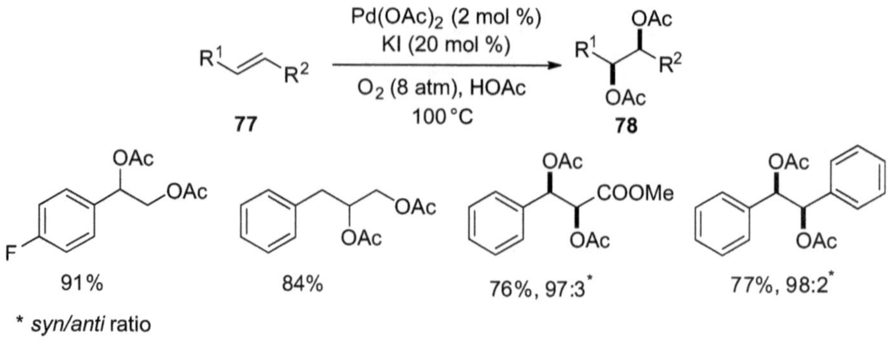

Scheme 11.32 Pd-catalyzed aerobic diacetoxylation reaction.

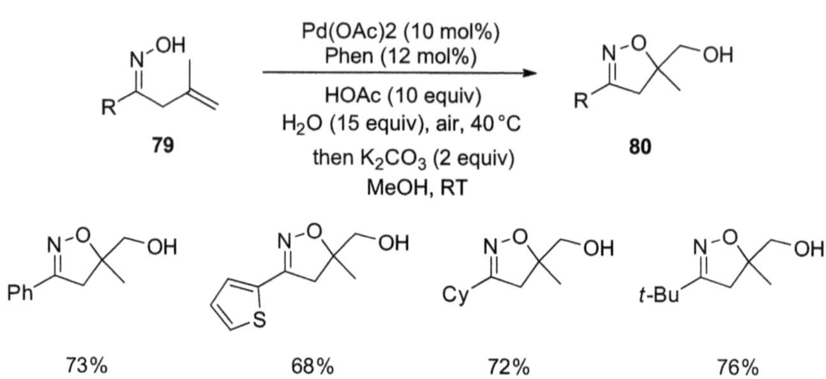

Scheme 11.33 Pd-catalyzed oxime-assisted intramolecular dioxygenation of alkenes.

C–O bond. However, a detailed mechanism on the formation of second C–O bond under dioxygen remained elusive. Loh and coworkers reported a similar Pd-catalyzed intramolecular dioxygenation of alkenes under dioxygen atmosphere (Scheme 11.33).[126] Isotopic labeling experiments with H_2O^{18} and O_2^{18} indicated the oxygen of the hydroxyl group was originated from O_2.

11.3.3 Pd-Catalyzed Oxidative Carbon–Carbon Bond Formation

11.3.3.1 Heck-Type Carbon–Carbon Bond Formation Initiated by Transmetalation

Heck-type reactions are one of the most powerful tools for carbon–carbon bond forming reactions in organic synthesis.[127–134] In addition to the use of conventional aryl halide and triflate, employment of organometallic reagents such as organosilanes and organostannanes as coupling partners have greatly expanded the reaction scope of traditional Heck reactions. These types of reactions proceed through initial transmetalation with the palladium catalyst, subsequent migratory insertion and final β-hydride elimination lead to the coupling products.

In 1994, Uemura reported a Pd(0)-catalyzed cross coupling of arylboronic acids and alkenes *via* oxidative addition of Pd(0) to a carbon–boron bond.[135] Fundamental work on Pd-catalyzed oxidative coupling of alkenes and organoboron reagents bearing various electronic properties was provided by Jung and coworkers.[136]

Alkenyl boronic compounds **81** with various functional groups are compatible for the aerobic oxidative Heck cross-coupling reactions. Exclusive (*E*,*E*)-dienes were obtained with excellent selectivity and yields (Scheme 11.34). Arylstannane reagents **85** as coupling element were also applied in a similar aerobic oxidative coupling reactions (Scheme 11.35).[137]

Lei and coworkers provided a strategy of synthesizing β-aryl ketones and aldehydes **89** from allylic alcohols **88** *via* Pd-catalyzed oxidative coupling reactions (Scheme 11.36).[138] Ligand-modulated oxidative Heck-type reactions have been documented during the pursuit for higher catalytic efficiency and minimization of side products. Larhed reported the first case using 2,9-dimethyl-phenanthroline as an effective ligand (Scheme 11.37).[139–142] Jung reported a base-free protocol for the oxidative Heck reactions with the same bidentate nitrogen ligand.[143]

Sigman published a unique strategy to achieve difunctionalization of terminal alkenes with organostannane reagents using a Pd-carbene catalyst.[144] The combination of palladium carbene catalyst, DMA

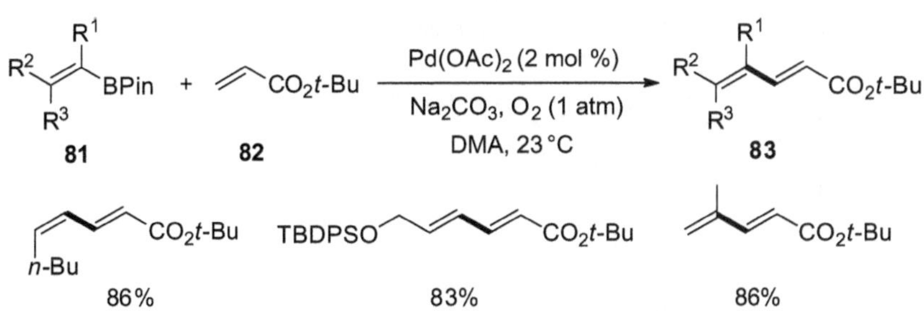

Scheme 11.34 Pd-catalyzed oxidative coupling of alkenes with alkenylboron reagents.

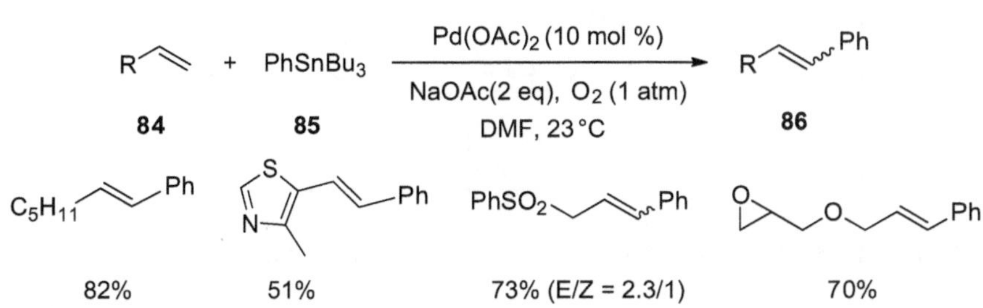

Scheme 11.35 Pd-catalyzed oxidative coupling of alkenes with organostannane reagents.

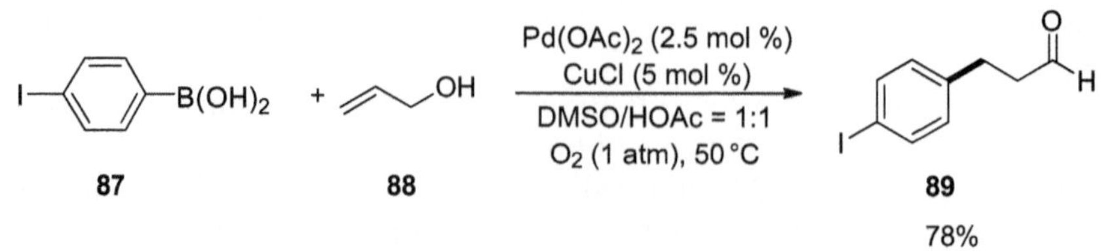

Scheme 11.36 Pd-catalyzed oxidative coupling of allylic alcohols and boronic acid.

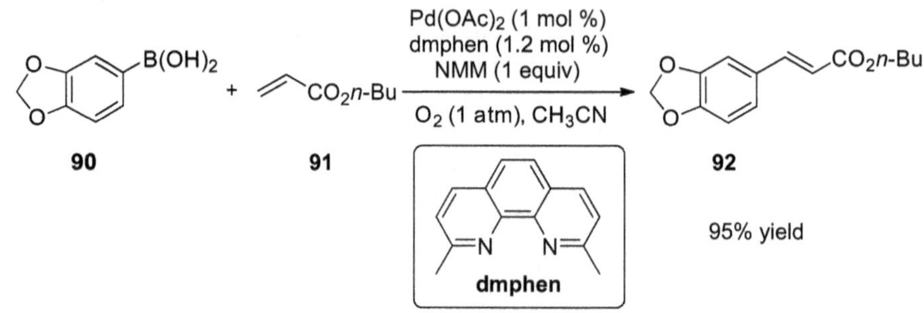

Scheme 11.37 Ligand-modulated oxidative Heck vinylation.

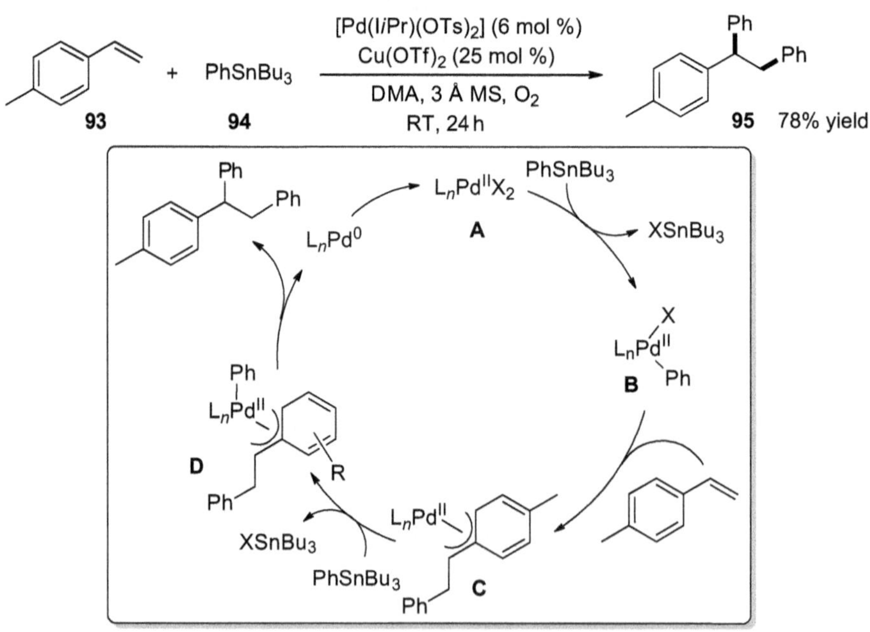

Scheme 11.38 Sigman's synthesis of 1,2-diarylation products.

(N,N-dimethyl-acetamide) as solvent and $Cu(OTf)_2$ as additive provided **95** with good yield and selectivity (Scheme 11.38). The author surmised two consecutive transmetalation steps are involved in the catalytic cycle. Initial transmetalation of palladium catalyst with aryl stannane to form species **B**, migratory insertion to the styrene leads to π-benzyl intermediate **C**, a second transmetalation step and reductive elimination give the 1,2-diarylation product **95**.

The emergence of asymmetric inter- and intramolecular oxidative Heck reactions using dioxygen as the terminal oxidants relied on the introduction of chiral ligands or preformed palladium-ligand complexes (Scheme 11.39).[42,130] Mikami reported the first intermolecular enantioselective oxidative Heck-type reactions. (*S*,*S*)-Chiralphos was chosen as optimized ligand to give a moderate enantiomeric excess (Scheme 11.40).[145] An intramolecular mode using (*S*,*S*)-chiralphos was also investigated, which provided improved results up to 86% ee.[146] Jung's group discovered the preformed palladium-ligand complexes gave better reactivity than *in-situ* generation of the catalyst in intermolecular Heck-type reaction (Scheme 11.41).[147–149]

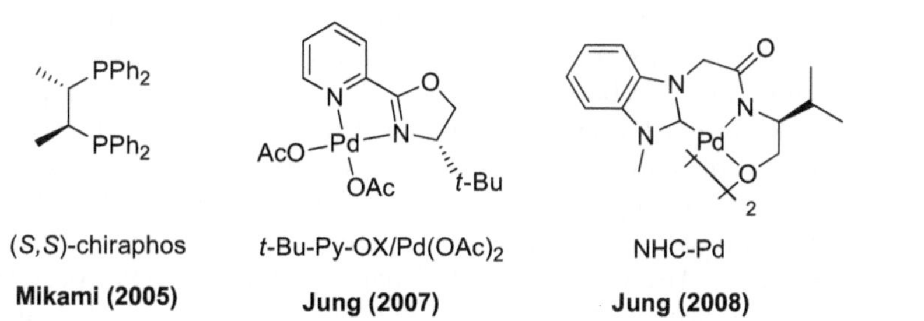

Scheme 11.39 Ligands and Pd complexes for asymmetric oxidative Heck reactions.

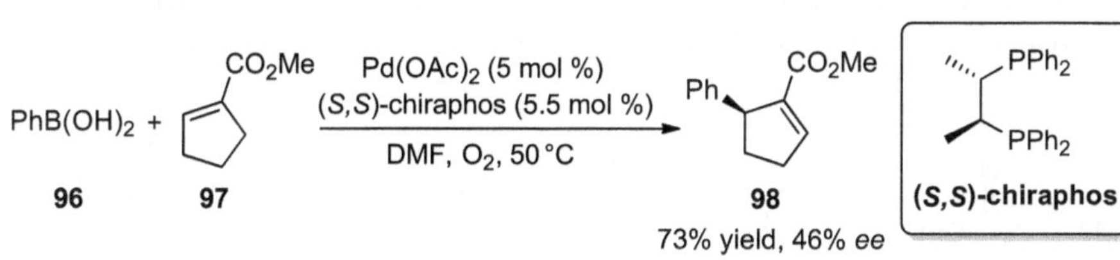

Scheme 11.40 Mikami's asymmetric oxidative Heck reaction.

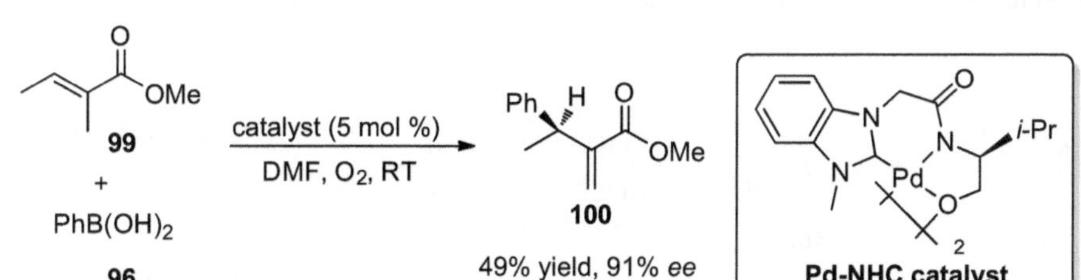

Scheme 11.41 Jung's catalytic system for the enantioselective Heck-type reaction.

11.3.3.2 Carbon–Carbon Bond Formation Initiated by C–H Bond Activation

During the last several decades, massive efforts have been devoted to developing efficient methods to realize C–H bond functionalization and further transformation by transition metals, including palladium catalysts.[150–161] Many cases have been reported on the functionalization of alkenes through sp^2 and sp^3 C–H bond cleavage-coupled reactions under aerobic conditions; an exhaustive review related to this issue was presented recently by Yeung and Dong.[44]

Shue and coworkers reported the early case of Pd-catalyzed oxidative C–C coupling reaction of benzene and alkenes.[162] Two possible mechanisms account for dehydrogenative cross-coupling regarding the carbopalladation pathways.[44] However, a drawback of this procedure is the requirement of large excess of benzene to achieve reasonable yields. Another example of this type of reaction using oxygen as the sole oxidant was reported by Jacobs and coworkers (Scheme 11.42).[163] Addition of catalytic benzoic acid is beneficial to promote arene sp^2 C–H bond activation. The challenge of using simple alkenes instead of electron-rich arenes as coupling elements has been achieved recently under mild aerobic conditions by Loh and coworkers (Scheme 11.43).[164]

Oxidative cross-coupling of acrylates with vinyl carboxylates was successfully achieved by using a $Pd(OAc)_2/HPMoV/O_2$ system reported by Ishii and coworkers.[165] The author indicated that the initial vinyl–Pd species formed from the vinyl acetate followed by migratory and β-hydride elimination provided the products (Scheme 11.44).

Miura reported oxidative cross-coupling of benzene-sulfonamides **107** or benzoic and naphthoic acids with alkenes using a Pd-Cu catalyst system under aerobic conditions (Scheme 11.45).[166] Beck and Gaunt provided a regioselective switch of pyrrole C–H alkenylation and annulation reactions under

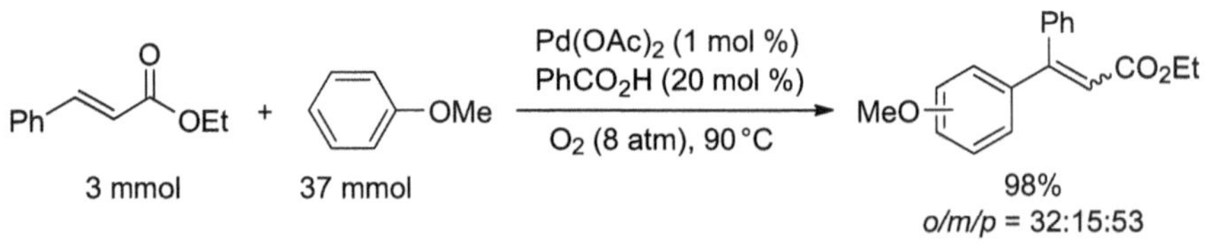

Scheme 11.42 Pd-catalyzed oxidative coupling of styrene and benzene derivatives.

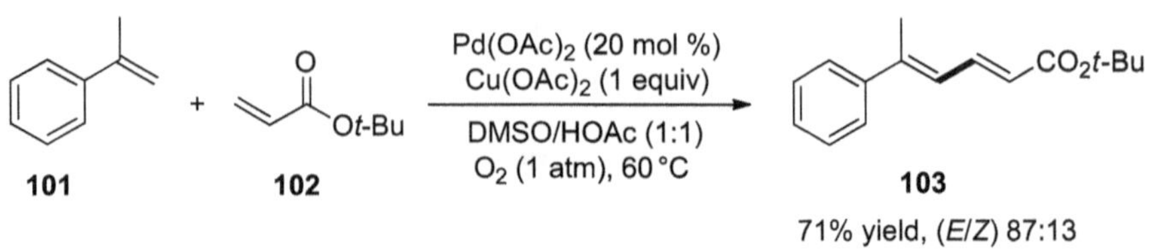

Scheme 11.43 Pd-catalyzed oxidative cross-coupling of simple alkenes with acrylates.

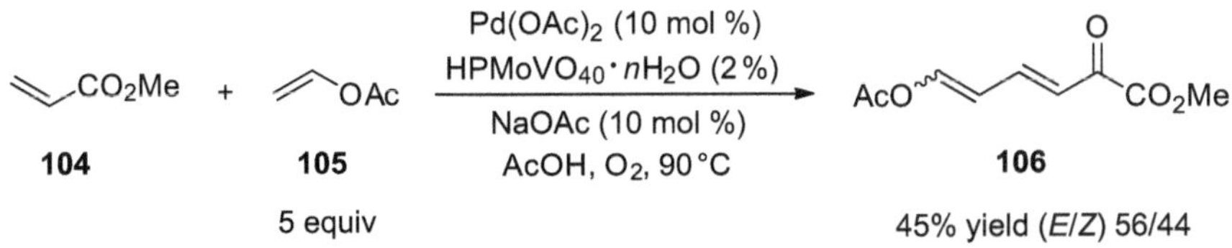

Scheme 11.44 Oxidative cross-coupling of acrylates with vinyl carboxylates.

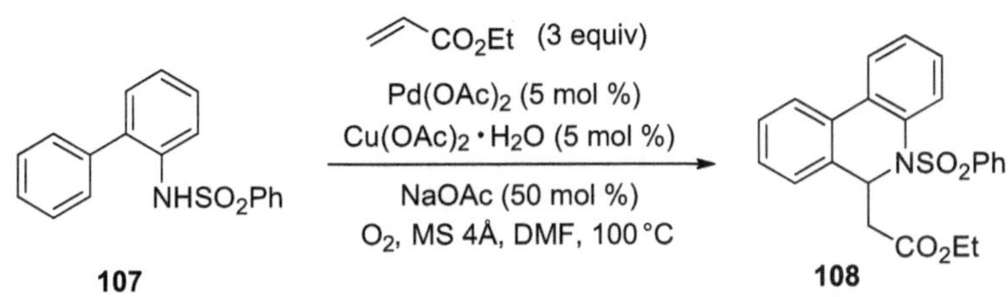

Scheme 11.45 Pd-catalyzed oxidative coupling of benzene-sulfonamides with alkenes.

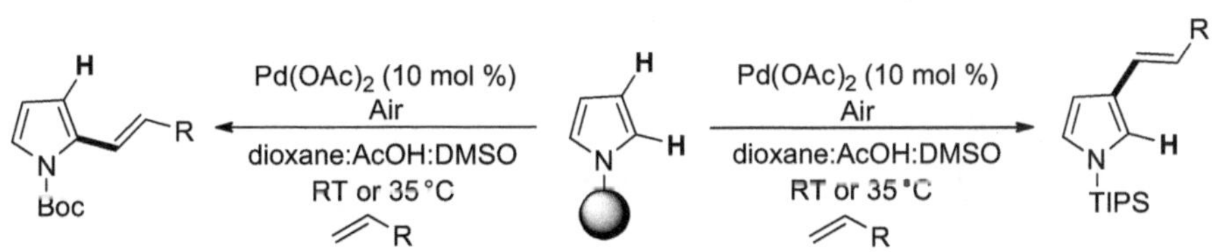

Scheme 11.46 Pd-catalyzed regioselective and switchable C–H alkenylation of pyrroles.

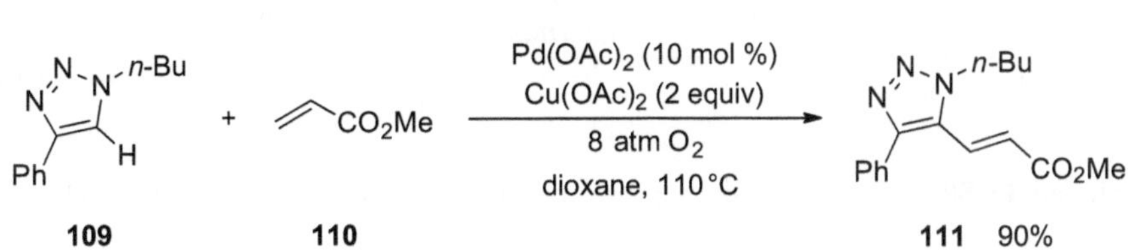

Scheme 11.47 Pd-catalyzed alkenylation of 1,2,3-trizoles under aerobic conditions.

aerobic conditions. Different protecting groups on the nitrogen influenced the regioselectivity of C–H functionalizations (Scheme 11.46).[167]

Jiang reported a Pd-catalyzed alkenylation of 1,2,3-triazoles **109** with terminal conjugated alkenes. 8 atm of oxygen and co-oxidant $Cu(OAc)_2$ are both necessary to achieve satisfactory yields (Scheme 11.47).[168] A straightforward access to biologically important carbazoloquinones by Pd-catalyzed cyclizations of arylaminoquinones under aerobic conditions was developed by Åkermark and coworkers (Scheme 11.48).[169]

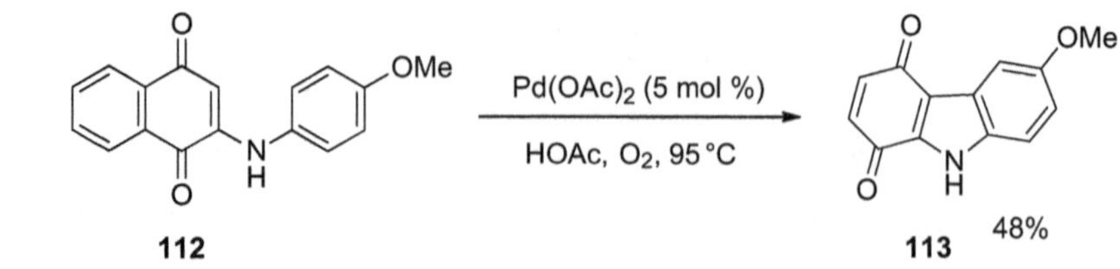

Scheme 11.48 Pd-catalyzed cyclizations of arylaminoquinones.

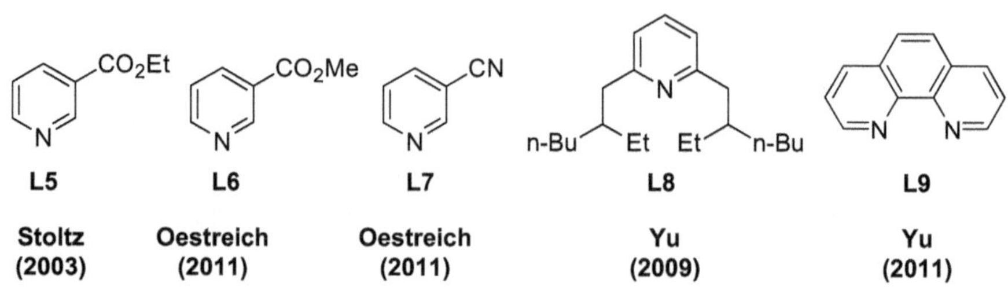

Scheme 11.49 Typical ligands for Pd-catalyzed C–H bond cleavage/C–C bond coupling.

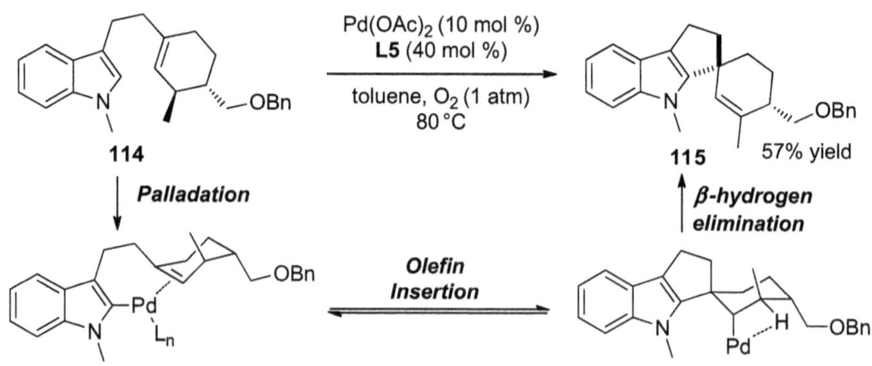

Scheme 11.50 Proposed mechanism of Pd-catalyzed indole carbocyclization reaction.

Ligands that were elaborated with multiple roles to facilitate cross coupling reactions have been utilized in C–H bond cleavage/C–C bond cross-coupling reactions. Pyridine derivatives were generally evaluated regarding the balance of coordination ability and electrophlicity alterations on the Pd centre. Introduction of steric factors on pyridine rings were also examined (Scheme 11.49).

Stoltz reported a novel case of C–H functionalization of indoles to achieve carbocyclization under aerobic conditions (Scheme 11.50).[170] Electron-deficient pyridine ligand **L5** (see Scheme 11.49) enhanced the electrophilicity of the Pd catalyst to promote the cyclization process. Oestreich reported a Pd-catalyzed aerobic cyclization of α,β-unsaturated amides from electron-rich anilines initiated by arene electrophilic palladation (Scheme 11.51).[171] A similar protocol has been applied to diastereoselective C-2 alkenylation of indoles with tri- and tetra-substituted double bonds (Scheme 11.52).[172]

Yu reported a Pd-catalyzed aerobic *meta*-olefination of highly electron-deficient arenes **121** by using a 2,6-dialkyl pyridine ligand.[173] Steric bulky side chains on the pyridine ring are essential to attenuate ligand-Pd coordination and promote association of electron-deficient arene with Pd catalyst (Scheme 11.53). A similar strategy has been applied on Pd-catalyzed C-3 selective C–H olefination of pyridines, using bidentate 1,10-phenanthroline ligand to weaken the coordination of substrates with Pd catalyst (Scheme 11.54).[174]

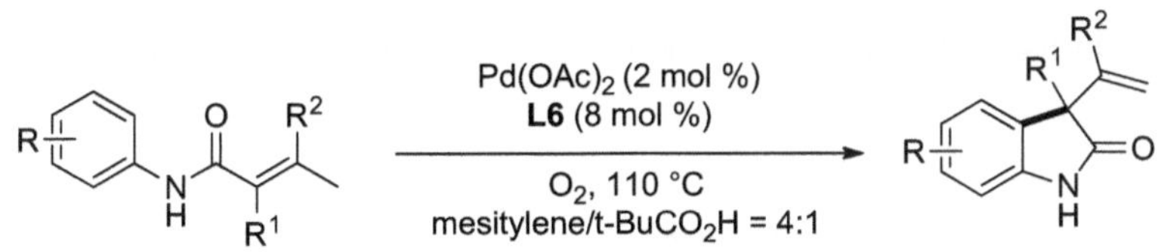

Scheme 11.51 Pd-catalyzed aerobic cyclization of α, β-unsaturated amides.

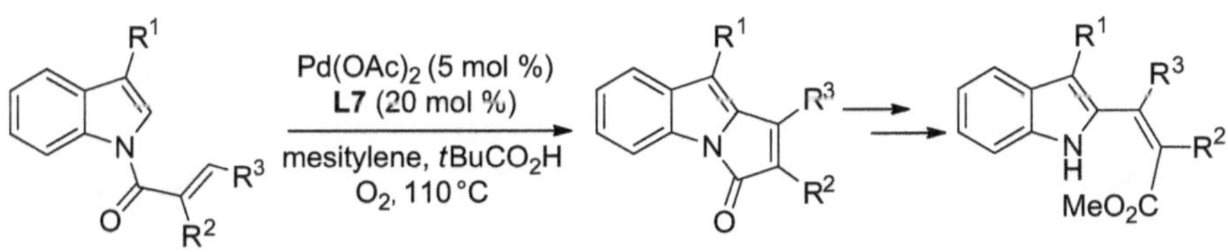

Scheme 11.52 Pd-catalyzed diastereoselective C-2 alkenylation of indoles.

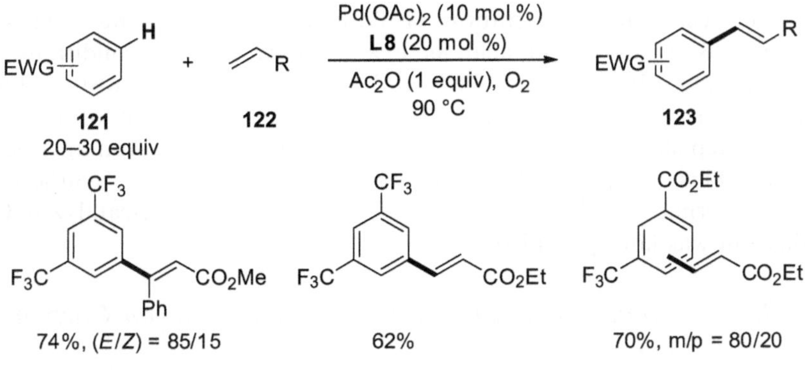

Scheme 11.53 Pd-catalyzed aerobic *meta*-olefination of highly electron-deficient arenes.

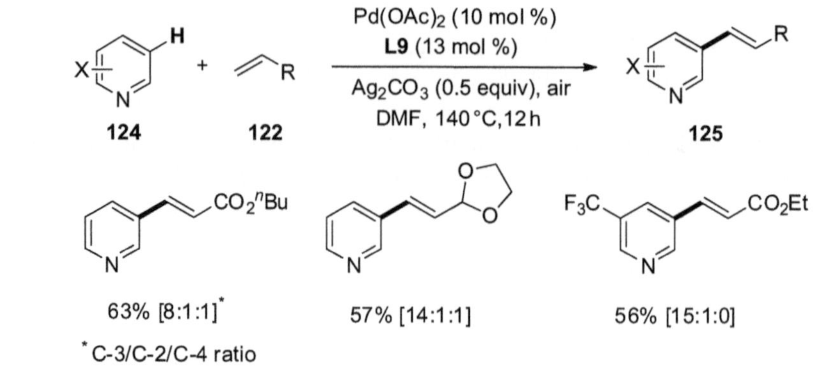

Scheme 11.54 Pd-catalyzed C-3 selective C–H olefination of pyridines.

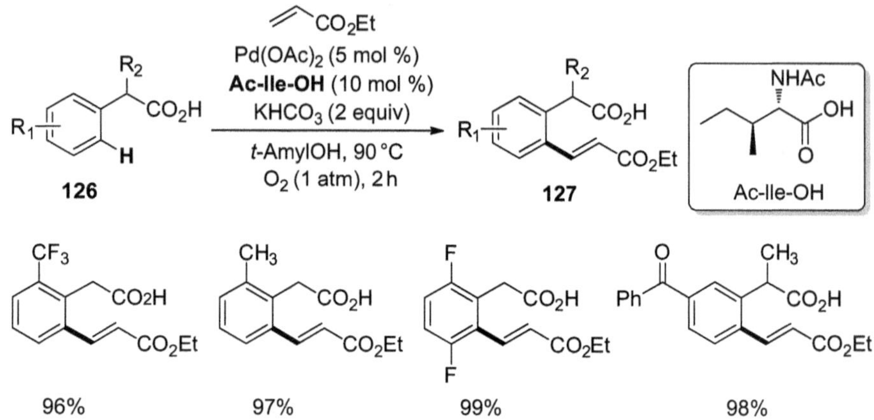

Scheme 11.55 Acceleration of Pd-catalyzed C–H bond olefination reaction.

An unprecedented acceleration of Pd-catalyzed C–H bond olefination reactions using an amino acid as ligand under aerobic conditions was disclosed recently (Scheme 11.55).[175,176] Detailed mechanistic studies indicated the boost effect may stem from the acceleration of the C–H cleavage step shifting from electrophilic palladation to a proton abstraction process. This unique method was later applied to the synthesis of (+)-lithospermic acid.[177] An enantioselective version of Pd-catalyzed C–H olefination was later provided.[178]

11.3.3.3 Aerobic Alcohol Oxidation Coupled Carbon-Carbon Bond Formation

Sigman and co-workers reported an excellent strategy to build the C–C bond utilizing a Pd^{II}–hydride species as the starting point.[179] They assumed the

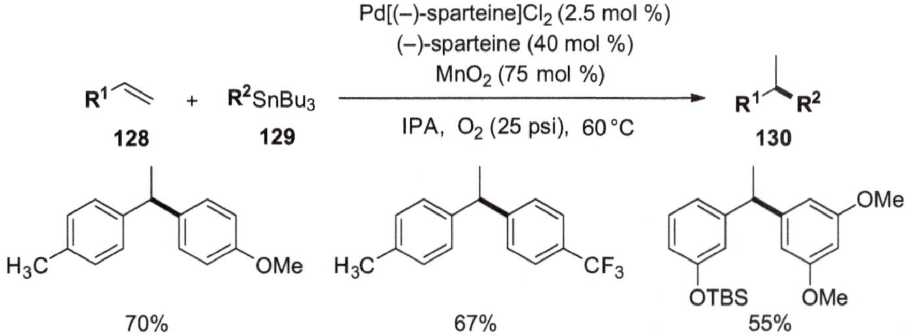

Scheme 11.56 Pd-catalyzed hydroarylation of alkenes with organostannane reagents.

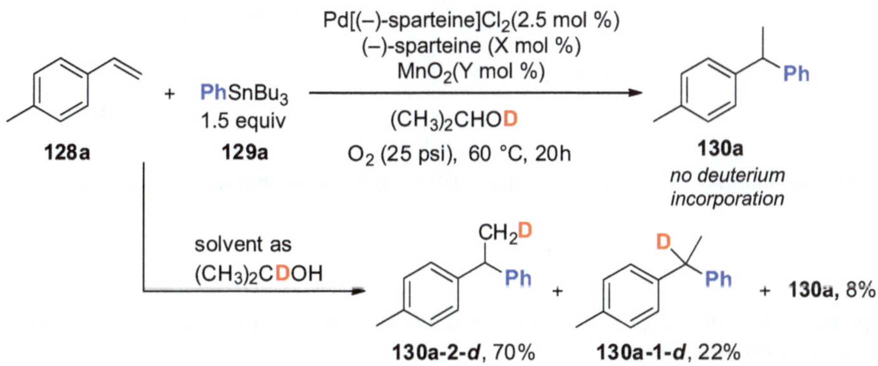

Scheme 11.57 Deuterium experiments to substantiate initial alcohol oxidation process.

Pd^{II}–hydride intermediate produced from alcohol oxidation could be inserted into the alkene moiety to provide a Pd^{II}–alkyl species; sequential transmetallation and reductive elimination finally constructs the C–C bond. This protocol has been put into practice with organostannane reagents[180] and organoboron reagents.[181] Addition of a H_2O_2 scavenger such as MnO_2 dramatically enhanced the yield of **130**. Extra addition of (–)-sparteine and escalated oxygen pressure and temperature are beneficial to the yields and selectivity (Scheme 11.56).

To substantiate the source of the Pd^{II}–hydride, two deuterated isopropyl alcohols were prepared and subjected to reaction condition (Scheme 11.57). When $(CH_3)_2CHOD$ was used as solvent, no deuterium is incorporated into the product which excluded the involvement of acidic protons in the reaction. However, when $(CH_3)_2CDOH$ was applied as solvent, 92% of the hydroarylation products containing a single deuterium atom as a mixture of isotopomers were obtained (Scheme 11.57). The observed phenomenon

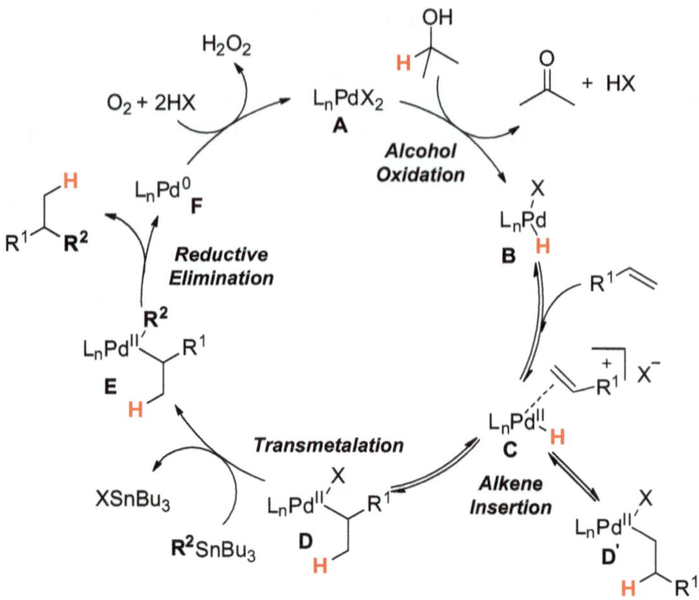

Scheme 11.58 Proposed mechanism of Pd-catalyzed alcohol-initiated hydroarylation.

justified the initial aerobic oxidation of alcohols to produce a Pd^{II}–hydride species (Scheme 11.58).

11.3.3.4 Carbopalladation Initiated Carbon–Carbon Bond Formation

Carbopalladation refers to the formation of a C–C bond and C–Pd bond in sequential steps. β-Hydride elimination is generally observed as the major termination mode of the resultant C–Pd bond among reported cases. Pd-catalyzed aerobic alkenylations to construct bicyclic structures *via* carbopalladation and rearomatization steps were reported (Scheme 11.59).[182]

Intramolecular oxidative carbocyclization of silyl enol ether substrates under identical condition afforded [3.2.1]-octane derivatives.[183,184] Tius reported a Pd-catalyzed Nazarov-type cyclization to form cyclopentenones under oxygen atmosphere following a similar reactions pathway (Scheme 11.60).[185]

Yang and coworkers provided a Lewis acid-assisted intramolecular aerobic oxidative coupling between β-ketoamide and alkenes *via* an enolate nucleophilic attack initiated palladation process (Scheme 11.61).[186] Bäckvall and coworkers reported an approach towards Pd-catalyzed aerobic allylic oxidative carbocyclization of allene-substituted olefins.[187] FePc as an

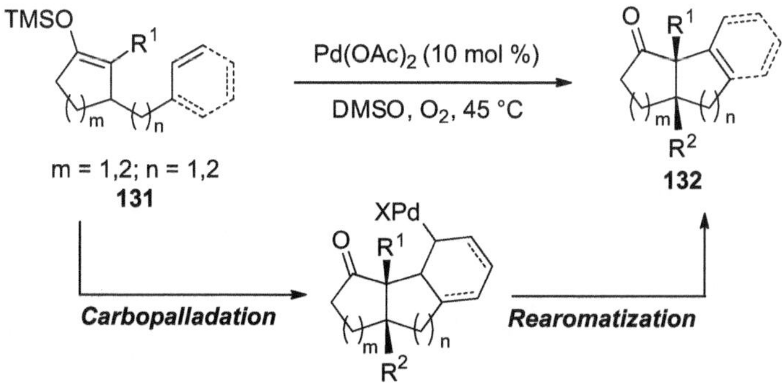

Scheme 11.59 Pd-catalyzed alkenylations to construct bicyclo-structures.

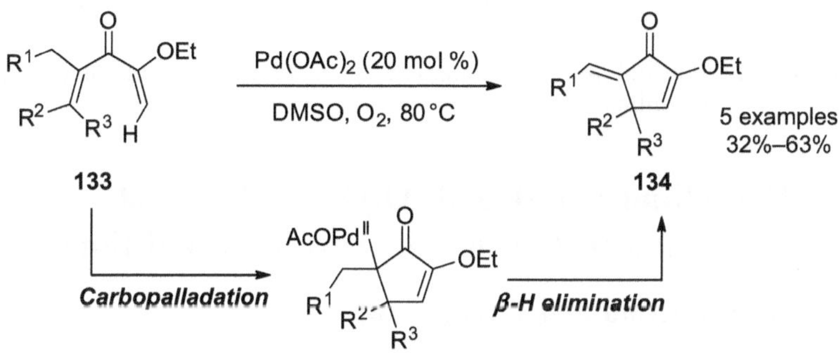

Scheme 11.60 Pd-catalyzed Nazarov-type cyclization towards cyclopentenones.

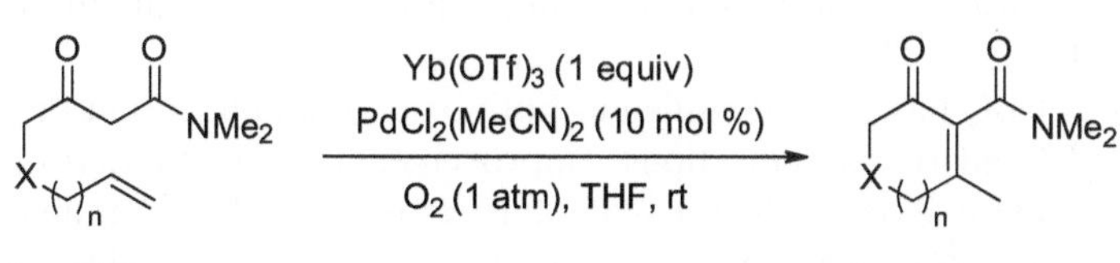

Scheme 11.61 Lewis acid-assisted Pd-catalyzed heterocycle synthesis.

electron-transfer mediator is required to achieve high catalytic efficiency. A mechanistic scenario involving allylic C–H bond activation and alkenylation could not be excluded. Loh and coworkers reported a Pd-catalyzed bis-olefination of C–C triple bonds under aerobic conditions initiated by alkene carbopalladation (Scheme 11.62).[188] Final β-hydride elimination or protonation provided alkene or alkyl substitution on the naphthalene ring.

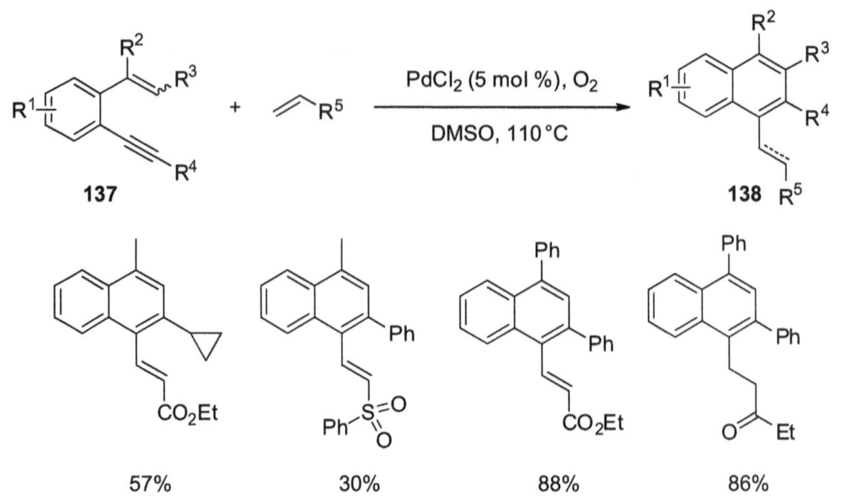

Scheme 11.62 Pd-catalyzed bis-olefination of C–C triple bonds under aerobic conditions.

11.4 Palladium Catalyzed Allylic C–H Bond Functionalization under Aerobic Conditions

11.4.1 Introduction and Background

Transition metal-catalyzed allylic C–H bond cleavage functionalization is a mechanistically and synthetically attractive process.[189–192] A large series of related reactions have been reported to transform alkene into π-allylpalladium complexes. Trost and coworkers reported two sets of useful preparation procedures using $Pd(O_2CCF_3)_2$ and $PdCl_2$ as palladium source (Scheme 11.63).[193,194]

The role of BQ has been identified as both oxidant and ligand in Pd-catalyzed allylic C–H bond functionalizations. Improvements on *modulation* of Pd^{II}/BQ systems to achieve selective allylic C–H bond acetoxylation,[195–201] amination[202,203] and alkylation[204,205] under aerobic conditions have been provided recently.

11.4.2 Allylic C–H Acetoxylation Reactions

Kaneda and co-workers reported a catalytic system for the dioxygen-coupled allylic acetoxylation of terminal alkenes (Scheme 11.64).[206] High pressure of O_2 (6 atm) was required to achieve good results due to the rate-determining reoxidation step of Pd^0 species.

Ligand modulation-based strategy to realize aerobic allylic C–H acetoxylation reaction have been achieved by Stahl and coworkers.[207] Screening of ligands indicated that 4,5-diazafluorenone (**L10**) significantly facilitated the allylic C–H acetoxylation reaction (Scheme 11.65). The author proposed that

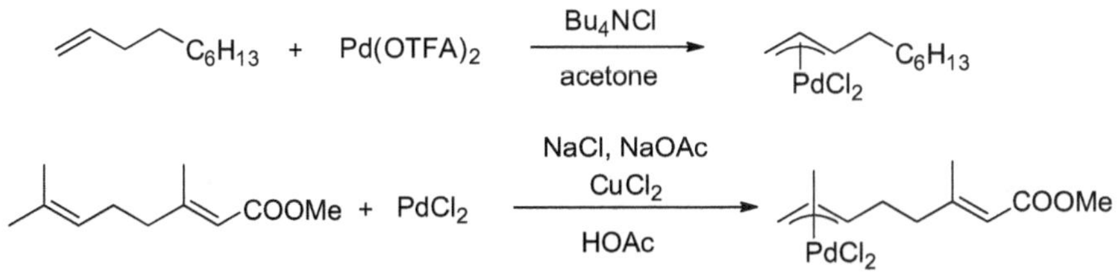

Scheme 11.63 Trost's strategy towards the synthesis of π-allylpalladium complexes.

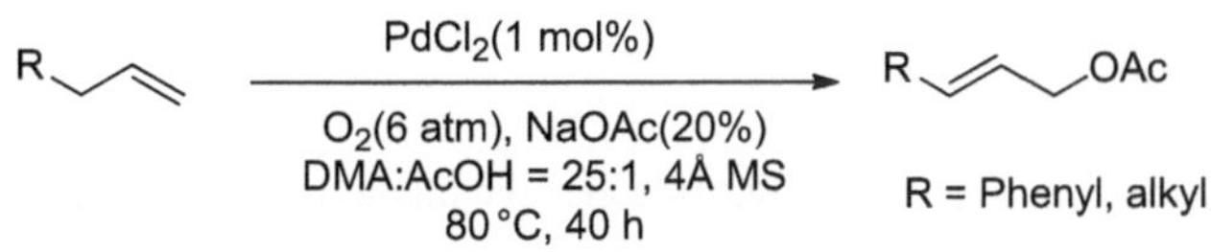

Scheme 11.64 Pd-catalyzed aerobic allylic C–H acetoxylation.

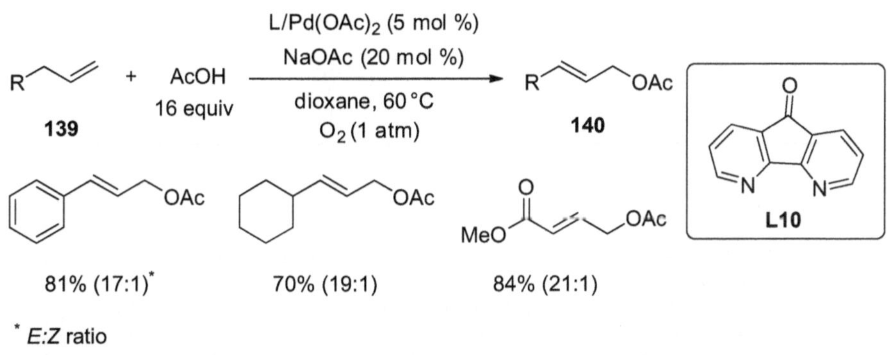

Scheme 11.65 Ligand-modulated aerobic allylic C–H bond acetoxylation.

the distorted bite angle of 4,5-diazafluorenone could destabilize the Pd^{II} center and allow the π-allylic intermediate to undergo nucleophilic attack in the absence of BQ. A recent account suggested that the ligand may participate in the C–H bond activation step.[208]

11.4.3 Allylic C–H Amination Reactions

Liu and coworkers reported a BQ-free aerobic catalytic system for the allylic amination reaction.[209] Two types of external nucleophiles, such as *N*-tosyl carbamate and saccharin, afforded the linear allylamine derivatives **143** with high regioselectivity, nonallylic isomers as minor products were also observed (Scheme 11.66). Addition of maleic anhydride (MA) increased the yield, indicating that the MA acted as a ligand similar to that of BQ, which promotes the nucleophilic attack. The amide solvent and high pressure of dioxygen are both essential to this reaction.

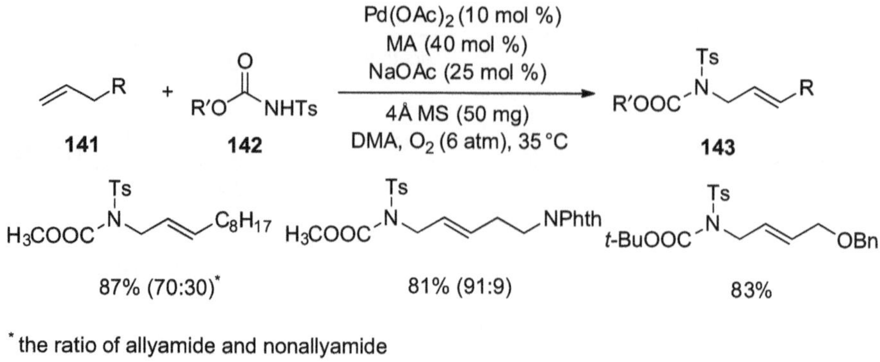

Scheme 11.66 Liu's report on BQ-free aerobic allylic C–H amination reaction.

Table 11.1 Regio-control of Pd-catalyzed aerobic intramolecular allylic C–H amination.

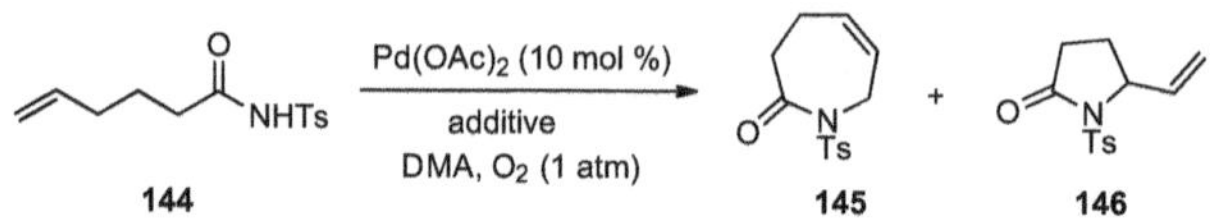

Entry	*Additive*	*Yield (145:146)*
1	none	20% (<3:97)
2	NaOBz (100 mol %)	44% (82:18)
3	NaOBz (100 mol %), 4 A MS (15 mg) MA (40 mol %) **[condition A]**	91% (86:14)
4	(salen)Cr^{III}Cl (10 mol %) **[condition B]**	65% (<3:97)
5	NaOBz (100 mol %), (salen)Cr(III)Cl (10 mol %)	54% (80:20)

Pd-catalyzed intramolecular aerobic oxidative allylic aminations of unactivated alkenes were also developed by the same group (Table 11.1).[210] In the presence of (salen)Cr^{III}Cl, the five-membered cyclic amide **146** was the major product (Table 11.1, entry 4). On the contrary, the addition of a Brønsted base significantly changed the regioselectivity to give seven-membered ring **145** as the major product (Table 11.1, entry 3). When a methyl group was placed at the β-position of the alkene moiety, a seven-membered ring amide **148** was produced as the sole product with good yield (Scheme 11.67).

Obara, Ishii and coworkers recently reported an intermolecular aerobic oxidative allylic amination of simple alkyl alkenes by a $Pd(OCOCF_3)_2$/vanadophosphate salt (NPMoV) catalytic system (Scheme 11.68).[211]

11.4.4 Allylic C–H Alkylation Reactions

In 2008, Shi and coworkers reported the intra/intermolecular coupling of carbon nucleophiles with terminal alkenes *via* allylic C–H activation reaction using bis

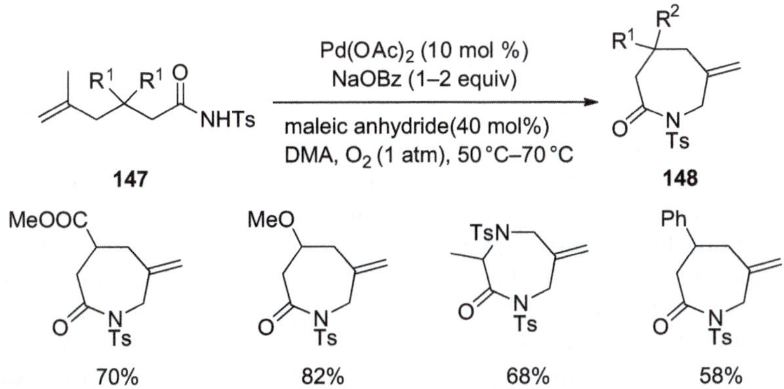

Scheme 11.67 Pd-catalyzed aerobic intramolecular allylic C–H amination.

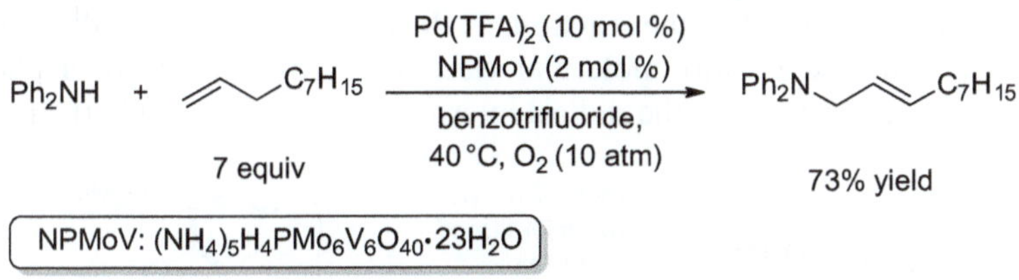

Scheme 11.68 NPMoV facilitated aerobic allylic C–H amination reaction.

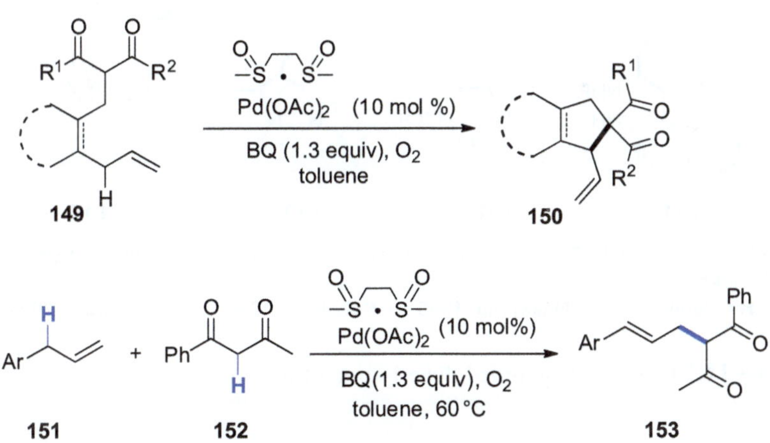

Scheme 11.69 Pd-catalyzed intra- and intermolecular aerobic allylic C–H alkylation.

(sulfoxide)/Pd^{II}/BQ catalyst system (Scheme 11.69).[205] Extension of the intramolecular version to an intermolecular reaction was then investigated. Only aromatic allylic substrates are compatible to this transformation. Aliphatic substrates give inferior results compared to their aromatic counterparts.

11.5 Copper Catalyzed Aerobic Aryl C–H Bond Functionalization

Copper(II) could participate in a wide range of oxidation reactions due to its diversified reaction pathways, which represent powerful tools in synthetic organic chemistry. Copper catalyzed aerobic arene C–H bond functionalizations have received extensive attention and a profound review by Stahl on this subject has been published recently.[48] We here present a brief introduction of this realm.

Electron rich arenes and heterocycles are susceptible to a copper mediated single electron transfer (SET) mechanism. Typical examples are oxidative dimerization of naphthol, using O_2 as terminal oxidant (Scheme 11.70).[212–219] The facile single-electron transfer protocol has also been applied in the oxidative bromination and chlorination of electron-rich arenes.[220–222] Gusevskaya and coworkers reported a highly selective method for the oxidative halogenation of phenols under aerobic copper-catalyzed conditions.[220,222] The phenolic OH group is proposed to be crucial to the reaction, as the formation of the Cu^{II}–phenolate complex is followed by intramolecular electron transfer to afford a phenoxyl radical. Halogen transfer to the radical intermediate would provide the product.

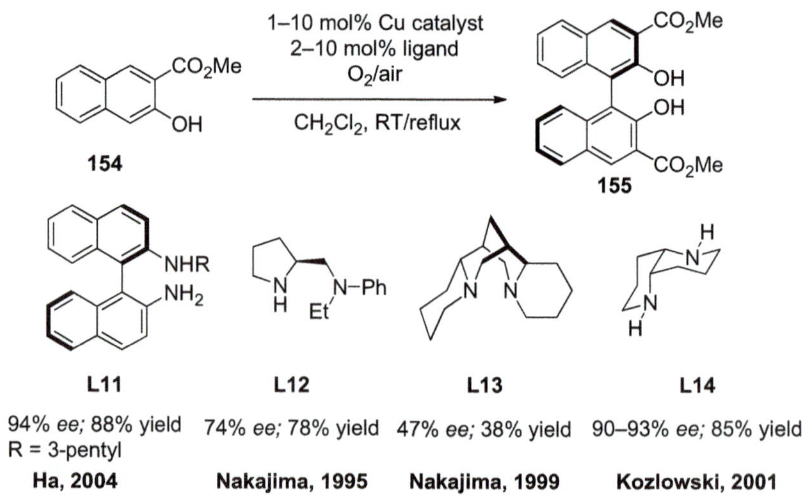

Scheme 11.70 Copper-catalyzed dimerization of naphthol.

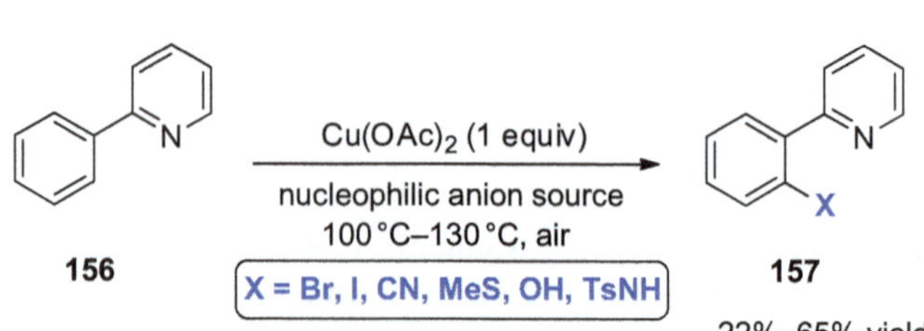

Scheme 11.71 Cu(II)-mediated diverse aryl C–H bond functionalizations.

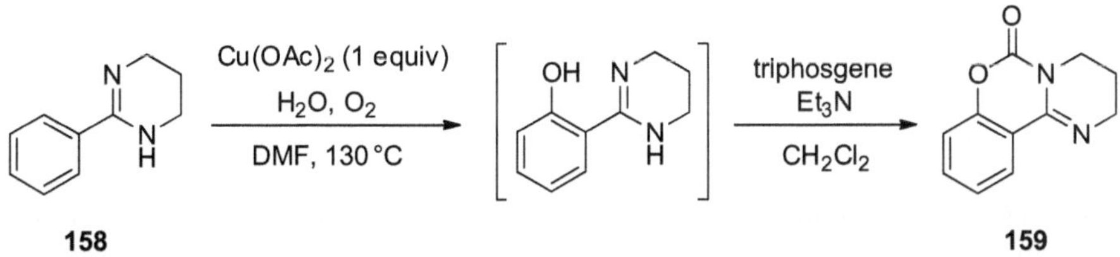

Scheme 11.72 Cu(II)-mediated oxidative intermolecular *ortho* C–H functionalization using tetrahydropyrimidine as the directing group.

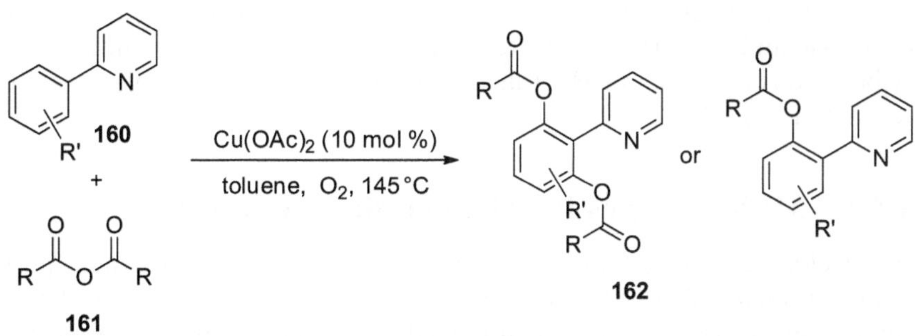

Scheme 11.73 Cu(II)-catalyzed *ortho*-acyloxylation of the 2-arylpyridine C–H bond.

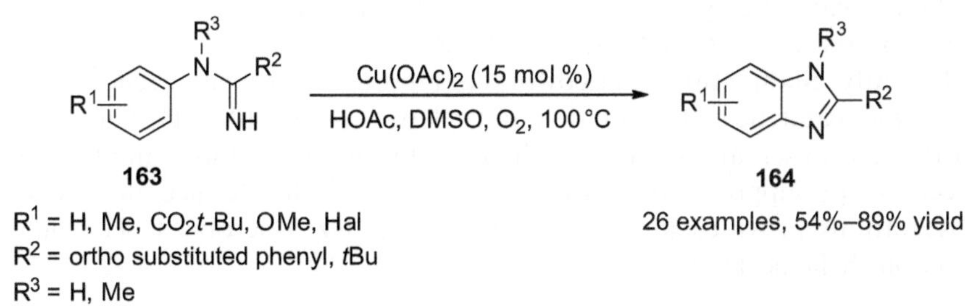

Scheme 11.74 Cu(II)-catalyzed C–H functionalization/C–N bond-forming process.

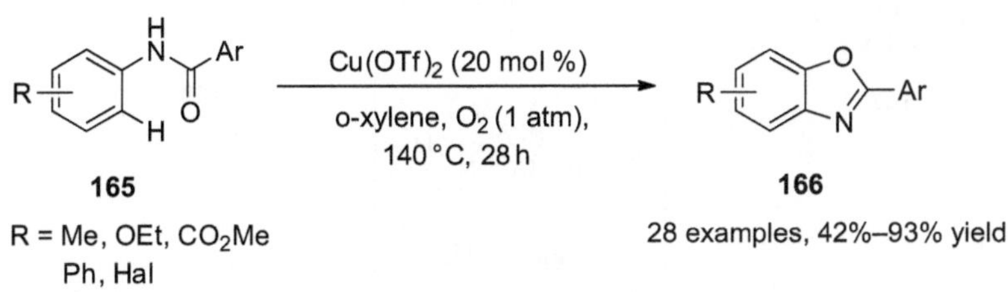

Scheme 11.75 Cu(II)-catalyzed aerobic C–H functionalization for the synthesis of 2-arylbenzoxazoles.

Yu and coworkers reported a Cu(II)-mediated diverse aryl C–H bond functionalizations using dioxygen as the oxidant (Scheme 11.71).[223] A pyridyl directing group is necessary for this transformation. A single electron transfer (SET) process was accounted for this reaction. Chatani and coworkers reported a novel case of a Cu(II)-mediated 2-phenylpyridine *ortho* C–H amination with anilines.[224] Fujii, Ohno and coworkers reported a Cu(II)-mediated oxidative intermolecular *ortho* C–H functionalization using tetrahydropyrimidine as the directing group (Scheme 11.72).[225] Cheng and coworkers provided an example of Cu(II)-catalyzed *ortho*-acyloxylation of the 2-arylpyridine **160** C–H bond under aerobic conditions (Scheme 11.73).[226] Buchwald and coworkers reported a novel synthesis of benzimidazoles **164** from amidines **163** through a Cu(II)-catalyzed C–H functionalization/C–N bond-forming process (Scheme 11.74).[227] Nagasawa and coworkers reported similar Cu(II)-catalyzed synthesis of 2-arylbenzoxazoles *via* C–H functionalization under aerobic condition (Scheme 11.75).[228,229]

11.5.1 Model Systems for Mechanistic Investigation

The development of Cu^{II} and Cu^{III} organometallic chemistry has been pushed forward dramatically due to a number of advances in recent years on the characterization of organocopper(III) intermediates in non-oxidative Cu-mediated reactions. In 2007, direct experimental evidence of Cu^{III} in organocuprate catalysis by low temperature NMR was reported.[230–233] Ullmann-type cross-coupling reactions of aryl halides are commonly proposed to proceed through a Cu^{I}/Cu^{III} catalytic cycle. The experimental evidence of the involvement of a Cu^{III} species was firstly substantiated by Ribas and Stahl.[234]

Recently, Stahl, Ribas and coworkers reported that macrocyclic arene **170** undergoes catalytic C–H oxidation under 1 atm O_2 with 10 mol% $Cu(ClO_4)_2$ or $CuBr_2$. The observation of intermediates by O_2-uptake methods and UV-visible spectroscopy supported the existence of aryl–Cu^{III}–Br intermediate **168**. These observations provided insight into the Cu-catalyzed aerobic C–H oxidation reaction (Scheme 11.76).[235]

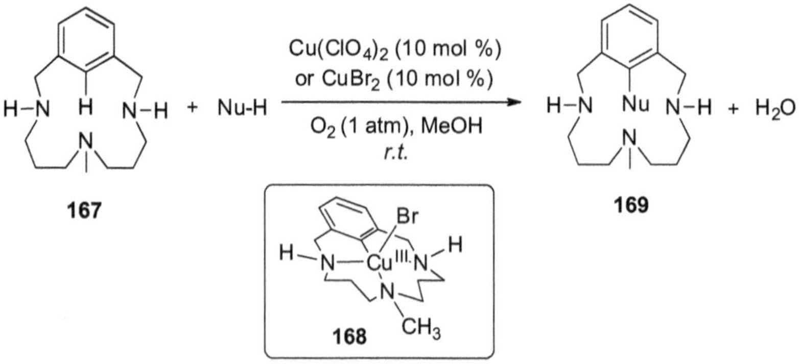

Scheme 11.76 Cu-catalyzed aerobic C–H oxidative functionalization.

11.6 Conclusion and Perspectives

The Pd-catalyzed aerobic reactions have been greatly developed during the last several decades. Plentiful efforts have been devoted to exploiting highly efficient and stereoselective catalyst systems to achieve Pd-catalyzed oxidation of different types of alcohols. Mechanistic studies on the aerobic alcohol oxidation process, especially the regeneration of resting-state Pd(II) catalyst by reaction of Pd(0) with dioxygen, have provided experimental and computational evidence to support the HX-reductive elimination pathway and hydrogen-atom transfer pathway in different catalyst system.

These types of Pd-catalyzed aerobic alcohol oxidation systems have been modified to realize aerobic alkene functionalizations, such as oxy- and aza-Wacker reactions to build C–O and C–N bond in both inter- and intra-molecular fashion. New protocols using C–H bond cleavage and alcohol oxidation-triggered C–C bond formation under aerobic conditions have also been developed. Mechanistic studies of these reactions corroborated the proposed pathways.

Pd-catalyzed allylic C–H bond functionalization is an arduous target for organometallic chemistry. Applications of oxygen as the sole oxidant or with co-oxidants to achieve allylic C–H bond functionalization have been developed recently by several groups.

Copper-catalyzed aerobic oxidative functionalizations are briefly reviewed in this chapter due to their emerging relevance. A review on the copper catalyzed oxidative reactions was published recently.[236]

Further improvement of the aerobic reactions could be anticipated by modulation of the metal catalyst and ligands to achieve diverse kinds of organic transformations.

Acknowledgement

We are grateful to the Financial Support provided by the National Natural Science Foundation of China, National Basic Research Program of China (973 Program 2009CB825300) and the Chinese Academy of Sciences.

References

1. S. S. Stahl, *Science*, 2005, **309**, 1824–1826.
2. T. Punniyamurthy, S. Velusamy and J. Iqbal, *Chem. Rev.*, 2005, **105**, 2329–2364.
3. C. N. Cornell and M. S. Sigman, *Inorg. Chem.*, 2007, **46**, 1903–1909.
4. M. S. Sigman and D. R. Jensen, *Acc. Chem. Res.*, 2006, **39**, 221–229.
5. S. Arita, T. Koike, Y. Kayaki and T. Ikariya, *Chem.–Asian J.*, 2008, **3**, 1479–1485.
6. M. J. Schultz and M. S. Sigman, *Tetrahedron*, 2006, **62**, 8227–8241.
7. T. Punniyamurthy, S. Velusamy and J. Iqbal, *Chem. Rev.*, 2005, **105**, 2329–2363.

8. I. W. C. E. Arends and R. A. Sheldon, Wiley-VCH Verlag GmbH & Co. KGaA: 2010, p 147.
9. B. Karimi and A. Zamani, *J. Iran. Chem. Soc.*, 2008, **5**, S1–S20.
10. M. Hayashi and H. Kawabata, *Adv. Chem. Res.*, 2006, **1**, 45–62.
11. J. Muzart, *Tetrahedron*, 2003, **59**, 5789–5816.
12. I. E. Marko, P. R. Giles, M. Tsukazaki, I. Chelle-Regnaut, A. Gautier, R. Dumeunier, F. Philippart, K. Doda, J.-L. Mutonkole, S. M. Brown and C. J. Urch, *Adv. Inorg. Chem.*, 2004, **56**, 211–240.
13. T. Nishimura and S. Uemura, *Synlett*, 2004, 201–216.
14. R. C. Larock, T. R. Hightower, L. A. Hasvold and K. P. Peterson, *J. Org. Chem.*, 1996, **61**, 3584–3585.
15. K. P. Peterson and R. C. Larock, *J. Org. Chem.*, 1998, **63**, 3185–3189.
16. B. A. Steinhoff, S. R. Fix and S. S. Stahl, *J. Am. Chem. Soc.*, 2002, **124**, 766–767.
17. B. A. Steinhoff and S. S. Stahl, *J. Am. Chem. Soc.*, 2006, **128**, 4348–4355.
18. T. Nishimura, T. Onoue, K. Ohe and S. Uemura, *Tetrahedron Lett.*, 1998, **39**, 6011–6014.
19. T. Nishimura, T. Onoue, K. Ohe and S. Uemura, *J. Org. Chem.*, 1999, **64**, 6750–6755.
20. T. Nishimura, N. Kakiuchi, T. Onoue, K. Ohe and S. Uemura, *J. Chem. Soc., Perkin Trans.*, 2000, **1**, 1915–1918.
21. B. A. Steinhoff and S. S. Stahl, *Org. Lett.*, 2002, **4**, 4179–4181.
22. B. A. Steinhoff, I. A. Guzei and S. S. Stahl, *J. Am. Chem. Soc.*, 2004, **126**, 11268–11278.
23. B. A. Steinhoff, A. E. King and S. S. Stahl, *J. Org. Chem.*, 2006, **71**, 1861–1868.
24. G.-J. ten Brink, I. W. C. E. Arends and R. A. Sheldon, *Science*, 2000, **287**, 1636–1639.
25. R. A. Sheldon, I. W. C. E. Arends, G.-J. ten Brink and A. Dijksman, *Acc. Chem. Res.*, 2002, **35**, 774–781.
26. G.-J. ten Brink, I. W. C. E. Arends and R. A. Sheldon, *Adv. Synth. Catal.*, 2002, **344**, 355–369.
27. G.-J. ten Brink, I. W. C. E. Arends, M. Hoogenraad, G. Verspui and R. A. Sheldon, *Adv. Synth. Catal.*, 2003, **345**, 497–505.
28. E. M. Ferreira and B. M. Stoltz, *J. Am. Chem. Soc.*, 2001, **123**, 7725–7726.
29. D. R. Jensen, J. S. Pugsley and M. S. Sigman, *J. Am. Chem. Soc.*, 2001, **123**, 7475–7476.
30. D. R. Jensen and M. S. Sigman, *Organic Letters*, 2002, **5**, 63–65.
31. S. K. Mandal, D. R. Jensen, J. S. Pugsley and M. S. Sigman, *J. Org. Chem.J. Org. Chem.*, 2003, **68**, 4600–4603.
32. J. A. Mueller, D. R. Jensen and M. S. Sigman, *J. Am. Chem. Soc.*, 2002, **124**, 8202–8203.
33. D. R. Jensen, M. J. Schultz, J. A. Mueller and M. S. Sigman, *Angew. Chem., Int. Ed.*, 2003, **42**, 3810–3813.
34. M. J. Schultz, S. S. Hamilton, D. R. Jensen and M. S. Sigman, *J. Org. Chem.*, 2004, **70**, 3343–3352.

35. D. R. Jensen and M. S. Sigman, *Org. Lett.*, 2003, **5**, 63–65.
36. L.-j. Liu, F. Wang and M. Shi, *Organometallics*, 2009, **28**, 4416–4420.
37. S.-J. Liu, L.-j. Liu and M. Shi, *Appl. Organometallic Chem.*, 2009, **23**, 183–190.
38. T. Chen, J.-J. Jiang, Q. Xu and M. Shi, *Org. Lett.*, 2007, **9**, 865–868.
39. C. Berini, D. F. Brayton, C. Mocka and O. Navarro, *Org. Lett.*, 2009, **11**, 4244–4247.
40. Y. Hou, X. Ji, G. Liu, J. Tang, J. Zheng, Y. Liu, W. Zhang and M. Jia, *Catal. Commun.*, 2009, **10**, 1459–1462.
41. L. Bettucci, C. Bianchini, W. Oberhauser, T.-H. Hsiao and H. M. Lee, *J. Mol. Catal. A: Chem.*, 2010, **322**, 63–72.
42. R. I. McDonald, G. Liu and S. S. Stahl, *Chem. Rev.*, 2011, **111**, 2981–3019.
43. E. M. Beccalli, G. Broggini, M. Martinelli and S. Sottocornola, *Chem. Rev.*, 2007, **107**, 5318–5365.
44. C. S. Yeung and V. M. Dong, *Chem. Rev.*, 2011, **111**, 1215–1292.
45. J. Le Bras and J. Muzart, *Chem. Rev.*, 2011, **111**, 1170–1214.
46. V. Kotov, C. C. Scarborough and S. S. Stahl, *Inorg. Chem.*, 2007, **46**, 1910–1923.
47. S. R. Chemler and P. H. Fuller, *Chem. Soc. Rev.*, 2007, **36**, 1153–1160.
48. A. E. Wendlandt, A. M. Suess and S. S. Stahl, *Angew. Chem., Int. Ed.*, 2011, **50**, 11062–11087.
49. B. V. Popp and S. S. Stahl, *Top. Organomet. Chem.*, 2007, **22**, 149–189.
50. G. Wilke, H. Schott and P. Heimbach, *Angew. Chem., Int. Ed. Engl.*, 1967, **6**, 92–93.
51. S. Otsuka, A. Nakamura, Y. Tatsuno and M. Miki, *J. Am. Chem. Soc.*, 1972, **94**, 3761–3767.
52. T. Yoshida and S. Otsuka, *J. Am. Chem. Soc.*, 1977, **99**, 2134–2140.
53. J. S. Valentine, *Chem. Rev.*, 1973, **73**, 235–245.
54. S. S. Stahl, J. L. Thorman, R. C. Nelson and M. A. Kozee, *J. Am. Chem. Soc.*, 2001, **123**, 7188–7189.
55. M. M. Konnick, B. A. Gandhi, I. A. Guzei and S. S. Stahl, *Angew. Chem., Int. Ed.*, 2006, **45**, 2904–2907.
56. M. M. Konnick and S. S. Stahl, *J. Am. Chem. Soc.*, 2008, **130**, 5753–5762.
57. B. V. Popp and S. S. Stahl, *J. Am. Chem. Soc.*, 2007, **129**, 4410–4422.
58. B. V. Popp, J. E. Wendlandt, C. R. Landis and S. S. Stahl, *Angew. Chem., Int. Ed.*, 2007, **46**, 601–604.
59. J. M. Keith, R. P. Muller, R. A. Kemp, K. I. Goldberg, W. A. Goddard and J. Oxgaard, *Inorg. Chem.*, 2006, **45**, 9631–9633.
60. M. C. Denney, N. A. Smythe, K. L. Cetto, R. A. Kemp and K. I. Goldberg, *J. Am. Chem. Soc.*, 2006, **128**, 2508–2509.
61. J. M. Keith, R. J. Nielsen, J. Oxgaard and W. A. Goddard, *J. Am. Chem. Soc.*, 2005, **127**, 13172–13179.
62. N. Decharin, B. V. Popp and S. S. Stahl, *J. Am. Chem. Soc.*, 2011, **133**, 13268–13271.

63. B. V. Popp and S. S. Stahl, *Chem.– Eur. J.*, 2009, **15**, 2915–2922.
64. P. M.Henry, *Palladium Catalyzed Oxidation of Hydrocarbons*, D. Reidel Publishing Co, Dordrecht, The Netherlands, 1980; vol. 2.
65. E.Negishi, *Handbook of Organopalladium Chemistry for Organic Synthesis*, Wiley, New York, 2002; vol. 2.
66. R.Jira, *Applied Homogeneous Catalysis with Organometallic Compounds*, Wiley-VCH, Weinheim, Germany, 2002.
67. J. A. Keith and P. M. Henry, *Angew. Chem., Int. Ed.*, 2009, **48**, 9038–9049.
68. J. E. Backvall, *Acc. Chem. Res.*, 1983, **16**, 335.
69. M. S. Sigman and M. J. Schultz, *Org. Biomol. Chem.*, 2004, **2**.
70. B. M. Stoltz, *Chem. Lett.*, 2004, **33**, 362–367.
71. L. F. Tietze, H. Ila and H. P. Bell, *Chem. Rev.*, 2004, **104**, 3453–3516.
72. A. Minatti and K. Muniz, *Chem. Soc. Rev.*, 2007, **36**, 1142–1152.
73. L. S. Hegedus, *J. Mol. Catal.*, 1983, **19**, 201.
74. L. S. Hegedus, *Angew. Chem.*, 1988, **100**, 1147.
75. M. Rönn, J.-E. Bäckvall and P. G. Andersson, *Tetrahedron Lett.*, 1995, **36**, 7749–7752.
76. M. A. Rönn, P. G. Andersson and J.-E. Bäckvall, *Acta. Chem. Scand.*, 1997, **51**, 773.
77. A. T. M. van Benthem, H. Hiemstra, G. R. Longarela and W. N. Speckamp., *Tetrahedron Lett.*, 1994, **35**, 9281.
78. E. M. Beccalli, G. Broggini, G. Paladino, A. Penoni and C. Zoni, *J. Org. Chem.*, 2004, **69**, 5627–5630.
79. S. R. Fix, J. L. Brice and S. S. Stahl, *Angew. Chem., Int. Ed.*, 2002, **41**, 164–166.
80. M. M. Rogers, J. E. Wendlandt, I. A. Guzei and S. S. Stahl, *Org. Lett.*, 2006, **8**, 2257–2260.
81. K.-T. Yip, M. Yang, K.-L. Law, N.-Y. Zhu and D. Yang, *J. Am. Chem. Soc.*, 2006, **128**, 3130–3131.
82. R. I. McDonald and S. S. Stahl, *Angew. Chem., Int. Ed.*, 2010, **49**, 5529–5532.
83. F. Jiang, Z. Wu and W. Zhang, *Tetrahedron Lett.*, 2010, **51**, 5124–5126.
84. W. He, K.-T. Yip, N.-Y. Zhu and D. Yang, *Org. Lett.*, 2009, **11**, 5626–5628.
85. R. I. McDonald, P. B. White, A. B. Weinstein, C. P. Tam and S. S. Stahl, *Org. Lett.*, 2011, **13**, 2830–2833.
86. K.-T. Yip, N.-Y. Zhu and D. Yang, *Org. Lett.*, 2009, **11**, 1911–1914.
87. K.-T. Yip and D. Yang, *Org. Lett.*, 2011, **13**, 2134–2137.
88. G. Yang and W. Zhang, *Org. Lett.*, 2011, **14**, 268–271.
89. G. Liu and S. S. Stahl, *J. Am. Chem. Soc.*, 2007, **129**, 6328–6335.
90. X. Ye, G. Liu, B. V. Popp and S. S. Stahl, *J. Org. Chem.*, 2011, **76**, 1031–1044.
91. P. B. White and S. S. Stahl, *J. Am. Chem. Soc.*, 2011, **133**, 18594–18597.
92. T. Hosokawa, M. Takano, Y. Kuroki and S.-I. Murahashi, *Tetrahedron Lett.*, 1992, **33**, 6643–6646.

93. F. Ragaini, T. Longo and S. Cenini, *Journal of Molecular Catalysis A: Chemical*, 1996, **110**, L171–L175.
94. J. M. Lee, D.-S. Ahn, D. Y. Jung, J. Lee, Y. Do, S. K. Kim and S. Chang, *J. Am. Chem. Soc.*, 2006, **128**, 12954–12962.
95. G. L. J. Bar, G. C. Lloyd-Jones and K. I. Booker-Milburn, *J. Am. Chem. Soc.*, 2005, **127**, 7308–7309.
96. V. I. Timokhin, N. R. Anastasi and S. S. Stahl, *J. Am. Chem. Soc.*, 2003, **125**, 12996–12997.
97. V. I. Timokhin and S. S. Stahl, *J. Am. Chem. Soc.*, 2005, **127**, 17888–17893.
98. J. L. Brice, J. E. Harang, V. I. Timokhin, N. R. Anastasi and S. S. Stahl, *J. Am. Chem. Soc.*, 2005, **127**, 2868–2869.
99. M. M. Rogers, V. Kotov, J. Chatwichien and S. S. Stahl, *Org. Lett.*, 2007, **9**, 4331–4334.
100. H. H. R. A. T. M. van Benthem, J. J. Michels and W. N. Speckamp, *J. Chem. Soc. Chem. Commun.*, 1994, 357.
101. R. M. Trend, Y. K. Ramtohul, E. M. Ferreira and B. M. Stoltz, *Angew. Chem. Int. Edit.*, 2003, **42**, 2892–2895.
102. R. M. Trend, Y. K. Ramtohul and B. M. Stoltz, *J. Am. Chem. Soc.*, 2005, **127**, 17778–17788.
103. T. Hosokawa, S. Miyagi, S.-I. Murahashi and A. Sonoda, *Journal of the Chemical Society, Chemical Communications*, 1978, 687.
104. T. Hosokawa, T. Uno, S. Inui and S. Murahashi, *J. Am. Chem. Soc.*, 1981, **103**, 2318–2323.
105. T. Hosokawa, C. Okuda and S. Murahashi, *J. Org. Chem.*, 1985, **50**, 1282–1287.
106. Y. Uozumi, H. Kyota, K. Kato, M. Ogasawara and T. Hayashi, *J. Org. Chem.*, 1999, **64**, 1620–1625.
107. T. Hosokawa, S. Miyagi, S. Murahashi and A. Sonoda, *J. Org. Chem.*, 1978, **43**, 2752–2757.
108. T. Hosokawa, M. Hirata, S.-I. Murahashi and A. Sonoda, *Tetrahedron Lett.*, 1976, **17**, 1821–1824.
109. S. Jabre-Truffert and B. Waegell, *Tetrahedron Lett.*, 1997, **38**, 835–836.
110. M. Reiter, S. Ropp and V. Gouverneur, *Org. Lett.Org. Lett.*, 2003, **6**, 91–94.
111. M. Reiter, H. Turner, R. Mills-Webb and V. Gouverneur, *J. Org. Chem.*, 2005, **70**, 8478–8485.
112. T. Hosokawa, F. Nakajima, S. Iwasa and S.-I. Murahashi, *Chem. Lett.*, 1990, **19**, 1387–1390.
113. T. M. Meulemans, N. H. Kiers, B. L. Feringa and P. W. N. M. v. Leeuwen, *Tetrahedron Lett.*, 1994, **35**, 455–458.
114. G. V. M. Sharma, A. S. Chander, K. Krishnudu and P. R. Krishna, *Tetrahedron Lett.*, 1997, **38**, 9051–9054.
115. T. Hosokawa, T. Ohta and S. I. Murahashi, *J. Chem. Soc., Chem. Commun.*, 1983, 848–849.

116. T. Hosokawa, T. Ohta, S. Kanayama and S. I. Murahashi, *J. Org. Chem.*, 1987, **52**, 1758–1764.
117. T. Hosokawa, Y. Ataka and S. I. Murahashi, *Bull. Chem. Soc. Jpn.*, 1990, **63**, 166–169.
118. T. Hosokawa and S. I. Murahashi, *Acc. Chem. Res.*, 1990, **23**, 49–54.
119. T. Hosokawa, S. Aoki and S.-I. Murahashi, *Synthesis*, 1992, **1992**, 558–561.
120. Y. Kawamura, Y. Kawano, T. Matsuda, Y. t. Ishitobi and T. Hosokawa, *J. Org. Chem.*, 2009, **74**, 3048–3053.
121. M. J. Schultz and M. S. Sigman, *J. Am. Chem. Soc.*, 2006, **128**, 1460–1461.
122. Y. Zhang and M. S. Sigman, *J. Am. Chem. Soc.*, 2007, **129**, 3076–3077.
123. K. H. Jensen, T. P. Pathak, Y. Zhang and M. S. Sigman, *J. Am. Chem. Soc.*, 2009, **131**, 17074–17075.
124. T. P. Pathak, K. M. Gligorich, B. E. Welm and M. S. Sigman, *J. Am. Chem. Soc.*, 2010, **132**, 7870–7871.
125. A. Wang, H. Jiang and H. Chen, *J. Am. Chem. Soc.*, 2009, **131**, 3846–3847.
126. M.-K. Zhu, J.-F. Zhao and T.-P. Loh, *J. Am. Chem. Soc.*, 2010, **132**, 6284–6285.
127. B. Karimi, H. Behzadnia, D. Elhamifar, P. F. Akhavan, F. K. Esfahani and A. Zamani, *Synthesis*, 2010, 1399–1427.
128. J. P. Knowles and A. Whiting, *Org. Biomol. Chem.*, 2007, **5**, 31–44.
129. A. B. Dounay and L. E. Overman, *Chem. Rev.*, 2003, **103**, 2945–2964.
130. D. Mc Cartney and P. J. Guiry, *Chem. Soc. Rev.*, 2011, **40**, 5122–5150.
131. J. Ruan and J. Xiao, *Acc. Chem. Res.*, 2011, **44**, 614–626.
132. G. P. McGlacken and I. J. S. Fairlamb, *Eur. J. Org. Chem.*, 2009, **2009**, 4011–4029.
133. C. Amatore and A. Jutand, *Acc. Chem. Res.*, 2000, **33**, 314–321.
134. G. T. Crisp, *Chem. Soc. Rev.*, 1998, **27**, 427–436.
135. C. S. Cho and S. Uemura, *J. Organometallic Chem.*, 1994, **465**, 85–92.
136. Y. C. Jung, R. K. Mishra, C. H. Yoon and K. W. Jung, *Org. Lett.Org. Lett.*, 2003, **5**, 2231–2234.
137. J. P. Parrish, Y. C. Jung, S. I. Shin and K. W. Jung, *J. Org. Chem.*, 2002, **67**, 7127–7130.
138. M. Chen, J. Wang, Z. Chai, C. You and A. Lei, *Adv. Synth. Catal.*, 2012, **354**, 341–346.
139. M. M. S. Andappan, P. Nilsson and M. Larhed, *Chem. Commun.*, 2004, 218–219.
140. M. M. S. Andappan, P. Nilsson, H. von Schenck and M. Larhed, *J. Org. Chem.*, 2004, **69**, 5212–5218.
141. P.-A. Enquist, P. Nilsson, P. Sjöberg and M. Larhed, *J. Org. Chem.*, 2006, **71**, 8779–8786.
142. J. Lindh, P.-A. Enquist, Å. Pilotti, P. Nilsson and M. Larhed, *J. Org. Chem.*, 2007, **72**, 7957–7962.
143. K. S. Yoo, C. H. Yoon and K. W. Jung, *J. Am. Chem. Soc.*, 2006, **128**, 16384–16393.

144. K. B. Urkalan and M. S. Sigman, *Angew. Chem., Int. Ed.*, 2009, **48**, 3146–3149.
145. K. Akiyama, K. Wakabayashi and K. Mikami, *Adv. Synth. Catal.*, 2005, **347**, 1569–1575.
146. K. A. a. K. Mikami, *Heterocycles*, 2007, **74**, 827.
147. K. S. Yoo, C. P. Park, C. H. Yoon, S. Sakaguchi, J. O'Neil and K. W. Jung, *Org. Lett.Org. Lett.*, 2007, **9**, 3933–3935.
148. K. S. Yoo, J. O'Neill, S. Sakaguchi, R. Giles, J. H. Lee and K. W. Jung, *J. Org. Chem.*, 2009, **75**, 95–101.
149. S. Sakaguchi, K. S. Yoo, J. O'Neill, J. H. Lee, T. Stewart and K. W. Jung, *Angew. Chem., Int. Ed.*, 2008, **47**, 9326–9329.
150. S. H. Cho, J. Y. Kim, J. Kwak and S. Chang, *Chem. Soc. Rev.*, 2011, **40**.
151. S. L. You and J. B. Xia, in *C–H Activation*, ed. J. Q. Yu and Z. Shi, Springer-Verlag Berlin, Berlin, 2010, vol. 292, pp. 165–194.
152. W. J. Yoo and C. J.Li, in *C–H Activation*, ed. J. Q. Yu and Z. Shi, Springer-Verlag Berlin, Berlin, 2010, vol. 292, pp. 281–302.
153. L.-M. Xu, B.-J. Li, Z. Yang and Z.-J. Shi, *Chem. Soc. Rev.*, 2010, **39**, 712.
154. F. Shi and R. C. Larock, in *C–H Activation*, ed. J. Q. Yu and Z. Shi, Springer-Verlag Berlin, Berlin, 2010, vol. 292, pp. 123–164.
155. O. Dagoulis, in *C–H Activation*, ed. J. Q. Yu and Z. Shi, Springer-Verlag Berlin, Berlin, 2010, vol. 292, pp. 57–84.
156. E. M. Beck and M. J. Gaunt, in *C–H Activation*, ed. J. Q. Yu and Z. Shi, Springer-Verlag Berlin, Berlin, 2010, vol. 292, pp. 85–121.
157. O. Daugulis, H.-Q. Do and D. Shabashov, *Acc. Chem. Res.*, 2009, **42**, 1074–1086.
158. Z. Shi, C. Zhang, S. Li, D. Pan, S. Ding, Y. Cui and N. Jiao, *Angew. Chem., Int. Ed.*, 2009, **48**, 4572–4576.
159. X. Chen, K. M. Engle, D.-H. Wang and J.-Q. Yu, *Angew. Chem., Int. Ed.*, 2009, **48**, 5094–5115.
160. D. Alberico, M. E. Scott and M. Lautens, *Chem. Rev.*, 2007, **107**, 174–238.
161. C. Jia, T. Kitamura and Y. Fujiwara, *Acc. Chem. Res.*, 2001, **34**, 633–639.
162. R. S. Shue, *J. Chem. Soc., Chem. Commun.*, 1971, 1510.
163. M. Dams, D. E. De Vos, S. Celen and P. A. Jacobs, *Angew. Chem., Int. Ed.*, 2003, **42**, 3512–3515.
164. Y.-H. Xu, J. Lu and T.-P. Loh, *J. Am. Chem. Soc.*, 2009, **131**, 1372–1373.
165. Y. Hatamoto, S. Sakaguchi and Y. Ishii, *Org. Lett.*, 2004, **6**, 4623–4625.
166. M. Miura, T. Tsuda, T. Satoh, S. Pivsa-Art and M. Nomura, *J. Org. Chem.*, 1998, **63**, 5211–5215.
167. E. M. Beck, N. P. Grimster, R. Hatley and M. J. Gaunt, *J. Am. Chem. Soc.*, 2006, **128**, 2528–2529.
168. H. Jiang, Z. Feng, A. Wang, X. Liu and Z. Chen, *Eur. J. Org. Chem.*, 2010, **2010**, 1227–1230.
169. H. Hagelin, J. D. Oslob and B. Åkermark, *Chem.–Eur. J.*, 1999, **5**, 2413–2416.

170. E. M. Ferreira and B. M. Stoltz, *J. Am. Chem. Soc.*, 2003, **125**, 9578–9579.
171. J. A. Schiffner and M. Oestreich, *Eur. J. Org. Chem.*, 2011, **2011**, 1148–1154.
172. S. R. Kandukuri, J. A. Schiffner and M. Oestreich, *Angew. Chem., Int. Ed.*, 2012, **51**, 1265–1269.
173. Y.-H. Zhang, B.-F. Shi and J.-Q. Yu, *J. Am. Chem. Soc.*, 2009, **131**, 5072–5074.
174. M. Ye, G.-L. Gao and J.-Q. Yu, *J. Am. Chem. Soc.*, 2011, **133**, 6964–6967.
175. D.-H. Wang, K. M. Engle, B.-F. Shi and J.-Q. Yu, *Science*, 2010, **327**, 315–319.
176. K. M. Engle, D.-H. Wang and J.-Q. Yu, *J. Am. Chem. Soc.*, 2010, **132**, 14137–14151.
177. D.-H. Wang and J.-Q. Yu, *J. Am. Chem. Soc.*, 2011, **133**, 5767–5769.
178. B.-F. Shi, Y.-H. Zhang, J. K. Lam, D.-H. Wang and J.-Q. Yu, *J. Am. Chem. Soc.*, 2009, **132**, 460–461.
179. K. M. Gligorich and M. S. Sigman, *Chemical Commun.*, 2009, 3854–3867.
180. K. M. Gligorich, S. A. Cummings and M. S. Sigman, *J. Am. Chem. Soc.*, 2007, **129**, 14193.
181. Y. Iwai, K. M. Gligorich and M. S. Sigman, *Angew. Chem., Int. Ed.*, 2008, **47**, 3219–3222.
182. M. Toyota, A. Ilangovan, R. Okamoto, T. Masaki, M. Arakawa and M. Ihara, *Org. Lett.*, 2002, **4**, 4293–4296.
183. M. Toyota, M. Rudyanto and M. Ihara, *J. Org. Chem.*, 2002, **67**, 3374–3386.
184. M. Toyota and M. Ihara, *Synlett*, 2002, 1211–1222.
185. C. Bee, E. Leclerc and M. A. Tius, *Org. Lett.*, 2003, **5**, 4927–4930.
186. K.-T. Yip, J.-H. Li, O.-Y. Lee and D. Yang, *Org. Lett.*, 2005, **7**, 5717–5719.
187. J. Piera, K. Närhi and J.-E. Bäckvall, *Angew. Chem., Int. Ed.*, 2006, **45**, 6914–6917.
188. C. Feng and T.-P. Loh, *J. Am. Chem. Soc.*, 2010, **132**, 17710–17712.
189. C. J. Engelin and P. Fristrup, *Molecules*, 2011, **16**, 951–969.
190. G. S. Liu and Y. C. Wu, in *C–H Activation*, ed. J. Q. Yu and Z. Shi, Springer-Verlag Berlin, Berlin, 2010, vol. 292, pp. 195–209.
191. T. Jensen and P. Fristrup, *Chem.–Eur. J.*, 2009, **15**, 9632–9636.
192. H. Li, B. J. Lia and Z. J. Shi, *Catal. Sci. Technol.*, 2011, **1**, 191–206.
193. B. M. Trost and P. J. Metzner, *J. Am. Chem. Soc.*, 1980, **102**, 3572–3577.
194. B. M. Trost, P. E. Strege, L. Weber, T. J. Fullerton and T. J. Dietsche, *J. Am. Chem. Soc.*, 1978, **100**, 3407–3415.
195. M. S. Chen and M. C. White, *J. Am. Chem. Soc.*, 2004, **126**, 1346–1347.
196. J. H. Delcamp and M. C. White, *J. Am. Chem. Soc.*, 2006, **128**, 15076–15077.
197. K. J. Fraunhoffer, N. Prabagaran, L. E. Sirois and M. C. White, *J. Am. Chem. Soc.*, 2006, **128**, 9032–9033.
198. M. S. Chen, N. Prabagaran, N. A. Labenz and M. C. White, *J. Am. Chem. Soc.*, 2005, **127**, 6970–6971.

199. D. J. Covell and M. C. White, *Angew. Chem., Int. Ed.*, 2008, **47**, 6448–6451.
200. J. E. Bercaw, N. Hazari, J. A. Labinger and P. F. Oblad, *Angew. Chem., Int. Ed.*, 2008, **47**, 9941–9943.
201. B.-L. Lin, J. A. Labinger and J. E. Bercaw, *Can. J. Chem.*, 2009, **87**, 264–271.
202. K. J. Fraunhoffer and M. C. White, *J. Am. Chem. Soc.*, 2007, **129**, 7274.
203. S. A. Reed and M. C. White, *J. Am. Chem. Soc.*, 2008, **130**, 3316–3318.
204. A. J. Young and M. C. White, *J. Am. Chem. Soc.*, 2008, **130**, 14090–14091.
205. S. Lin, C.-X. Song, G.-X. Cai, W.-H. Wang and Z.-J. Shi, *J. Am. Chem. Soc.*, 2008, **130**, 12901–12903.
206. T. Mitsudome, T. Umetani, N. Nosaka, K. Mori, T. Mizugaki, K. Ebitani and K. Kaneda, *Angew. Chem., Int. Ed.*, 2006, **45**, 481–485.
207. A. N. Campbell, P. B. White, I. A. Guzei and S. S. Stahl, *J. Am. Chem. Soc.*, 2010, **132**, 15116–15119.
208. A. N. Campbell and S. S. Stahl, *Acc. Chem. Res.*, 2012, **45**, 851–863.
209. G. S. Liu, G. Y. Yin and L. Wu, *Angew. Chem., Int. Edit.*, 2008, **47**, 4733–4736.
210. L. Wu, S. F. Qiu and G. S. Liu, *Org. Lett.Org. Lett.*, 2009, **11**, 2707–2710.
211. Y. Shimizu, Y. Obora and Y. Ishii, *Org. Lett.Org. Lett.*, 2010, **12**, 1372–1374.
212. M. Nakajima, I. Miyoshi, K. Kanayama, S.-i. Hashimoto, M. Noji and K. Koga, *J. Org. Chem.*, 1999, **64**, 2264–2271.
213. K. K. M. Nakajima, I. Miyoshi and S.-I. Hashimoto, *Tetrahedron Lett.*, 1995, **36**, 9519.
214. M. N. M. Noji, K. Koga and Kenji, *Tetrahedron Lett.*, 1994, **35**, 7983.
215. D.-W. L. K. H. Kim, Y.-S. Lee, D.-H. Ko and D.-C. Ha, *Tetrahedron*, 2004, **60**, 9037.
216. X. Li, J. Yang and M. C. Kozlowski, *Org. Lett.*, 2001, **3**, 1137–1140.
217. G. B. G. A. Caselli, G. Palmisano, M. Sisti and T. Pilati, *Tetrahedron: Asymmetry*, 2003, **14**, 1451.
218. T. Sakamoto, H. Yonehara and C. Pac, *J. Org. Chem.*, 1997, **62**, 3194–3199.
219. T. Sakamoto, H. Yonehara and C. Pac, *J. Org. Chem.*, 1994, **59**, 6859–6861.
220. L. Menini and E. V. Gusevskaya, *Chem. Commun.*, 2006.
221. Y.-F. Song, G. A. van Albada, J. Tang, I. Mutikainen, U. Turpeinen, C. Massera, O. Roubeau, J. S. Costa, P. Gamez and J. Reedijk, *Inorg. Chem.*, 2007, **46**, 4944–4950.
222. L. Menini, L. A. Parreira and E. V. Gusevskaya, *Tetrahedron Lett.*, 2007, **48**, 6401–6404.
223. X. Chen, X.-S. Hao, C. E. Goodhue and J.-Q. Yu, *J. Am. Chem. Soc.*, 2006, **128**, 6790–6791.
224. T. Uemura, S. Imoto and N. Chatani, *Chem. Lett.*, 2006, **35**, 842–843.

225. T. Mizuhara, S. Inuki, S. Oishi, N. Fujii and H. Ohno, *Chem. Commun.*, 2009, 3413–3415.
226. W. Wang, F. Luo, S. Zhang and J. Cheng, *J. Org. Chem.*, 2010, **75**, 2415–2418.
227. G. Brasche and S. L. Buchwald, *Angew. Chem., Int. Ed.*, 2008, **47**, 1932–1934.
228. S. Ueda and H. Nagasawa, *Angew. Chem., Int. Ed.*, 2008, **47**, 6411–6413.
229. S. Ueda and H. Nagasawa, *J. Org. Chem.*, 2009, **74**, 4272–4277.
230. S. H. Bertz, S. Cope, M. Murphy, C. A. Ogle and B. J. Taylor, *J. Am. Chem. Soc.*, 2007, **129**, 7208–7209.
231. H. Hu and J. P. Snyder, *J. Am. Chem. Soc.*, 2007, **129**, 7210–7211.
232. T. Gärtner, W. Henze and R. M. Gschwind, *J. Am. Chem. Soc.*, 2007, **129**, 11362–11363.
233. S. H. Bertz, S. Cope, D. Dorton, M. Murphy and C. A. Ogle, *Angew. Chem., Int. Ed.*, 2007, **46**, 7082–7085.
234. A. Casitas, A. E. King, T. Parella, M. Costas, S. S. Stahl and X. Ribas, *Chem. Sci.*, 2010, **1**, 326–330.
235. A. E. King, L. M. Huffman, A. Casitas, M. Costas, X. Ribas and S. S. Stahl, *J. Am. Chem. Soc.*, 2010, **132**, 12068–12073.
236. A. E. Wendlandt, A. M. Suess and S. S. Stahl, *Angew. Chem., Int. Ed.*, 2011, **50**, 11062–11087.

CHAPTER 12

Catalytic C–H/C–X Bond Functionalisation of Nucleosides, Nucleotides, Nucleic Acids, Amino Acids, Peptides and Proteins

SARA DE ORNELLAS,[a,b] THOMAS J. WILLIAMS,[a] CHRISTOPH G. BAUMANN[b] AND IAN J. S. FAIRLAMB*[a]

[a] Department of Chemistry, University of York, York, United Kingdom, YO10 5DD; [b] Department of Biology, University of York, York, United Kingdom, YO10 5DD
*Email: ian.fairlamb@york.ac.uk

12.1 Introduction

Palladium-catalysed cross-coupling (hereafter referred to as 'coupling') reactions have been used in the synthesis of modified nucleosides for many years. Modified nucleosides, nucleotides, amino acids, peptides and proteins have a huge range of applications, from biologically active compounds for use in pharmaceuticals, to materials applications and uses as biological probes and reporter molecules.[1] Their huge range of applications, and the ease and versatility with which they can be modified using palladium-catalysed coupling methodology, has made this an exceptionally diverse and active field.

RSC Catalysis Series No. 11
C–H and C–X Bond Functionalization: Transition Metal Mediation
Edited by Xavi Ribas

Published by the Royal Society of Chemistry, www.rsc.org

12.2 Palladium-Catalysed Modification of Nucleosides, Nucleotides and Nucleic Acids

12.2.1 Introduction

Generally, in the Pd-catalysed coupling of nucleosides and nucleotides, the nucleoside or nucleotide is used as the halide component of the reaction, and coupled with an organometallic reagent. Nucleosides and nucleotides can usually be halogenated using standard iodination and bromination reagents.[2–4] The challenges for coupling chemistry of these halogenated nucleosides are (i) their poor solubility in many solvents classically used for coupling, (ii) the often difficult purification of the products, (iii) the highly sensitive functional groups they contain such as amines, hydroxyls and pyrimidones, and (iv) the acid, base and temperature sensitive glycosyl bond. Nucleotides, containing triphosphate groups, are even more sensitive and decompose at room temperature. However, much work in this area, driven by the highly useful and versatile products, has given rise to robust cross-coupling chemistry for the functionalisation of these complex molecules. The limitations imposed by these substrates has required the development of improved coupling conditions such as milder conditions (*e.g.* lower temperatures) and advances in aqueous coupling conditions and water-soluble catalysts. The application of palladium coupling to these substrates highlights the utility of cross-coupling, and more recently direct arlyation reactions for the synthesis of highly complex and sensitive compounds.

Modified nucleosides based on all five canonical bases (adenine, guanine, cytosine, thymine and uracil), as well as many unnatural or unusual bases, have been functionalised using palladium-catalysed chemistry. In this way, palladium-catalysed coupling allows easy access to a wide variety of previously difficult to synthesise nucleoside analogues. Primarily, palladium-catalysed coupling has been used to functionalise the C-8, C-2 and C-6 positions of purine bases (and C-7 of 7-deazapurines), and the C-5 position of pyrimidine bases. A detailed review of palladium-catalysed functionalisation of nucleosides was published by Agrofoglio *et al.* in 2003,[5] and therefore this section will focus primarily on significant developments since this publication.

The major advance in coupling on nucleosides and nucleotides has come with the development of conditions which allow the coupling of unprotected nucleosides. For many years, both the sugar hydroxyls and free N–Hs were usually protected prior to coupling. This improves solubility and reduces side-reactions and catalyst deactivation pathways. However, in 1990, Casalnuovo and Calabrese published a variety of coupling reactions on unprotected nucleosides and nucleotides under aqueous conditions.[6] Both Sonogashira and Suzuki–Miyaura couplings were reported on 5-iodopyrimidine nucleosides and nucleotide mono- and triphosphates. The reactions were carried out using the novel water-soluble catalyst $Pd(tppms)_3$ (TPPMS = sodium triphenylphosphine monosulfonate) in water/acetonitrile or water/methanol mixtures. In most instances, the yields were moderate but even the Sonogashira coupling of

5-iodo-dideoxyuridine 5′-triphosphate to give T-505, a fluorescently labelled nucleotide, was achieved in 50% isolated yield after purification by ion exchange chromatography (Scheme 12.1). Although applied to some syntheses of modified nucleosides,[7] the full potential of this work was not realised for several years.

12.2.2 Suzuki–Miyaura Coupling

In 2003, Shaughnessy and co-workers published the arylation of unprotected halonucleosides by Suzuki–Miyaura coupling.[8] 8-Bromopurine and 5-iodouracil nucleosides were coupled with arylboronic acids under aqueous conditions using the water soluble phosphine ligand TPPTS (trisodium tri(3-sulfonatophenyl)phosphine) to solubilise the palladium catalyst (Scheme 12.2). The conditions used were not dissimilar to those published by Casalnuovo, however the range of substrates was expanded upon and the reaction conditions optimised, so that the active Pd-phosphine species is formed *in situ* ($Pd(OAc)_2$ (2.7 mol%), TPPTS (5.4 mol%), $Ar\text{-}B(OH)_2$ (1.2 eq.), Na_2CO_3 (2 eq.), $MeCN/H_2O$ (1 : 2), 80 °C). 8-Bromoadenosine, 8-bromo-2′-deoxyadenosine (8-Br-dA) and 5-iodouridine (5-I-U) were arylated with excellent yields under these conditions (68–99%), and even the more challenging substrates 8-bromoguanosine and 8-bromo-2′-deoxyguanosine gave high yields (64–95%). Shaughnessy and Western subsequently showed that the guanine moiety has an inhibitory effect on Pd-catalysed coupling reactions due to the readily ionisable amine group and multiple binding modes to palladium.[9] Further optimisation of the reaction found that TXPTS (trisodium tri(4,6-dimethyl-3-sulfonatophenyl)phosphine) was an even more effective ligand, giving arylated nucleosides quantitatively at room temperature in 30 minutes.

Interestingly, the Pd^{II}-TXPTS pre-catalyst has been shown to be remarkably stable to the base, solvent and aryl boronic acid used in this reaction and is reduced very slowly (*ca.* 24 h) to the Pd^0 active catalyst (thought to be a Pd monophosphine complex), leaving the mechanism for the reduction of Pd^{II} to Pd^0 unclear under these conditions.[10]

Collier and Wagner have improved the synthesis of 8-arylguanosine mono- and triphosphates by using a pyridine additive and acetonitrile solvent for the phosphorylation of the halogenated nucleoside.[11] The 8-bromoguanosine mono- and triphosphates can then undergo Suzuki–Miyaura coupling using conditions similar to those above. Their process is an efficient two step synthesis of 8-substituted guanosine nucleotide triphosphates from 8-bromoguanosine (8-Br-G). They have also developed a one-pot synthesis of 8-aryl-NAD derivatives from NAD by first brominating the adenine residue, and then carrying out an *in situ* Suzuki-Miyaura coupling of the brominated adenine with a variety of aryl boronic acids (Scheme 12.3).[12] The resulting arylated products were found to inhibit sirtuin proteins.

Hocek and co-workers have also used conditions similar to the Shaughnessy work to synthesise a range of substituted nucleosides. Specifically, they are

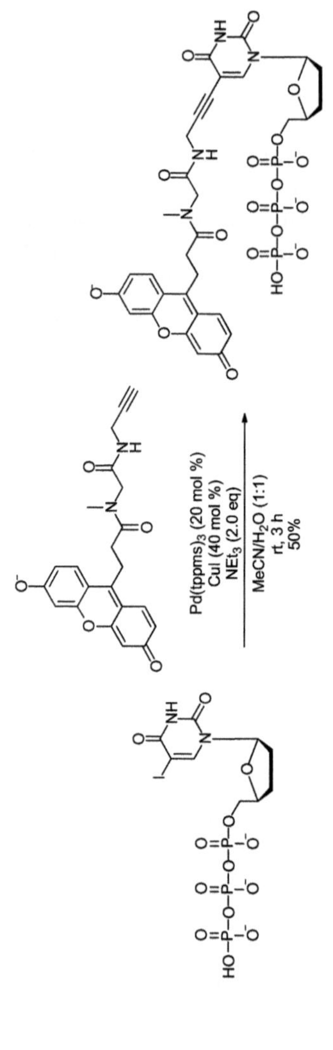

Scheme 12.1

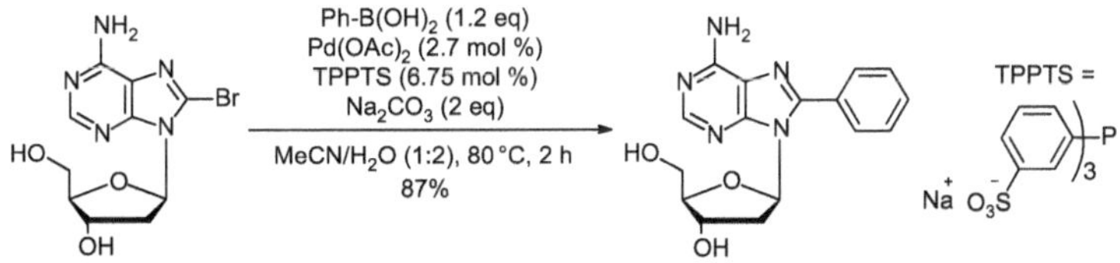

Scheme 12.2

interested in electrochemically-labelled modified nucleosides for possible analytical applications, for example, DNA or RNA nucleotide sequencing or in detection of single-point mutations. They optimised the conditions for use in the microwave, by increasing the catalyst loading and the ligand : metal ratio, showing that a variety of halogenated nucleosides could be coupled effectively in only 5 minutes at 150 °C.[13] Using these, and similar thermal conditions, they have synthesised a variety of labelled adenine, 7-deazaadenine and uracil nucleosides bearing electrochemically active groups. Utilising these conditions they were able to couple organoboronic acids containing Ru-bpy, Ru-tpy and Ru-phen complexes with phenylene linkers successfully, initially with 8-bromo-*N*-benzyladenine,[14] and later with 8-bromo-2′-deoxyadenosine[15] and 7-iodo-7-deaza-2′-deoxyadenosine (7-I-c^{7}dA) (Scheme 12.4).[16]

Hocek and co-workers have also optimised the aqueous Suzuki-Miyaura coupling conditions to directly functionalise unstable halogenated nucleotide mono- and triphosphates. In this case some decomposition of the nucleotides was observed, but it could be minimised by employing Cs_2CO_3 as the base and minimising the reaction time (Scheme 12.5).[13] Many organic co-solvents were investigated but did not appear to affect the stability of the nucleotides. This methodology allows access to a series of electrochemically active nucleotides in an efficient and divergent manner. 7-Iodo-7-deaza-2′-deoxyadenosine, 5-iodo-2′-deoxycytidine and 5-iodo-2′-deoxyuridine triphosphates were functionalised with labels containing amino- and nitro-substituted phenyl groups, and incorporated into nucleic acids using the polymerase chain reaction (PCR).[17]

One of the most remarkable Pd-catalysed functionalisations of nucleosides was published recently by Manderville and co-workers.[18] Using solid-phase synthesis, Manderville and co-workers incorporated 8-bromo-2′-deoxyguanosine into oligonucleotides. The oligonucleotide containing the brominated guanine could then be arylated using Suzuki-Miyaura coupling conditions almost identical to those originally published by Shaughnessy and Western (Scheme 12.6). A large excess of aryl boronic acid (10 eq.) was required, and the arylated DNA (up to 15mers) was produced in 45% yield. A variety of organoboronic acids, including phenol and benzothiophene boronic acids could be successfully coupled in this way. Using these arylated oligonucleotides, Manderville and co-workers intend to study the effect of chemical carcinogens, such as phenol radicals, on DNA. Phenol radicals can

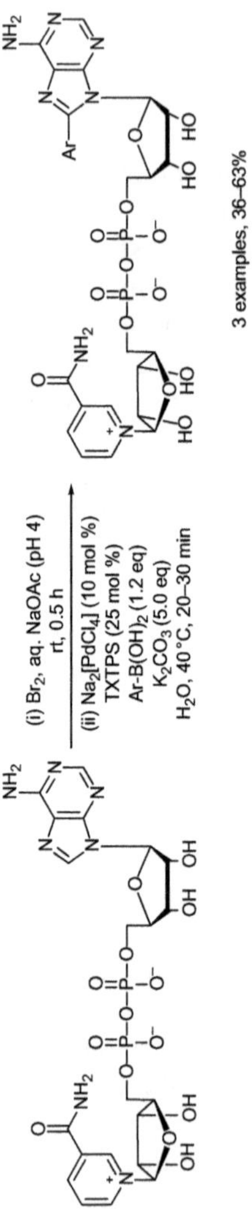
(i) Br2, aq. NaOAc (pH 4)
rt, 0.5 h
(ii) Na2[PdCl4] (10 mol %)
TXTPS (25 mol %)
Ar-B(OH)2 (1.2 eq)
K2CO3 (5.0 eq)
H2O, 40°C, 20–30 min
3 examples, 36–63%

Scheme 12.3

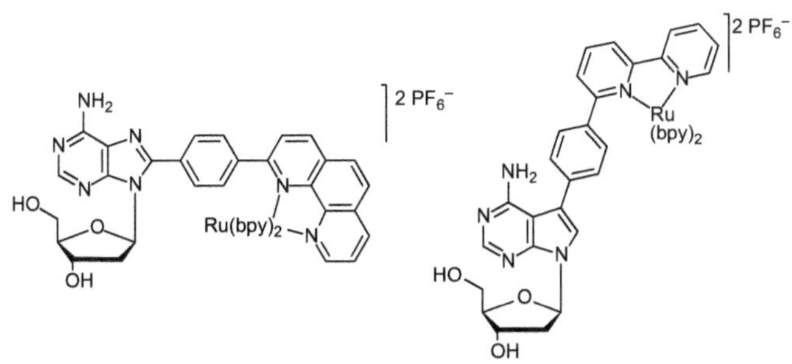

Scheme 12.4

cause mutagenic and carcinogenic damage to DNA, possibly by inducing a conformational change from B- to Z-type DNA.

12.2.3 Sonogashira Coupling

As described above, Sonogashira coupling of unprotected nucleosides and nucleotides under aqueous conditions was carried out as early as 1990.[6] Sonogashira coupling is generally more compatible with nucleosides, as often it requires polar solvents such as DMF, moderate temperatures and mild bases (*e.g.* Et_3N). Consequently, Sonogashira coupling of unprotected nucleosides was adopted as a synthetic tool much earlier than the analogous Suzuki–Miyaura couplings, although couplings on purine bases remained rare. One limitation in the application of Sonogashira coupling to the synthesis of 5-alkynylated uracil derivatives is the cyclisation of the alkynylated product to form a furan derivative.[19] This was observed in an early example of Sonogashira coupling on an unprotected nucleoside published in 1993 by Crisp *et al.* (Scheme 12.7).[20] They showed that Sonogashira coupling could be carried out on 5-trifloxy-2′-deoxyuridine with good yields and trace amounts of the bicyclic by-product. This can be minimised by reducing reaction times or temperatures, optimising reaction conditions or protection of N-3.

A few examples of Sonogashira coupling of unprotected 8-bromoadenosine with terminal acetylenes were reported by Cristalli and co-workers in 2001, although no yields or characterisation of the products was given.[21] Unlike the conditions reported by Casalnuovo and co-workers who used a water-soluble catalyst under aqueous conditions, Cristalli used more typical Sonogashira conditions ($PdCl_2(PPh_3)_2$, CuI, Et_3N, DMF).

Liebscher and co-workers also synthesised a variety of lipophilic 8-substituted adenosines by Sonogashira coupling, however synthesis of the 8-substituted guanosine analogues required protection of the sugar hydroxyls and the exocyclic amine.[22] The Baumann and Fairlamb groups investigated the Sonogashira coupling of unprotected purine nucleosides in 2006.[23] A library of 8-substituted adenosines and guanosines was synthesised by Sonogashira coupling, including thienylacetylene- and ferrocenylacetylene-substituted

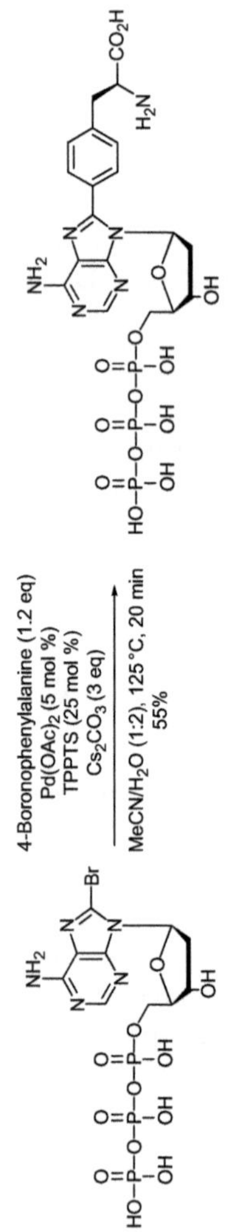
4-Boronophenylalanine (1.2 eq)
Pd(OAc)2 (5 mol %)
TPPTS (25 mol %)
Cs2CO3 (3 eq)
MeCN/H2O (1:2), 125 °C, 20 min
55%

Scheme 12.5

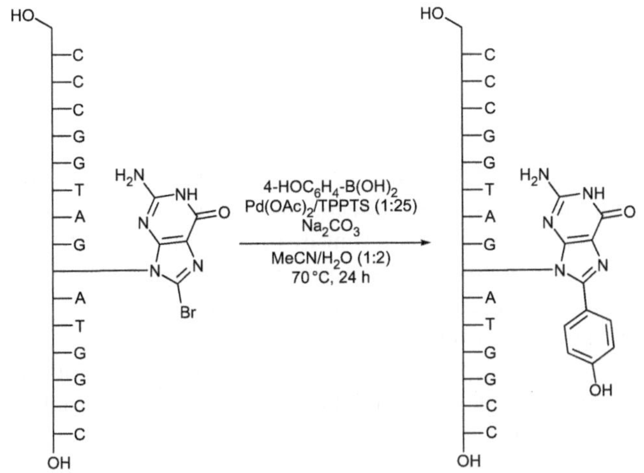

Scheme 12.6

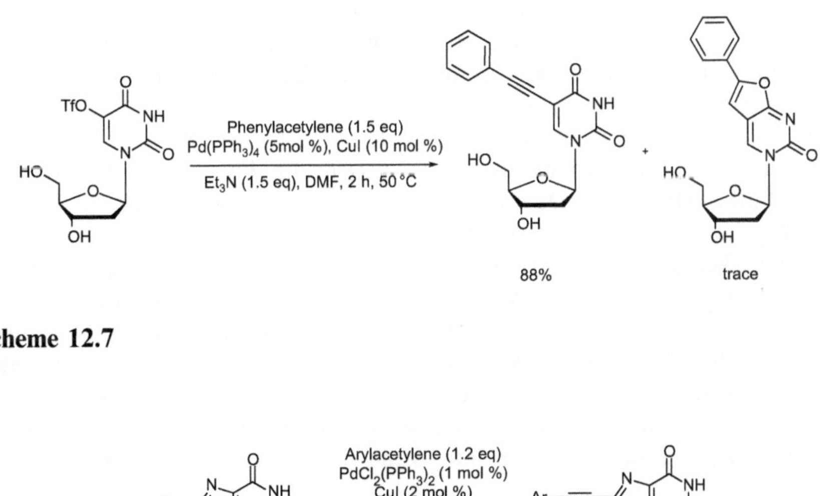

Scheme 12.7

Scheme 12.8

products (Scheme 12.3). The effect of the Pd : Cu ratio in the Sonogashira coupling of guanosines was investigated, and the optimal conditions were found to require 2 : 1 Cu : Pd, with lower Pd loadings (1 mol%) significantly enhancing the coupling (Scheme 12.8).

An investigation into the application of Amberlite IRA-67 resin as a heterogeneous base and solid support for Pd for Sonogashira coupling of uracil, cytosine, adenine and guanine nucleosides was carried out by Quake, Stoltz and co-workers.[24] They employed Pd/C as the catalyst for the uridines and $Pd(PPh_3)_4$ for the other, more challenging nucleosides. The advantages of this system include easier purification, low reaction temperatures and greater reactivity compared with the equivalent homogeneous system. A detailed investigation into the heterogeneous *vs.* homogeneous nature of the Amberlite IRA-67 system in the Sonogahira coupling of 8-bromoguanosine with phenylacetylene by $PdCl_2(PPh_3)_2$ was carried out by Fairlamb, Baumann and co-workers.[25] They used XPS (X-ray photoelectron spectroscopy), SEM (scanning electron microscopy) and TEM (transmission electron microscopy) to observe the aggregation of metal particles in the Amberlite resin, but conclude that the reaction is actually catalysed by homogeneous metal species leached from the resin.

The catalyst system usually employed for Suzuki–Miyaura coupling of halonucleosides ($Pd(OAc)_2$/TPPTS or $Pd(OAc)_2$/TXPTS) has also been used for Sonogashira coupling of the same substrates.[15,16] In particular, coupling of highly unstable nucleoside triphosphates has been effectively carried out by Hocek and co-workers, using the Pd/TPPTS catalyst with CuI in water/acetonitrile (as for Suzuki–Miyaura couplings, *vide supra*).[26] In this manner, they successfully synthesised amino acid-labelled nucleotides triphosphates with alkyne linkers from 8-Br-dA, 7-I-c^7dA and 5-I-U. Nucleotides with acetylene, rather than phenylene, linkers were found to be more suitable substrates for DNA polymerase enzymes, presumably due to the decreased steric bulk. The same chemistry was used in the synthesis of (cytosin-5-yl)ethynyl deoxynucleotide triphosphates (dNTPs) from 5-I-dU, 5-I-dC and 7-I-c^7dA (Scheme 12.9).[27] In the case of the Sonogashira coupling on the uridine nucleoside, the base was changed from diisoproylethylamine to triethylamine, and the temperature lowered from 80 °C to 50 °C to avoid cyclisation of the alkyne onto the carbonyl group, however the furopyrimidine product was not observed when the nucleotide triphosphate substrate was used.

Shaughnessy and co-workers subsequently optimised conditions for aqueous Sonogashira couplings on 8-bromopurine nucleosides and 5-iodopyrimidine

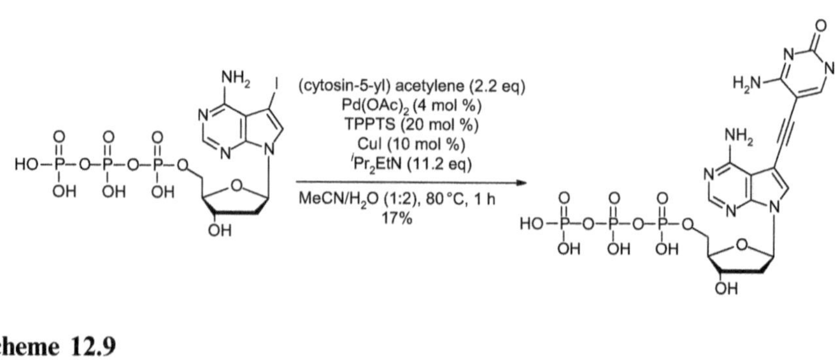

Scheme 12.9

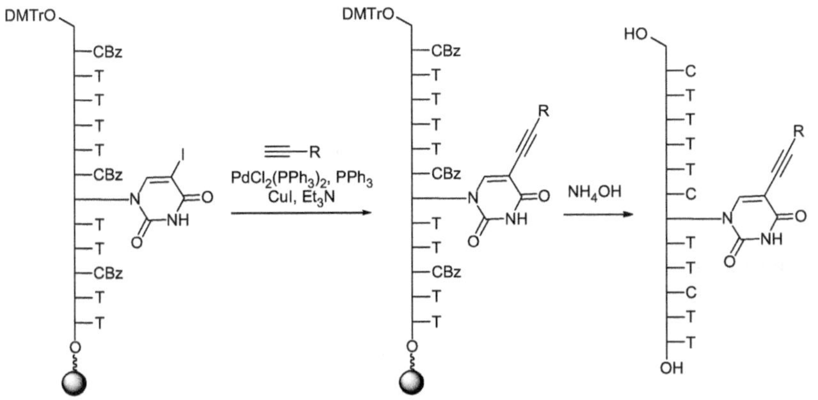

Scheme 12.10

nucleosides, and found that, as with Suzuki–Miyaura couplings, TXPTS was the most efficient of the sulfonated ligands tested.[28]

Several years before the remarkable application of Pd-catalysed coupling to oligonucleotides by Manderville and co-workers,[17] Richert and co-workers reported Sonogashira coupling on a solid-supported oligonucleotide.[29] This work followed a report in 2001 of a method for synthesising base-modified oligonucleotides by stopping the solid-phase synthesis after incorporation of a halogenated nucleoside, subjecting the incomplete oligonucleotide to Sonogashira conditions, and then continuing the solid phase synthesis[30] Richert's method differs in that it does not require interruption of the solid-phase synthesis. Instead, the completed oligonucleotide is subjected to the Sonogashira conditions prior to deprotection and cleavage from the solid support (Scheme 12.10). A wide variety of alkynyl side chains were incorporated into oligonucleotides (12mers) at the 5-position of uracil and their effect on duplex stability investigated.

12.2.4 Stille Coupling

Whilst the Stille coupling has become less popular in recent years, due in part to the toxic reagents and byproducts involved, there are still numerous applications for which it is the most efficient coupling reaction. The Stille coupling can be particularly beneficial where the substrates are base-sensitive.

Krattz and co-workers developed the synthesis of ferrocenyl pyrimidine derivatives protected at the 5′-position by 4,4′-dimethoxytrityl (DMTr) using the Stille coupling.[31] The DMTr group aids solubility of the nucleoside, as well as being the protecting group required for solid-phase synthesis of oligonucleotides. They used conditions based on those developed by Farina and co-worker, with a copper co-catalyst and triphenylarsine ligand, to synthesise 5-ferrocenyl 2′-deoxyuridine and cytidine in excellent yields (95% and 80%) (Scheme 12.11).

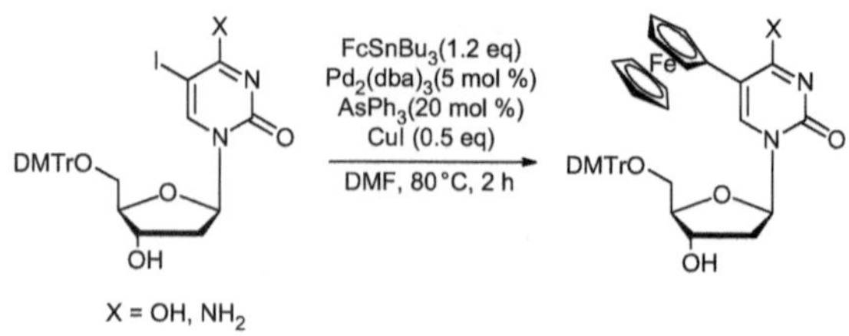

Scheme 12.11

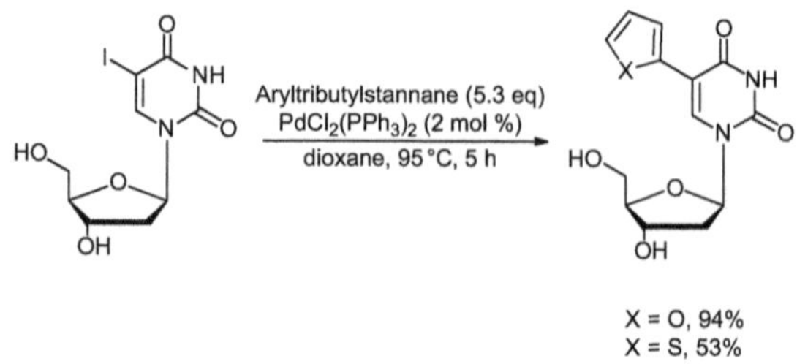

Scheme 12.12

Greco and Tor used Stille coupling to synthesise pyrimidine nucleosides with 5-membered heterocycles at the 5-position.[32] 5-Furanyl- and 5-thienyl 2′-deoxyuridine were synthesised from the unprotected 5-iodonucleoside using catalytic $PdCl_2(PPh_3)_2$ in dioxane in moderate or good yield (Scheme 12.12). Coupling with oxazole and thiazole stannanes, however, required initial protection of the hydroxyl groups as toluoyl esters for successful Stille coupling. The uridines could then be converted to the corresponding cytidines in good yields. 8-Modified purines were also synthesised by this route, albeit in extremely low yields. The furan-containing nucleosides were shown to have biophysically useful fluorescence properties.

An alternative approach to Stille couplings with nucleosides has been to use the nucleoside as the organometallic reagent. This is particularly useful in cases where modification of the nucleoside leads to a less electrophilic aryl halide (so the nucleoside is therefore less reactive towards oxidative addition), such as that reported by Agrofoglio and coworkers (Scheme 12.13).[33] Although the yields are quite low, this route gives access to a useful class of compounds.

12.2.5 Mizoroki–Heck Coupling

The Mizoroki–Heck coupling has been underutilised in the synthesis of C-modified nucleosides. In the report of the "heterogeneous" Amberlite system reported by Quake, Stoltz and co-workers, they also included a single example of this system applied in a Mizoroki–Heck coupling.[22] 5-Iodo-2′-deoxyuridine was coupled with ethyl acrylate in the presence of Amberlite IRA-67 to give the corresponding alkenyl-derivative, which was then converted to the potent

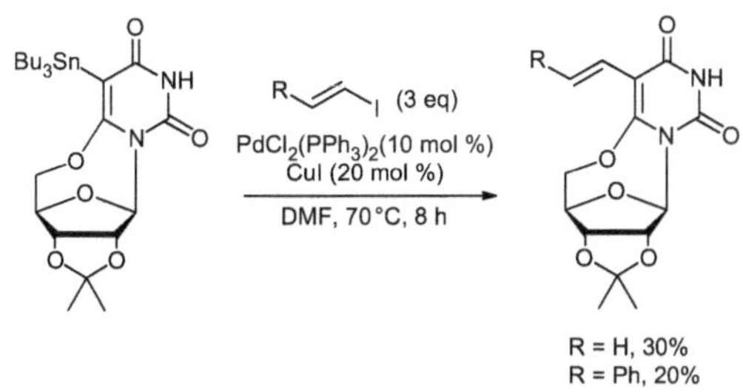

Scheme 12.13

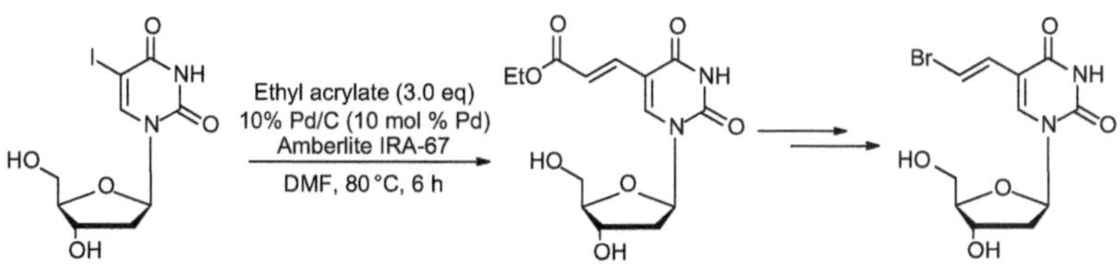

Scheme 12.14

antihepetic agent BVdU (Scheme 12.14). BVdU was synthesised by Mizoroki–Heck coupling of unprotected 5-iodo-2′-deoxyuridine as far back as 1987 by Jones and co-workers using a more traditional catalyst system ($Pd(OAc)_2$ (0.5 eq.), PPh_3 (1.0 eq.), methyl acrylate (2.0 eq.), Et_3N/dioxane (1 : 13), reflux, 40 minutes, 70% yield).[34]

Lakshman and co-workers found that 8-bromoadenosine was not a suitable substrate for Mizoroki–Heck couplings under a variety of reaction conditions. However when protected 8-vinyladenosine was used as a substrate for Mizoroki–Heck couplings with aryl halides, the desired styrene-derivative products were formed in excellent yields, predominantly as the *trans*-product (Scheme 12.15).[35] Despite some pioneering Mizoroki–Heck chemistry being carried out on unprotected pyrimidine nucleosides by Bergstrom and co-workers,[36–38] to our knowledge there has not been significant subsequent development of this chemistry to expand the substrate scope or extend it to unprotected purine nucleosides.

12.2.6 Direct Arylation

Pd/Cu direct arylation systems were first developed by Miura and co-workers in the 1990s for the functionalisation of nitrogen-containing heterocycles.[39] They found that the selectivity of Pd-catalysed arylations was affected by the presence of Cu^I salts. When $Pd(OAc)_2$ was used as the sole catalyst, the arylation of 1-methylimidazole with iodobenzene was selective for C-5, however when a stoichiometric amount of CuI was added, the selectivity was pushed towards C-2 arylation. Subsequently, much research has been carried out into Pd(cat.)/Cu(stoich.) direct arylation systems.[40]

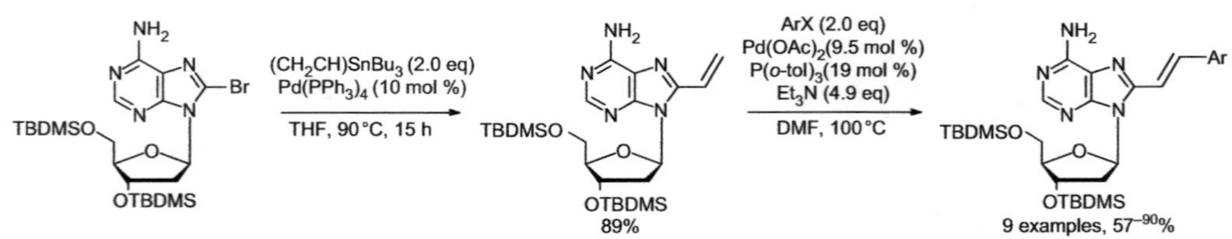
NH2
N
N
N
N
Br
TBDMSO
O
OTBDMS
(CH2CH)SnBu3 (2.0 eq)
Pd(PPh3)4 (10 mol %)
THF, 90 °C, 15 h
89%
ArX (2.0 eq)
Pd(OAc)2(9.5 mol %)
P(o-tol)3(19 mol %)
Et3N (4.9 eq)
DMF, 100 °C
Ar
9 examples, 57–90%

Scheme 12.15

The direct arylation of 9-benzylpurines was accomplished by Hocek and co-workers in 2006.[41] 2,6-Diaryl-9-benzylpurines were substituted at C-8-H by coupling with aryl iodides using catalytic $Pd(OAc)_2$ and excess CuI (3 eq.) and Cs_2CO_3. The yields for the direct arylation were generally very good (42–95%, 10 examples), however the conditions (160 °C, 60 h) were far too harsh for thermally unstable nucleosides. Hocek and co-workers developed the direct arylation of unprotected adenine nucleosides simultaneously with the Fairlamb group.[42,43] Both processes employ a palladium catalyst and excess CuI. The Fairlamb conditions require Cs_2CO_3 as a base (Scheme 12.16), whereas Hocek uses piperidine. In both cases, diarylated products, where both the 8-position and the exocyclic amine group were arylated, were by-products of the reaction, however this was more problematic under the Hocek conditions (Fairlamb – 4%, Hocek – 18%). The higher temperatures used by Hocek also led to increased deglycosylation, and therefore lower yields. Fairlamb and co-workers also investigated the reaction mechanism. Although Pd^0 nanoparticles were formed under the reaction conditions, a negative mercury drop test suggested the reaction to be homogenous. However, the efficacy of this test as a probe for the homogeneity of a reaction has not been fully determined, particularly in very polar solvents such as DMF.[44] The role of CuI was not fully understood despite these investigations. A Sonogashira-like mechanism was proposed, however this does not account for the requirement for three equivalents of copper iodide (Scheme 12.17). Further investigation into this reaction by Fairlamb and co-workers has suggested that coordination of the nucleobase to Cu^I plays a key role in this reaction, and may account for the need for an excess of copper iodide.[45]

Fairlamb, Baumann and co-workers have subsequently developed conditions for the direct arylation of 2′-deoxyadenosine. This is a more challenging substrate due to the increased instability of the glycosyl bond (10^2 times more unstable than adenosine[46]). Lowering the reaction temperature from 120 to 80 °C was required to prevent deglycosylation (Scheme 12.18). It was found that the thermal degradation of DMF to dimethylamine[47] played a key role in the reaction. When the temperature was lowered to 80 °C, dimethylamine was not produced in the reaction. After screening of a variety of amines, it was found that sub-stoichiometric piperidine (40 mol%) was an adequate dimethylamine mimic, giving good, repeatable yields. The piperidine additive

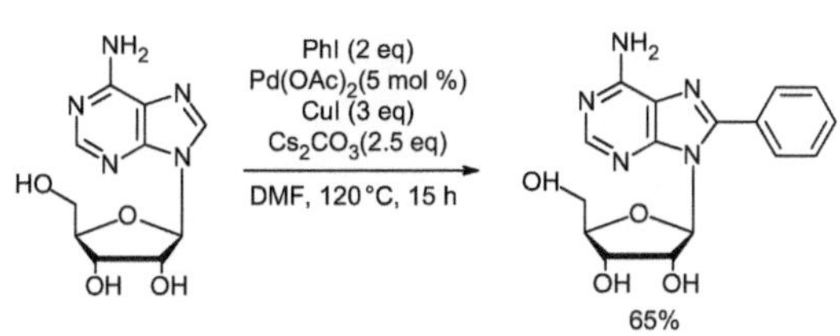

Scheme 12.16

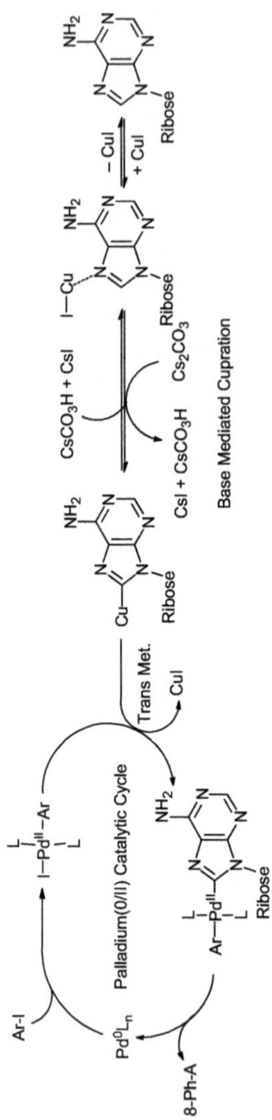
Palladium(0/II) Catalytic Cycle
Base Mediated Cupration
Trans Met.
CuI
Ar-I
8-Ph-A
Ribose
− CuI
+ CuI
CsCO$_3$H + CsI
CsI + CsCO$_3$H
Cs$_2$CO$_3$

Scheme 12.17

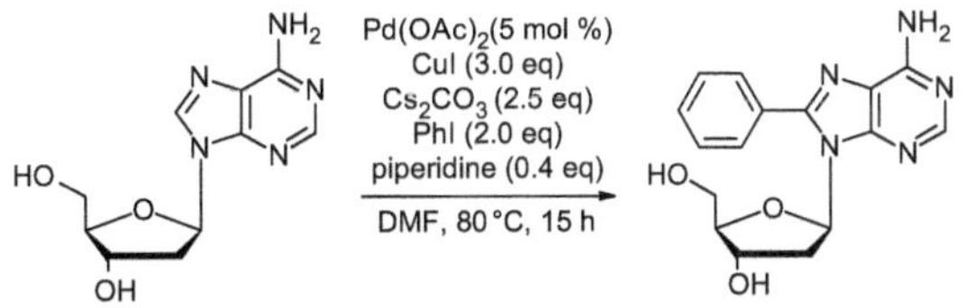

Scheme 12.18

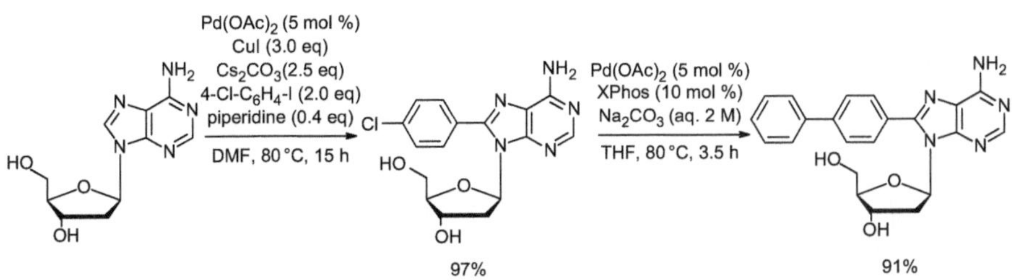

Scheme 12.19

was found to aid the reduction of $Pd(OAc)_2$ to Pd^0 *via* $[Pd(HNC_5H_{10})_2(OAc)_2]$ under these more mild reaction conditions.

A wide variety of aryl halides, including heteroaryl halides, can be coupled to 2′-deoxyadenosine under these conditions with moderate to excellent yields (32–99%, 12 examples). Highly fluorescent nucleosides can be synthesised in this way by first carrying out the direct arylation reaction with 3- or 4-chloro- or bromo- iodobenzene, selectively functionalising the iodide over the other halogens. These fluorophores can be extended further by Suzuki–Miyaura coupling the second halogen to give biphenyl or terphenyl products with useful fluorescent properties (Scheme 12.19).[48]

The direct arylation of adenine nucleosides has been developed for solid-phase synthesis by Hlaváč and co-workers.[49] In this instance, piperidine was required as the base as Cs_2CO_3 was not compatible with the solid-phase conditions. It is notable that although Suzuki–Miyaura coupling is a more established methodology, and the direct arylation has a stoichiometric requirement for Cu^I, the direct arylation methodology is arguably more cost-efficient than the equivalent Suzuki–Miyaura coupling due to the need for bromination of the nucleobase prior to cross-coupling, which is capricious and low-yielding. Based on a 1 mmol scale reaction, the direct arylation reaction is *ca.* 30% cheaper (based on prices from Sigma Aldrich, catalogue accessed online 12/3/12).

Hocek and co-workers have also reported an intramolecular direct arylation strategy for the synthesis of purino[8,9-*f*]phenanthridines and 5,6-dihydro-purino[8,9-*a*]isoquinolines.[50] The appropriate linker, with an aryl halide, is attached to the N9 position of purines *via* Chan–Lam methodology, and then treated with $Pd(OAc)_2$, $PCy_3{\cdot}HBF_4$ and base to yield the corresponding fused

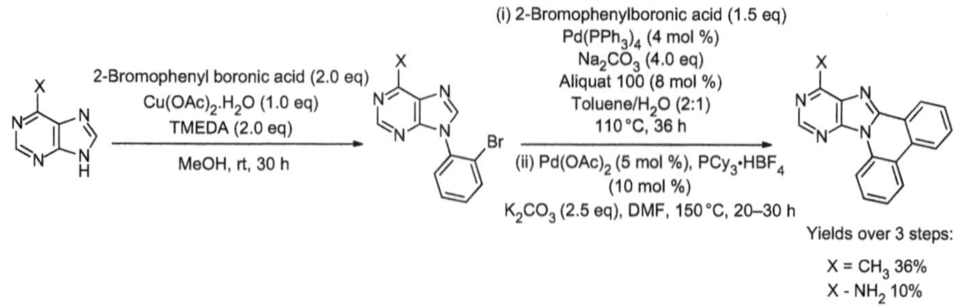

Scheme 12.20

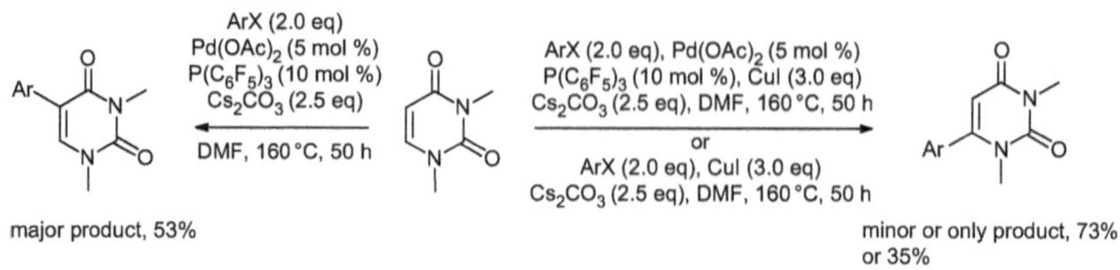

Scheme 12.21

tricyclic or tetracyclic system (Scheme 12.20). Attempts to synthesise these types of structures using two C–H activations (from 1,2-diiodobenzene) or by double C–H activation (coupling of two C–H bonds) resulted in little or no yield.

In 2009, the direct C–H arylation of 1,3-dimethyluracil was reported, where the regioselectivity could be switched between C-5 and C-6 depending on the conditions used.[51] Using $Pd(OAc)_2/P(C_6F_5)_3$ as the catalyst, with Cs_2CO_3 as base in DMF, both aryl iodides and bromides can be coupled to give 5-arylated-1,3-dimethyluracils as the major product in 50 h at 160 °C (C-5 : C-6 86 : 14, 62% conversion). Addition of 3 equivalents of CuI to the reaction leads to a switch in selectivity, giving predominantly C-6 arylated product (C-5 : C-6 6 : 94, 77% conversion) (Scheme 12.21). Addition of CuI is proposed to induce a change in mechanism: the Pd-only catalysed reaction is thought to proceed *via* a CMD-type mechanism,[52] whilst addition of CuI causes a switch to a deprotonation–cupration mechanism (similar to that suggested for adenine nucleosides). Interestingly, when CuI is used with no palladium catalyst, the yield decreases but the C-6 arylated product is formed exclusively (C-5 : C-6 0 : 100, 35% conversion).

This work was extended to the examination of a variety of removable protecting groups.[53] 1,3-Dibenzyluracil undergoes the regioselective palladium-catalysed direct arylation reaction in moderate yield, giving either isomer depending on whether CuI is added. The benzyl groups can then be removed to yield the free nucleobases in good yields.

12.3 Palladium-Catalysed Modification of Amino Acids, Peptides and Proteins

12.3.1 C–X Functionalisation of Amino Acids, Peptides and Proteins

Modifications to amino acids by the activation of C–X bonds were pioneered in an industrial rather than academic setting. Early work by Petrakis and co-workers had indicated the possibility of forming C–P bonds from tyrosine *para*-triflate residues by palladium-mediated catalysis as a step towards alkene formation in pharmaceutical synthesis (Scheme 12.22).[54]

Casalnuovo was the first to apply cross-coupling methodology to the formation of C–C bonds in amino acids. In his pioneering paper on coupling in aqueous media, he demonstrated that the Sonogashira coupling (mediated by $Pd(tppms)_3$) could be used to functionalise 4-iodotyrosine (Scheme 12.23).[6] This was the only example of a functionalised amino acid in the paper. However, the product cyclised *in situ*. This benzofuran formation is common in *ortho*-OH phenylalkynes.[55]

Crisp and co-workers later developed a wide range of low temperature Sonogashira conditions in DMF, DMSO and piperidine solvents, and applied them to the synthesis of amino acids labelled with common, commercially available fluorescent tags.[56]

At a similar time, Tilley and co-workers published work that extended coupling on amino acids into Mizoroki–Heck and Stille couplings (Schemes 12.24–26).[57] Tyrosine *para*-triflate was successfully functionalised with stannyl acrylates or *tert*-butyl acrylates. Couplings on

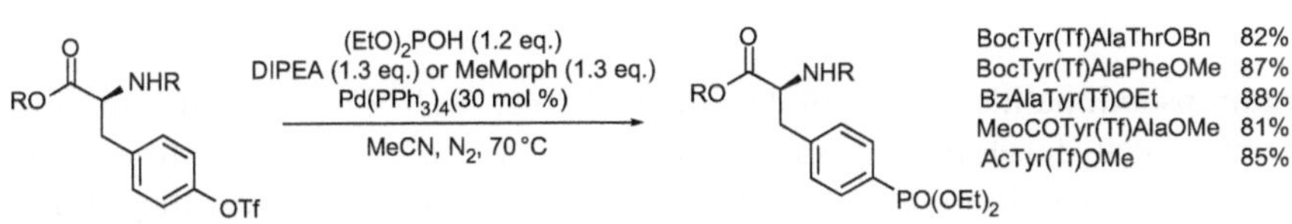

Scheme 12.22

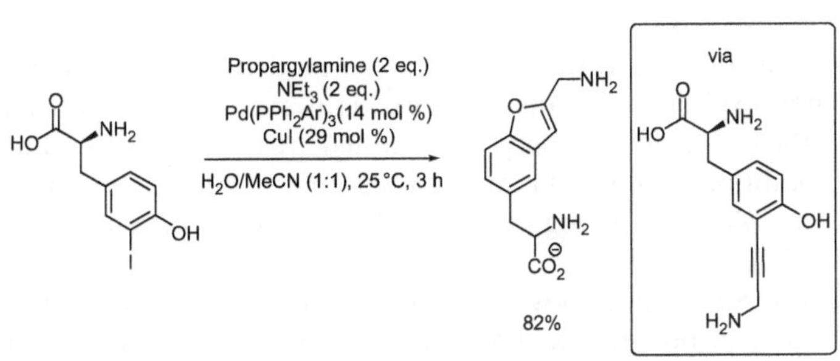

Scheme 12.23

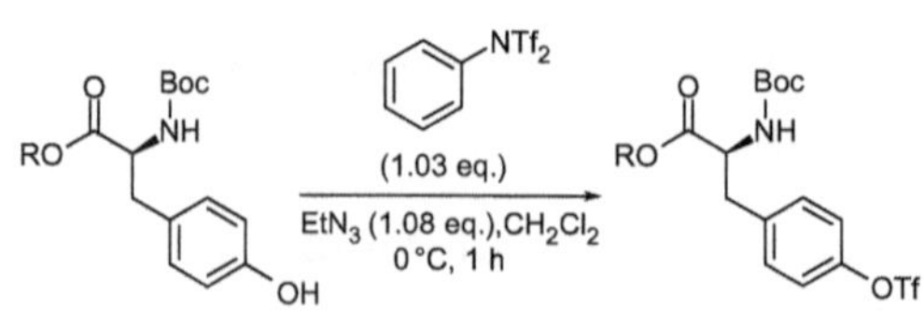

Scheme 12.24

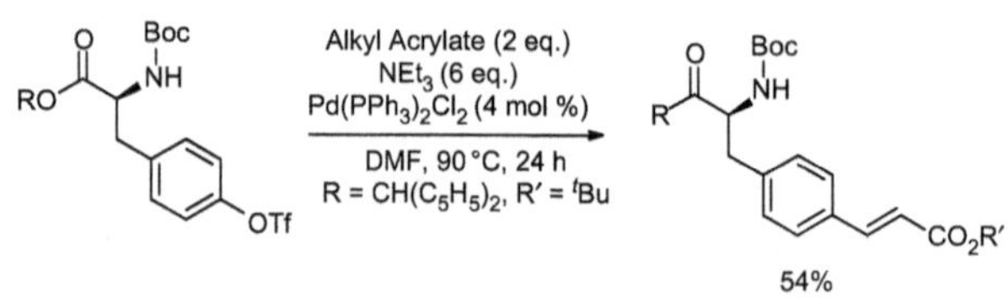

Scheme 12.25

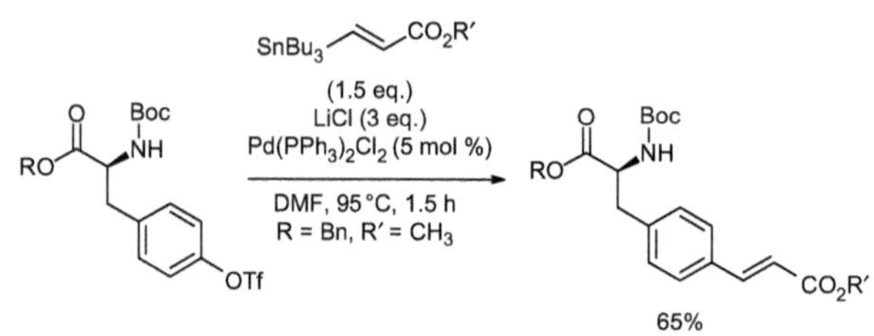

Scheme 12.26

organopseudohalides like triflates were already well known in the literature (having been developed by Echavarren and Stille),[58] and Tilley found these more accessible than traditional organohalides, primarily as they utilised functionality already present in natural amino acids. A modest yield of 54% was achieved in the Mizoroki–Heck coupling, whilst in the Stille coupling a more respectable 65% was obtained after a significantly reduced reaction time (1.5 hours, down from 24 hours in the Mizoroki–Heck coupling). It is worth noting that various protecting groups were used for the different couplings, which is not explained by the authors. However, unlike the Sonogashira products, they did not undergo unwanted rearrangements. Later, using aqueous conditions, Sengupta and co-workers published a one-pot Heck-Matsuda coupling,[59] giving the product in an excellent 84% yield (Scheme 12.27).[60]

In addition, Sheih and co-workers demonstrated that the Suzuki–Miyaura coupling was also possible on tyrosine *para*-triflate substrates.[61] The authors initially trialled catalytic $Pd(PPh_3)_4$ with K_2CO_3 in DMF at 90 °C, and obtained high yields, offset by an erosion of enantiomeric excess. No racemisation was

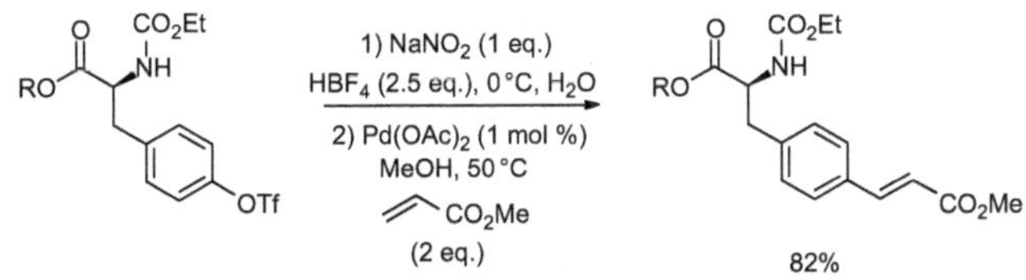

Scheme 12.27

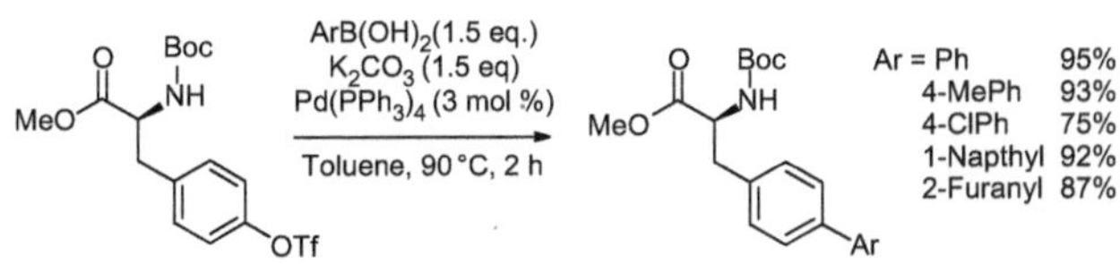

Scheme 12.28

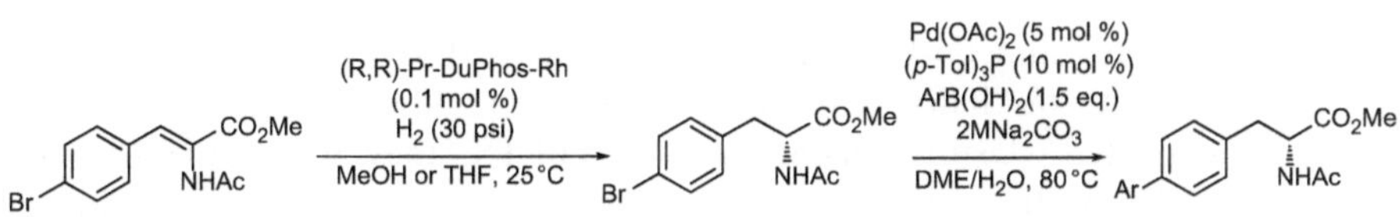

Scheme 12.29

observed when an organic base (triethylamine) was substituted for K_2CO_3, although the rate of reaction was significantly reduced. Following further reaction optimisation (including, adventurously, assessing the use of TlOH as a possible base),[62] it was concluded that exchanging DMF for toluene gave both high yield (93%) and retained enantiomeric excess (>99% by HPLC) (Scheme 12.28).

Burk and co-workers performed the first amino acid Suzuki–Miyaura coupling on a halide substrate, using 2-, 3-, and 4-bromophenylalanine produced by their novel DuPHOS-Rh catalysts.[63] Cross coupling conditions initially employed catalytic $Pd(PPh_3)_4$, but long reaction times (greater than 16 hours) resulted in the authors switching to $Pd(OAc)_2$ with 2 M Na_2CO_3, $P(p\text{-Tol})_3$ in DME at 80 °C. Unlike Sheih's, these conditions are bi-phasic and utilised an additional triaryl phosphine. Although reaction times were slightly greater, excellent yields (75–99%) were observed with no racemisation. These conditions are tolerant to a wide range of aryl and heteroaryl boronic acids. The authors applied their conditions to a tri-peptide (*N*-Boc-Met-Leu-*o*-BrPhe-OMe), isolating product in an excellent 71% yield (Scheme 12.29).

Satoh and co-workers have developed a complementary methodology to Sheih and Burk, in which (4-pinacolylborono)phenylalanine ethyl ester is cross-coupled with aryl halides with both palladium and nickel catalysts in NMP or dioxane.[64,65]

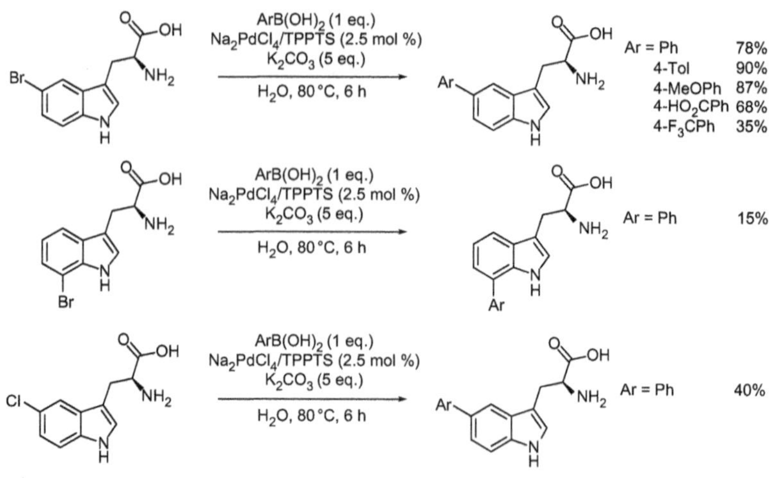

Scheme 12.30

Goss, Wagner and co-workers developed a methodology for an aqueous Suzuki–Miyaura on halotryptophans (Scheme 12.30),[66] synthesised *via* an enzymatic route.[67] Using Na_2PdCl_4 with TPPTS, they were able to functionalise 5-bromotryptophan with a variety of organoboronic acids in reasonable yields. 5-Chlorotryptophan was also functionalised, albeit with a substantially depressed yield compared to the 5-bromo compound. 7-Bromotryptophan gave a poor yield. Using methodology that had been applied to nucleosides, reaction temperature was successfully reduced to 40 °C (from 80 °C) by switching to TXPTS as ligand.[8]

A simple dipeptide (*N*-Boc-Ala-5-BrTrp-OMe) was subjected to the reaction conditions and, although deprotection was observed at the methyl ester, 62% functionalised product was isolated by HPLC.

The work by Tilley, Shieh, Burk and others demonstrated the possibilities for coupling on amino acids and established triflates and organohalides as accessible starting materials for such reactions. However, in some ways they represented a step backward relative to the aqueous conditions proposed by Casalnuovo. The use of non-polar organic solvents (such as toluene and benzene) make conditions unsuitable for application to unprotected amino acids, or even longer chained peptides and proteins. DMF brings its own synthetic, mechanistic and environmental challenges, and well as toxicity issues.[47]

Schmittchen and co-workers began to address the problems associated with coupling in proteins and peptides.[68] They were especially interested in bioconjugating peptides encoded with aryl iodide and propargyl amino acids. Under the basic conditions of many coupling reactions, peptides and proteins are likely to carry an overall negative charge. Therefore, the water-soluble phosphine ligands of Casalnuovo and co-workers (which contain sulfonates) would face an additional electrostatic barrier as compared to coupling on neutral substrates, although this had not previously been shown as a problem on nucleotides.[6] To overcome this problem, the authors developed positively charged analogues: the guanadinophosphines (Scheme 12.31).[27,69] Catalysts were prepared by combining $Pd(OAc)_2$, ligand, and CuI in acetonitrile in a ratio 1 : 5 : 10, respectively, and allowing the solution to 'condition' over a five day period under N_2 at 4 °C.

To simulate the peptide charge environment, Schmittchen performed Sonogashira couplings on 4-iodobenzoate with propagyl glycine (Scheme 12.32) under aqueous conditions (water/acetonitrile, 7 : 3). After three hours, a 75% yield of product was obtained, with no evidence of homo-coupling.[70] The reaction of 3-iodotyrosine with propiolic acid gave 86% after 25 hours (Scheme 12.33). Longer reaction times, or exposure to acid, resulted in rearrangements similar to those in Casalnuovo's chemistry. Use of buffered solution (3-[tris(hydroxymethyl)methylamino]-1-propanesulfonic acid (TAPS) buffer (Scheme 12.34), pH 8, as opposed to $MeCN/H_2O$, gave similar results. On free, multi-functional peptides a 75% yield of product was obtained (Scheme 12.35). Further, the authors demonstrated it was possible to label the same protein with a biotin derivative, commonly used in biological chemistry (Scheme 12.36).[71]

Ghadhiri and co-workers challenged Schmittchen's assertion on the suitability of sulfonated phosphines for the functionalisation of peptides by

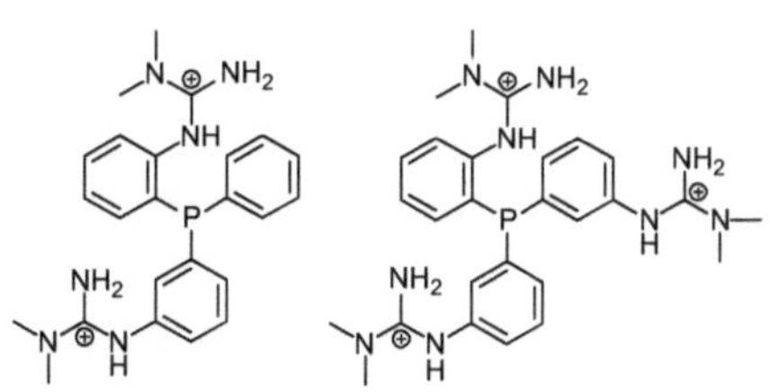

Scheme 12.31

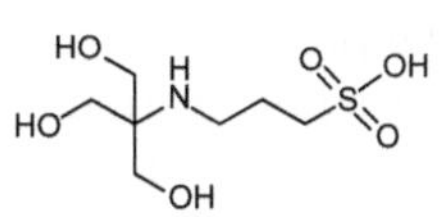

Scheme 12.32

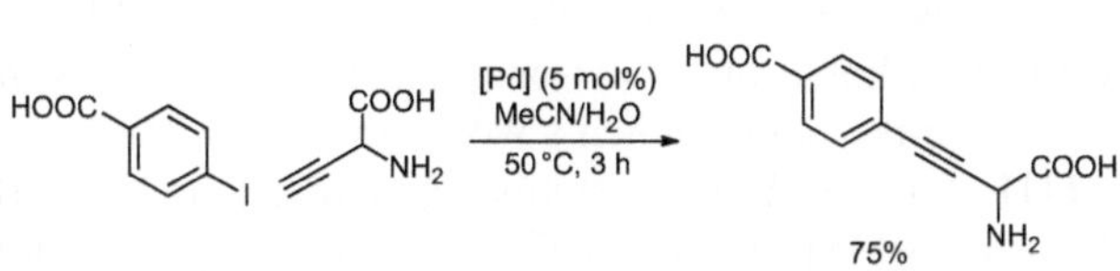

Scheme 12.33

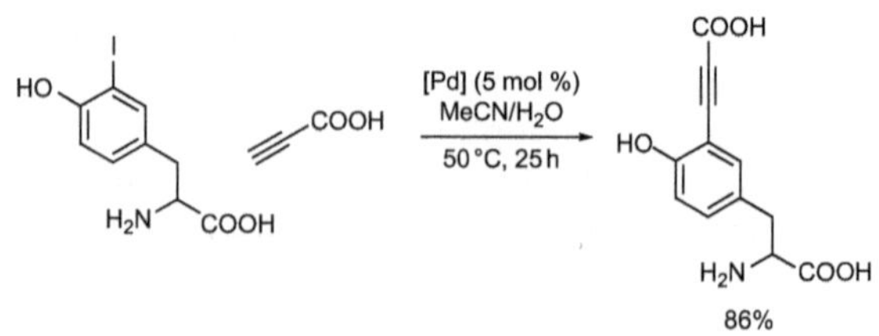

Scheme 12.34

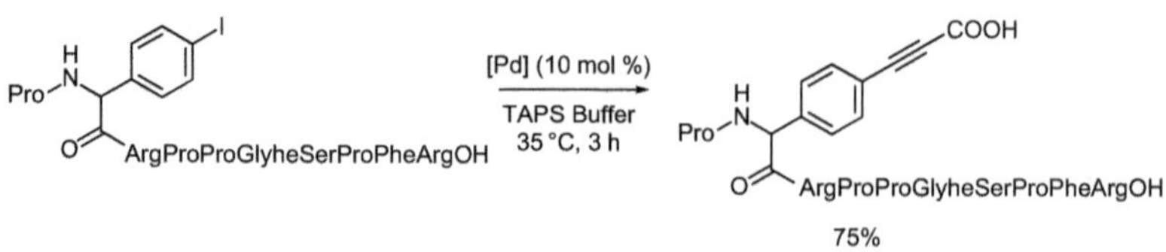

Scheme 12.35

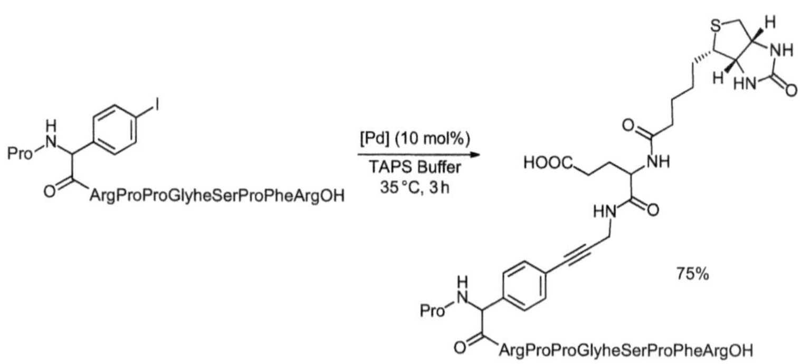

Scheme 12.36

demonstrating that quantitative yields could be achieved with such systems.[72] They further indicated that the presence of thiol, thioether or bipyridyl functionality in the peptide chain was not tolerated by the reaction, presumably as such groups compete as ligands for palladium and copper. This finding is surprising as Schmittchen showed that the reaction could occur in the presence of biotin, a cyclic thioether. Biotin has been shown to be tolerated by small molecule coupling reactions.[73]

Kotha and co-workers have also developed conditions for the Suzuki–Miyaura couplings on short-chained peptides, in generally good yields.[74] However, unlike Schmittchen's work, these reactions were performed in bi-phasic (THF/Toluene/H_2O) systems. Hamachi and co-workers demonstrated that it was possible to modify a 33-residue peptide (WW domain of Pin1 protein), which was utilised as a model for protein modification.[75] Using Na_2PdCl_4 and no phosphine, an initial yield of 40% was obtained. Higher temperatures increased the yield and with the addition of glycerol (10–50 mol%, depending on substrate) yields of between 54 and 90% were obtained. Glycerol likely enhanced the yield by increasing peptide solubility, as well as having a demonstrated role in the formation and stabilisation of Pd nanoparticles.[76] Further, they demonstrated that the peptides modified by the Suzuki–Miyaura couplings maintained their native structure and ligand binding function.

Fournier and co-workers have demonstrated that it is possible to modify peptides in this way whilst they remain tethered to solid-phase resins used for peptide synthesis. $Pd(PPh_3)_4$ was utilised as catalyst, and higher temperatures, in DMF, were found to be optimal. Such conditions would not be ideal for more sensitive substrates, have a high environmental impact, and are likely to be complicated by problems with solvent stability.[47] However, for a combinatorial chemist the advantages of solid-phase synthesis may outweigh these disadvantages.[77]

Although post-synthetic modification of the amino acid tyrosine to yield triflates as coupling partners was reported relatively early, post-synthetic halogenation remained elusive. Work by Valencia, Barluenga and co-workers combined aqueous iodination of tyrosine and phenylalanine residues in peptides with a Suzuki–Miyaura arylation (Scheme 12.37).[78] The coupling conditions differ from Casalnuovo's work mainly in their use of potassium aryl trifluoroborates (known to be only partially-soluble in water), as opposed to aryl boronic acids.[6,79] The authors had previously reported work into the iodination of biomolecules,[80] however, the innovation in this paper was the 'one-pot', sequential use of these reactions for what amounts to an entirely post-synthetic modification of a peptide.

The first coupling on a protein was published by Tachibana, Yokoyama and co-workers.[81] iF32-Ras-His was used, in TAPS buffer (pH 8.3), to $Pd(OAc)_2$, TPPTS, 12% DMSO, $MgCl_2$, and vinylated biotin (Scheme 12.38). $MgCl_2$ was used to suppress biotinylation at cysteine residues, presumably by co-ordinating the sulfur. This reaction occurred at low temperature (5 °C), although long (50 hours) reaction times were required. An estimated 2% yield was observed, coupled with significant amounts of dehalogenation (28% of the crude mixture). Sodium stearate was added to the crude reaction mixture in work-up to remove magnesium and palladium, and experiments indicated that the use of transition metals did not affect the subsequent biological assays.

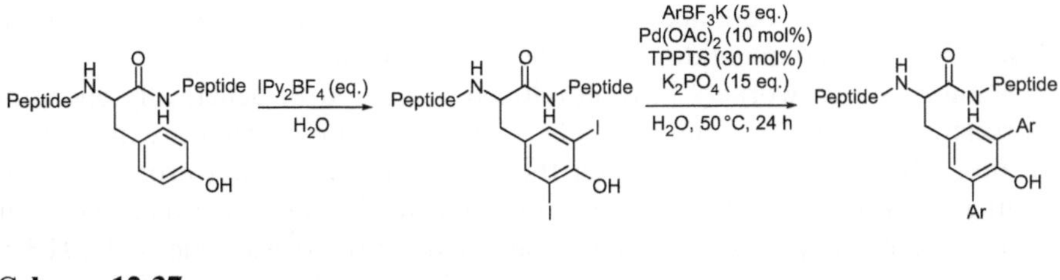

Scheme 12.37

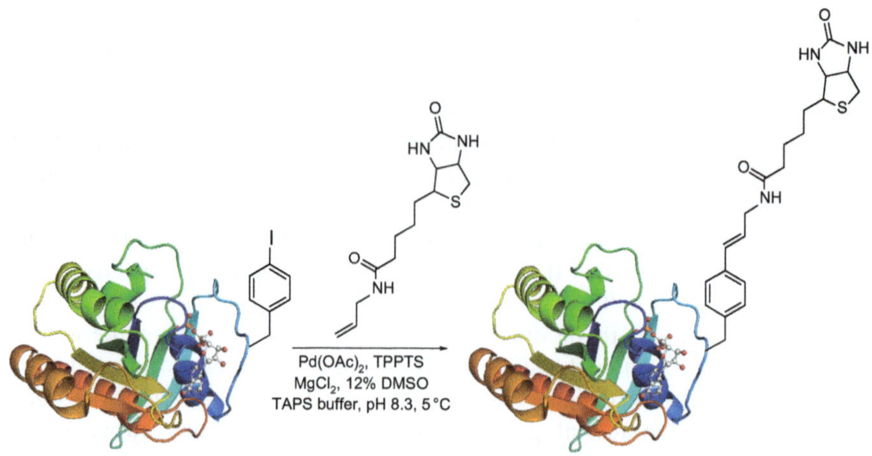

Scheme 12.38

Tachibana, Yokoyama and co-workers went on to publish a significant investigation into these findings.[82] Confining these studies to a model peptide substrate, an extensive optimisation found conditions that gave 100% conversion to product, as assayed by HPLC. This screen included reagent concentration, reaction time and temperature, and the presence of other additives. Optimal conditions (1 : 7.14 : 1.45 : 1 : 43 ratio of peptide, vinylated biotin, decyl-β-D-glucopyranoside (a mild surfactant) and Pd-TPPTS, respectively, at 22 °C for 2 hours) showed 76% conversion to product, and after purification by HPLC a 26% yield was obtained.

When the authors extended this methodology to the Sonogashira coupling, they found the reactions proceeded at higher rates than the equivalent Suzuki–Miyaura couplings, typically finishing within 4 hours. They noted, perhaps unsurprisingly, that the efficacy of this reaction was dependent on the concentration of palladium and copper in the reaction mixture. They further demonstrated that the reaction continued to proceed in the presence of contaminants known to result from the purification of proteins using nickel-affinity chromatography (Ni^{2+}-NTA agarose, Triton X-100 and imidazole hydrochloride).

The Sonogashira coupling conditions, with propargylated biotin, were applied to Ras protein containing a 4-iodophenylalanine residue (Scheme 12.39). A 25% yield was obtained, which was a significant improvement on the Mizoroki–Heck coupling done with the same substrates.

Schultz and co-workers have developed a method of genetically encoding boronate-containing amino acids into proteins. They explored a Suzuki–Miyaura coupling between their boronate-containing protein and an iodinated fluorescent BODIPY reporter molecule.[83] Catalysis was achieved in EPPS buffer (3-[4-(2-hydroxyethyl)-1-piperazinyl]propanesulfonic acid, pH 8.5) at an elevated temperature of 70 °C over 12 hours with, unusually for aqueous

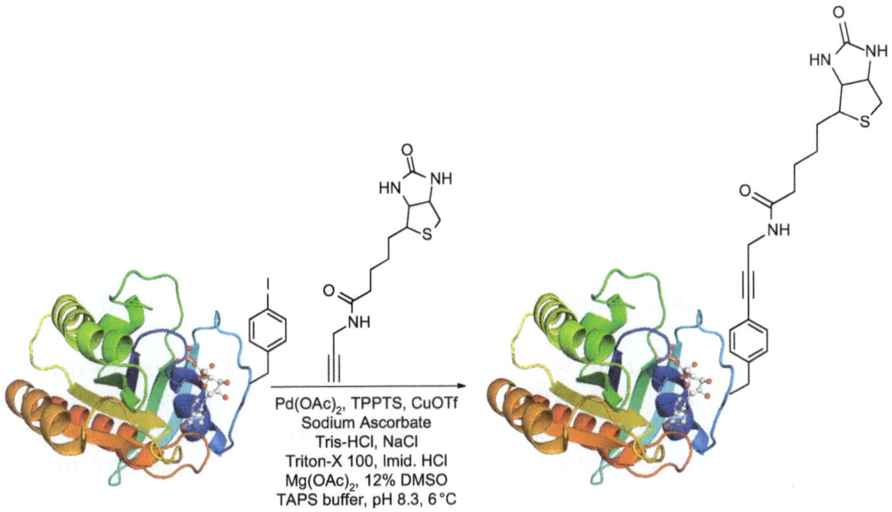

Scheme 12.39

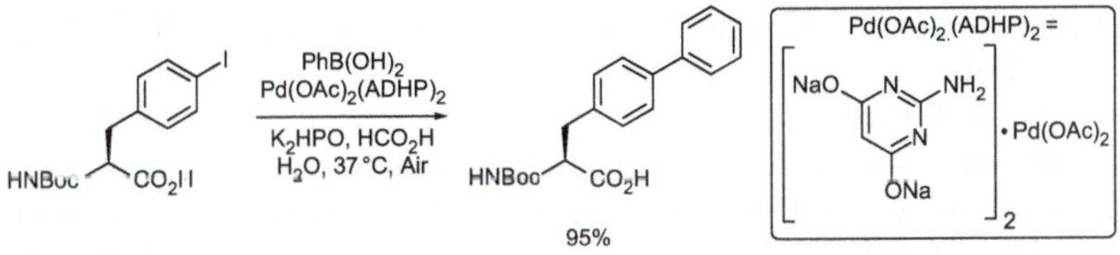

Scheme 12.40

chemistry, $Pd_2(dba)_3$ as a palladium source.[84] It should be noted that water-soluble hydroxylated dba has been reported.[85] An approximate 30% yield was reported, however the authors noted that the high temperature would limit the usefulness of this reaction for routine protein labelling.

These pioneering methodologies demonstrated how metal-catalysed coupling can be used on extremely complicated substrates. However, they suffer from low conversion to product or conditions that denature proteins. Davis and co-workers have published ground-breaking research which begins to overcome these problems.[86] The authors were keen to avoid phosphines, which they considered too sensitive to oxidation and often too expensive for everyday use in the modification of proteins, and therefore chose the sodium salt of 2-amino-4,6-dihydroxypyrimidine as substitute ligands. Similar pyrimidines had previously been used as ligands in Mizoroki–Heck alkynylation reactions in organic solvents.[87] Initially performing reactions on model amino acids and short-chain peptides, near quantitative yields were obtained for the coupling of aryl iodides and aryl bromides with aryl boronic acids. However, no conversion was observed with aryl chlorides (Scheme 12.40). The reaction with terminal cysteine-derived residues inhibited catalysis, but internal sulfur-containing *p*-iodobenzyl cysteine (Pic) residues gave the coupled product in >90% yield.

Building on this work, the authors chemically incorporated Pic into the serine protease subtilisin *Bacillus lentus* (SBL) mutant S156C. The Suzuki–Miyaura conditions were applied to this modified protein, at first with phenylboronic acid (Scheme 12.41). After 30 minutes, >95% conversion to product was obtained as assayed by LC-MS. A wide variety of aryl- and vinylboronic acids were used, and all gave similarly excellent yields. This represented a significant breakthrough in the field, and was the first example of full conversion in a coupling reaction occurring on a modified protein substrate.

Davis later extended this work to genetically encoded phenylalanine-derived aryl iodides.[88] The protein MBP (maltose binding protein), chosen for its high stability and solubility,[89,90] was modified at the solvent accessible E13 residue by UAG stop-codon suppression to generate MBP-E13*. This protein was expressed in high yields (15 mg L^{-1}) and with good fidelity.

Initially, it was found that non-specific metal binding to the protein resulted in mass spectral signal suppression, making monitoring of these reactions difficult.[91] As a solution to this, Davis trialled EDTA, DTT, cysteine and 3-mercaptopropionic acid[92] as potential palladium scavengers. The latter was found to be most effective in low concentrations (4.4 μmol mL^{-1}, 3 eq. with respect to Pd).

The Suzuki–Miyaura coupling (Scheme 12.42) was found to be effective with furan-3-boronic acid (at 680 equivalents), with 50% conversion observed after one hour at 37 °C, and full conversion after two hours (slower than their cysteine-derived methodology). The catalyst, $Pd(OAc)_2{\cdot}(ADHP)_2$, was loaded at 50 equivalents with respect to protein. Control reactions without palladium and organoboronic acid were performed, confirming that this reaction was, indeed, a palladium-mediated coupling reaction.

In a natural extension to this methodology, Davis and co-workers demonstrated that it was possible not only to perform Suzuki–Miyaura couplings on proteins in solution, but also those residing in cell membranes (Scheme 12.43).[93] Several *Escherichia coli* strains were grown which contained outer membrane OmpC proteins with pIPhe (p-iodophenylalanine) residues incorporated at different positions *via* amber stop codon suppression.[94] These cells (with $OD_{600} = 0.2$, OD_{600} defined as the optical density as measured at 600 nm) were treated with a fluorescein derived organoboronic acid and $Pd(OAc)_2{\cdot}(ADHP)_2$ in phosphate buffer at 37 °C. After one hour, the fluorescent labelling was clearly visible (excitation at 488 nm, emission between 500–640 nm). At lower temperatures, labelling was significantly decreased, even in the presence of high organoboronic acid and palladium loadings. Control reactions with wild-type OmpC proteins exhibited no fluorescence. Similarly, the absence of palladium or organoboronic acid gave the same result, discounting non-specific uptake or binding of the organoboronic acid. In the presence of reducing sugars, such as D-glucose, there was no difference in the efficacy of the reaction.

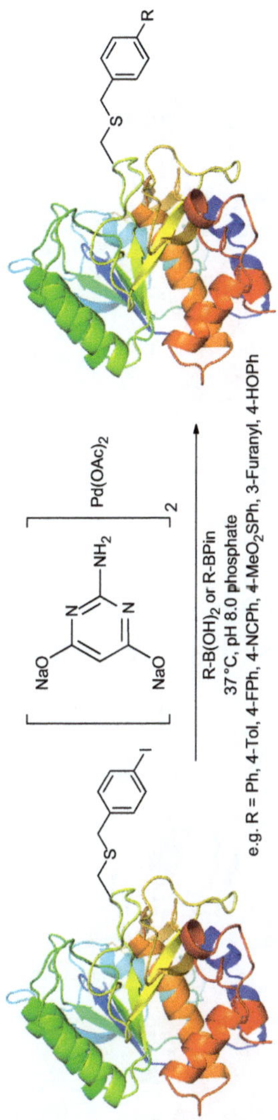
R-B(OH)2 or R-BPin
37 °C, pH 8.0 phosphate
e.g. R = Ph, 4-Tol, 4-FPh, 4-NCPh, 4-MeO2SPh, 3-Furanyl, 4-HOPh
Pd(OAc)2
NaO
NaO
NH2
S
S
I
R

Scheme 12.41

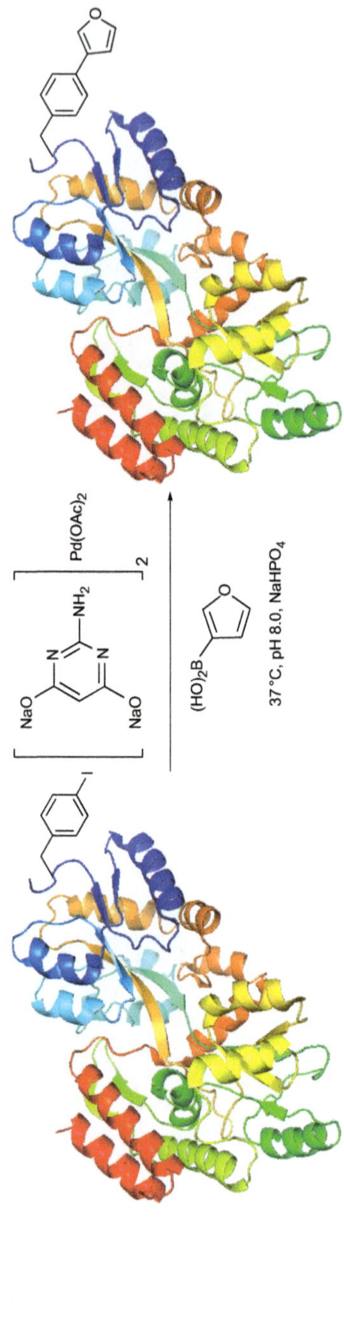

Scheme 12.42

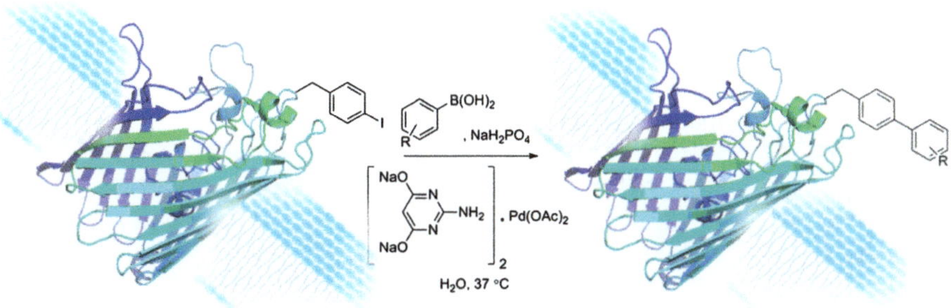

Scheme 12.43 Reprinted (adapted) with permission from C. D. Spicer, T. Triemer and B. G. Davis, *J. Am. Chem. Soc.*, 2012, **134**, 800–803. Copyright 2012 American Chemical Society.

Stoichiometrically it is difficult to compare these 'cell-surface' reactions with traditional coupling methodologies. However, Davis showed that the optimal organoboronic acid concentration was achieved at approximately 5.7 equivalents relative to the palladium loading. No functionalisation was observed at low palladium concentrations. However, a critical threshold for labelling was reached at 330 μM. No further benefit was observed on increasing the palladium concentration above 450 μM. Davis has suggested this effect could be used for a palladium-based titration system. Mindful of the toxicity observed by other metal-mediated synthetic modifications in biology (for example, copper-catalysed 'click' chemistry),[95] the Davis group demonstrated that at optimised palladium concentration cell death was identical to the control (< 3%).

12.3.2 C–H Functionalisation of Amino Acids and Peptides

Over the years, many methods for C–C bond formation by the functionalisation of C–H bonds in amino acids have been developed which utilise Grignard, Mannich or enolate chemistries. An amino acid example was reported by Li and co-workers, in which a Mannich-type reaction was used for forming a C–C bond at the α-position of protected glycine.[96] Other examples for the formation of similar bonds included the reaction of BuLi-generated enolates quenched with alkyl tin reagents,[97] or low-temperature Grignard chemistry.[98]

Palladium-catalysed direct C–H functionalisation reactions have remained limited. There has been significant interest in the synthesis of *ortho*-cyclopalladated complexes of amino acids, in particular phenylglycine and phenylalanine.[99–104] Pd^{II} palladacycles have proved of interest as pre-catalysts for palladium-mediated coupling reactions,[105] and for their possible cytotoxic activity.[106–108] Vicente and Urriolabeitia have separately demonstrated the utility of these *ortho*-palladated amino acid complexes in the synthesis of organic compounds.

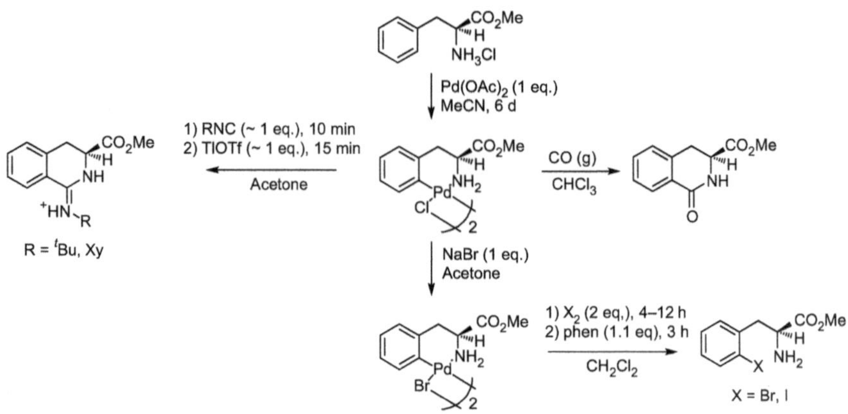

Scheme 12.44

Vicente and co-workers successfully formed a phenylalanine palladacycle (Scheme 12.44) by reacting methyl protected phenylalanine with palladium(II) acetate in MeCN.[99] From this intermediate, it proved possible to carbonylate (forming a tetrahydroisoquinoline heterocycle) and react with isocyanide (to form isoquinolinium salts). In addition, swapping to a bromide (as opposed to chloride) bridge allows for halogenation reactions *via* a proposed palladium(IV) intermediate (although no experimental evidence for such an intermediate was presented in the paper).[109] Phenanthroline hydrate (phen) was required to release the organic product from the complex.

Urriolabeitia and co-workers have demonstrated that a cyclopalladated phenylglycine[101] (synthesised in an optically pure form) can be functionalised in a similar way, and also used as an intermediate for C–O and C–N bond formation (Scheme 12.45).[104]

Halogenation proceeds from the chloride dimer (as opposed to the bromide dimer which is required in Vicente's reaction). Urriolabeitia also proposed an alternative mechanism, albeit *via* a similar Pd^{IV} intermediate. Enantiomeric excess, determined by the Mosher amide method,[110,111] indicated that there was a slight erosion of stereochemistry (*ee* 82% and 87% for bromo- and iodo-compounds, respectively). As with the phenylalanine system, carboxylation allows cyclisation to give an isoindolinone. However, complete epimerisation is observed. No insertion into the Pd–C bond is observed with isocynides, which instead formed simple co-ordination complexes.

Iodine(III) species[112,113] have long been used to access Pd^{IV}/Pt^{IV} oxidation states. In the presence of $PhI(OAc)_2$ and a large excess of alcohol (MeOH or EtOH), etherification of the palladacycle was observed. Significant erosion in enantiomeric excess was observed (*ee* 20% and 50% for MeOH and EtOH, respectively), but this does demonstrate a possible route to structural motifs which would have otherwise been generally unavailable.

The reaction of cyclopalladated phenylglycine with dimethyl acetylenedicarboxylate (DMAD) resulted in simultaneous C–C bond formation and an

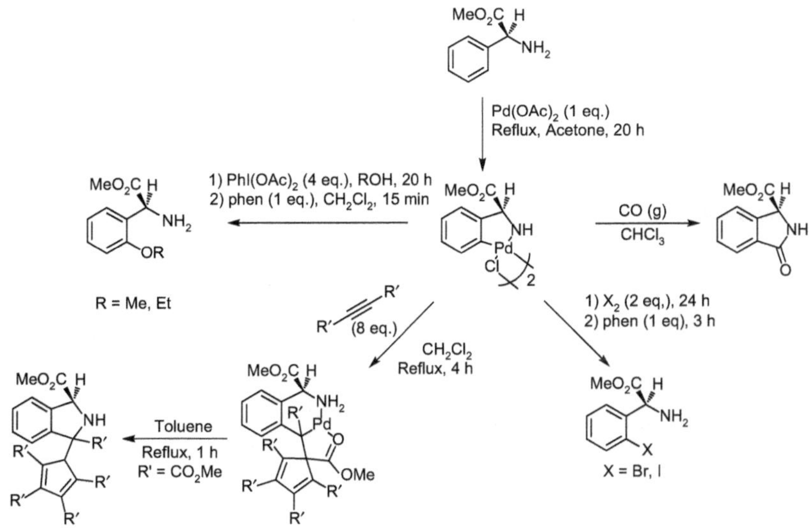

Scheme 12.45

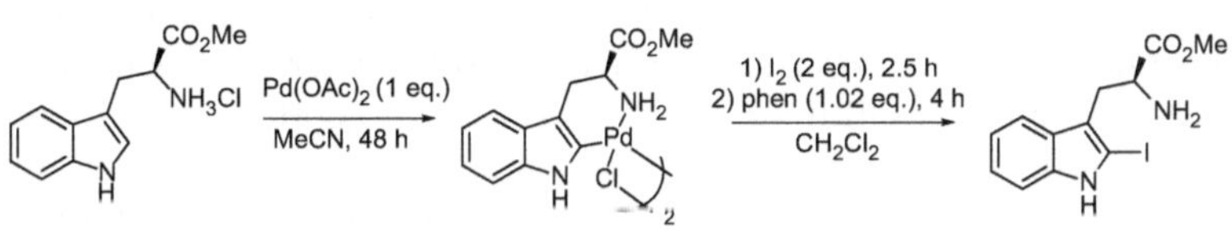

Scheme 12.46

interesting spiro-palladacycle system resulting from the insertion of three DMAD molecules. Reductive elimination of the isolated intermediate resulted in C–N bond formation, forming a novel isoindoline heterocycle. Optical activity indicated some stereospecificity; however as the Mosher amide approach could not be used, an exact *ee* remained undetermined from this reaction.

The cyclopalladated-tryptophan complex (Scheme 12.46) has been synthesised by Vicente and co-workers,[114] and reacted with I_2 to give iodination at the C2-position of the indole. No comment was made as to whether racemisation had occurred.

Examples of catalytic C–H functionalisations of amino acids are even more rare than their stoichiometric counterparts. Lavilla and co-workers have shown that it is possible to selectively arylate the C2-position of tryptophan using a palladium-mediated system similar to that initially developed by Larrosa (Scheme 12.47).[115,116] Excellent yields, and good stereochemical retention, were obtained with a protected amino acid. In the presence of other halogens, the reaction selectively proceeded at the iodine. Interestingly, a significantly higher temperature was required than in Larrosa's methodology for indoles.

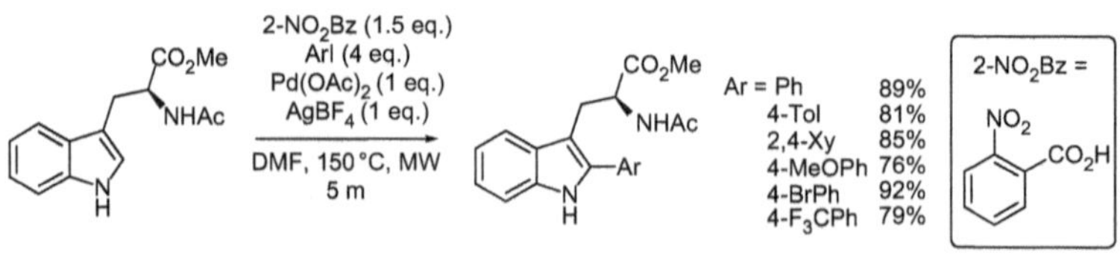

Scheme 12.47

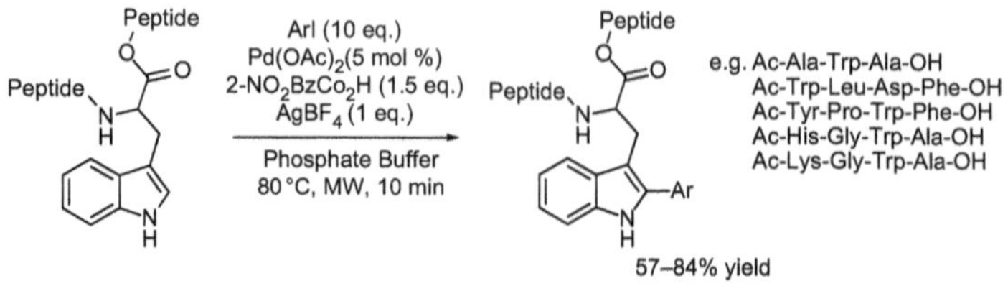

Scheme 12.48

High temperatures, and microwave irradiation, were required making these specific conditions unsuitable for more sensitive substrates. However, Lavilla successfully adopted this methodology to the C–H functionalisation of peptides in phosphate buffer, lowering the temperature to 80 °C (Scheme 12.48). Peptides containing sulfur residues were not tolerated due to selective hydrolysis of the peptide bond. Peptides containing histidine residues were also susceptible to this side reaction. The application of these conditions to full proteins remains limited due to the high reaction temperature. Despite these drawbacks, this work represents a significant step forward in C–H functionalisation of amino acids and peptides.

12.4 Summary

The catalytic functionalisation of C–H and C–X bonds in biomolecules remains a rapidly evolving field. The challenges these compounds present, including their intrinsic sensitivity to non-aqueous environments and high temperature, and structural complexity, arguably places this chemistry at the forefront of transition metal-catalysed organic chemistry.

Although functionalisation by Suzuki–Miyaura and Sonogashira couplings has been extensively developed over recent years, the Mizoroki–Heck reaction remains under-explored by comparison. This is particularly the case in the functionalisation of nucleosides and nucleotides. The driving force behind the development of much of this chemistry has been through synthetic need. The literature scarcity of Mizoroki–Heck couplings on biomolecules may be best explained by a simple lack of interest in their products for biological applications, rather than an incompatibility of the coupling with these substrates. Mechanistic studies will continue to provide useful information on suitable catalytic processes which are compatible with highly complex biomolecule structures.

Direct C–H functionalisation is an exciting and fast-developing methodology. The many advantages it presents over traditional cross-coupling methods makes it ideal for application to sensitive biomolecules. The next few years are likely to see significant advances in the site-specific C–H functionalisation of ever more complex biomolecular targets.

References

1. M. Sameiro and T. Goncalves, *Chem. Rev.*, 2009, **109**, 190–212.
2. R. E. Holmes and R. K. Robins, *J. Am. Chem. Soc.*, 1964, **86**, 1242–1245.
3. N. Ramzaeva and F. Seela, *Helv. Chim. Acta.*, 1995, **78**, 1083–1090.
4. W. W. Sy, *Synth. Comm.*, 1990, **20**, 3391–3394.
5. L. A. Agrofoglio, I. Gillaizeau and Y. Saito, *Chem. Rev.*, 2003, **103**, 1875–1916.
6. A. L. Casalnuovo and J. C. Calabrese, *J. Am. Chem. Soc.*, 1990, **112**, 4324–4330.
7. T. R. Battersby, D. N. Ang, P. Burgstaller, S. C. Jurczyk, M. T. Bowser, D. D. Buchanan, R. T. Kennedy and S. A. Benner, *J. Am. Chem. Soc.*, 1999, **121**, 9781–9789.
8. E. C. Western, J. R. Daft, E. M. Johnson, P. M. Gannett and K. H. Shaughnessy, *J. Org. Chem.*, 2003, **68**, 6767–6774.
9. E. C. Western and K. H. Shaughnessy, *J. Org. Chem.*, 2005, **70**, 6378–6388.
10. L. R. Moore, E. C. Western, R. Craciun, J. M. Spruell, D. A. Dixon, K. P. O'Halloran and K. H. Shaughnessy, *Organometallics*, 2008, **27**, 576–593.
11. A. Collier and G. Wagner, *Org. Biomol. Chem.*, 2006, **4**, 4526–4532.
12. T. Pesnot, J. Kempter, J. Schemies, G. Pergolizzi, U. Uciechowska, T. Rumpf, W. Sippl, M. Jung and G. K. Wagner, *J. Med. Chem.*, 2011, **54**, 3492–3499.
13. P. Capek, R. Pohl and M. Hocek, *Org. Biomol. Chem.*, 2006, **4**, 2278–2284.
14. M. Vrabel, M. Hocek, L. Havran, M. Fojta, I. Votruba, B. Klepetarova, R. Pohl, L. Rulisek, L. Zendlova, P. Hobza, I. h. Shih, E. Mabery and R. Mackman, *Eur. J. Org. Chem.*, 2007, 1752–1769.
15. M. Vrabel, R. Pohl, B. Klepetarova, I. Votruba and M. Hocek, *Org. Biomol. Chem.*, 2007, **5**, 2849–2857.
16. M. Vrabel, R. Pohl, I. Votruba, M. Sajadi, S. A. Kovalenko, N. P. Ernsting and M. Hocek, *Org. Biomol. Chem.*, 2008, **6**, 2852–2860.
17. H. Cahova, L. Havran, P. Brazdilova, H. Pivonkova, R. Pohl, M. Fojta and M. Hocek, *Angew. Chem., Int. Ed.*, 2008, **47**, 2059–2062.
18. A. Omumi, D. G. Beach, M. Baker, W. Gabryelski and R. A. Manderville, *J. Am. Chem. Soc.*, 2011, **133**, 42–50.
19. C. J. Yu, H. Yowanto, Y. J. Wan, T. J. Meade, Y. Chong, M. Strong, L. H. Donilon, J. F. Kayyem, M. Gozin and G. F. Blackburn, *J. Am. Chem. Soc.*, 2000, **122**, 6767–6768.

20. G. T. Crisp and B. L. Flynn, *J. Org. Chem.*, 1993, **58**, 6614–6619.
21. R. Volpini, S. Costanzi, C. Lambertucci, S. Vittori, K. N. Klotz, A. Lorenzen and G. Cristalli, *Bioorg. Med. Chem. Lett.*, 2001, **11**, 1931–1934.
22. W. Flasche, C. Cismas, A. Herrmann and J. Liebscher, *Synthesis*, 2004, 2335–2341.
23. A. G. Firth, I. J. S. Fairlamb, K. Darley and C. G. Baumann, *Tetrahedron Lett.*, 2006, **47**, 3529–3533.
24. N. K. Garg, C. C. Woodroofe, C. J. Lacenere, S. R. Quake and B. M. Stoltz, *Chem. Commun.*, 2005, 4551–4553.
25. A. G. Firth, K. Wilson, C. G. Baumann and I. J. S. Fairlamb, *Nucleosides, Nucleotides Nucleic Acids*, 2011, **30**, 168–184.
26. P. Capek, H. Cahova, R. Pohl, M. Hocek, C. Gloeckner and A. Marx, *Chem.–Eur. J.*, 2007, **13**, 6196–6203.
27. A. Hessler, O. Stelzer, H. Dibowski, K. Worm and F. P. Schmidtchen, *J. Org. Chem.*, 1997, **62**, 2362–2369.
28. J. H. Cho, C. D. Prickett and K. H. Shaughnessy, *Eur. J. Org. Chem.*, 2010, 3678–3683.
29. T. Kottysch, C. Ahlborn, F. Brotzel and C. Richert, *Chem. Eur. J.*, 2004, **10**, 4017–4028.
30. T. Strube, O. Schiemann, F. MacMillan, T. Prisner and J. W. Engels, *Nucleosides, Nucleotides, Nucleic Acids*, 2001, **20**, 1271–1274.
31. H. Song, X. Li, Y. Long, G. Schatte and H.-B. Kraatz, *Dalton Trans.*, 2006, 4696–4701.
32. N. J. Greco and Y. Tor, *Tetrahedron*, 2007, **63**, 3515–3527.
33. A. Mieczkowski, J. Blu, V. Roy and L. A. Agrofoglio, *Tetrahedron*, 2009, **65**, 9791–9796.
34. M. Ashwell, A. S. Jones, A. Kumar, J. R. Sayers, R. T. Walker, T. Sakuma and E. Declercq, *Tetrahedron*, 1987, **43**, 4601–4608.
35. P. Lagisetty, L. Zhang and M. K. Lakshman, *Adv. Synth. Catal.*, 2008, **350**, 602–608.
36. D. E. Bergstrom and J. L. Ruth, *J. Am. Chem. Soc.*, 1976, **98**, 1587–1589.
37. D. E. Bergstrom and M. K. Ogawa, *J. Am. Chem. Soc.*, 1978, **100**, 8106–8112.
38. J. L. Ruth and D. E. Bergstrom, *J. Org. Chem.*, 1978, **43**, 2870–2876.
39. S. Pivsa-Art, T. Satoh, Y. Kawamura, M. Miura and M. Nomura, *Bull. Chem. Soc. Jpn.*, 1998, **71**, 467–473.
40. S. De Ornellas, T. E. Storr, T. J. Williams, C. G. Baumann and I. J. S. Fairlamb, *Curr. Org. Synth.*, 2011, **8**, 79–101.
41. I. Cerna, R. Pohl, B. Klepetarova and M. Hocek, *Org. Lett.*, 2006, **8**, 5389–5392.
42. I. Cerna, R. Pohl and M. Hocek, *Chem. Commun.*, 2007, 4729–4730.
43. T. E. Storr, A. G. Firth, K. Wilson, K. Darley, C. G. Baumann and I. J. S. Fairlamb, *Tetrahedron*, 2008, **64**, 6125–6137.
44. J. A. Widegren and R. G. Finke, *J. Mol. Catal. A: Chem.*, 2003, **198**, 317–341.

45. T. E. Storr, C. G. Baumann, R. J. Thatcher, S. De Ornellas, A. C. Whitwood and I. J. S. Fairlamb, *J. Org. Chem.*, 2009, **74**, 5810–5821.
46. R. Saladino, C. Crestini, F. Ciciriello, E. Di Mauro and G. Costanzo, *J. Biol. Chem.*, 2006, **281**, 5790–5796.
47. J. Muzart, *Tetrahedron*, 2009, **65**, 8313–8323.
48. T. E. Storr, J. A. Strohmeier, C. G. Baumann and I. J. S. Fairlamb, *Chem. Commun.*, 2010, **46**, 6470–6472.
49. B. Vankova, V. Krchnak, M. Soural and J. Hlavac, *ACS. Comb. Sci.*, 2011, **13**, 496–500.
50. I. Cerna, R. Pohl, B. Klepetarova and M. Hocek, *J. Org. Chem.*, 2010, **75**, 2302–2308.
51. M. Cernova, R. Pohl and M. Hocek, *Eur. J. Org. Chem.*, 2009, 3698–3701.
52. S. I. Gorelsky, D. Lapointe and K. Fagnou, *J. Am. Chem. Soc.*, 2008, **130**, 10848–10849.
53. M. Cernova, I. Cerna, R. Pohl and M. Hocek, *J. Org. Chem.*, 2011, **76**, 5309–5319.
54. K. S. Petrakis and T. L. Nagabhushan, *J. Am. Chem. Soc*, 1987, **109**, 2831–2833.
55. N. G. Kundu, M. Pal, J. S. Mahanty and S. K. Dasgupta, *J. Chem. Soc., Chem. Commun.*, 1992, 41–42.
56. G. T. Crisp and J. Gore, *Tetrahedron*, 1997, **53**, 1523–1544.
57. J. W. Tilley, R. Sarabu, R. Wagner and K. Mulkerins, *J. Org. Chem.*, 1990, **55**, 906–910.
58. A. M. Echavarren and J. K. Stille, *J. Am. Chem. Soc.*, 1987, **109**, 5478–5486.
59. J. G. Taylor, A. V. Moro and C. R. D. Correia, *Eur. J. Org. Chem.*, 2011, 1403–1428.
60. S. Sengupta and S. Bhattacharyya, *Tetrahedron Lett.*, 1995, **36**, 4475–4478.
61. W. C. Shieh and J. A. Carlson, *J. Org. Chem.*, 1992, **57**, 379–381.
62. J. Uenishi, J. M. Beau, R. W. Armstrong and Y. Kishi, *J. Am. Chem. Soc.*, 1987, **109**, 4756–4758.
63. M. J. Burk, J. R. Lee and J. P. Martinez, *J. Am. Chem. Soc.*, 1994, **116**, 10847–10848.
64. F. Firooznia, C. Gude, K. Chan and Y. Satoh, *Tetrahedron Lett.*, 1998, **39**, 3985–3988.
65. Y. Satoh, C. Gude, K. Chan and F. Firooznia, *Tetrahedron Lett.*, 1997, **38**, 7645–7648.
66. A. D. Roy, R. J. M. Goss, G. K. Wagner and M. Winn, *Chem. Commun.*, 2008, 4831–4833.
67. R. J. M. Goss and P. L. A. Newill, *Chem. Commun.*, 2006, 4924–4925.
68. H. Dibowski and F. P. Schmidtchen, *Angew. Chem., Int. Ed.*, 1998, **37**, 476–478.
69. H. Dibowski and F. P. Schmidtchen, *Tetrahedron*, 1995, **51**, 2325–2330.
70. A. S. Batsanov, J. C. Collings, I. J. S. Fairlamb, J. P. Holland, J. A. K. Howard, Z. Y. Lin, T. B. Marder, A. C. Parsons, R. M. Ward and J. Zhu, *J. Org. Chem.*, 2005, **70**, 703–706.

71. E. A. Bayer and M. Wilchek, *Methods Enzymol.*, 1990, **184**, 138–160.
72. D. T. Bong and M. R. Ghadiri, *Org. Lett.*, 2001, **3**, 2509–2511.
73. C. Corona, B. K. Bryant and J. B. Arterburn, *Org. Lett.*, 2006, **8**, 1883–1886.
74. S. Kotha and K. Lahin, *Biopolymers*, 2003, **69**, 517–528.
75. A. Ojida, H. Tsutsumi, N. Kasagi and I. Hamachi, *Tetrahedron Lett.*, 2005, **46**, 3301–3305.
76. Y. W. Lee, S. B. Han, A. R. Ko, H. S. Kim and K. W. Park, *Catal. Commun.*, 2011, **15**, 137–140.
77. N. D. Doan, S. Bourgault, M. Letourneau and A. Fournier, *J. Comb. Chem.*, 2008, **10**, 44–51.
78. M. Vilaro, G. Arsequell, G. Valencia, A. Ballesteros and J. Barluenga, *Org. Lett.*, 2008, **10**, 3243–3245.
79. G. A. Molander and B. Biolatto, *Org. Lett.*, 2002, **4**, 1867–1870.
80. G. Espuna, D. Andreu, J. Barluenga, X. Perez, A. Planas, G. Arsequell and G. Valencia, *Biochemistry*, 2006, **45**, 5957–5963.
81. K. Kodama, S. Fukuzawa, H. Nakayama, T. Kigawa, K. Sakamoto, T. Yabuki, N. Matsuda, M. Shirouzu, K. Takio, K. Tachibana and S. Yokoyama, *ChemBioChem*, 2006, **7**, 134–139.
82. K. Kodama, S. Fukuzawa, H. Nakayama, K. Sakamoto, T. Kigawa, T. Yabuki, N. Matsuda, M. Shirouzu, K. Takio, S. Yokoyama and K. Tachibana, *ChemBioChem*, 2007, **8**, 232–238.
83. E. Brustad, M. L. Bushey, J. W. Lee, D. Groff, W. Liu and P. G. Schultz, *Angew. Chem., Int. Ed.*, 2008, **47**, 8220–8223.
84. A. G. Jarvis and I. J. S. Fairlamb, *Curr. Org. Chem.*, 2011, **15**, 3175–3196.
85. G. Deepa, R. Balamurugan and P. Kannan, *J. Mol. Struct.*, 2010, **963**, 219–227.
86. J. M. Chalker, C. S. C. Wood and B. G. Davis, *J. Am. Chem. Soc.*, 2009, **131**, 16346–16347.
87. J. H. Li, X. D. Zhang and Y. X. Xie, *Eur. J. Org. Chem.*, 2005, 4256–4259.
88. C. D. Spicer and B. G. Davis, *Chem. Commun.*, 2011, **47**, 1698–1700.
89. M. B. Zwick, I. L. C. Bonnycastle, K. A. Noren, S. Venturini, C. F. Barbas, C. J. Noren and J. K. Scott, *Anal. Biochem.*, 1998, **264**, 87–97.
90. J. D. Fox, R. B. Kapust and D. S. Waugh, *Protein Sci.*, 2001, **10**, 622–630.
91. J. M. Antos and M. B. Francis, *Curr. Opin. Chem. Biol.*, 2006, **10**, 253–262.
92. L. M. Rossi, L. L. R. Vono, F. P. Silva, P. K. Kiyohara, E. L. Duarte and J. R. Matos, *Appl. Catal., A.*, 2007, **330**, 139–144.
93. C. D. Spicer, T. Triemer and B. G. Davis, *J. Am. Chem. Soc.*, 2012, **134**, 800–803.
94. D. R. Liu, T. J. Magliery, M. Pasternak and P. G. Schultz, *Proc. Nat. Acad. Sci. U. S.A.*, 1997, **94**, 10092–10097.
95. V. Hong, N. F. Steinmetz, M. Manchester and M. G. Finn, *Bioconjugate Chem.*, 2010, **21**, 1912–1916.
96. L. Zhao and C. J. Li, *Angew. Chem., Int. Ed.*, 2008, **47**, 7075–7078.

97. R. M. Williams, D. J. Aldous and S. C. Aldous, *J. Org. Chem.*, 1990, **55**, 4657–4663.
98. A. L. Castelhano, S. Horne, G. J. Taylor, R. Billedeau and A. Krantz, *Tetrahedron*, 1988, **44**, 5451–5466.
99. J. Vicente, I. Saura-Llamas, J. Antonio Garcia-Lopez, B. Calmuschi-Cula and D. Bautista, *Organometallics*, 2007, **26**, 2768–2776.
100. J. Albert, J. M. Cadena and J. Granell, *Tetrahedron: Asymmetry*, 1997, **8**, 991–994.
101. Y. Fuchita, K. Yoshinaga, Y. Ikeda and J. KinoshitaKawashima, *J. Chem. Soc., Dalton Trans.*, 1997, 2495–2499.
102. J. Albert, J. Granell and G. Muller, *J. Organomet. Chem.*, 2006, **691**, 2101–2106.
103. A. R. Hajipour, K. Karami, A. Pirisedigh and A. E. Ruoho, *Amino Acids*, 2009, **37**, 537–541.
104. S. Nieto, P. Arnau, E. Serrano, R. Navarro, T. Soler, C. Cativiela and E. P. Urriolabeitia, *Inorg. Chem.*, 2009, **48**, 11963–11975.
105. J. L. Serrano, L. Garcia, J. Perez, E. Perez, J. Garcia, G. Sanchez, P. Sehnal, S. De Ornellas, T. J. Williams and I. J. S. Fairlamb, *Organometallics*, 2011, **30**, 5095–5109.
106. C. Navarroranninger, I. Lopezsolera, J. M. Perez, J. Rodriguez, J. L. Garciaruano, P. R. Raithby, J. R. Masaguer and C. Alonso, *J. Med. Chem.*, 1993, **36**, 3795–3801.
107. C. Navarroranninger, I. Lopezsolera, V. M. Gonzalez, J. M. Perez, A. Alvarezvaldes, A. Martin, P. R. Raithby, J. R. Masaguer and C. Alonso, *Inorg. Chem.*, 1996, **35**, 5181–5187.
108. F. Zamora, V. M. Gonzalez, J. M. Perez, J. R. Masaguer, C. Alonso and C. Navarroranninger, *Appl. Organomet. Chem.*, 1997, **11**, 659–666.
109. P. Sehnal, R. J. K. Taylor and I. J. S. Fairlamb, *Chem. Rev.*, 2010, **110**, 824–889.
110. J. A. Dale, D. L. Dull and H. S. Mosher, *J. Org. Chem.*, 1969, **34**, 2543–2549.
111. J. M. Seco, E. Quinoa and R. Riguera, *Chem. Rev.*, 2004, **104**, 17–117.
112. A.Varvoglis, *Hypervalent Iodine in Organic Synthesis*, Academic Press, London, 1997.
113. P. J. Stang and V. V. Zhdankin, *Chem. Rev.*, 1996, **96**, 1123–1178.
114. J. Vicente, I. Saura-Llamas and D. Bautista, *Organometallics*, 2005, **24**, 6001–6004.
115. J. Ruiz-Rodriguez, F. Albericio and R. Lavilla, *Chem.–Eur. J.*, 2010, **16**, 1124–1127.
116. N. Lebrasseur and I. Larrosa, *J. Am. Chem. Soc.*, 2008, **130**, 2926–2927.

Subject Index

References to tables and charts are in **bold** type